Higher PHYSICS

For SQA 2019 and beyond

Student Book

David McLean

001/25032020

10 9 8 7 6 5 4 3 2 1

ISBN 978-0-00-838439-5

Published by
Leckie & Leckie Ltd
An imprint of HarperCollins Publishers
Westerhill Road, Bishopbriggs, Glasgow, G64 2QT
T: 0844 576 8126 F: 0844 576 8131
leckiescotland@harpercollins.co.uk
www.leckiescotland.co.uk

Publisher: Sarah Mitchell
Project managers: Janice McNeillie and Gillian Bowman

Special thanks to
Jouve (layout and illustration)
Louise Robb (proofread)

Printed in Italy by GRAFICA VENETA S.p.A.

A CIP Catalogue record for this book is available from the British Library.

Acknowledgements
Fig 1.73 © Action Sports Photography / Shutterstock.com; Fig 1.138 © VanderWolf Images / Shutterstock.com; Fig 2.26 © sciencephotos / Alamy; Fig 2.27 © Dorling Kindersley / UIG / SCIENCE PHOTO LIBRARY; PP354-355 © D-VISIONS / Shutterstock.com

All other images © Shutterstock.com, © HarperCollins Publishers or public domain.

Whilst every effort has been made to trace the copyright holders, in cases where this has been unsuccessful, or if any have inadvertently been overlooked, the Publishers would gladly receive any information enabling them to rectify any error or omission at the first opportunity.

AREA 1, AREA 2 and AREA 3 exam-style questions and answers, plus answers to exercises can be found at www.collins.co.uk/pages/Scottish-curriculum-free-resources

Introduction

About this book

This book provides a resource to practise and assess your understanding of the physics covered for the Higher qualification. The book has been organised to map to the course specifications and is packed with worked examples, experiments and exercises to deepen your understanding of Physics and help you prepare for the assessments and final exam.

Features

YOU SHOULD ALREADY KNOW

Each chapter begins with a list of topics you should already know before you start the chapter. Some of these topics will have been covered at National 5, while others may depend on other chapters in this book.

You should already know (National 5):

- The wavelength of a wave is measured between two identical points, e.g. from peak to peak
- The frequency of the wave is the number of complete waves in one second

IN THIS CHAPTER

After the list of things you should be familiar with, there is a list of the topics covered in the chapter. This tells you what you should be able to do when you have worked your way through the chapter.

Learning intentions

- The Doppler effect is observed for both sound and light
- The Doppler effect links the change in observed frequency and wavelength of a wave to the relative motion of the source and observer

EXPERIMENT

Topics are usually introduced by way of an experiment that aims to familiarise you with the apparatus and methods that you are likely to encounter. Each includes a list of things to do to perform the experiment as well as questions that the experiments will raise.

GO! Experiment 6.3 Expansion of the universe

This experiment demonstrates the link between Hubble's law and the expansion of the universe.

Apparatus

- A balloon
- A marker pen

Instructions

1. Blow up the balloon to a very small size. Mark an X on the balloon to represent the Earth. Mark dots on the balloon at different distances away from the X to represent near and far celestial bodies.
2. Blow up the balloon slowly, watching how the near and far dots move relative to the X. Which dots are moving more quickly away from Earth?

WORKED EXAMPLES

New topics involving calculations are introduced with at least one worked example, which shows how to go about tackling the questions and activities. Each example breaks the question and solution down into steps, so you can see what calculations are involved, what kind of rearrangements are needed and how to work out the best way of answering the question.

Worked example

Calculate the kinetic energy of a Border Collie dog, of mass 18 kg, when it is moving with a velocity of 10 m s^{-1}.

Use the equation for kinetic energy here:

$$E_K = \frac{1}{2}mv^2$$

Substitute what you know and solve for the kinetic energy:

$$E_K = \frac{1}{2} \times 18 \times 10^2$$

$$E_K = 900\ J$$

Fig 1.122

KEY POINTS

Where appropriate, key point features provide useful summaries of critical information.

Key point

Acceleration is the rate of change of velocity, given by:

$$a = \frac{\Delta v}{t} = \frac{v - u}{t}$$

It is a vector quantity, so it has both magnitude and direction.

ASSESSMENTS

In each chapter, exercises provide topic-linked questions. End-of-area assessments are available online at: www.collins.co.uk/pages/Scottish-curriculum-free-resources. These assessments contain a number of exam-style questions and cover the minimum competence for the unit content and are a good preparation for your unit assessment.

Answers to exercises and to the end-of-area assessment questions are available online at: www.collins.co.uk/pages/Scottish-curriculum-free-resources.

THE CFE HIGHER

The Higher qualification is split into two sections that make up your final award:

- the exam (worth 80% of the final award)
- the assignment (worth 20% of your final award).

THE EXAM

The examination is split into two papers and lasts a total of three hours.

Paper 1

Paper 1 consists of 25 multiple choice questions with five options per question.

Each question is worth one mark so the total for the first paper is 25 marks.

Paper 2

Paper 2 of the examination is by far the longest, and worth 130 marks. It consists of longer questions where the focus is on application of knowledge through problem solving. Practice makes perfect with this, so it is vital to attempt many exam-level questions before the examination to give you practice in applying your knowledge to typical problem solving questions.

Open-ended questions

There are also open-ended questions in the examination. These questions give you the opportunity to share your knowledge of Physics in questions where there is not just one distinct correct answer! Instead, you must apply what you know to a specific problem and explain your answer. These questions are worth three marks each.

THE ASSIGNMENT

In addition to the 155-mark two examination papers, which are worth 80% of your overall award, there is a research task (assignment) worth 20% of your overall mark. The purpose of the research task is to give you experience with researching a topic in Physics through reading literature and conducting experiments.

The assignment consists of a research stage and an experimental stage. It is assessed by means of a written report that is written in school under exam conditions (you are allowed any notes or results you have found during your research stage). This report is then marked and the result combined with your examination papers to give the final mark.

1. **Motion equations and graphs**
    - Properties of motion
    - Motion–time graphs
    - The equations of motion
2. **Force, energy and power**
    - Forces in one dimension
    - Forces in two dimensions
    - Energy and power
3. **Collisions, explosions and impulse**
    - Momentum
    - Conservation of momentum
    - Impulse
4. **Gravitation**
    - Horizontally launched projectiles
    - Oblique projectiles
    - Newton's universal law of gravitation
    - Gravitation and the universe
5. **Special relativity**
    - Galilean invariance and Newtonian relativity
    - Special relativity
6. **The expanding universe**
    - The Doppler effect
    - Hubble's law
    - Expansion of the universe

AREA 1

Our Dynamic Universe

1 Motion equations and graphs

You should already know (National 5)

- Definition of quantities relating to motion:
  - Displacement as the distance travelled from the start point to the end point in one particular direction
  - Velocity as the rate of change of displacement with time
  - Acceleration as the rate of change of velocity with time
- Vector and scalar quantities, including:
  - Velocity as a vector quantity
  - Displacement as a vector quantity
  - Addition of vectors in one or two dimensions
- Velocity–time graphs:
  - Plot and interpret velocity–time graphs for motion with a constant velocity or constant acceleration
  - Calculation of acceleration from a velocity–time graph
  - Calculation of displacement from a velocity–time graph

Learning intentions

- Definition of the quantities relating to the motion of an object:
  - Displacement
  - Velocity
  - Acceleration
- Use of motion–time graphs to represent the motion of an object with a constant acceleration
- Link between the three main graphs of motion
- Derivation of equations to represent the motion of an object with a constant acceleration (for example, motion under gravity)
- Analysis of the motion of an object using the equations of motion

1.1 Properties of motion

There are three main quantities used to describe motion as functions of time: displacement, velocity and acceleration. It is important to remember that these quantities are vectors – they have both magnitude and direction!

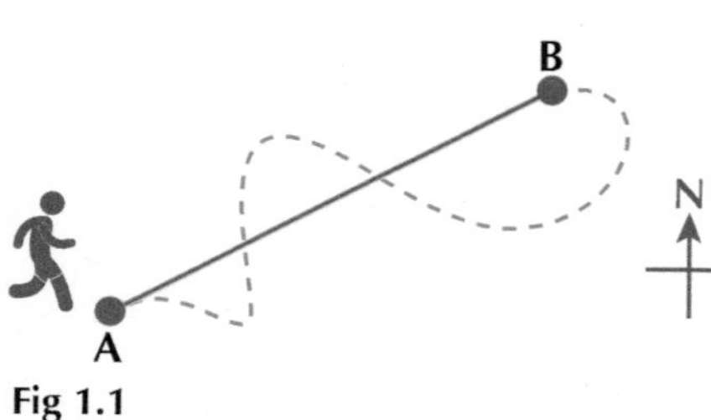

Fig 1.1

1.1.1 Displacement

The concept of displacement has been introduced at National 5. The magnitude of the displacement of an object is measured in metres (m). It is the straight line that connects your starting position to your finishing position as shown in Figure 1.1.

It can be thought of as the route travelled 'as the crow flies'. The actual route taken between the start and finish point does not matter – the displacement will always be the same straight line connecting both points. As displacement is a vector quantity, it is essential that a direction is also given! In Figure 1.1, the length of the grey dashed line represents the distance travelled.

At National 5, it was shown that the resultant displacement of an object can be found by adding together the individual displacements as vectors. Vectors are added 'tip-to-tail' using either a scale drawing or Pythagoras' theorem and trigonometry. For more detail, refer to National 5 Physics. As well as working with displacements at right angles to each other, we must also be able to add displacements (or any vector) that are not at right angles. For this, we can use either a scale drawing or trigonometry, or we can split each vector into rectangular components as described in National 5 Mathematics. The best option will depend on the problem you are faced with.

Key point

The resultant displacement is found by adding individual displacements together as vectors using the 'tip-to-tail' method shown in the diagram.

The resultant can be found by either scale drawing or trigonometry.

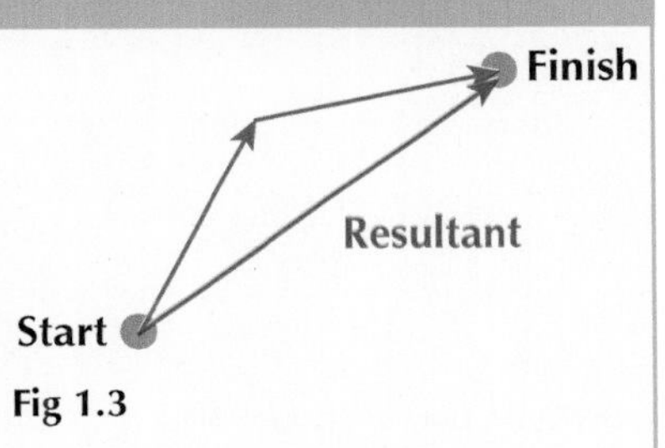

Fig 1.3

Key point: Pythagoras' theorem and trigonometry

The following rules are from National 5 Mathematics and are applied to a right-angled triangle:

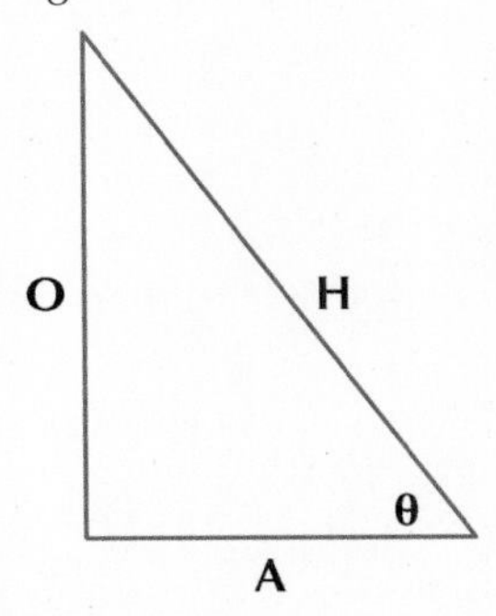

Fig 1.2

$$H^2 = O^2 + A^2$$

$$\sin\theta = \frac{O}{H}$$

$$\cos\theta = \frac{A}{H}$$

$$\tan\theta = \frac{O}{A}$$

The following rules are from National 5 Mathematics and are applied to all triangles, regardless of internal angles:

$$\frac{a}{\sin A} = \frac{b}{\sin B} = \frac{c}{\sin C}$$

$$a^2 = b^2 + c^2 - 2bc\cos A$$

Worked example

A plane flies 140 km due north before turning and flying 100 km on a bearing of 040. Calculate the resultant displacement of the plane.

This solution uses trigonometry. Scale drawing could also be used.

Start by drawing a diagram of the flight path (not to scale) that adds the vectors tip-to-tail:

The black dashed line represents the northerly direction (same as the 140 km vector). You can use the fact that all the angles on a straight line intersected by other straight line(s) add up to 180°.

$$\theta = 180 - 40 = 140°$$

Fig 1.4

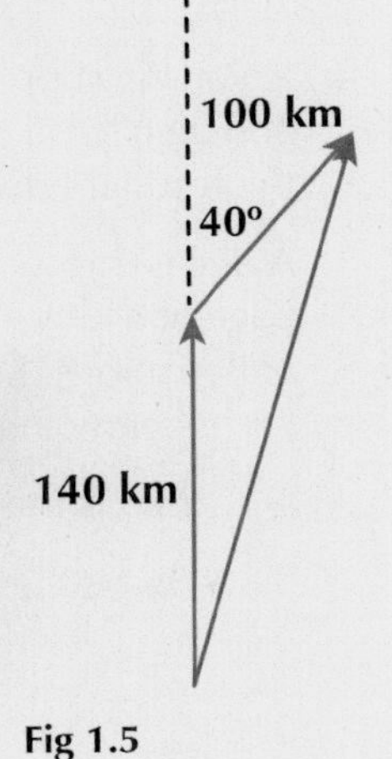

Fig 1.5

(continued)

This is displayed in Figure 1.6.

You need to find the missing length that gives the magnitude of the resultant displacement and the angle, θ, that corresponds to the bearing.

The missing length can be found using the cosine rule:

$$a^2 = b^2 + c^2 - 2bc\cos A$$
$$a^2 = 140^2 + 100^2 - 2(140)(100)\cos 140°$$
$$a^2 = 29600 - 28000\cos 140°$$
$$a^2 = 29600 - -21449$$
$$a^2 = 51049$$
$$a = 226\ km$$

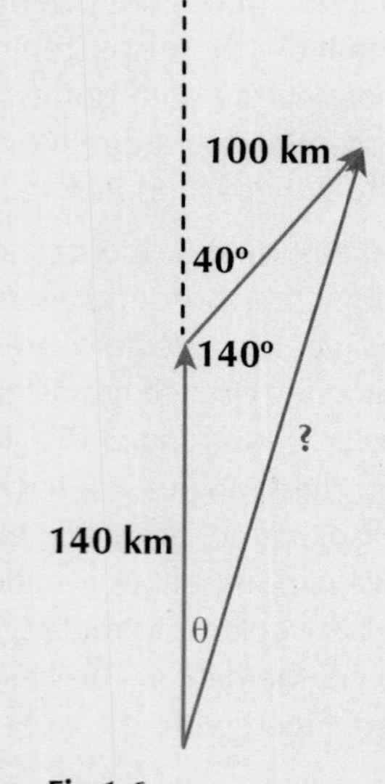

Fig 1.6

The bearing, θ, can then be found using the sine rule:

$$\frac{a}{\sin 140°} = \frac{b}{\sin B}$$
$$\frac{226}{\sin 140°} = \frac{100}{\sin\theta}$$
$$100\sin 140° = 226\sin\theta$$
$$226\sin\theta = 64$$
$$\sin\theta = \frac{64}{226}$$
$$\theta = \sin^{-1}\left(\frac{64}{226}\right)$$
$$\theta = 16$$

Hence, the aircraft's resultant displacement was 226 km on a bearing of 016.

Exercise 1.1.1 Displacement

1. Explain the difference between displacement and distance.
2. A runner runs a total distance of 12 km due north and then turns and runs 5 km due east. Find the resultant displacement of the runner (magnitude and direction).
3. During an orienteering competition, a team navigate a distance of 4.6 km on a bearing of 050. They then turn and walk 3.4 km on a bearing of 140. Find the resultant displacement of the team.
4. An oil-rig supply vessel sails from port to oil rig A and then to oil rig B as shown in Fig 1.7. Calculate the resultant displacement of the supply vessel.
5. A ball is thrown vertically upwards 0.5 m from the edge of a cliff. The ball rises to a maximum height of 8 m above the cliff before then falling to the sea 45 m below the edge of the cliff.
   a) What is the distance travelled by the ball?
   b) What is the resultant displacement of the ball?

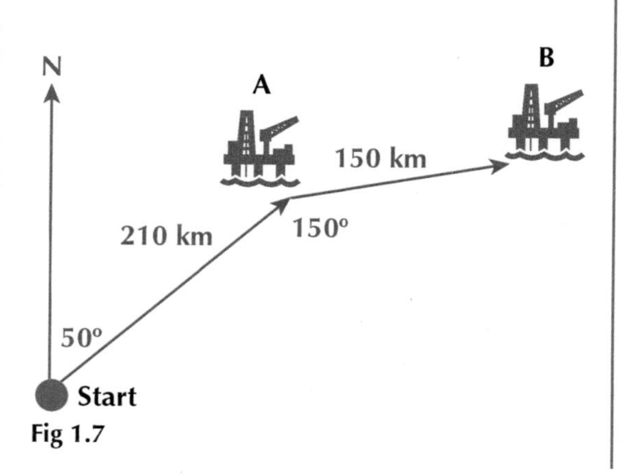

Fig 1.7

6 A passenger jet is carrying passengers to a remote island in the Pacific on holiday. The jet flies with a constant velocity of 600 km h^{-1} throughout the flight. If the jet flies due south for 2 hours and then on a bearing of 210 for 3 hours, find:

a) The total distance flown.

b) The resultant displacement of the aircraft.

7 A hiker walks 4 km due east, then 7 km due north and then 9 km due east. Calculate the total distance walked and the resultant displacement of the hiker.

1.1.2 Velocity

Velocity is a vector quantity that describes the rate of change of displacement of an object. The greater the velocity of an object, the greater the change in displacement in a given time. The velocity of an object can be found using,

$$\bar{v} = \frac{s}{t}$$

where s is the displacement of the object. The direction of the velocity is the same as the direction of the displacement used in the calculation.

Worked example: displacement and velocity

A boat sails 90 km due north before changing direction and sailing 120 km on a bearing of 320 as shown in Figure 1.8a. The ship sails this distance in a time of 12 hours.

Find:

a) The resultant displacement of the ship.

b) The resultant average velocity of the ship.

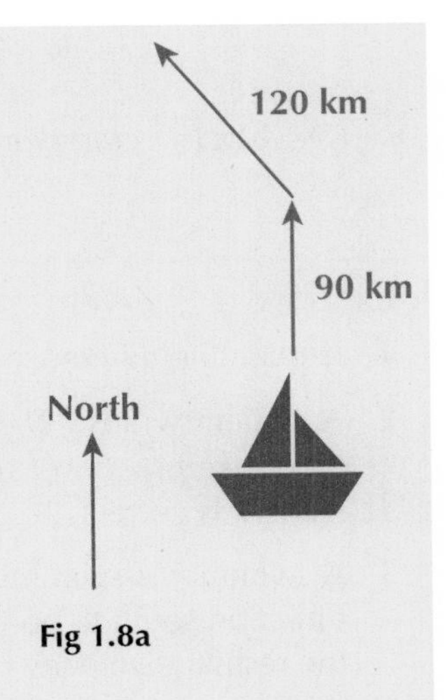

Fig 1.8a

a) *To answer this question, add the individual displacement vectors tip-to-tail. The angle between them can be calculated by considering the bearing. A bearing of 320 is 140° further round than due south, giving the angle between the vectors shown in Figure 1.8b.*

The magnitude of the resultant displacement can be found using the cosine rule:

$$a^2 = b^2 + c^2 - 2bc\cos A°$$
$$a^2 = 90^2 + 120^2 - 2(90)(120)\cos 140°$$
$$a^2 = 22500 - 21600\cos 140°$$
$$a^2 = 22500 - -16546$$
$$a^2 = 39046$$
$$a = 198\ km$$

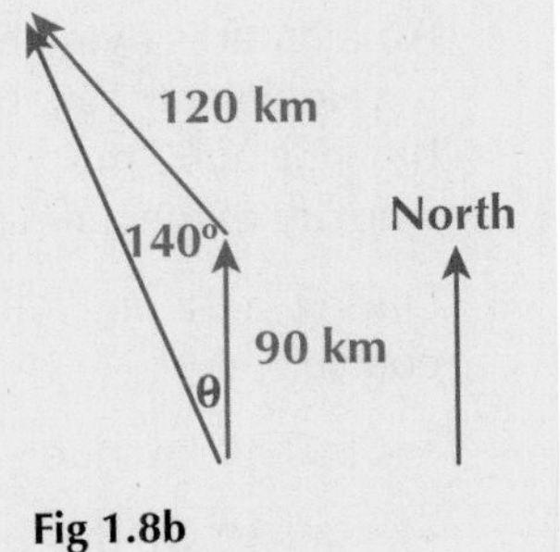

Fig 1.8b

(continued)

The angle, θ, can then be found using the sine rule:

$$\frac{a}{\sin A} = \frac{b}{\sin B}$$

$$\frac{198}{\sin 140°} = \frac{120}{\sin\theta}$$

$$120\sin 140° = 198\sin\theta$$

$$198\sin\theta = 77$$

$$\sin\theta = \frac{77}{198}$$

$$\theta = \sin^{-1}\left(\frac{77}{198}\right)$$

$$\theta = 23°$$

The bearing can be seen to be 360–θ, hence the bearing here is 337. This means the resultant displacement of the ship is:

$$s = 198 \text{ km on a bearing of } 337$$

b) *The resultant velocity is given by,*

$$v = \frac{s}{t}$$

Remember that the direction of the resultant velocity is the same as the direction of the resultant displacement calculated above. So, just find the magnitude of the resultant velocity by calculation:

$$v = \frac{s}{t}$$

$$v = \frac{198}{12}$$

$$v = 16.5 \text{ km h}^{-1}$$

The velocity is therefore: $v = 16.5$ *km* h^{-1} *on a bearing of* 337

Exercise 1.1.2 Velocity

1 Explain the difference between the velocity and the speed of an object.

2 A student writes, *'The average velocity of an object depends on the distance travelled and the time taken to travel this distance.'* Comment on this statement, explaining whether or not the student is correct.

3 A swimmer is swimming from south to north across a river. The swimmer swims at 4 m s^{-1} in the direction south to north. The current of the river is flowing from west to east at 3 m s^{-1}. Calculate the resultant velocity of the swimmer.

4 An aircraft is flying across Europe. The plane's engines provide a velocity of 150 m s^{-1} in an easterly direction. A wind is blowing at 60 m s^{-1} on a bearing of 140. Calculate the resultant velocity of the aircraft.

Fig 1.9

5 During a sailing race, a boat must complete the following course:

- Sail 40 km due east.
- Sail 30 km due north.
- Sail 10 km due west.

One boat completes the course in a time of 6 hours.

a) By scale drawing or otherwise, calculate the magnitude and direction of the resultant displacement of the boat.

b) Find the average velocity of the boat as it completes the above route.

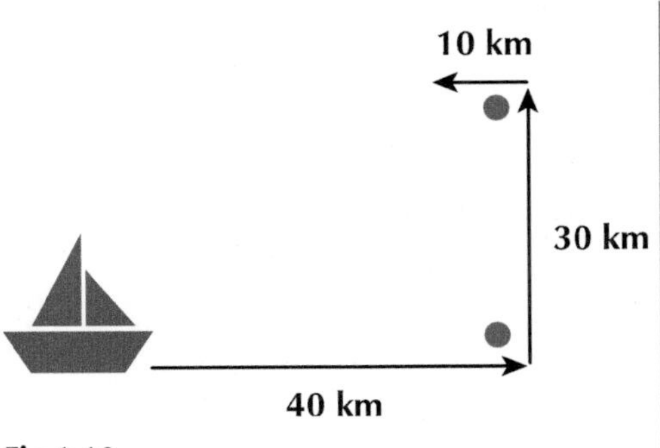

Fig 1.10

6 The velocity of an arrow fired by an archer at the Olympic Games is analysed. The arrow was found to be moving horizontally with a velocity of 80 $m s^{-1}$ while simultaneously moving vertically downwards at 10 $m s^{-1}$. Find the resultant velocity of the arrow.

7 While navigating across a glacier, a group of polar explorers walk the following route:

- An average velocity of 3 $m s^{-1}$ on a bearing of 070 for a time of 30 minutes.
- An average velocity of 2.5 $m s^{-1}$ on a bearing of 190 for a time of 3 h.

a) Find the total distance walked by the explorers.

b) Find the resultant displacement of the explorers.

c) Calculate the resultant average velocity of the explorers.

Fig 1.11

1.1.3 Acceleration

Acceleration is a vector quantity that describes the rate of change of velocity as a function of time. The greater the acceleration, the greater the change in velocity in a given time. Acceleration can be found using:

$$a = \frac{\Delta v}{t}$$

where Δv is the change in velocity. This leads to the familiar equation for acceleration:

$$a = \frac{v - u}{t}$$

where u is the initial velocity and v is the final velocity.

As acceleration is a vector quantity, it is necessary to quote both magnitude and direction. Usually in Higher Physics, acceleration is only considered in one dimension – so use a positive or negative sign to represent the direction. For example, you would assume that downwards would be negative, so the acceleration caused by gravity would be: $a = -9.8\ m s^{-2}$

Key point

Acceleration is the rate of change of velocity, given by:

$$a = \frac{\Delta v}{t} = \frac{v - u}{t}$$

It is a vector quantity, so it has both magnitude and direction.

Worked example

Calculate the acceleration of a car that starts with an initial velocity of 15 m s^{-1} and accelerates to 25 m s^{-1} in a time of 4 s.

- $a = ?$
- $u = 15\ m s^{-1}$
- $v = 25\ m s^{-1}$
- $t = 4\ s$

Use the equation for acceleration, substitute what is known and solve:

$$a = \frac{v-u}{t}$$

$$a = \frac{25-15}{4}$$

$$a = \frac{10}{4} = 2.5\ m s^{-2}$$

Exercise 1.1.3 Acceleration

1 Calculate the acceleration of a lorry that starts from rest and accelerates to a velocity of 25 m s^{-1} in a time of 40 seconds.

2 A train is travelling at 40 m s^{-1} and needs to slow down when approaching a station. The train has 2 minutes to come to rest. Calculate its deceleration.

3 A sprinter starts from rest and accelerates at 3 m s^{-2} for a time of 2 s. Calculate the final speed of the sprinter.

4 A plane must reach a certain velocity in order to take off (known as 'v2'). A Boeing 747-400 must reach a velocity of 90 m s^{-1} in order to take off. If it can accelerate at 8 m s^{-2}, what is the minimum time required for the take-off velocity to be reached, assuming it starts from rest?

5 A dart hits a dart board and comes to rest in a time of 0.05 seconds. The deceleration of the dart was determined to be 450 m s^{-2}. What was the velocity of the dart just before it hit the dart board?

6 Two cars are racing. The first car accelerates from rest with an acceleration of 6 m s^{-2} for 8 s. The second car accelerates from rest with an acceleration of 7 m s^{-2} for 7 s. Which car reaches the greatest top speed?

1.2 Motion–time graphs

So far, three different quantities linked to motion have been considered: displacement, velocity and acceleration. Each of these quantities can be plotted on a graph to quickly display the motion of an object. As the quantities are all linked (for example, acceleration is the *rate of change* of velocity), the graphs representing them are linked too.

All of the graphs of motion are plotted as functions of time; for example, a plot of the velocity of an object against time, which is called a velocity–time graph.

1.2.1 Displacement–time graphs

A displacement–time graph shows the displacement of an object at any given time. As displacement is a vector quantity, the graph can be either positive or negative to represent the direction of the displacement in one dimension.

Displacement–time graphs are difficult to plot for examples where the speed is not constant.

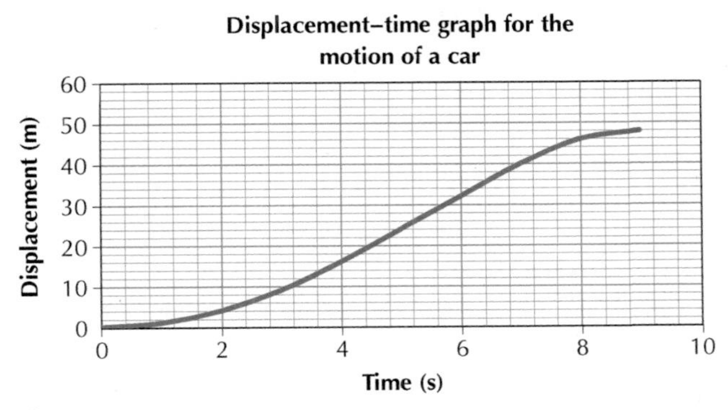

Fig 1.12

Notice that this graph has two curved sections with a straight line in between them. The initial curved section is where the car is accelerating and the final one is where it is decelerating. The straight line in the middle is where the car is travelling at a constant speed. The straight line highlights that the displacement is increasing at a constant rate – as expected for a constant speed.

1.2.2 Velocity–time and acceleration–time graphs

A velocity–time graph plots the velocity of an object as a function of time. Remember that velocity is a vector quantity, so the graph must also show the direction of the motion. When working in one dimension, the direction is represented using a positive or negative sign – a velocity–time graph can therefore be either positive or negative.

An example of a velocity–time graph is shown below. It is plotted for the motion of a car during a short journey. The car starts from rest ($u = 0\ \mathrm{m\,s^{-1}}$) and accelerates with a constant acceleration to a velocity of $v = 8\ \mathrm{m\,s^{-1}}$ in a time of 4 seconds. The velocity of the car remains constant for the next 3 seconds. The car then decelerates at a constant rate back to rest in a time of 2 seconds.

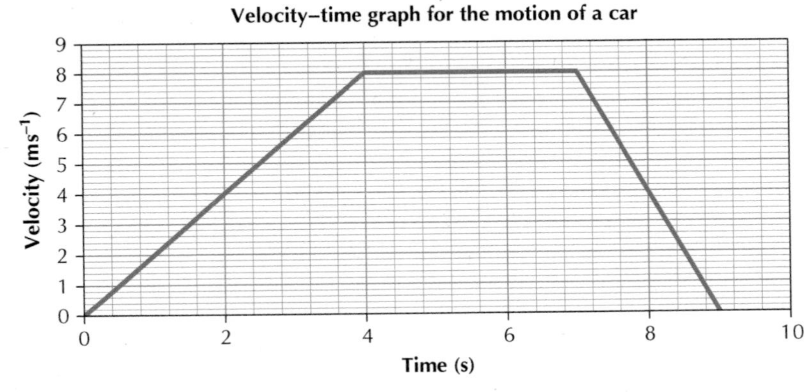

Fig 1.13

The velocity is always positive. This means that the car is always going in the same direction.

Calculating the acceleration for the first half of the journey:

$$a = \frac{v - u}{t}$$

$$a = \frac{8 - 0}{4}$$

$$a = 2\ m s^{-2}$$

Similarly, to calculate the acceleration when the car slows down:

$$a = \frac{0 - 8}{2} = -4\ m s^{-2}$$

When the car is travelling at a constant speed, the acceleration is zero. This gives the following acceleration–time graph.

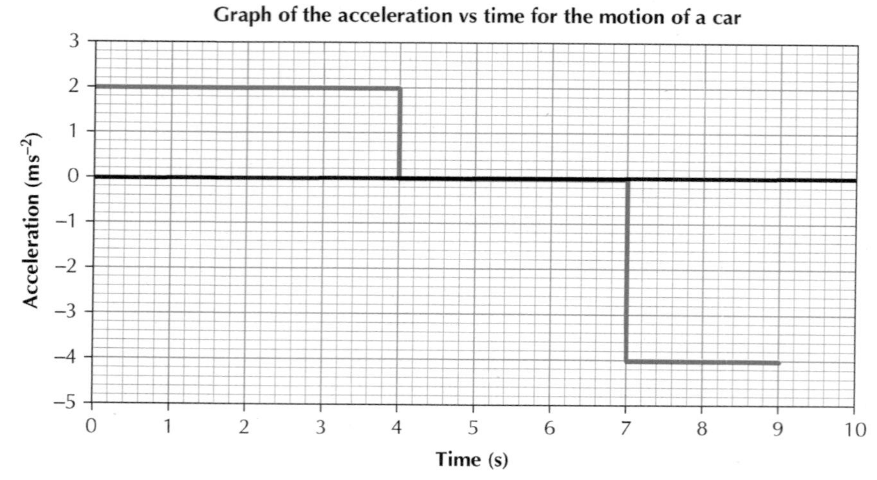

Fig 1.14

Exercise 1.2.2 Velocity–time and acceleration–time graphs

1 A train leaves a station and accelerates from rest to a speed of 15 m s^{-1} in a time of 50 s. The train then cruises at this speed for a further 200 s.

Plot a velocity–time graph for the motion of the train.

Fig 1.15

2 A runner in a race runs as follows:

- 0–2 s: constant acceleration from rest to 5 m s^{-1}
- 2–10 s: constant velocity of 5 m s^{-1}
- 10–11 s: constant deceleration to rest
- 11–15 s: remains at rest
- 15–17 s: constant acceleration to a velocity of 5 m s^{-1} in the opposite direction
- 17–26 s: constant velocity of 5 m s^{-1}

Plot a velocity–time graph for the motion of the runner.

Fig 1.16

3 The velocity–time graph below represents the motion of a dynamics cart during a lab experiment.

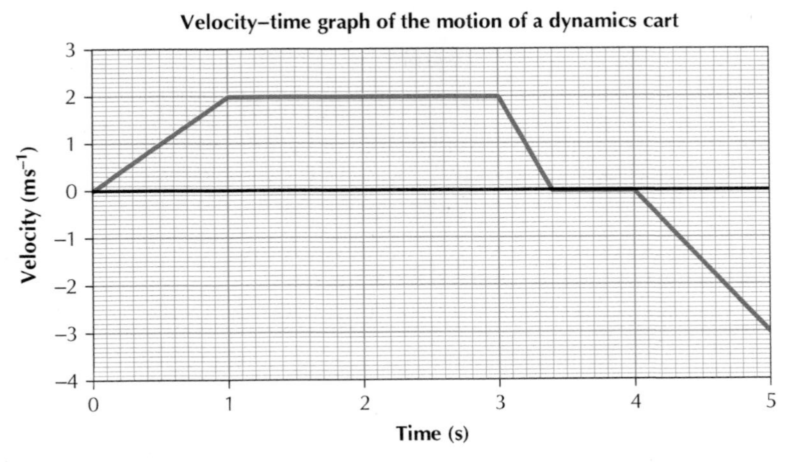

Fig 1.17

a) Describe the motion of the cart during the first second of the motion.

b) Between what times is the cart moving with a constant velocity?

c) Between what times is the cart stationary?

d) Explain the significance of the negative velocity between 4 and 5 s.

4 The acceleration–time graph shown below shows the acceleration for a train as it travels between two stations.

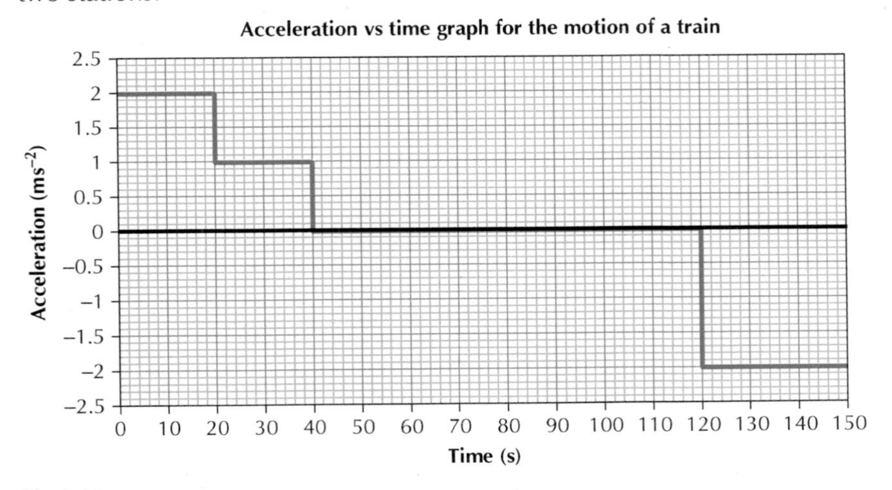

Fig 1.18

a) Between what times of the journey is the train accelerating?

b) Between what times of the journey is the train decelerating?

c) Between what times of the journey is the train travelling at a constant speed?

5 The journey of a car during a hill-climb race was as follows:

- 0–10 s: constant acceleration of 2 m s^{-2}
- 10–15 s: constant deceleration of 1 m s^{-2}
- 15–25 s: constant speed
- 25–30 s: constant acceleration of 4 m s^{-2}
- 30–45 s: constant deceleration of –6 m s^{-2}

Plot an acceleration–time graph to represent the motion of the car.

Fig 1.19

6 A bus accelerates uniformly from rest to 10 m s^{-1} in a time of 5 s. It then travels at a constant speed of 10 m s^{-1} for 1 minute before slowing down to rest in a time of 20 s.

a) Calculate the acceleration of the bus during the first 5 s of the journey.

b) Calculate the deceleration of the bus during the last 20 s of the journey.

c) Plot an acceleration–time graph to represent the motion of the bus.

1.2.3 Linking the motion–time graphs

Displacement, velocity and acceleration are all linked. Therefore, the graphs representing them must also be linked. Consider these velocity–time and acceleration–time graphs for the motion of the car referred to in section 1.2.2:

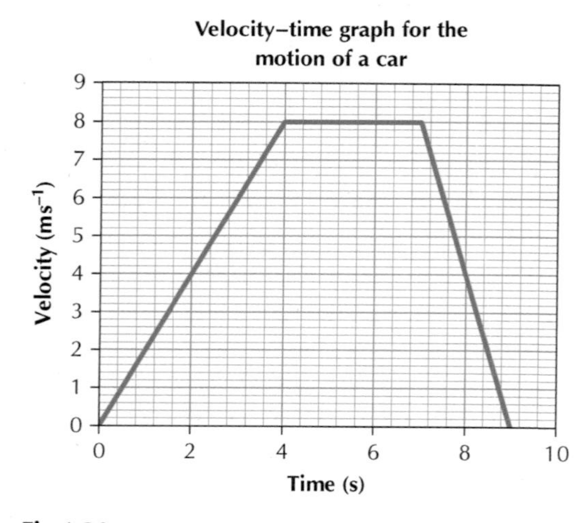

Fig 1.20

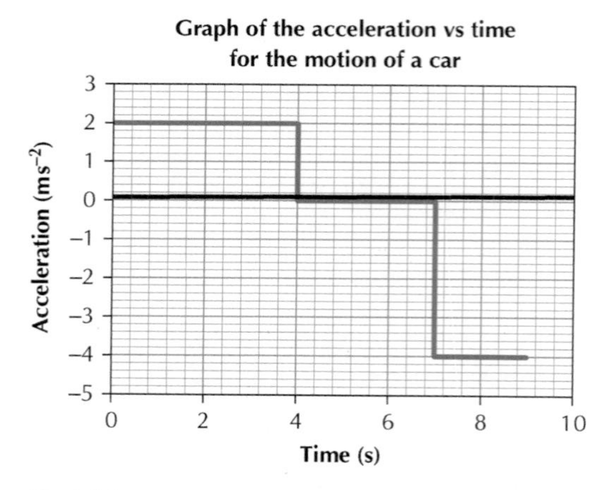

Fig 1.21

The gradient of the velocity–time graph can be calculated as the difference in y divided by the difference in x:

$$m = \frac{y_2 - y_1}{x_2 - x_1}$$

For the example above, the gradient in the first 4 seconds is:

$$m = \frac{8 - 0}{4 - 0}$$

$$m = 2$$

This corresponds to the acceleration of the car during the first 4 seconds of the journey. The acceleration can be found by finding the gradient of a velocity–time graph.

Consider these velocity–time and displacement–time graphs for the motion of the car:

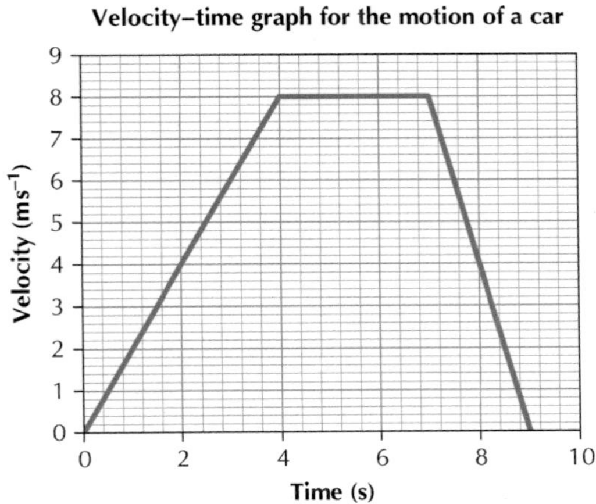

Fig 1.22

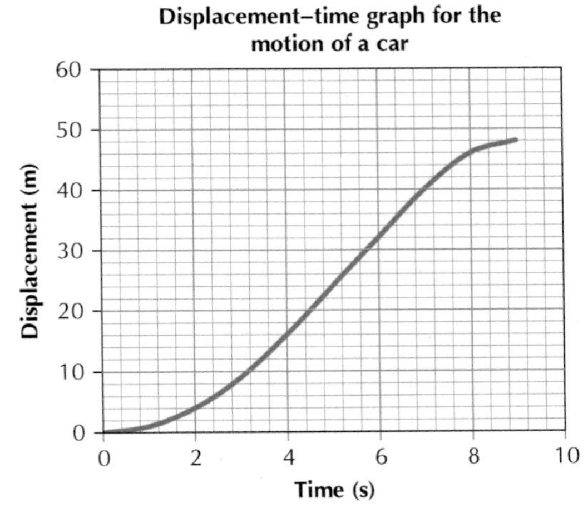

Fig 1.23

The displacement–time graph shows that after a time of 4 s, the displacement of the car is 16 metres. The area under the velocity–time graph over the first 4 seconds can be calculated by finding the area of the triangle:

$$A = \frac{1}{2}bh$$

$$A = \frac{1}{2}(4)(8)$$

$$A = 16$$

This corresponds to the displacement of the car over the first 4 seconds of the journey. The displacement can therefore be found by finding the area under a velocity–time graph. It is important to note that when the velocity is negative, the displacement will also be negative. Therefore, any areas that are calculated below the x-axis should be taken as negative.

The **gradient** of a displacement–time graph provides the velocity. The **area** under an acceleration–time graph provides the velocity.

Key point

The links between the three graphs are shown in the diagram below.

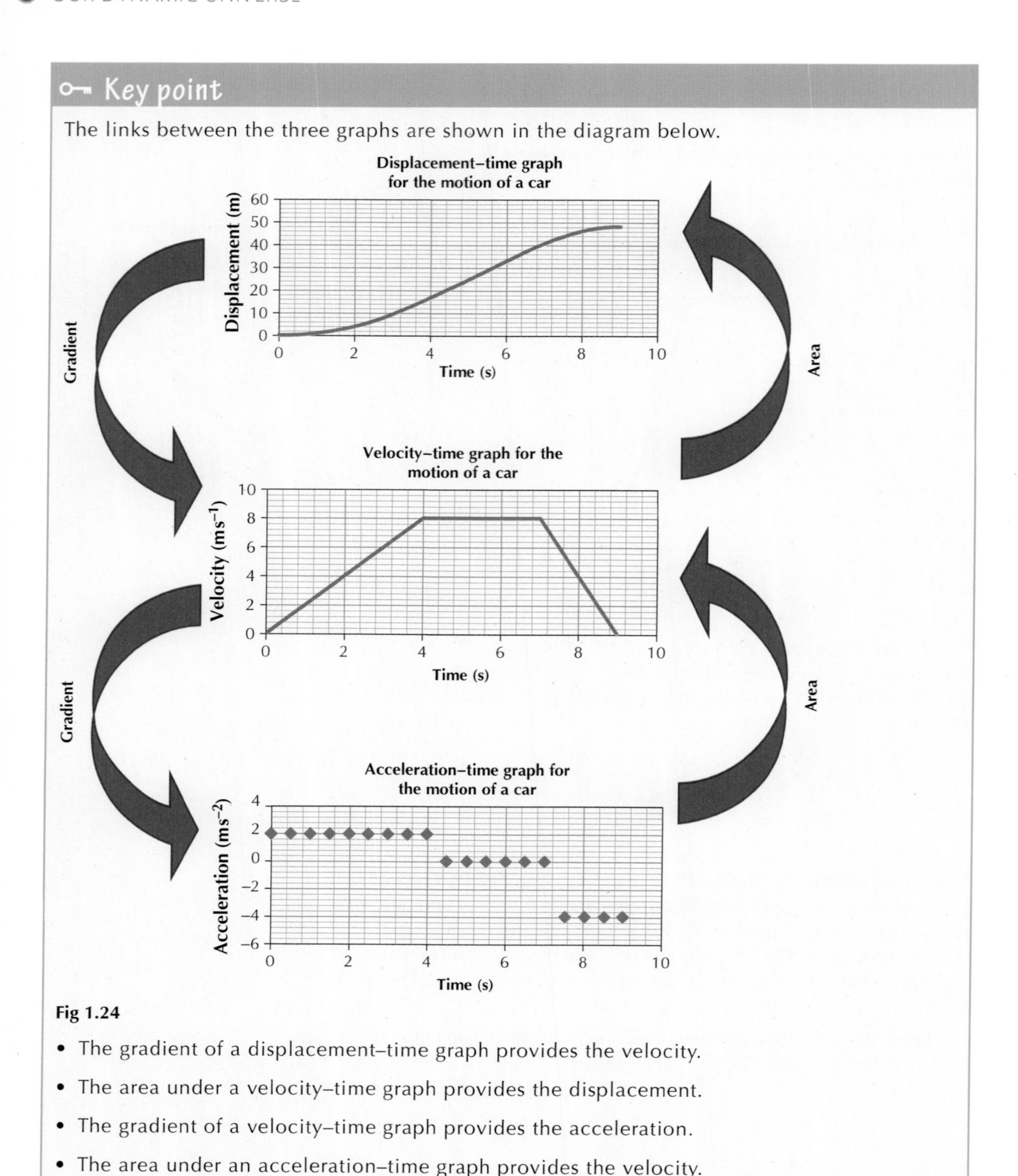

Fig 1.24

- The gradient of a displacement–time graph provides the velocity.
- The area under a velocity–time graph provides the displacement.
- The gradient of a velocity–time graph provides the acceleration.
- The area under an acceleration–time graph provides the velocity.

Worked example

The velocity–time graph for a car is shown in Figure 1.25:

a) Find the velocity of the car at 10 s into the journey.

b) Calculate the acceleration of the car in the first 10 s of the journey.

c) Calculate the deceleration of the car in the last 4 s of the journey.

d) Calculate the total displacement travelled over the 20 s journey.

e) Describe the motion of the car between 10 s and 16 s.

f) Plot an acceleration–time graph for the motion of the car over the whole journey.

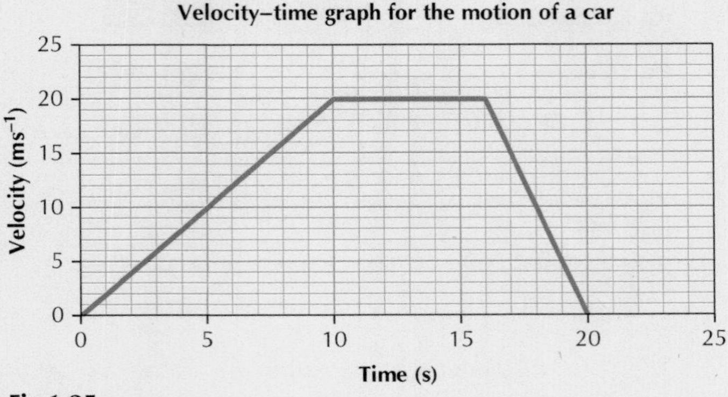

Fig 1.25

a) *Using the velocity–time graph directly, the velocity can be seen as:*

$$v = 20\ ms^{-1}$$

b) *The acceleration can be found by working out the gradient of the velocity–time graph:*

$$a = \frac{y_2 - y_1}{x_2 - x_1}$$

$$a = \frac{20 - 0}{10 - 0}$$

$$a = 2\ ms^{-2}$$

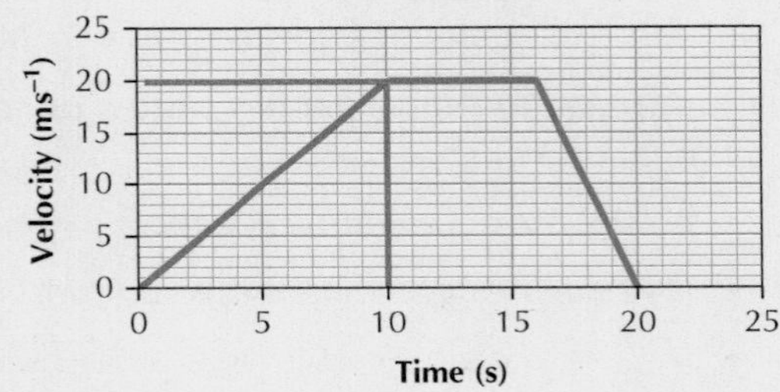

Fig 1.26

**The acceleration can also be found by using the equation* $a = \frac{\Delta v}{t} = \frac{v - u}{t}$.

This equation is exactly the same as the one for the gradient of the line, it is just specific to the example of velocity, acceleration and time!

c) *The deceleration of the car is found in a similar way to the acceleration of the car above. Note that the deceleration of the car will be negative because the car is slowing down.*

$$a = \frac{y_2 - y_1}{x_2 - x_1}$$

$$a = \frac{0 - 20}{20 - 16}$$

$$a = -5\ ms^{-2}$$

d) *The displacement is found by working out the area under a velocity–time graph. First split the area up into manageable shapes that you can easily calculate the area of (rectangles and triangles):*

Now calculate the area of each of the shapes individually:

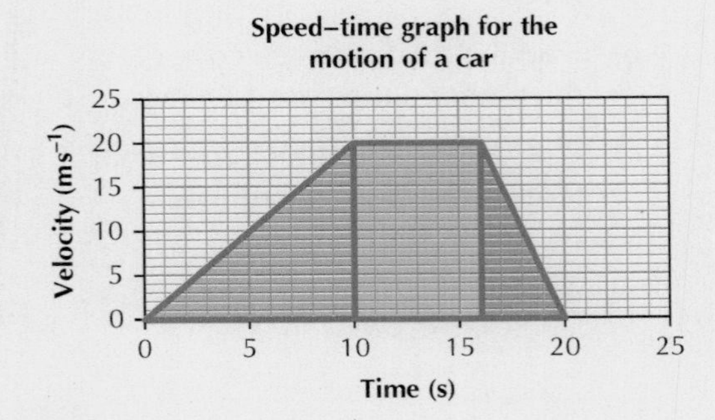

Fig 1.27

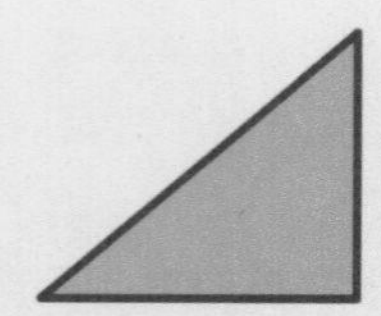

Fig 1.28

0–10 seconds:

$A = \frac{1}{2}b \times h$

$A = \frac{1}{2} \times 10 \times 20$

$A = 100\ m$

Fig 1.29

10–16 seconds:

$A = l \times b$

$A = 6 \times 20$

$A = 120\ m$

Fig 1.30

16–20 seconds:

$A = \frac{1}{2}b \times h$

$A = \frac{1}{2} \times 4 \times 20$

$A = 40\ m$

The total displacement can then be found by adding together all the individual areas:

$s = 100 + 120 + 40$

$s = 260\ m$

e) *The car is moving with a constant velocity of 20 $m s^{-1}$ (or zero acceleration) between 10 s and 16 s.*

f) *The acceleration–time graph is plotted using the following information from the parts above:*

- *0–10 s, the acceleration is 2 $m s^{-2}$.*
- *10–16 s, the acceleration is zero (constant speed).*
- *16–20 s, the acceleration is –5 $m s^{-2}$ (deceleration).*

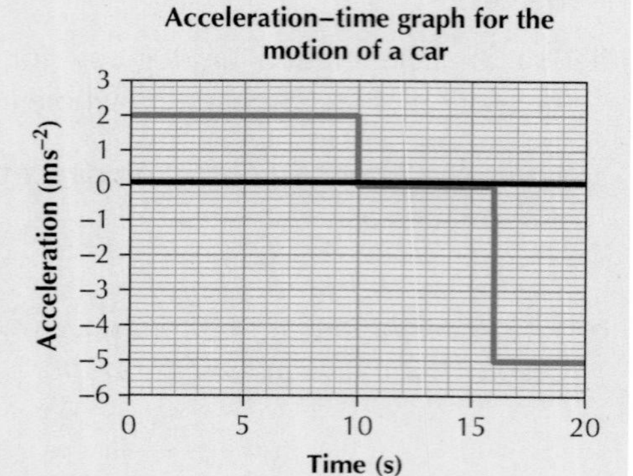

Fig 1.31

Exercise 1.2.3 Linking the motion–time graphs

1 The velocity–time graph for the motion of a runner during a race is shown in Figure 1.32.

a) Calculate the acceleration of the runner during the first 4 s of the race.

b) Find the total distance covered by the runner during the 12 s race.

c) Plot an acceleration–time graph for the motion of the runner.

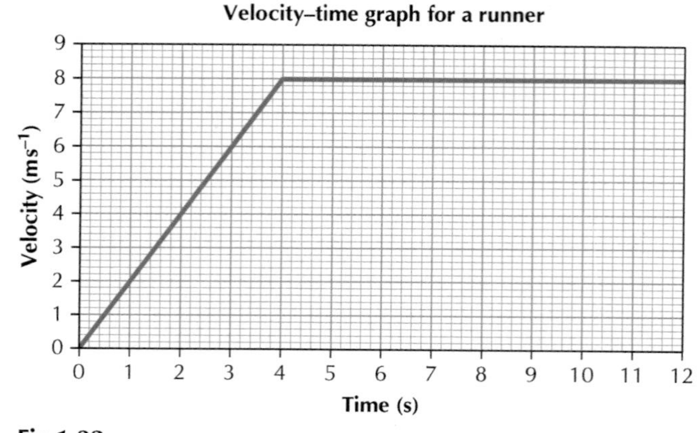

Fig 1.32

2 The acceleration–time graph in Fig 1.33 represents the motion of a dog chasing a frisbee. The dog starts from rest.

a) Calculate the velocity of the dog after 2 s.

b) Calculate the velocity of the dog after 5 s.

c) Plot a velocity–time graph for the motion of the dog.

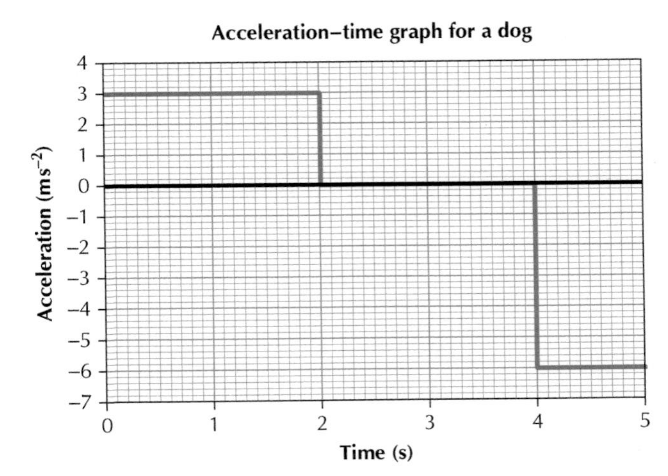

Fig 1.33

3 When taking off, a plane must reach a certain speed in order to generate enough lift to become airborne. This velocity is known as 'V2'.

A particular plane has a constant acceleration of 4 ms^{-2}. It must reach a velocity of 80 ms^{-1} to be able to take off.

a) How long does it take the plane to reach 80 ms^{-1}?

b) Plot a velocity–time graph for the motion of the plane during its acceleration to 80 ms^{-1}.

c) Calculate the minimum length of runway required to allow the plane to take off safely.

Fig 1.34

4 A man is late for his train. He runs along the platform to try to catch it. After 8 s, he catches up with the train and jumps on. The graph representing his motion is shown in Figure 1.35.

a) Calculate the acceleration of the man during the first 4 s of the motion.

b) Calculate the acceleration of the man when he jumps onto the train.

c) Between what times is the man moving at a constant velocity?

d) How far did the man have to run before jumping on the train?

e) Calculate the total displacement of the man over the 20 s.

f) Plot an acceleration–time graph for the motion of the man over the 20 s.

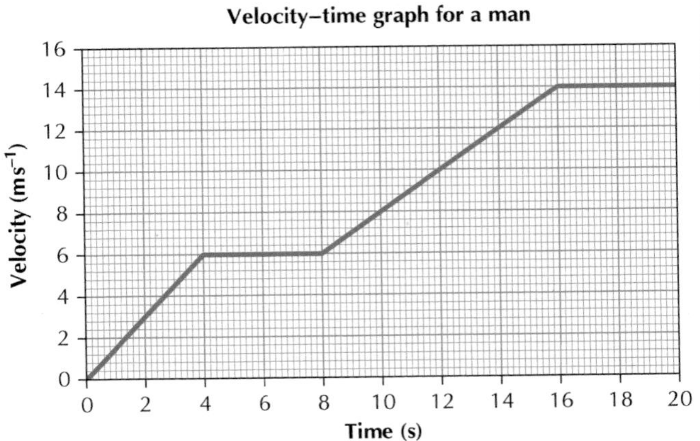

Fig 1.35

5 A car is travelling at a constant velocity of 15 ms^{-1} for a time of 7 s.

a) Plot a velocity–time graph to represent the motion of the car.

b) Plot a displacement–time graph to represent the motion of the car.

c) Calculate the displacement of the car using the equation $s = vt$. Is there a link between the area of a rectangle on the velocity–time graph and the equation used here? Explain you answer.

1.2.4 Motion–time graphs: special cases

In the previous section, the properties of motion–time graphs and how the different graphs are linked to each other are considered. Now look at two special cases and their resulting motion–time graphs.

Experiment 1.2.4 The bouncing ball

This experiment investigates the motion–time graphs of a bouncing ball.

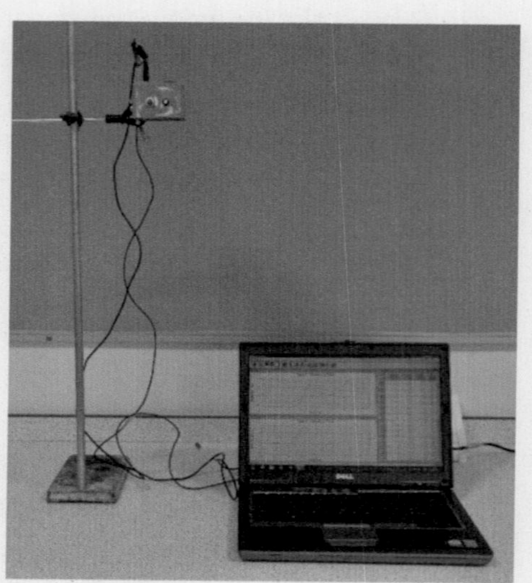

Fig 1.36

Apparatus

- A motion sensor connected to a computer
- A ball

Instructions

1. Connect the motion sensor to a computer with appropriate software to plot the motion–time graphs for the bouncing ball.
2. Hold the ball underneath the motion sensor. Drop the ball and use the motion sensor to record the velocity, displacement and acceleration. Don't throw the ball – allow it to fall freely under gravity.
3. Plot three graphs of the motion:
   (a) a displacement–time graph of the motion
   (b) a velocity–time graph of the motion
   (c) an acceleration–time graph of the motion.
4. Explain the links between the above graphs of motion. Highlight on the graphs the points where the ball hits the ground and where it reaches its maximum height.
5. Is the maximum height reached by the ball the same after each bounce? Explain this result.

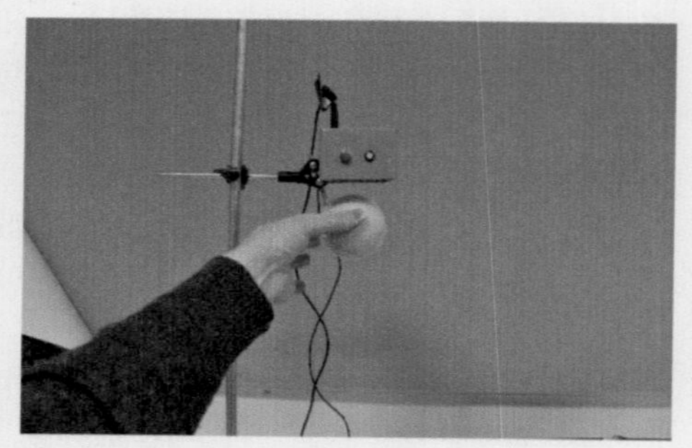

Fig 1.37

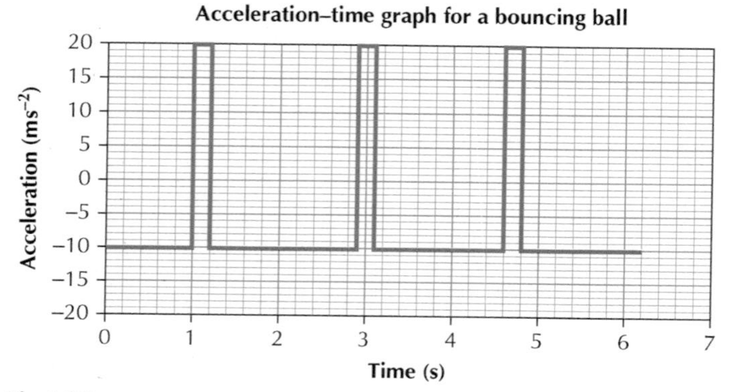

Fig 1.38

When a ball is dropped, it is subject to a constant acceleration downwards due to gravity. When the ball is moving downwards, gravity will make the ball travel faster. When the ball is moving upwards, gravity will act to slow the ball down. The graphs of a bouncing ball are shown and explained. For simplicity, the acceleration due to gravity is taken to be $-9.8\ m\,s^{-2}$.

Acceleration–time graph

The acceleration–time graph for a bouncing ball may look a little strange at first! The ball is subject to a constant acceleration of $-9.8\ \text{m s}^{-2}$ due to gravity: this gives the horizontal line (between 0 and 1 s). The peaks are given when the ball bounces and changes direction quickly. The fast direction change requires a large acceleration.

Velocity–time graph

The velocity time graph of the bouncing ball has two main parts. The first is velocity that is always decreasing – with a constant acceleration due to gravity. The second is when the ball changes direction and switches from a negative velocity (downwards) to a positive velocity (upwards). This represents the bounce of the ball.

The gradient of the line provides the acceleration. The area under the graph provides the displacement, which is shown in Figure 1.42.

Displacement–time graph

The displacement of the ball is measured using the sensor, with the downward direction taken to be negative. The graph here shows that the ball falls 5 m downwards, hits the ground and returns upwards to a maximum height of 1 m below the sensor. The sensor must therefore be 5 m above the ground.

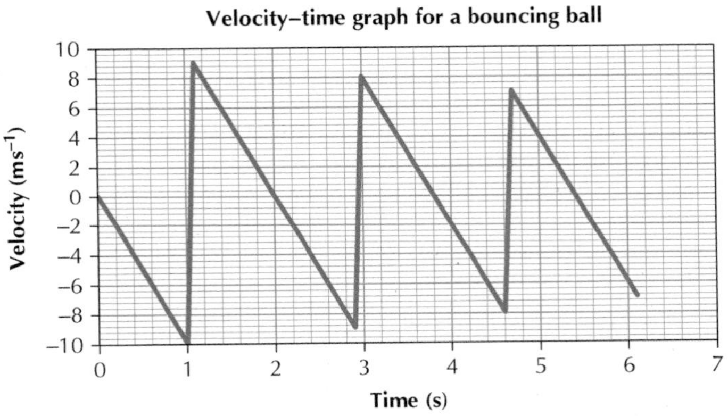

Velocity–time graph for a bouncing ball

Velocity (ms^{-1})

Time (s)

Fig 1.39

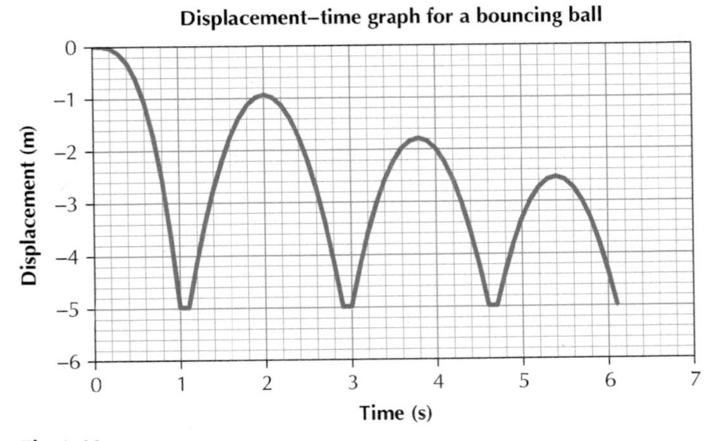

Fig 1.40

Energy losses result in the maximum height the ball reaches decreasing after each bounce.

Consider the section of the velocity–time graph for the bouncing ball shown here:

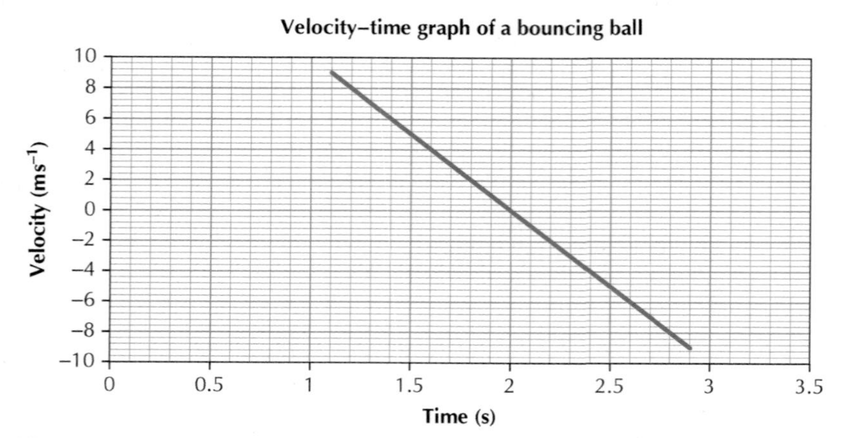

Fig 1.41

The acceleration is given by the gradient of the line, which here is given by:

$$m = \frac{y_2 - y_1}{x_2 - x_1}$$

$$m = \frac{-8 - 8}{2.8 - 1.2}$$

$$m = \frac{-16}{1.6}$$

$$m = -10$$

$$m = -10\ m s^{-2}$$

The displacement is given by the area under a velocity–time graph. Direction is important!

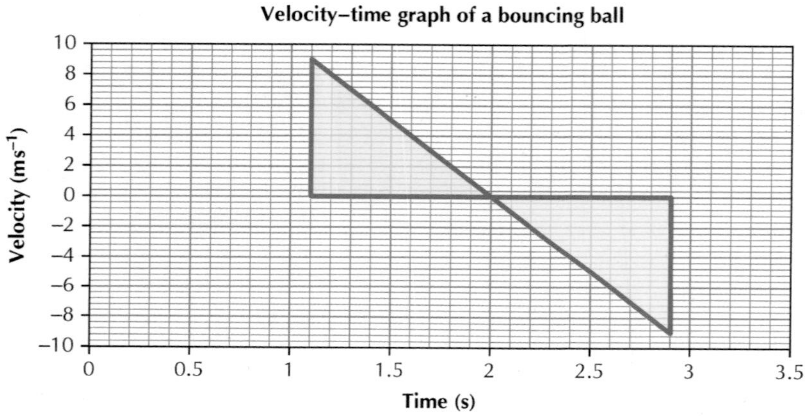

Fig 1.42

Up to 2 s, velocity is positive because the ball is moving upwards. This means that the area will be positive, giving a positive displacement (upward motion). After 2 s, the velocity is negative, so the displacement will be negative (downward motion). The resultant displacement between 1.1 and 2.9 s is found by calculating two separate areas, highlighted in different colours on the graph.

$$A = \frac{1}{2}bh \qquad A = \frac{1}{2}bh$$

$$A = \frac{1}{2} \times 0.9 \times 9 \qquad A = \frac{1}{2} \times 0.9 \times -9$$

$$A = 4.05\ m \qquad A = -4.05\ m$$

$$s = 4.05 - 4.05 = 0\ m$$

Here, the positive area is cancelled out by the negative area – the overall displacement is zero, meaning that the ball has started from the ground and then returned to the ground.

1.3 The equations of motion

This section considers equations that can be used to represent the motion of objects that are moving with a constant acceleration.

1.3.1 Deriving the equations of motion

Three main equations of motion are considered here. The derivations of these equations are shown below. For the Higher course, you do not need to be able to derive these equations, but you do need to be able to understand and apply them!

Equation 1

The first equation of motion follows from the definition of acceleration:

$$a = \frac{v - u}{t}$$

If you rearrange this equation, you get:

$$at = v - u$$

$$v = u + at$$

The first equation of motion is:

$$v = u + at$$

Equation 2

The second equation of motion can be derived using the velocity–time graph for the motion of an object. Consider the graph below for an object starting with initial velocity u and accelerating to a final speed v in a time t.

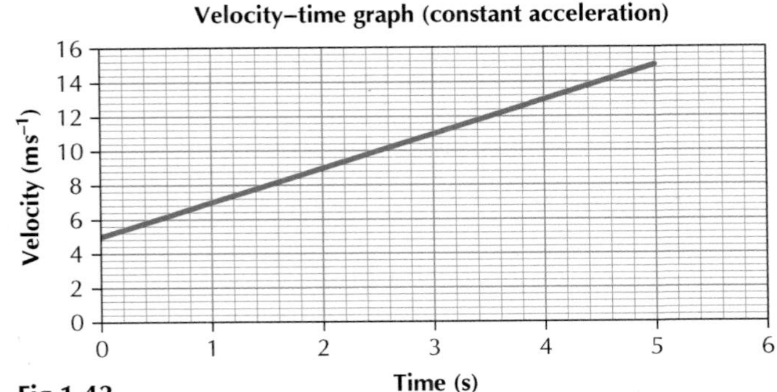

Fig 1.43

As shown in the previous section, finding the area under the graph will provide the displacement, *s*. Here, you can divide the graph into easy-to-work-with shapes to get the following:

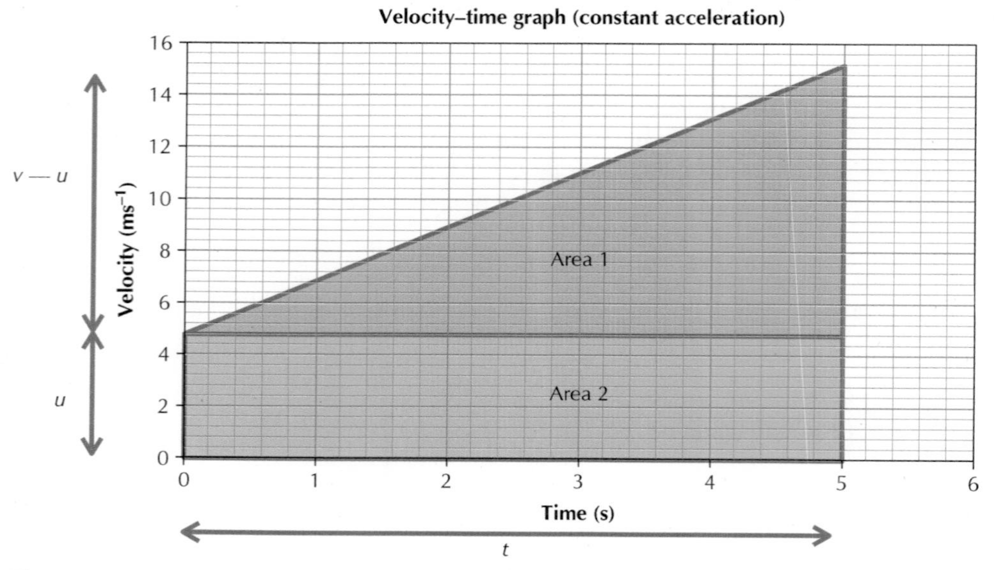

Fig 1.44

The displacement is therefore given by the area of the triangle plus the area of the rectangle. Working this out gets:

$$s = \text{Area 1} + \text{Area 2}$$

$$s = (l \times b) + \left(\frac{1}{2} \times b \times h\right)$$

$$s = ut + \frac{1}{2}(v - u)t$$

The derived first equation is:

$$v - u = at$$

and if you substitute this into the previous equation for displacement you get:

$$s = ut + \frac{1}{2}(at)t$$

which is the second equation of motion, written as:

$$s = ut + \frac{1}{2}at^2$$

Equation 3

The third equation of motion comes from taking the first equation of motion, squaring both sides and rearranging:

$$v = u + at$$

$$v^2 = (u + at)^2$$

$$v^2 = u^2 + 2uat + a^2t^2$$

This equation can also be rearranged as follows:

$$v^2 = u^2 + 2a\left(ut + \frac{1}{2}at^2\right)$$

As the term in the bracket is the equation for displacement (equation 2), the third equation of motion is:

$$v^2 = u^2 + 2as$$

Experiment 1.3.1 Proving an equation of motion

This experiment measures the time taken for a ball to fall through known heights to prove the equation of motion:

$$s = ut + \frac{1}{2}at^2$$

Apparatus

- Metal ball bearing
- Electromagnet
- 'Trap door'*
- Timing device
- Metre stick

* The trap door can be made from aluminium foil stretched over the landing area. The ball will then break this foil to break the circuit as required.

Instructions

1. Set up the experiment as shown in Figure 1.45. The timing device should be set up to measure the time difference between two events:
   - The current to the electromagnet being broken (causing the ball to start falling).
   - The current through the trap-door series circuit being broken (by the ball breaking the trap door).
2. Measure the height, h, between the electromagnet and the trap door.
3. With the electromagnet switched on, attach the ball and check that a current is flowing through the trapdoor circuit.
4. Trigger the electromagnet to switch off and release the ball. Use the timing device to measure the time taken for the ball to reach the trap door, t.
5. Record your results in a table similar to the one shown below:

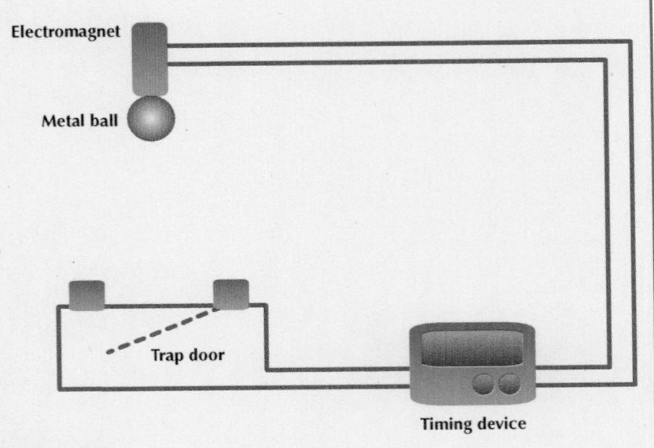

Fig 1.45

Height, h (m)	Time, t (s)	Time squared, t^2 (s^2)

Table 1.1

6 Plot a graph of the height vs the time taken to fall.
7 Plot a graph of the height vs the time squared – this should produce a straight line graph. Calculate the gradient of the straight line you have produced.
8 Consider the equation of motion:

$$s = ut + \frac{1}{2}at^2$$

Here, the initial velocity is zero and the acceleration is the acceleration due to gravity, so

$$h = \frac{1}{2}gt^2$$

where the displacement is equal to the height and the downward direction taken to be positive. This can be compared with the equation of a straight line,

$$y = mx$$

to prove the equation of motion and find the acceleration due to gravity as the gradient of the line,

$$m = \frac{1}{2}g$$

1.3.2 Applying the equations of motion

The three equations of motion above can be used when solving problems that involve a constant acceleration; for example, motion under gravity. The derivations need not be memorised, but you should be familiar with the three equations:

$$v = u + at$$

$$s = ut + \frac{1}{2}at^2$$

$$v^2 = u^2 + 2as$$

There are five possible variables that you may encounter when solving motion problems:

- Initial velocity, u
- Final velocity, v
- Acceleration, a
- Displacement, s
- Time, t

When solving a motion problem, writing down a table of these five values and filling in what you know and what you are trying to find out can help with choosing the correct equation from the three. Remember you are dealing with vector quantities so take care to get the sign correct! A useful convention is 'motion to the right or motion upwards is positive'.

Worked example 1

A lorry starts accelerating from 20 ms^{-1} to a speed of 35 ms^{-1}. A car covers a distance of 100 m during this acceleration. Calculate the acceleration of the lorry.

Fig 1.46

Write down what you know and what you are trying to find out in the suvat table:

s	u	v	a	t
100	20	35	?	X

Table 1.2

You can now choose the correct equation – you do not have a value for time and you are not looking for time, so choose the equation that does not have time:

$$v^2 = u^2 + 2as$$

Substitute what you know and then solve for the acceleration:

$$35^2 = 20^2 + 2(a)(100)$$
$$1225 = 400 + 200a$$
$$200a = 825$$
$$a = \frac{825}{200}$$
$$a = 4.13\ ms^{-2}$$

Worked example 2

When a cannon is fired, the cannonball travels the length of the cannon. The cannonball is initially at rest and accelerates along the length of the cannon in a time of 0.2 s with an acceleration of 150 ms^{-2}. Determine:

(a) The length of the cannon.

(b) The final velocity of the cannonball when it is launched.

(a) Start by writing down what you know and what you are trying to find out in the suvat table:

s	u	v	a	t
?	0	X	150	0.2

Table 1.3

Here the appropriate equation is:

$$s = ut + \frac{1}{2}at^2$$

Substitute what you know and then solve for the displacement, which is the length of the cannon:

$$s = (0)(0.8) + \frac{1}{2}(150)(0.2)^2$$
$$s = \frac{1}{2}(150)(0.04)$$
$$s = 3\ m$$

(b) Again, write down the suvat table, filling in what you know and what you are trying to find out (the final velocity, v):

s	u	v	a	t
3	0	?	150	0.2

Table 1.4

You can choose any equation here:

$$v = u + at$$

Fill in what you know and solve for v:

$$v = (0) + (150)(0.2)$$
$$v = 30\ ms^{-1}$$

Exercise 1.3.2 Applying the equations of motion

Note: ignore air resistance throughout.

Fig 1.47

1. A train accelerates from 5 ms^{-1} with an acceleration of 4 ms^{-2}. Calculate the velocity of the train after accelerating for 8 s.
2. A bus is initially stationary. It accelerates with an acceleration of 2.5 ms^{-2} for a time of 6 s. Calculate the distance the bus has travelled while accelerating and what final speed it reached.
3. Engineers are working out the minimum acceleration of an aircraft such that it can reach its take-off speed of 65 ms^{-1}. It must accelerate from rest and the length of the runway is 1500 m. Calculate the minimum acceleration of the aircraft so that it reaches its take-off speed before the end of the runway.

Fig 1.48

4. During an emergency stop, a car decelerates from 25 ms^{-1} to rest in a time of 4 s. Calculate the distance taken for the car to come to a stop.
5. A train is travelling with a velocity of 35 ms^{-1}. In order to stop before entering a station ahead, it must slow down over a distance of 300 m. Calculate the minimum deceleration the train must achieve to stop before entering the station.

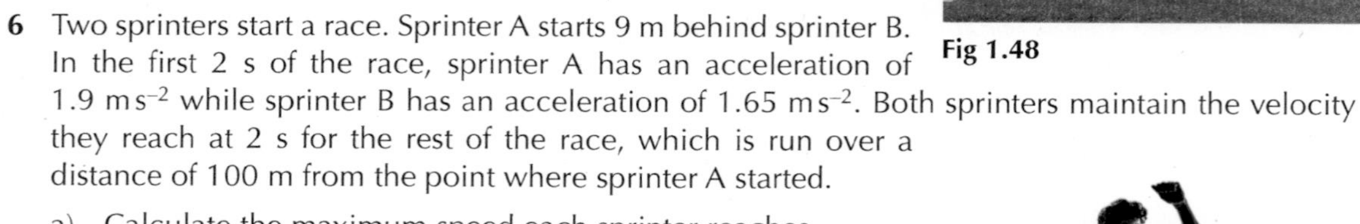

6. Two sprinters start a race. Sprinter A starts 9 m behind sprinter B. In the first 2 s of the race, sprinter A has an acceleration of 1.9 ms^{-2} while sprinter B has an acceleration of 1.65 ms^{-2}. Both sprinters maintain the velocity they reach at 2 s for the rest of the race, which is run over a distance of 100 m from the point where sprinter A started.

   a) Calculate the maximum speed each sprinter reaches.

   b) State which sprinter wins the race.

   c) Determine the time between the first sprinter and the second sprinter.

Fig 1.49

7. Derive the three equations of motion shown below where the symbols have their usual meaning:

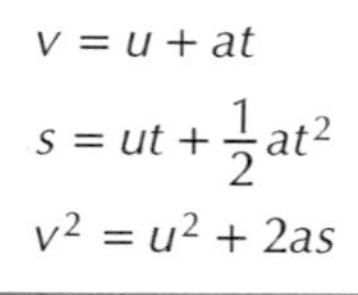

$$v = u + at$$
$$s = ut + \frac{1}{2}at^2$$
$$v^2 = u^2 + 2as$$

1.3.3 Equations of motion applied to gravity

At National 5 level, we learned that gravity can affect the motion of an object falling vertically. Motion under gravity is an important application of the equations of motion. To study the motion of a projectile, split it into horizontal and vertical components. The vertical component of the motion is subject to gravity and therefore the equations of motion apply.

At Higher level, gravity is taken to be $a = g = -9.8\ m\,s^{-2}$.

The negative sign highlights that the upward direction is assumed to be positive and that gravity acts downwards. Remember, you can have the acceleration due to gravity as positive if you define the downward direction to be positive.

Key point

When dealing with vertical motion under gravity (such as the velocity of a stone thrown vertically upwards), the acceleration will always be equal to the gravitational acceleration, $a = g = -9.8\ m\,s^{-2}$.

Key point

An object reaches its maximum height when its vertical velocity is 0.

Worked example 1

A man throws a stone upwards into the air with an initial velocity of 18 $m\,s^{-1}$. Calculate the maximum height reached by the stone.

Fig 1.51

First of all define the positive direction. Take the positive direction to be upward:

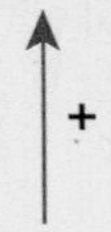

Fig 1.50

Now construct a suvat table, filling in what you know and what you are looking for. Remember that the final velocity at the maximum height is 0.

s	u	v	a	t
?	18	0	–9.8	X

Table 1.5

Choose the appropriate equation, which in this case does not involve time:

$$v^2 = u^2 + 2as$$

Substitute what you know and solve for the displacement that corresponds to the maximum height:

$$\frac{a}{\sin A} = \frac{b}{\sin B} = \frac{c}{\sin C}$$

$$a^2 = b^2 + c^2 - 2bc \cos A$$

Exercise 1.3.3 Equations of motion applied to gravity

Note: ignore air resistance throughout.

1 A ball is thrown vertically upwards. It reaches a height of 35 m. Calculate the following:

a) The velocity the ball was thrown with.

b) The time taken for it to reach its maximum height.

c) The velocity of the ball when it returns to the thrower's hand.

2 a) A stone is dropped off of the edge of a cliff that is 29 m above sea level. Calculate how long it takes the stone to reach the water.

b) A catapult is used to launch a stone vertically upwards with an initial velocity of 45 $m\,s^{-1}$. Ignoring the effects of air resistance, calculate the maximum height reached by the stone and the length of time taken for it to reach this maximum height.

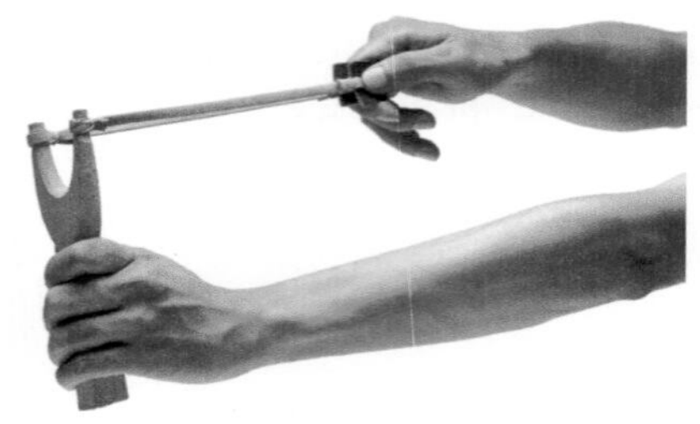

Fig 1.52

3 A ball is thrown vertically into the air with an initial speed of 9.8 $m\,s^{-1}$.

a) Calculate:

(i) The maximum height the ball will reach.

(ii) The time taken for it to reach the maximum height.

b) Plot a velocity–time graph to represent the motion. Use this graph to calculate the maximum height. Do your answers agree?

Fig 1.53

4 A hot air balloon, travelling at a constant altitude, needs to gain height to avoid colliding with transmission lines. To do this, sandbags are thrown over the side to reduce the weight of the balloon. A sandbag takes 4.7 s to hit the ground. Calculate what height the balloon was flying at when the sandbag was released.

5 A ball is thrown vertically off the edge of a cliff with an initial speed of 15 $m\,s^{-1}$ upwards. The cliff is 45 m high.

a) Calculate the maximum height above sea level that the ball reaches.

b) Find the total time of the flight of the ball.

c) Calculate the total distance travelled by the ball.

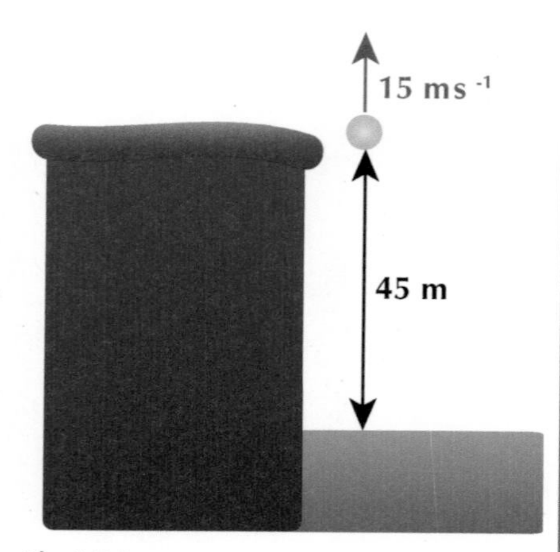

Fig 1.54

2 Force, energy and power

You should already know (National 5)

- Newton's laws of motion:
  - First law
  - Second law *($F = ma$)*
  - Third law
- How to find the unbalanced force when more than one force is acting
- Potential energy and kinetic energy of moving objects
- Power is the rate of change of energy

Learning intentions

- Understand balanced and unbalanced forces in one dimension:
  - Effects of friction
  - Effects of air resistance on falling objects
  - Terminal velocity
- Analysis of motion using Newton's laws of motion
- Tension in systems
- Rectangular components of vectors to resolve forces
- Forces and motion on an inclined plane
- Balanced and unbalanced forces in two dimensions
- Energy:
  - Work done
  - Potential energy
  - Kinetic energy
  - Conservation of energy

2.1 Forces in one dimension

The concept of balanced and unbalanced forces has already been studied in previous Physics courses.

2.1.1 Resultant force in one dimension

When more than one force is acting on an object, these forces can be added together as vectors to find the resultant force. If this resultant force is zero, the forces acting on the object are balanced. The unbalanced force (magnitude and direction) can be calculated in one dimension by adding or subtracting the individual forces acting on the object. Figure 1.55 shows the direction of the force; one of these directions is defined as positive and the other negative, so the sign of the resultant force shows the direction in which it is acting.

The resultant force, which has previously been called the 'unbalanced force', is given by:

F_{un} = (forces acting in positive direction) – (forces acting in negative direction)

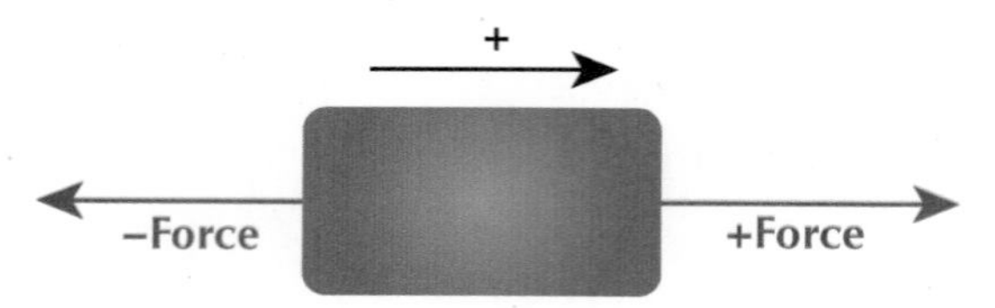

Fig 1.55

Weight

Sometimes we need to calculate the individual forces acting on an object. An example of this would be the weight of an object, which is the downward force acting due to gravity. National 5 Physics has shown that the weight of an object, W, is given by:

$$W = mg$$

where m is the mass (kg) and g is the gravitational field strength in $\mathrm{Nkg^{-1}}$. Any object on Earth is subject to a downward acceleration due to gravity. Objects on the ground or on a table have an equal and opposite upward force so they don't fall downwards.

Worked example

A rocket is accelerated by a force of 2500 N upwards. The mass of the rocket is 100 kg. Calculate the resultant force acting on the rocket.

Start by drawing a diagram that shows all of the forces acting on the rocket. Define the upwards direction as positive:

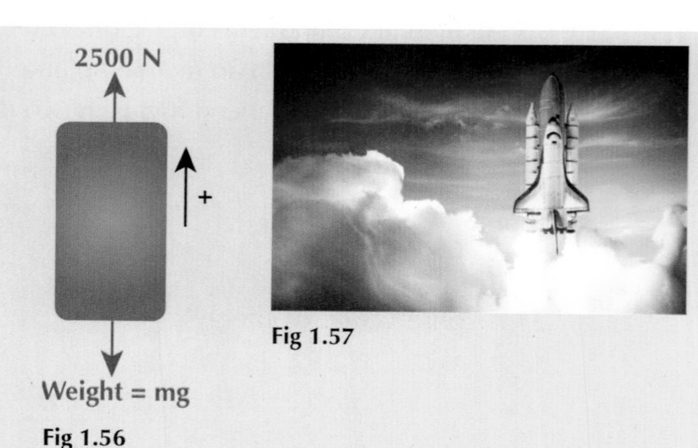

Fig 1.56

Fig 1.57

The weight of the rocket is calculated using the equation:

$$W = mg$$
$$W = 100 \times 9.8$$
$$W = 980\ N$$

The weight is acting in the negative direction. The resultant force can then be calculated:

$$F = 2500 - 980$$
$$F = 1520\ N$$

The positive sign here shows that the resultant force is acting upwards.

Exercise 2.1.1 Resultant force in one dimension

1 Calculate the resultant force acting on each of the objects below.

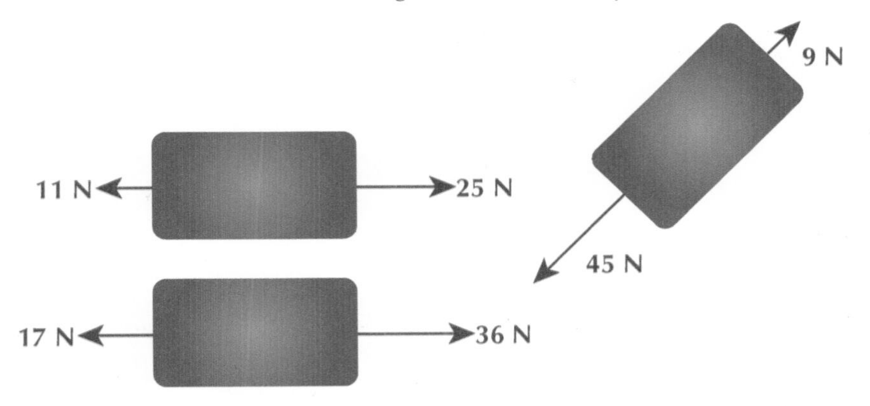

Fig 1.58

2 A bag of shopping has a mass of 2.7 kg. It is being lifted upwards with an upward force of 40 N. Calculate the resultant force acting on the shopping bag.

3 A rocket of mass 2700 kg is accelerating upwards into space from the Earth. The resultant force acting on the rocket is 11,000 N. Find the size of the upward force acting on the rocket.

4 A weightlifter applies a force of 900 N to a mass. The unbalanced force acting on the mass is 50 N upwards. Find the mass of the weight being lifted.

Fig 1.59

2.1.2 Newton's first law and friction

In the 17th century, Issac Newton published three main laws of motion. The first of these laws is, **'An object will remain at rest or continue moving with a constant velocity unless acted on by an unbalanced force.'**

This law is stating something that on the face of it looks to be common sense: an object will not move unless you apply a force to move it; an object will not stop unless you apply a force to stop it.

Thinking a little deeper, this law has many important consequences in everyday life. A prime example would be the seatbelt in a car. In the event of a car crash, the car is stopped by an external object. However, this object does not make contact with the occupants of the car, so according to Newton's first law they keep moving. To keep the occupants in their seat requires an unbalanced force. This force is provided by the seatbelt.

This law is also important to space travel. Spacecraft have to travel very large distances in space. If they required their engines to be on for the whole journey, it would demand a massive amount of fuel. However, the spacecraft do not need to use their engines once in space, except to change speed and/or direction – there are no unbalanced forces acting, so the rocket keeps moving with a constant speed.

Fig 1.60

Returning to Earth, if an object is given an initial velocity, why does it not keep moving indefinitely? The fact that an object slows down means that it must be acted on by an unbalanced force. There is a force that is opposing the motion of the object. This force is friction. As force is a vector quantity, and friction always acts in the opposite direction to the motion: friction is a *negative vector quantity*.

Key point

Friction is a force that opposes the motion of an object. It always acts in the opposite direction to the motion of the object. Friction is a negative vector quantity.

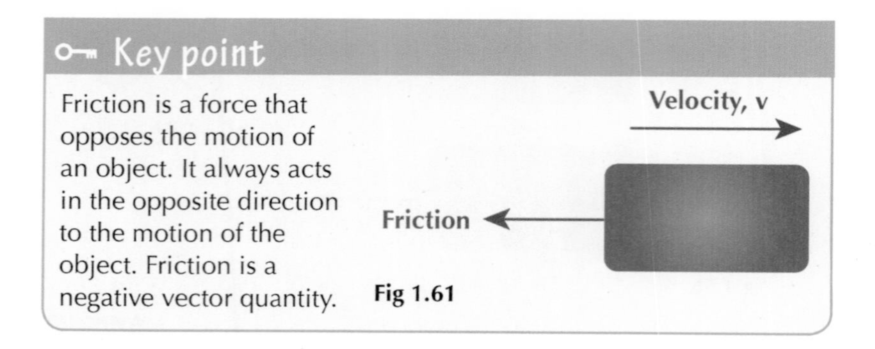

Fig 1.61

Exercise 2.1.2 Newton's first law and friction

1 State Newton's first law of motion.

2 Using your knowledge of Newton's first law of motion, explain how a seatbelt works to keep a passenger in their seat in the event of a car accident. Why is it vital that you wear a seatbelt when travelling in a car?

3 Friction can be described as a negative vector quantity. Explain what is meant by this statement.

4 Using your knowledge of Newton's first law and friction, explain why a ball that is rolled across the table comes to rest and does not keep moving indefinitely.

5 For each of the objects in Figure 1.62, the direction of the motion (velocity) has been labelled. Copy and complete the diagrams to show the direction in which friction will be acting.

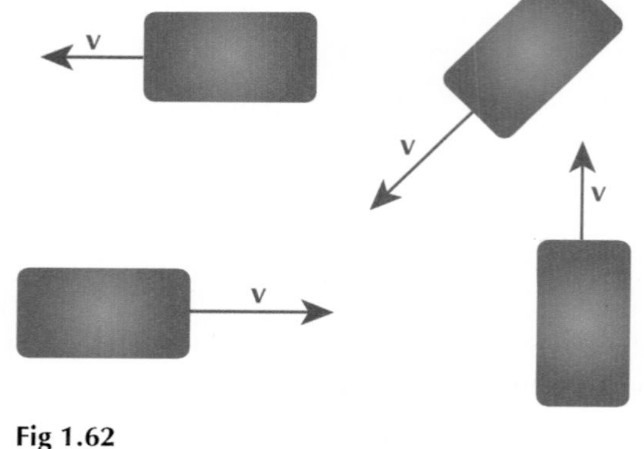

Fig 1.62

2.1.3 Newton's second law

Fig 1.63

Newton's second law of motion builds on the first law – it explains the effects of the unbalanced force that acts on an object. **An unbalanced force will cause an object to accelerate or decelerate.** Consider applying an unbalanced force to a small pen lying on your desk. The pen has a small mass and will therefore accelerate quickly. Now consider applying the same force to a more massive object, such as the desk itself. The desk will accelerate more slowly because it has a greater mass. Furthermore, if you change the size of the force applied to the pen, you will also change the acceleration; the greater the force, the greater the acceleration.

In summary, the magnitude of this acceleration, a, will depend on:

- The magnitude of the unbalanced force, F (N).
- The mass of the object, m (kg).
- The direction of the applied force relative to its motion (see below).

The direction of the acceleration will be the same as the direction of the unbalanced force. Hence, the familiar equation for Newton's second law:

$$F = ma$$

If you work out the size and direction of the unbalanced force, using a vector diagram as described above, you can work out the size and direction of the acceleration.

Key point

Newton's second law states that an unbalanced force, F, will cause an object of mass m to accelerate with an acceleration, a:

$$F = ma$$

Worked example

A car's engine produces a force of 8000 N. If friction of 6000 N acts, calculate the acceleration of the car if it has a mass of 1700 kg.

In order to work out the acceleration of the car, use Newton's second law:

$$F = ma$$

Start off by drawing a diagram of the car and the forces that are acting:

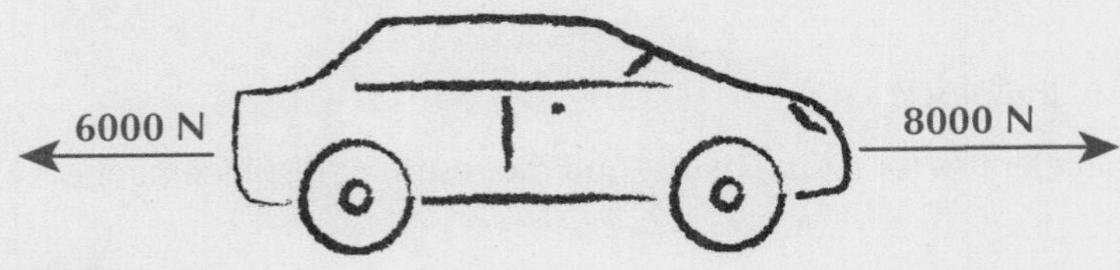

Fig 1.64

The resultant force can then be calculated:

$$F = 8000 - 6000$$
$$F = 2000\ N$$

Now apply Newton's second law to work out the acceleration:

$$F = ma$$
$$2000 = 1700\ a$$
$$a = \frac{2000}{1700}$$
$$a = 1.2\ ms^{-2}$$

Exercise 2.1.3 Newton's second law

1 Calculate the acceleration of each of the blocks shown in the diagram below. Include a direction. The masses of the blocks are given on the diagram.

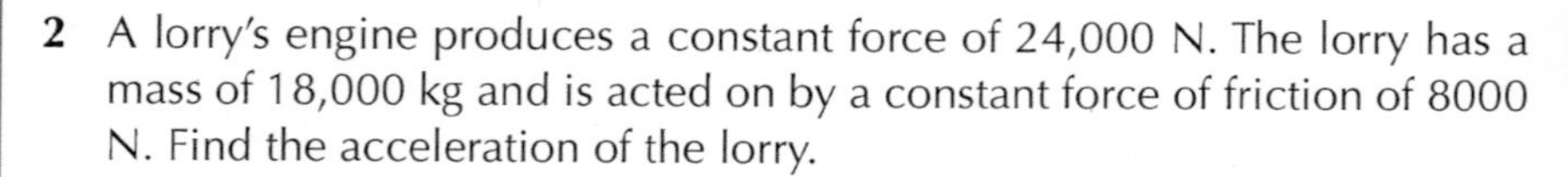

Fig 1.65

2 A lorry's engine produces a constant force of 24,000 N. The lorry has a mass of 18,000 kg and is acted on by a constant force of friction of 8000 N. Find the acceleration of the lorry.

Fig 1.66

3 A block of mass is being pulled along a horizontal surface by a constant force of 25 N. It has an acceleration of 1.4 m s^{-2}.

a) Find the force of friction acting on the block.

b) If the block starts from rest and accelerates for a time of 5 s, find the resultant displacement of the block.

4 A rocket of mass 4000 kg is lifting off from the surface of the Moon (where g = 1.6 N/kg). Its engines provide a constant thrust of 60 kN. Calculate the upward acceleration of the rocket from the surface of the Moon.

5 A crane is used to lift a crate upwards to the roof of a building. The crate has a mass of 240 kg and is subject to a constant frictional force downwards of 1600 N. If the crane supplies an upward force of 5000 N, calculate the upward acceleration of the crate.

6 A hot air balloon is accelerating vertically upwards with a constant acceleration of 0.8 m s^{-2}. The mass of the balloon is 450 kg and it is subject to a constant frictional force of 720 N downwards.

a) Calculate the upward force acting on the balloon required to produce the acceleration.

b) The balloon rises from the ground to a height of 450 m. Calculate the time taken for the balloon to reach this height.

c) What is the upward velocity of the balloon when it reaches 450 m?

d) Plot a velocity–time graph showing the velocity of the balloon between the ground and 450 m.

Fig 1.67

2.1.4 Terminal velocity

So far, examples have shown where the frictional force acting on an object is constant regardless of the velocity. In practice, this is not the case. Friction, such as air resistance and rolling resistance, increases as the velocity increases. This means that the faster your velocity, the greater the force of resistance acting against you.

Fig 1.68

Consider a squirrel that is enjoying an afternoon snack of an acorn high up in a tree canopy. Caught by surprise, the squirrel drops the acorn from the top of the tree. Assuming the air resistance is negligible, or constant with respect to velocity, the acorn would have a constant acceleration downwards. A constant acceleration would suggest that the velocity of the acorn would continue to increase indefinitely until it smashed into the ground. However, in practice this does not happen. The acorn reaches a maximum speed, known as the **terminal velocity**.

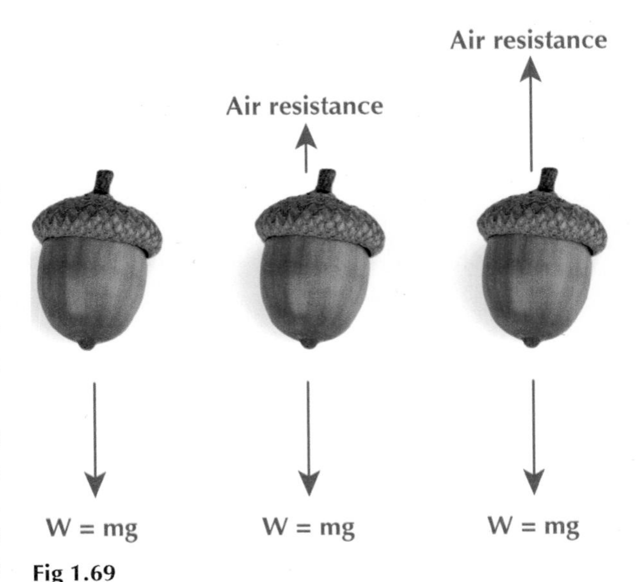

Fig 1.69

As the acorn falls to the ground, its velocity increases. As the velocity increases, the force of air resistance acting on the acorn increases as shown in Figure 1.69.

The weight of the acorn acting downwards remains constant. Air resistance is a form of friction, so acts in the opposite direction to the downwards motion. The magnitude of the air resistance increases as the velocity increases. Once the velocity is high enough, the magnitude of the air resistance equals the weight: the forces acting on the acorn are balanced. Newton's first law says that when the forces are balanced, the acorn will stop accelerating and continue moving with a constant velocity. This velocity is the maximum velocity the acorn will reach – the terminal velocity.

The increasing force of air resistance acts to reduce the unbalanced force acting on the acorn as its speed increases. This reduces the acceleration of the acorn. If you plot the velocity of the acorn on a speed–time graph, we would see the graph in Figure 1.70:

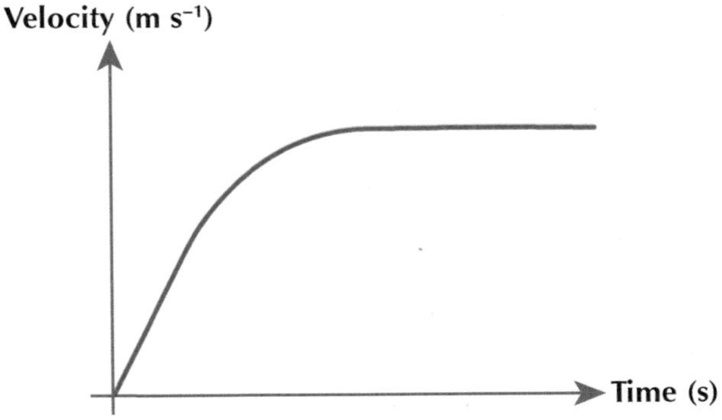

Velocity (m s^{-1})

Time (s)

Fig 1.70

The concept of terminal velocity also applies to objects moving horizontally. For example, a car has got a maximum speed that it can reach. This maximum speed (the terminal velocity) depends on the force the engine can provide and the friction acting. When the friction force equals the maximum force the engine can provide, the car will be moving with the maximum speed. Friction increases as the speed increases as described above.

Friction acting on an object can be reduced by a variety of methods, including streamlining and reduction of rolling resistance. The lower the friction, the greater the terminal velocity.

Fig 1.71

Exercise 2.1.4 Terminal velocity

1 Consider dropping a ball from the side of a building. The ball falls under gravity and hits the ground a short time later.

a) It is found that the velocity of the ball reaches a maximum. Using your knowledge of the forces acting on the ball, explain this observation.

b) Suggest a method that could be used to increase the terminal velocity of the ball.

Fig 1.72

2 A parachute is used to slow down a skydiver so that he can make a safe landing.

a) Draw a diagram that shows the vertical forces acting on the skydiver at one point in the descent.

b) Explain why the unbalanced force of the skydiver reduces as his velocity increases.

c) Describe how the parachute allows the skydiver to make a safe landing.

3 Use your knowledge of forces to explain the concept of the terminal velocity of a car.

4 A drag racing car uses parachutes to slow down at the end of the run.

a) After deployment, why should the parachutes be retracted to ensure the car reaches as high a speed as possible? Give other examples of how the maximum speed of the car could be increased.

b) Describe the effect of deploying the parachutes on the forces acting on the car. Explain why the parachutes cause the car to slow down.

Fig 1.73

2.1.5 Tension and towing

If you attach a rope to a box and pull it along the ground, you are supplying an unbalanced force to make the box move. The force you apply is transferred to the box through the rope. The rope is pulled 'tight' and then the box will begin to move. There is a tension in the rope connecting you to the box. This tension is equal to the force supplied to the box.

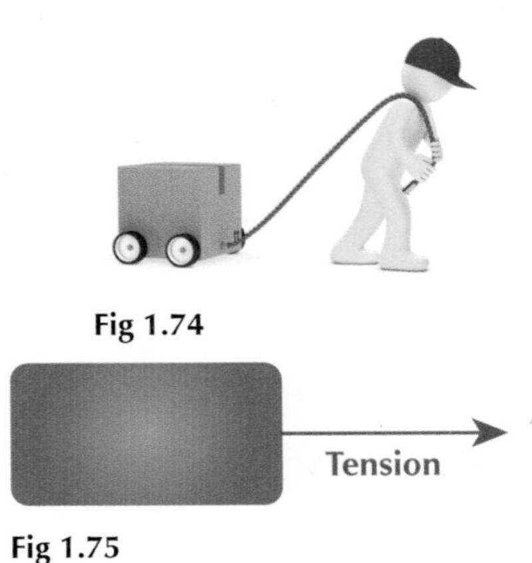

Fig 1.74

The same concept applies to an object that is being lifted; for example, by the cable of a crane. There will be a tension in the cable of the crane equal to the force supplied to the object.

Fig 1.75

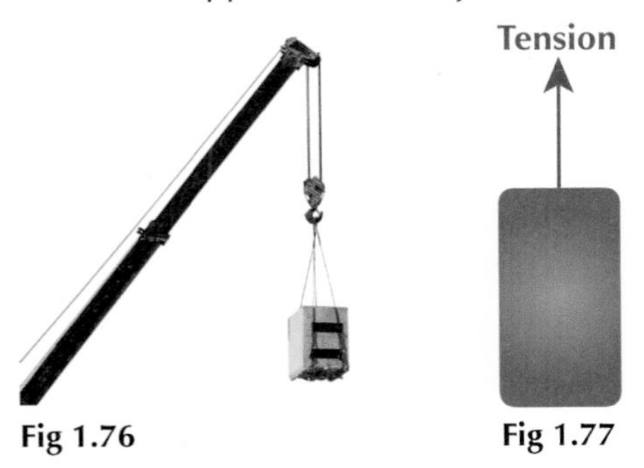

Fig 1.76 Fig 1.77

Horizontal tension

Consider pulling a box along a horizontal surface using a rope. There is a tension in the rope equal to the force you are supplying to the box, as shown in Figure 1.78.

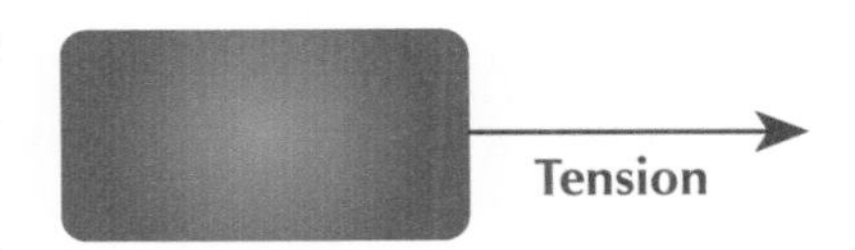

Fig 1.78

If the tension is known, then the acceleration of the block can be calculated using Newton's second law:

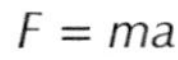

$$F = ma$$

Just as with multiple forces acting, the unbalanced force can be worked out by adding or subtracting depending on the direction of the forces. Tension would be one of the forces acting. Typically, on a level surface, the forces acting would be tension and friction. The unbalanced force acting on the block would then be:

$$F = \text{Tension} - \text{Friction}$$

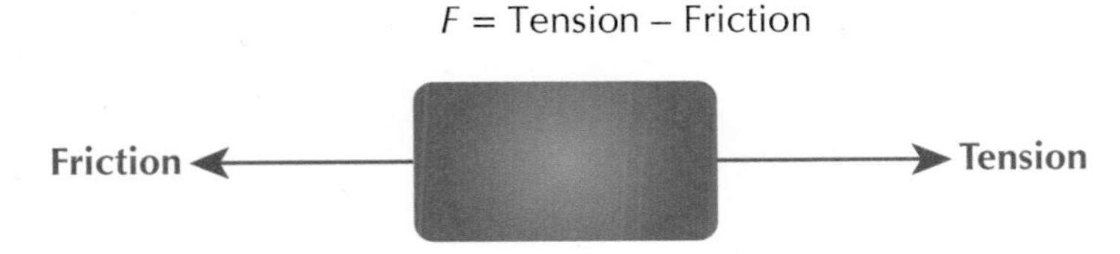

Fig 1.79

Towing on horizontal ground

Consider the train in the picture (Figure 1.80).

Each of the carriages of the train is pulled by the power car at the front. The engine in the power car provides the force required to pull the train. This force is transferred to the carriages via tension in the links as shown in Figure 1.81.

Fig 1.80

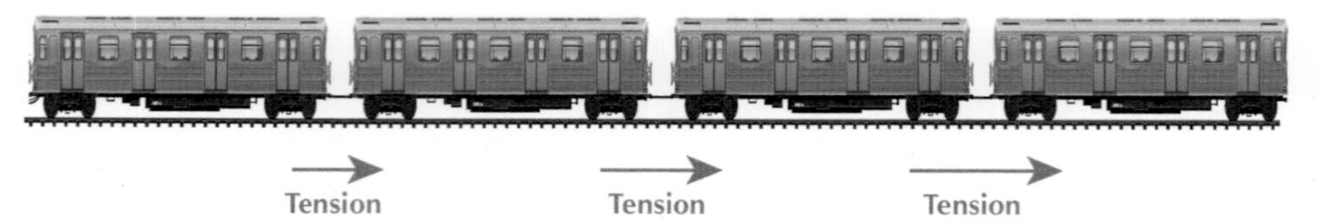

Fig 1.81

The tension in each of the links will not be equal. The link at the front of the train is responsible for pulling all the carriages behind it, while the link at the back is only responsible for pulling the last carriage.

Experiment 2.1.5 Train carriage tension

This experiment investigates how the tension in the links between objects being towed changes.

Apparatus

- Four Newton balances
- Four identical masses
- String
- Light gate and timer

Instructions

1. Make a 'train' using the four identical masses, using the Newton balances and string to form the links between the carriages, as shown in the photograph.
2. Fit a double mask to one of the carriages so that the acceleration can be measured – this carriage should start the run close to the light gate.
3. Using the front Newton balance, apply a pulling force horizontally as shown in Figure 1.83 and measure the acceleration of the train using the light gate. Also, measure the force recorded on each Newton balance.
4. Write a short report that describes how the force on each Newton Balance compares as you move along the train from front to back. Can you explain the results?

Fig 1.82

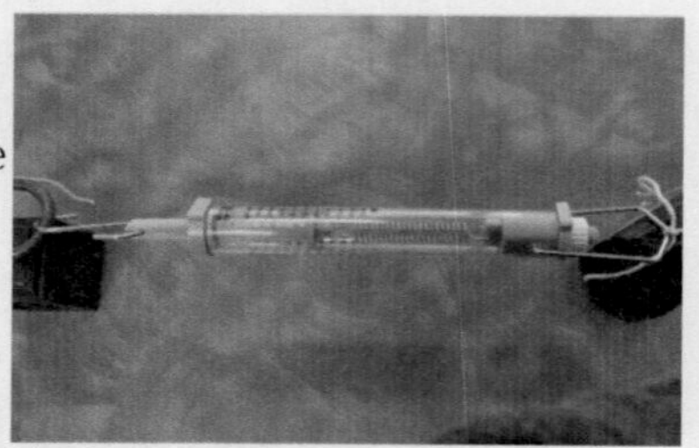
Fig 1.83

In examples such as the train above where an engine provides a total force to the system, the tension in the links depends on the acceleration and the mass being pulled by that link. This means that links further down the train will have a smaller tension as they only need to supply an unbalanced force to pull a small mass compared to links near the front of the train. This is illustrated in Figure 1.84.

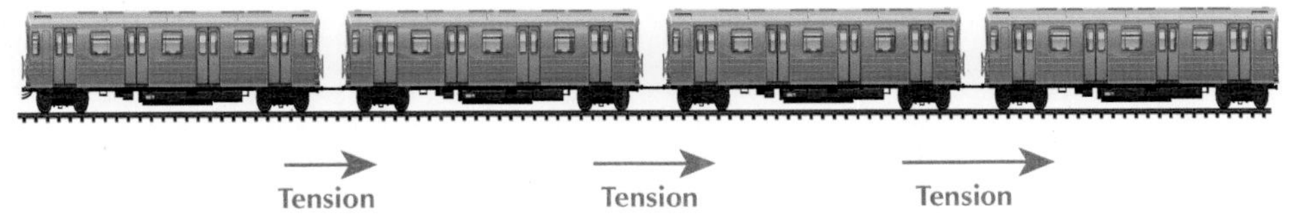

Fig 1.84

Calculating the tension in a towing link

The tension in a towing link depends on the acceleration of the system and the mass that the link is responsible for towing.

To find the tension in a link, first of all find the acceleration of the system as a whole using Newton's second law of motion:

$$F = ma$$

where F is the total unbalanced force supplied to the system and m is the total mass of the system (i.e. the mass of the power car and all of the carriages together).

Newton's second law can then be applied to find the unbalanced force and hence the tension in the link:

$$F = ma$$

where m is the mass that the link is responsible for towing and a is the acceleration of the system calculated above.

Worked example: no friction

In Australia, road trains are long lorries consisting of a 'tractor' at the front and three trailers behind. The tractor's engine supplies a force of 250,000 N. The mass of the tractor is 12,000 kg and the mass of each of the trailers is 25,000 kg.

a) Calculate the acceleration of the road train.

b) Find the tension in the link between the first trailer and the second trailer.

a) *The acceleration of the road train is the acceleration in the whole system, so apply Newton's second law:*

$$F = ma$$

Fig 1.85

(continued)

The total mass of the road train is:

$$m = 12{,}000 + 25{,}000 + 25{,}000 + 25{,}000$$
$$m = 87{,}000\ kg$$

Applying Newton's second law gives:

$$250{,}000 = 87{,}000\ a$$
$$a = \frac{250{,}000}{87{,}000}$$
$$a = 2.87\ ms^{-2}$$

b) *To find the tension in the link between the first and second trailers, you need to know the mass that this link is responsible for towing. The diagram shows that this link is responsible for towing the last two trailers:*

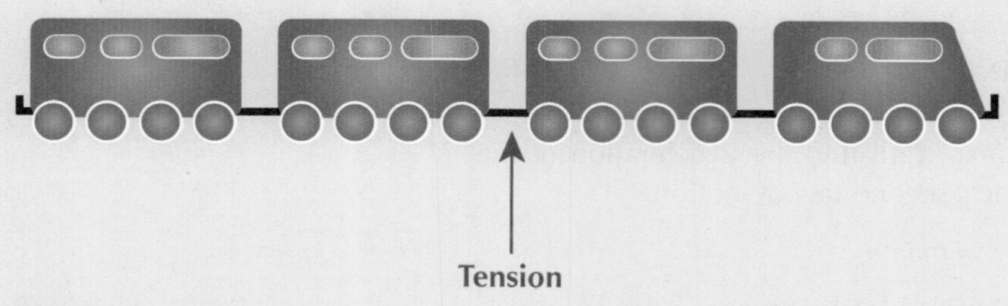

Fig 1.86

Newton's second law is applied to find the tension: $F = ma$

Here, the mass is the mass of the two trailers behind the link:

$$m = 25{,}000 + 25{,}000$$
$$m = 50{,}000\ kg$$

The acceleration has been calculated in part a) above, so you have:

$$F = 50{,}000 \times 2.87$$
$$F = 143{,}500\ N$$

Worked example: including friction

A Class 47 locomotive is used to pull two Class 3 carriages at a heritage railway.

The size of the force of friction acting on the locomotive is 4000 N and the size of the force of friction acting on each carriage is 2000 N. The mass of the locomotive is 35,000 kg and the mass of each carriage is 9000 kg. The locomotive's engine produces a force of 80,000 N.

a) Calculate the acceleration of the train.

b) Calculate the magnitude of the force in the link between the locomotive and the first of the two carriages.

Fig 1.87

a) *There are multiple forces acting on the train and its carriages, so draw a diagram to show these forces and their direction.*

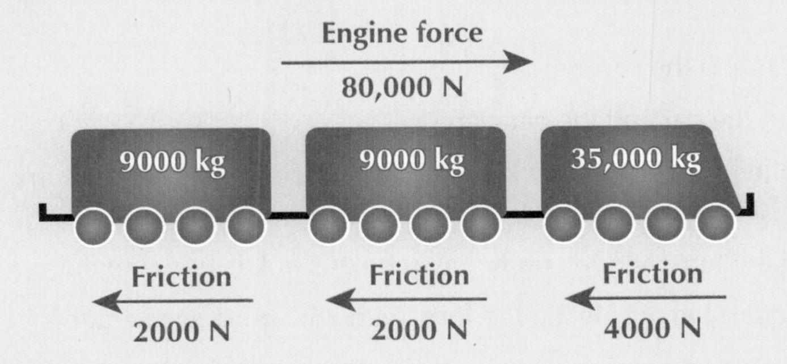

Fig 1.88

The acceleration of the train will be found by using Newton's second law:

$$F = ma$$

You need to find the resultant force acting on the train as a whole, using the diagram in Fig 1.88.

$$F = 80{,}000 - 4000 - 2000 - 2000$$
$$F = 72{,}000 \text{ N}$$

The total mass of the train:

$$m = 35{,}000 + 9000 + 9000$$
$$m = 53{,}000 \text{ kg}$$

The acceleration of the train is then given by:

$$72{,}000 = 53{,}000\ a$$
$$a = \frac{72{,}000}{53{,}000}$$
$$a = 1.36 \text{ ms}^{-2}$$

b) *The force in the link between the engine and the carriages is force required to accelerate these carriages with the acceleration calculated above. This requires an unbalanced force given by Newton's second law:*

$$F = ma$$
$$F = 18{,}000 \times 1.36$$
$$F = 24{,}480 \text{ N}$$

Friction is acting on both of the carriages, so the total resistive force acting on the carriages is:

$$F = 2000 + 2000$$
$$F = 4000 \text{ N}$$

The force in the link must be equal to the unbalanced force required to produce the acceleration plus the resistive force that must be overcome, hence the force in the link is:

$$F = 24{,}480 + 4000$$
$$F = 28{,}480 \text{ N}$$

Exercise 2.1.5 Tension and towing

1 A car of mass 2100 kg tows a caravan of mass 900 kg along a level road. The car's engine produces a force of 14 kN.

a) Calculate the acceleration of the car and the caravan.

b) Find the tension in the link between the car and the caravan.

2 A lorry of mass 12,000 kg pulls a trailer of mass 24,000 kg. Frictional forces acting on the lorry can be neglected. It is found that the lorry accelerates with an acceleration of 1.9 $m\,s^{-2}$.

a) Find the tension in the link joining the lorry and the trailer (assume this link is horizontal).

b) Calculate the total engine force required to accelerate the lorry with the above acceleration.

3 Students set up an experiment in the lab to demonstrate the forces in links between the carriages of a train. They fit a small Newton balance between each of the carriages of a model train set. Each carriage has a mass of 1.3 kg and the engine at the front has a mass of 1.7 kg. Use your knowledge of physics to explain the results the students will see on the Newton balances when the engine is pulling the train forwards with a total force of 10 N. You may ignore frictional forces in your discussion.

4 At a heritage railway, an old steam train is used to pull three carriages. The steam train can supply a force of 90 kN and has a mass of 55,000 kg. Each of the carriages has a mass of 12,000 kg.

a) Calculate the acceleration of the train.

b) Find the tension in the link joining the second and third carriages.

c) When the train begins to climb a hill, its acceleration reduces. Explain this observation.

5 A van with a mass of 2100 kg tows a trailer with a mass of 1000 kg. Frictional forces of 2000 N act on both the van and the trailer. The van's engine produces a force of 10 kN.

a) Find the resultant force acting on the van.

b) Calculate the acceleration of the van.

c) Calculate the tension in the link between the trailer and the van.

6 An Australian road train is used to carry gravel across Austrialia's Northern Territory. It consists of a front tractor with a mass of 45,000 kg, and three separate trailers, each with a mass of 30,000 kg fully loaded. The engine of the road train provides a total force of 250 kN.

a) The road train has an acceleration of 1.3 $m\,s^{-2}$. If the force of friction acting on each of the sections of the road train is equal, calculate the force of friction acting on one of the trailers.

b) Calculate the tension in the link between the first and second trailers.

2.2 Forces in two dimensions

In previous studies of vector quantities, it was shown that a vector can be split into two rectangular components. An example of this is the velocity of a projectile, which is split into a horizontal and a vertical component (see Projectiles sections on page 85 for more detail, and National 5 Physics).

2.2.1 Resolving forces into components

Any vector can be split into two rectangular components (components acting at right angles to each other) as shown in Figure 1.89. The choice of these components depends on the problem being tackled. For example, when dealing with the velocity of a projectile, the velocity is split into horizontal and vertical components that are easier to analyse separately.

Trigonometry can be used to work out the magnitude of the two components.

When the horizontal and vertical components are added tip-to-tail, the resultant is the force *F*.

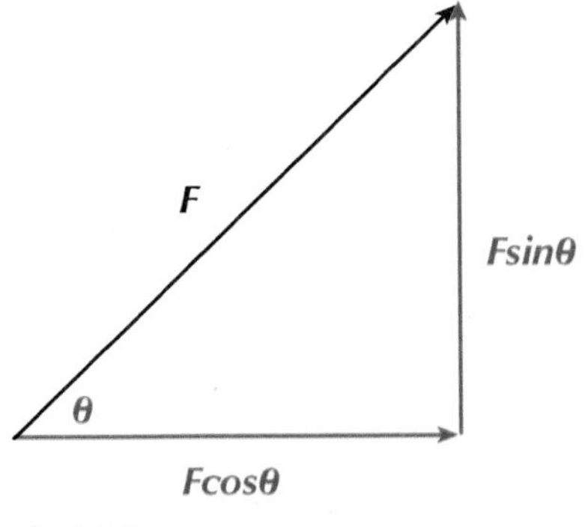

Fig 1.89

Key point

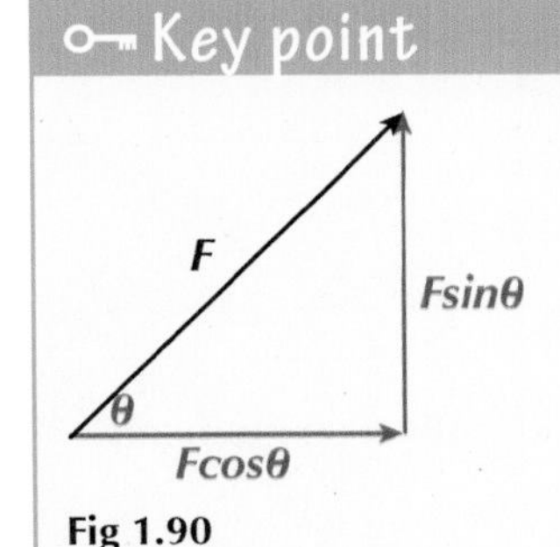

Fig 1.90

A force can be split into horizontal and vertical components as shown opposite. The magnitudes of the components are given by:

$$F_{Horizontal} = F\cos\theta$$
$$F_{Vertical} = F\sin\theta$$

Worked example

A crate is pulled along a horizontal floor by a force of 70 N at an angle of 40° above the horizontal. Find the horizontal and vertical components of the pulling force.

Sketch a diagram of the force (Figure 1.92).

The horizontal component of the force is given by:

$$F_H = 70\cos 40°$$
$$F_H = 53.6\ N$$

The vertical component of the force is also given by:

$$F_V = 70\sin 40°$$
$$F_V = 45.0\ N$$

Fig 1.91

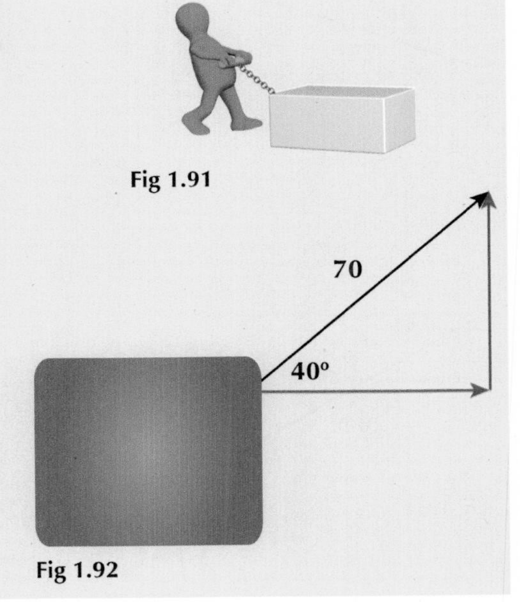

Fig 1.92

Make the link: vector components in mathematics

In the National 5 Mathematics course, vectors are represented in terms of their components. In two dimensions, a vector is represented as follows:

$$F = \begin{pmatrix} x \\ y \end{pmatrix}$$

where x is the horizontal component and y is the vertical component. This is illustrated in Figure 1.93.

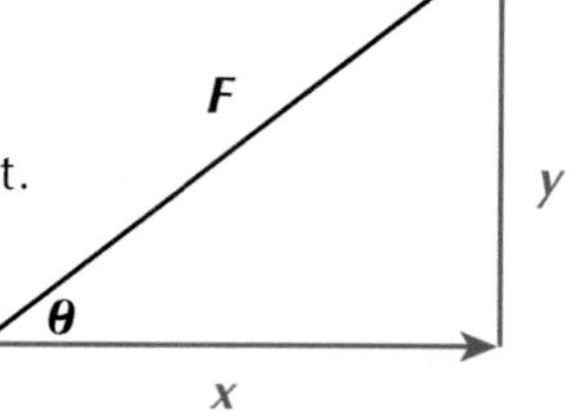

Fig 1.93

The magnitude of the vector is found using Pythagoras' theorem:

$$|F|^2 = \sqrt{x^2 + y^2}$$

The angle can be found using trigonometry:

$$\theta = \tan^{-1}\frac{y}{x}$$

Applying trigonometry to the notation above, the vector can be represented in terms of components as follows:

$$F = \begin{pmatrix} x \\ y \end{pmatrix} = \begin{pmatrix} F\cos\theta \\ F\sin\theta \end{pmatrix}$$

Exercise 2.2.1 Resolving forces into components

1 For each of the forces shown below, find the magnitude of the horizontal and vertical components.

a)

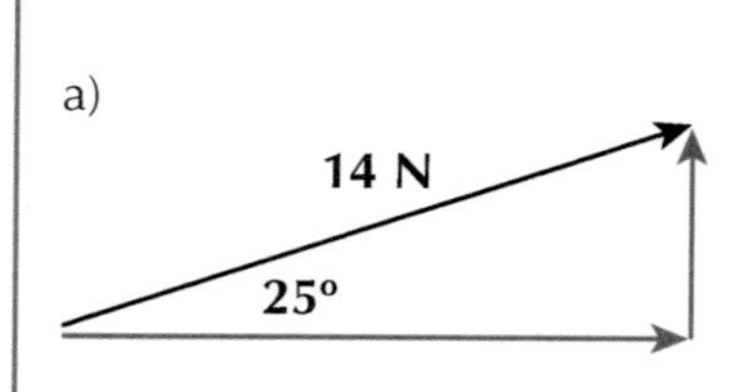

b)

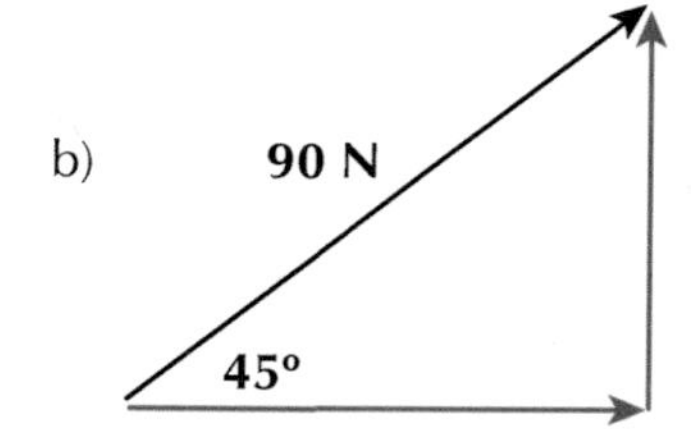

c)

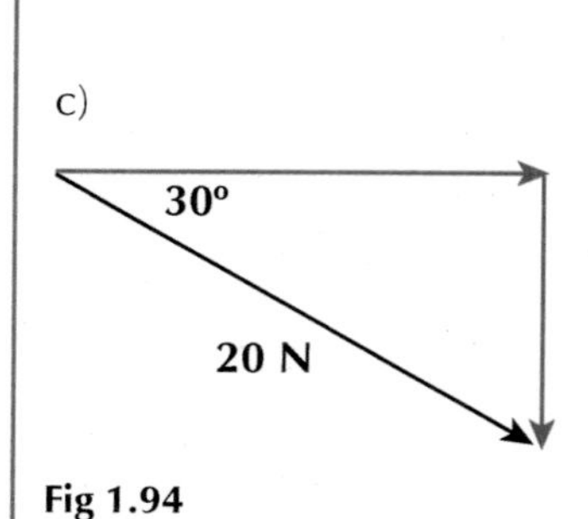

d)

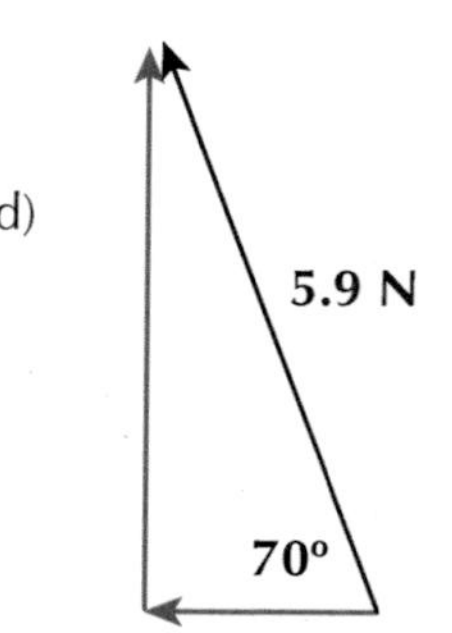

Fig 1.94

2 To examine the forces acting on objects, the following vectors have been resolved into their horizontal and vertical components. The magnitudes of these are shown. Find the magnitude of the resultant vector and the angle, q, of the resultant vector to the horizontal component.

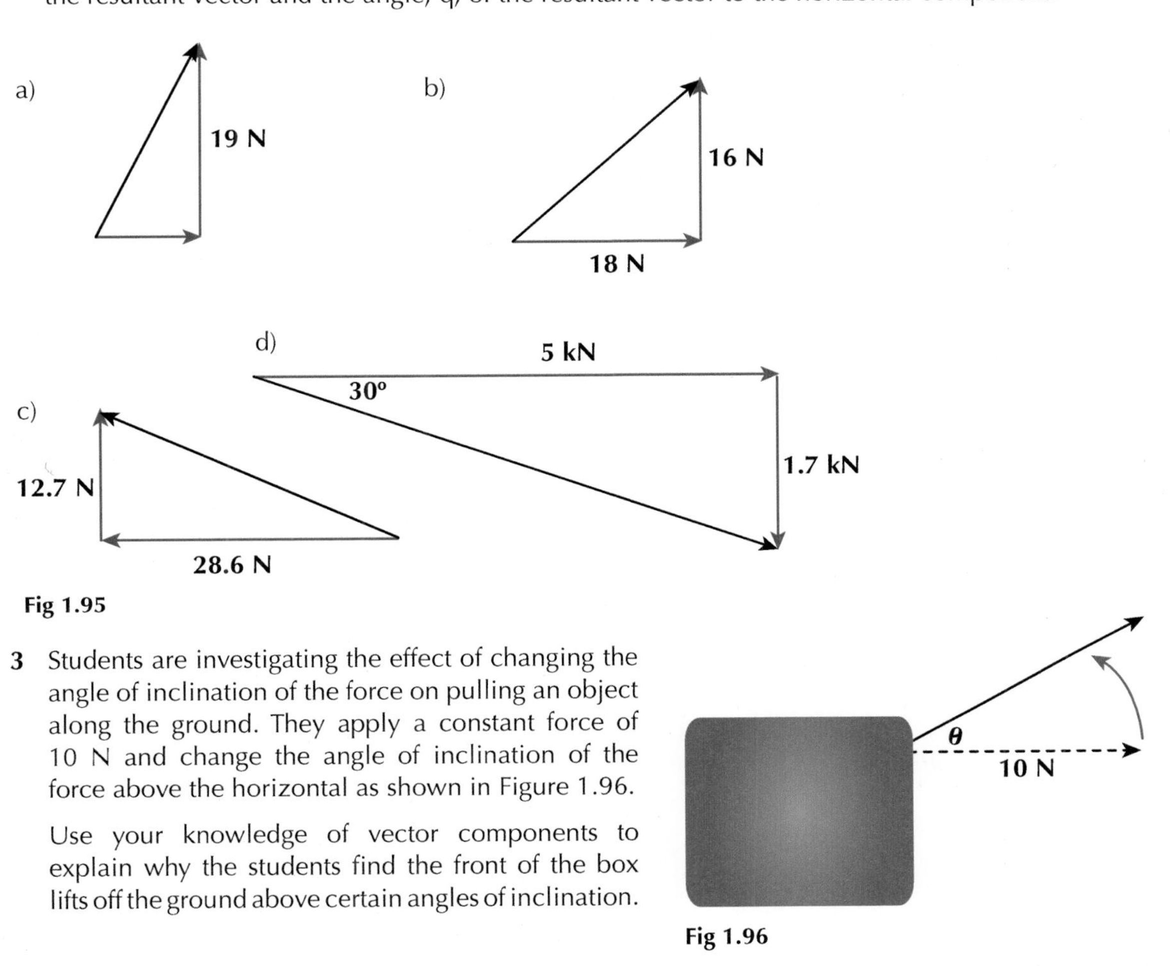

Fig 1.95

3 Students are investigating the effect of changing the angle of inclination of the force on pulling an object along the ground. They apply a constant force of 10 N and change the angle of inclination of the force above the horizontal as shown in Figure 1.96.

Use your knowledge of vector components to explain why the students find the front of the box lifts off the ground above certain angles of inclination.

Fig 1.96

2.2.2 Resultant force in two dimensions

In two dimensions, the vectors are added 'tip-to-tail' as before. This can be done by using scale drawing, Pythagoras' theorem and trigonometry, or resolving to components (similar to National 5 Mathematics).

Vectors at right angles

When the vectors are acting at right angles to each other, the resultant force, F_R can be readily found by applying Pythagoras' theorem and SOH-CAH-TOA as seen in Figure 1.97.

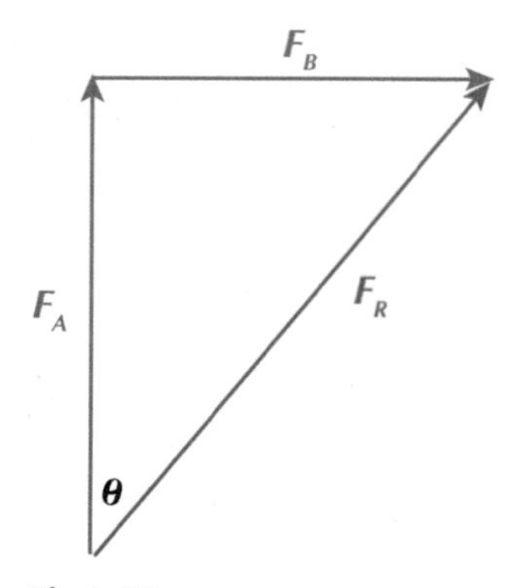

Fig 1.97

The magnitude of the resultant force is found using Pythagoras' theorem:

$$F_R = \sqrt{F_A^2 + F_B^2}$$

The direction of the force is found using SOH-CAH-TOA:

$$\theta = \tan^{-1}\frac{F_B}{F_A}$$

This technique is covered in detail at National 5 level.

Worked example

Two people are pulling a heavy crate of mass 50 kg along a frictionless floor. They apply forces of 20 N and 70 N as shown in Figure 1.98.

a) Calculate the resultant force acting on the crate.

b) Calculate the resultant acceleration (magnitude and direction) of the crate.

Fig 1.98

a) *To find the resultant force acting on the crate, add the vectors above tip-to-tail. Construct a diagram showing the vectors tip-to-tail (Fig 1.99).*

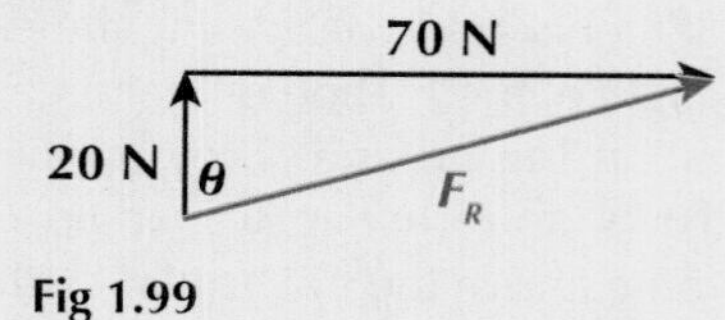

Fig 1.99

Pythagoras' theorem is used to find the magnitude of the resultant force:

$$F_R = \sqrt{20^2 + 70^2}$$
$$F_R = \sqrt{5300}$$
$$F_R = 73\ N$$

SOH-CAH-TOA is used to find the direction of the unbalanced force:

$$\theta = \tan^{-1}\frac{70}{20}$$
$$\theta = 74°$$

Hence, the resultant force is 73 N acting 74° to the right of the 20 N force.

b) *The unbalanced force is responsible for producing an acceleration according to Newton's second law:*

$$F = ma$$

The above equation can be used to work out the magnitude of the acceleration by substituting the magnitude of the resultant force and the mass of the crate:

$$73 = 50\ a$$
$$a = 1.46\ m s^{-2}$$

The direction of the acceleration is the same as the direction of the unbalanced force, hence the acceleration of the crate is 1.46 $m s^{-2}$ on 74° to the right of the 20 N force.

Vectors not at right angles

At Higher you are expected to find the resultant force when the individual forces are not acting at right angles. In these situations, Pythagoras' theorem and SOH-CAH-TOA will no longer work! Instead,you must turn to two rules of trigonometry for triangles – the sine rule and the cosine rule discussed in Chapter 1. Alternatively, you can use scale drawings. The examples here will be considered using trigonometry. You must ensure that the vectors added tip-to-tail from a common reference point. This may mean that it is necessary to calculate angles between the vectors. This is highlighted by the worked example below.

Worked example

Two forces are acting on a crate of mass 20 kg as shown in Figure 1.100. Calculate the resultant force acting on the crate.

Start by drawing a diagram of the vectors acting tip-to-tail, labelling the resultant triangle (Figure 1.101).

The angle inside the triangle is found by subtracting the 40° angle from 180° as both angles lie on the same straight line.

Now, apply the cosine rule to find the magnitude of the resultant force:

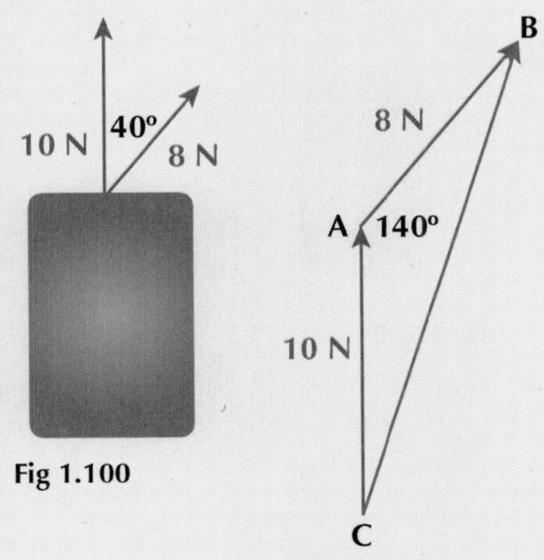

Fig 1.100

Fig 1.101

$$a^2 = b^2 + c^2 - 2bc\cos A°$$
$$a^2 = 10^2 + 8^2 - 2(10)(8)\cos 140°$$
$$a^2 = 100 + 64 - 160\cos 140°$$
$$a^2 = 164 - (-122.6)$$
$$a^2 = 286.6$$
$$a = 16.9$$

Hence, the magnitude of the resultant force is:

$$F = 16.9\ N$$

The direction of the force is found using the sine rule. Include the calculated magnitude on the diagram (Figure 1.102).

Apply the sine rule to find the size of angle C:

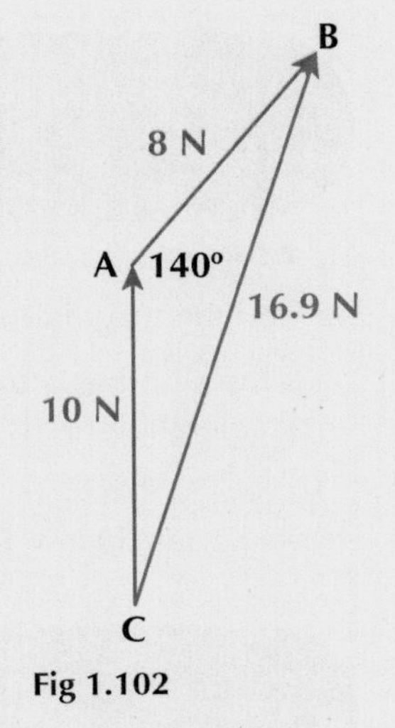

Fig 1.102

$$\frac{a}{\sin A} = \frac{b}{\sin B} = \frac{c}{\sin C}$$
$$\frac{16.9}{\sin 140°} = \frac{8}{\sin C}$$
$$16.9\sin C = 8\sin 140°$$
$$16.9\sin C = 5.1$$
$$\sin C = \frac{5.1}{16.9}$$
$$C = \sin^{-1}\left(\frac{5.1}{16.9}\right)$$
$$C = 17.6°$$

Hence, the resultant force is: F = 16.9 N acting 17.6° to the right of the vertical force.

Exercise 2.2.2 Resultant force in two dimensions

1 Calculate the resultant force acting on each of the crates in the diagram below.

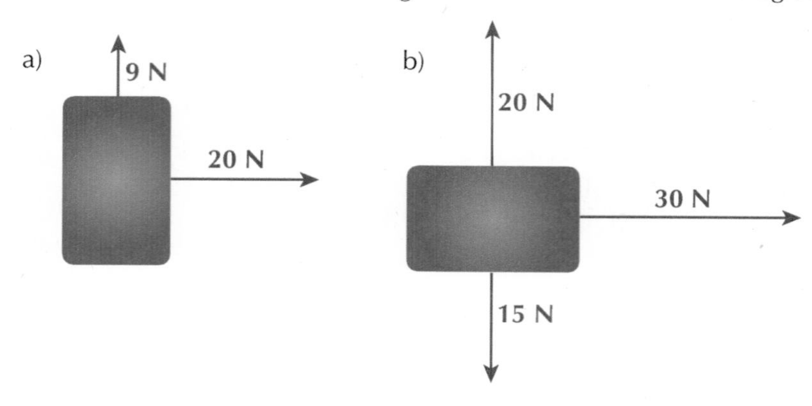

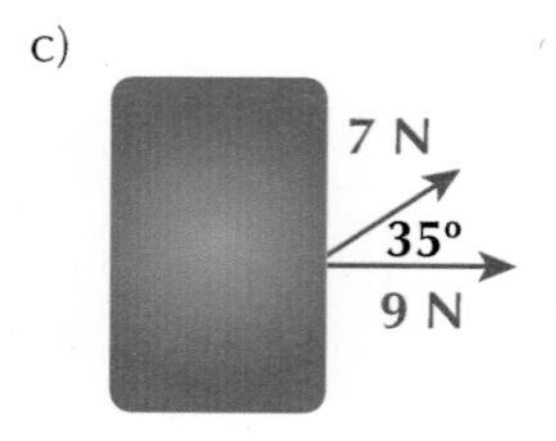

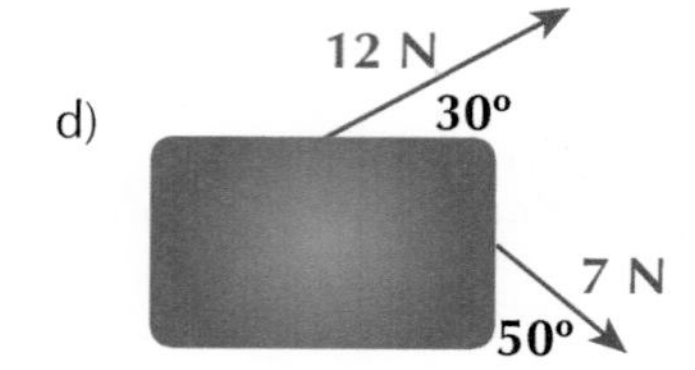

Fig 1.103

2 A large container ship is being towed out of port by two tugs. Each tug supplies a force of 250 kN to the ship as shown in the diagram below.

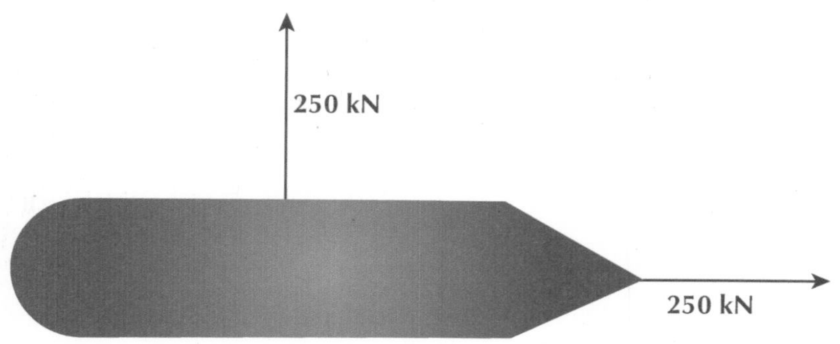

Fig 1.104

a) Find the magnitude and direction of the resultant force applied to the container ship.

b) If the ship has a total mass of 300,000 kg, find the magnitude and direction of the resultant acceleration of the ship.

3 A barge is pulled along a canal by two horses. They each exert a force of 1500 N on the barge as shown in the diagram below. Find the magnitude and direction of the resultant force acting on the barge.

Fig 1.105

2.2.3 Forces at an angle to motion

In the above examples, an unbalanced force produced an acceleration in the same direction as the force acted. In many cases in Physics, the force acts in the same direction as the motion. However, there are times when the force acts at an angle to the motion. An example of this would be pulling a crate along the floor using a rope inclined at an angle above the horizontal, as shown in Figure 1.106.

Fig 1.106

In this case, only the force that is acting in the direction of the motion (horizontally for the crate moving along the floor) contributes to the acceleration of the crate (and the work done moving the crate). The vertical force is 'wasted'. When calculating acceleration along the floor, only the horizontal component should be used.

2.2.4 The inclined plane

An inclined plane is a surface that has been lifted at one end to form a slope, as shown in Figure 1.107. Driving up a hill could therefore be considered as driving up an inclined plane. The plane has an angle of inclination, θ, that shows how steep the plane is.

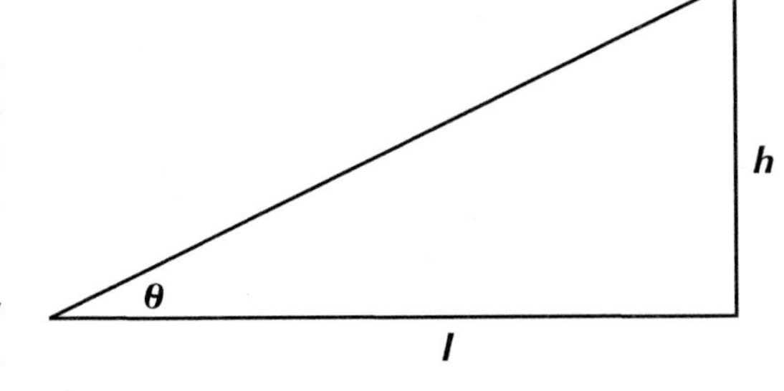

Fig 1.107

The angle of inclination can be found using trigonometry for a right-angled triangle if the length, l, and the height, h, are known:

$$\theta = \tan^{-1}\frac{h}{l}$$

Fig 1.108

An object placed on an inclined plane will slide down when the angle of inclination is big enough. This means that as the angle of inclination (steepness) is increased, the force acting down the plane to make the object move must be increasing. Once this force exceeds the force of friction, the object will accelerate down the slope. If you wish to push an object up an inclined plane, it is more difficult on a steeper slope. This is because you must work against a greater force when the plane is steeper.

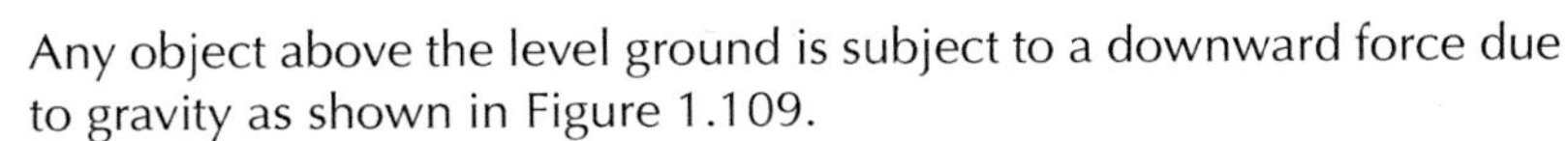

Any object above the level ground is subject to a downward force due to gravity as shown in Figure 1.109.

This downward force is the weight of the box and it acts straight downwards:

$$W = mg$$

Force is a vector quantity, so it can be split into two rectangular components, as shown in Figure 1.110.

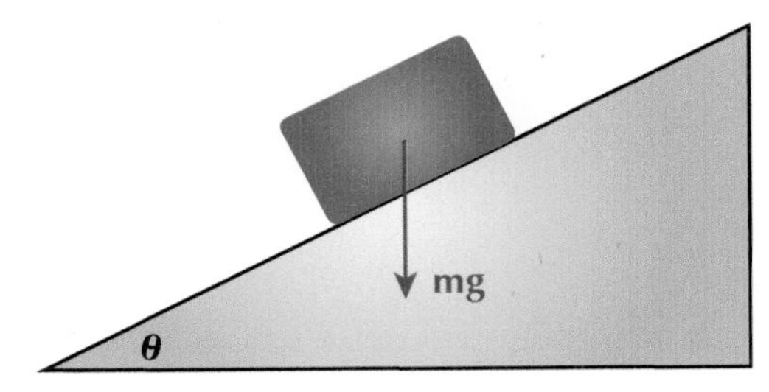

Fig 1.109

There is a component that is acting perpendicular to the slope shown in green. The other component is acting parallel to the slope, which is shown in blue. The component of the force acting parallel to the slope is responsible for making the object slide. When the components are analysed using trigonometry, the results show:

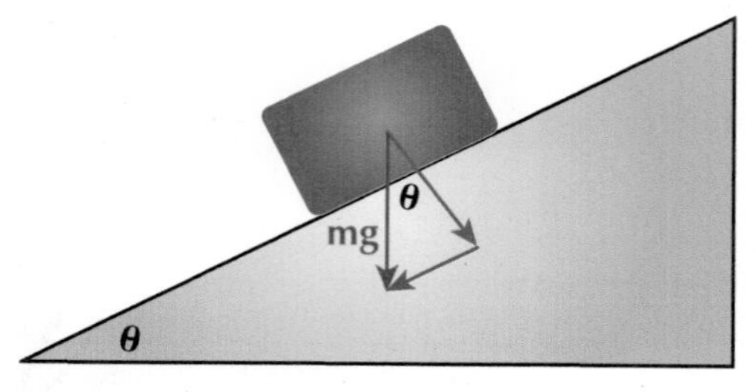

Fig 1.110

$$\sin\theta = \frac{o}{h}$$

$$\sin\theta = \frac{o}{mg}$$

$$o = mg\sin\theta$$

So the component of force acting down the slope is given by:

$$F = mg\sin\theta$$

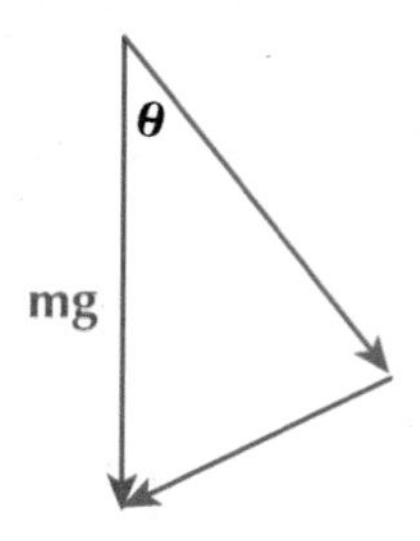

Fig 1.111

Key point

For a plane inclined at angle q above the horizontal, there will be a force,

$$F = mg\sin\theta$$

acting down the plane as shown in the diagram.

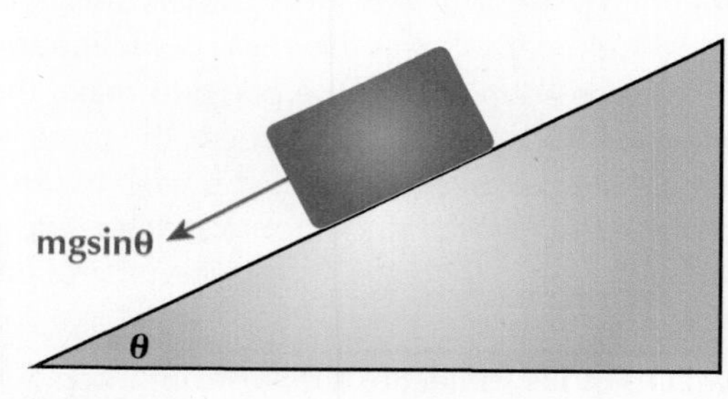

Fig 1.112

Acceleration down an inclined plane

Newton's second law states that when there is an unbalanced force, an object will accelerate in the direction of the unbalanced force. The greater the force, the greater the acceleration:

$$F = ma$$

If friction is neglected, you can calculate the acceleration of the object using the equation above:

$$ma = F$$

$$ma = mg\sin\theta$$

$$a = g\sin\theta$$

As you can see, the larger the angle of inclination, θ, the greater the acceleration. This agrees with the knowledge about forces acting on an inclined plane.

GO! Experiment 2.2.4 Acceleration on an inclined plane

This experiment investigates the effect of angle of inclination on the acceleration down an inclined plane in the absence of friction. The results are used to calculate the gravitational field strength on Earth.

Apparatus

- a linear air track (can be substituted for a dynamics track and carts with low friction bearings)
- dynamics cart
- light gate
- timing device
- metre stick

Instructions

1 Set up the experiment using the above apparatus. The light gate connected to the timing device can be used to measure the acceleration of the dynamics cart. Prop up the ramp at one end to form an inclined plane.

2 Measure the horizontal (l) and vertical (h) lengths of the ramp in order to work out the angle of inclination θ (Figure 1.113).

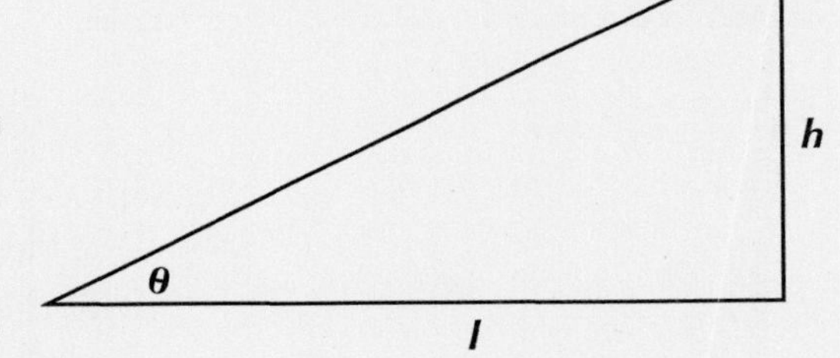

Fig 1.113

3 With the air track switched on to reduce the effects of friction as much as possible, release the dynamics cart from just in front of the light gate to measure the acceleration down the plane. Record your results in a table similar to the one below:

Angle of inclination, θ (°)	Acceleration ($m\,s^{-2}$)

Table 1.6

4 Plot a graph of $\sin\theta$ vs acceleration – this should give a straight line. Find the gradient of this straight line.

5 The analysis above shows that the acceleration down an inclined plane in the absence of friction is given by:

$$a = g\sin\theta$$

When you plot the acceleration on the y-axis and $\sin\theta$ on the x-axis, you can see the above is the equation of a straight line,

$$y = mx$$

Hence, the gradient of the line gives the gravitational field strength, g.

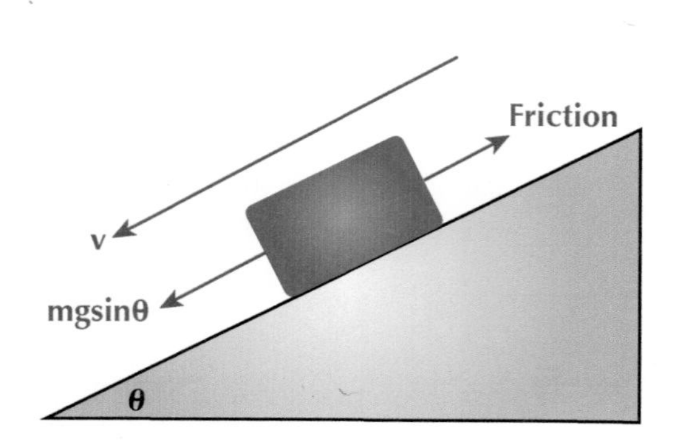

Fig 1.114

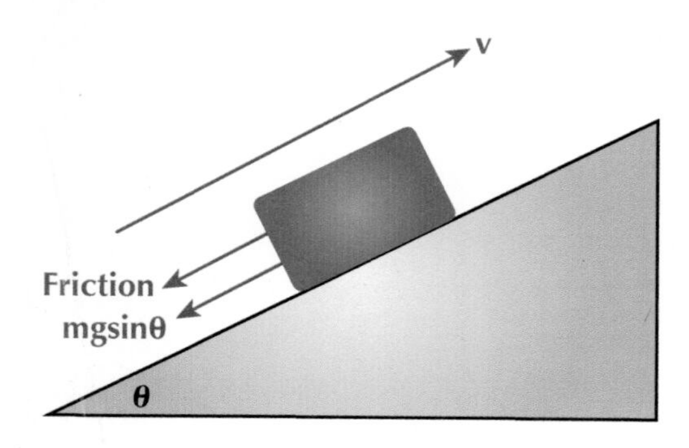

Fig 1.115

Friction on an inclined plane

If friction is acting, then the situation is more complicated. In this case, the acceleration will be given unbalanced force, F_{un}. For an object sliding down the slope, friction will be acting upwards because friction always acts to oppose the motion. This is shown in Figure 1.114.

The unbalanced force is therefore equal to the difference between force acting down the plane and the force of friction:

$$F_{un} = F_{slope} - F_{friction}$$

$$F_{un} = mg\sin\theta - F_{friction}$$

Once the unbalanced force has been found, it is then possible to find the acceleration down the slope using:

$$F_{un} = ma$$

However, if an object is being pushed up the slope, then friction will be acting down the slope (in the same direction as the component of weight)!

This is because friction always opposes the direction of the motion. In this case, the total force acting down the slope would be:

$$F_{un} = F_{slope} + F_{friction}$$

$$F_{un} = mg\sin\theta + F_{friction}$$

As with all problems involving multiple forces, drawing a diagram to show the direction of the motion and the forces acting is invaluable to ensuring you solve the problems correctly!

Worked example

A skier is waiting at the top of a downhill slope. Due to the icy nature of the snow, the friction acting on the skier is negligible. If the skier has a mass of 70 kg and the plane is inclined at an angle of 25°, calculate the acceleration of the skier down the slope.

a) Calculate the component of force acting down the plane.

b) Find the acceleration of the skier down the plane.

a) *The component of weight acting down an inclined plane is given by:*

$$F = mg\sin\theta$$

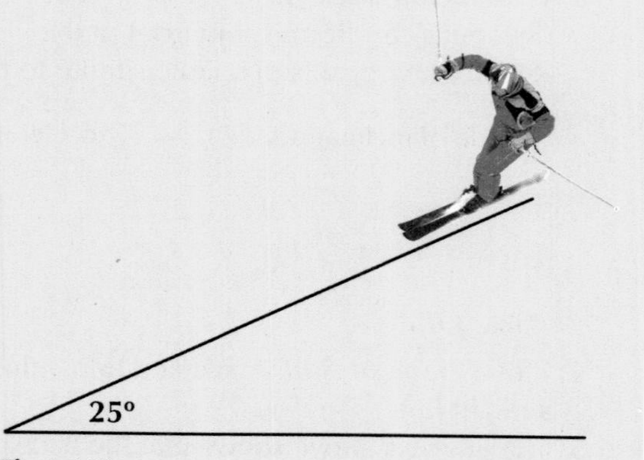

Fig 1.116

The mass of the skier and the inclination angle are known, so it is possible to work out the component of the force down the slope by substituting into the above equation and solving for force:

$$F = (70)(9.8)\sin 25°$$

$$F = 289.9\ N$$

b) *To find the acceleration, use Newton's second law:*

$$F = ma$$

There is no friction acting on the skier, so the unbalanced force is equal to the force acting down the plane. The mass of the skier is known, so it is possible to work out the acceleration:

$$289.9 = 70a$$

$$a = \frac{289.9}{70}$$

$$a = 4.1\ ms^{-2}$$

Worked example

A large crate of mass 40 kg is placed onto an inclined plane. A constant force of friction of 200 N acts on the crate. Calculate the minimum angle of inclination required to cause the crate to slide down the plane.

This problem can be solved by drawing a diagram to illustrate the situation and the forces acting (Fig 1.117).

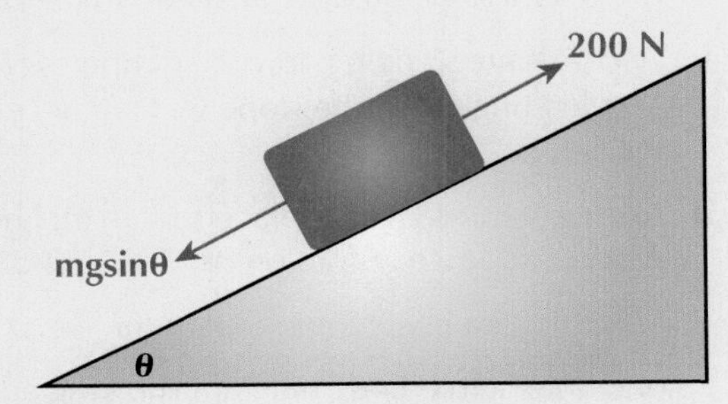

Fig 1.117

The crate will begin to slide down the plane when the downward force is greater than the force of friction acting upwards. Hence, the minimum angle can be found by setting the two forces equal to each other and then solving for the angle:

$$mg\sin\theta = 200$$

$$(40)(9.8)\sin\theta = 200$$

$$392\sin\theta = 200$$

$$\sin\theta = \frac{200}{392}$$

$$\theta = \sin^{-1}\left(\frac{200}{392}\right)$$

$$\theta = 30.7°$$

Exercise 2.2.4 Inclined plane

1 For each of the masses and inclination angles shown in the diagrams below, calculate the component of the weight acting parallel to the slope.

a)

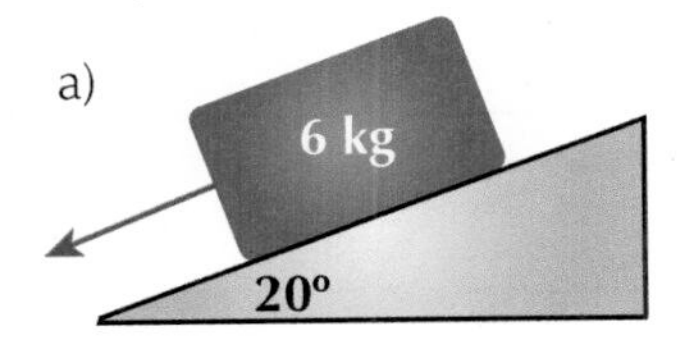

b)

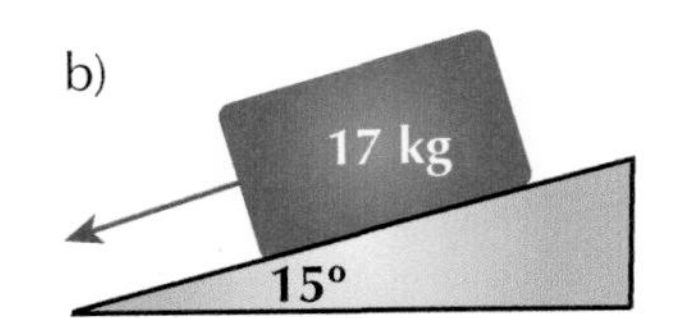

c)

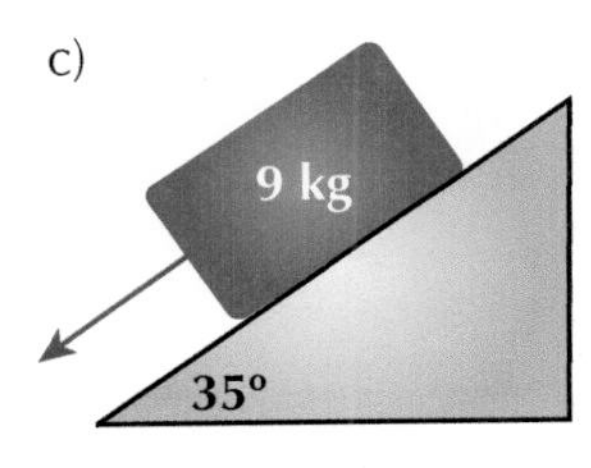

d)

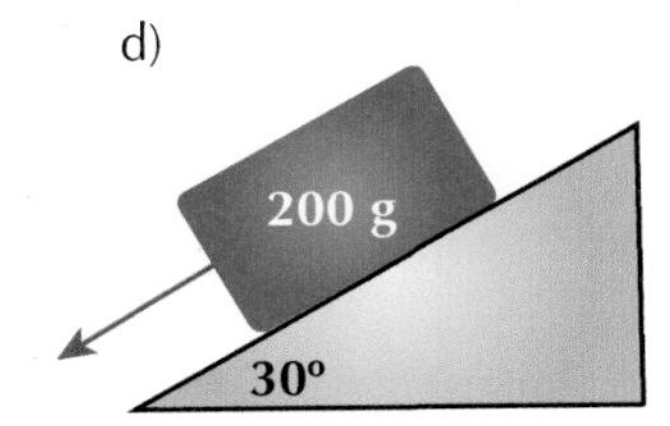

Fig 1.118

(continued)

2 A box of mass 25 kg is placed on a frictionless slope that is inclined at an angle of 25° to the horizontal. Calculate:

a) The component of the box's weight acting down the slope.

b) The acceleration of the box down the slope.

3 In an experiment to study the component of weight acting parallel to a slope, a block is placed on the slope at a given angle of inclination. Masses are attached via a pulley onto the block to provide a force up the plane ($W = mg$). You may assume the slope is frictionless.

a) Explain why attaching a mass to the block via a pulley keeps it stationary on the slope.

b) The mass of the block is 2 kg and the plane is inclined at an angle of 30° to the horizontal. Calculate the component of the weight of the block acting parallel to the slope.

c) Find the mass required to be attached to the pulley to hold the block stationary on the slope.

d) If a mass of 1.6 kg is attached to the pulley, find the acceleration of the block up the slope.

4 Two girls are sledging down a frictionless icy slope. They have a combined mass of 110 kg and are accelerating down the slope with an acceleration of 2 $m s^{-2}$. Calculate the angle of inclination of the slope.

5 A crate of mass 24 kg is placed on an inclined plane that makes an angle of 20° with the horizontal. A constant force of friction of 70 N acts on the crate.

a) Calculate the component of the weight of the crate acting parallel to the slope.

b) Draw a free body diagram showing all of the forces acting on the crate.

c) Find the unbalanced force acting on the crate.

d) Calculate the acceleration of the crate on the slope.

6 A car of mass 900 kg is sliding down a slope that is covered in ice. The slope is inclined at an angle of 25°. The acceleration of the car is found to be 4 $m s^{-2}$. Calculate the force of friction that is acting against the car on the slope.

7 A crate of mass 29 kg is being pulled up a slope by a force F with a constant acceleration of 1.5 $m s^{-2}$. A constant force of friction of 150 N acts on the crate as it moves up the slope. The slope is inclined at 15° above the horizontal as shown in Fig 1.119.

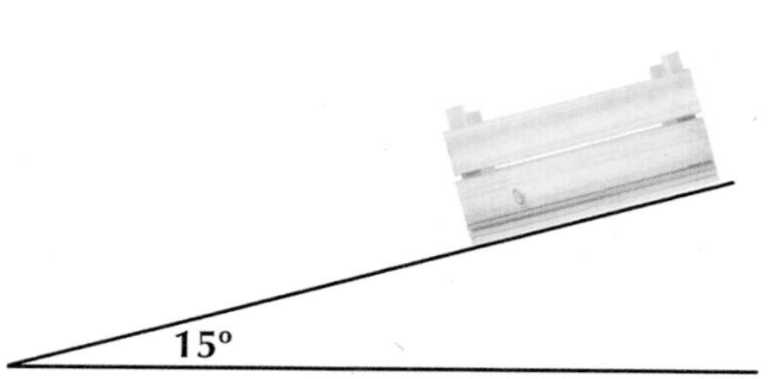

Fig 1.119

a) State the direction that friction acts when the crate is being pulled up the slope.

b) Calculate the component of weight acting parallel to the slope.

c) Draw a free body diagram showing all of the forces acting on the crate as it is pulled up the slope.

d) Calculate the magnitude of the force acting to pull the crate up the slope.

e) The crate starts from rest at the bottom of the slope and is pulled to the top of the slope in a time of 8 s.

(i) Find the speed of the crate at the top of the slope.

(ii) Calculate the length of the slope.

(iii) What is the maximum height of the slope?

(iv) Calculate the gain in potential energy of the crate moving up the slope.

2.3 Energy and power

There are many different forms of energy, from heat energy to light energy, sound energy to kinetic energy. However, one thing always remains the same – the total amount of energy in a system. Energy cannot be created or destroyed, only changed in form. Two types of energy linked to moving objects are considered here: kinetic energy, which any moving object possesses, and gravitational potential energy, which any object lifted above the ground has.

2.3.1 Work, gravitational potential and kinetic energy

Work, gravitational potential and kinetic are all forms of energy measured in Joules. This section considers mechanical work. Energy is a scalar quantity.

Work

Work is done whenever a force is applied to an object and the object is made to move. The amount of work done depends on the size of the force applied, F, and the distance, d, over which the force is applied. If you push a lawnmower across a lawn, you are doing work in order to make it move. You apply a force, F, for a distance, d, and the work done is given by:

$$E_W = Fd$$

Crucially, if the object does not move, you have done no work! You can apply a very large force but in order to do work, the object has to move.

Fig 1.120

Key point

Work is done when a force, F, is applied over a distance, d,

$$E_W = Fd$$

Worked example

Calculate the amount of work done pushing a crate across the floor using a constant force of 50 N for a distance of 12 m.

Here, apply the equation for work done:

$$E_W = Fd$$

Substitute what you know and solve for the work done:

$$E_W = 50 \times 12$$

$$E_W = 600\ J$$

Gravitational potential energy

When an object is lifted from the ground up to a height, h, it gains gravitational potential energy, E_P. When you lift an object, you have to do work against gravity. You have to supply energy to the box. If you lower the box, you get this energy back as kinetic energy.

The equation for potential energy can be derived using the equation for work done:

$$E_W = Fd$$

Key point

The gravitational potential energy, E_P, of an object of mass, m, lifted to a height, h, is given by:

$$E_P = mgh$$

When lifting an object at constant speed, you have to apply a force:

$$F = mg$$

If you apply this force for a distance (height) h, then the work done lifting an object is given by:

$$E_W = Fd$$
$$E_W = (mg)(h)$$
$$E_W = mgh$$

Hence, the gravitational potential energy of an object of mass, m, being lifted to a height, h, is given by:

$$E_P = mgh$$

Worked example

A crane lifts a crate of mass 340 kg at constant speed from the ground to a height of 20 m. Calculate the gravitational potential energy of the crate.

Apply the equation for gravitational potential energy:

$$E_P = mgh$$

Substitute what you know and solve for the potential energy:

$$E_P = (340)(9.8)(20)$$
$$E_P = 66640\ J$$

Fig 1.121

Kinetic energy

All moving objects have kinetic energy that depends on both the mass of the object and its velocity. The equation for kinetic energy, E_K, is:

$$E_K = \frac{1}{2}mv^2$$

Worked example

Calculate the kinetic energy of a Border Collie dog, of mass 18 kg, when it is moving with a velocity of 10 m s^{-1}.

Use the equation for kinetic energy here:

$$E_K = \frac{1}{2}mv^2$$

Substitute what you know and solve for the kinetic energy:

$$E_K = \frac{1}{2}\times 18 \times 10^2$$
$$E_K = 900\ J$$

Fig 1.122

Exercise 2.3.1 Work, gravitational potential and kinetic energy

1 In an experiment to investigate friction, students apply a constant force of 3 N to a wooden block of mass 1.2 kg. The block is found to move at a constant speed and covers a distance of 6 m. Calculate the work done to move the block.

2 While trying to move a very heavy crate, a student applies a constant force of 210 N. This is not enough to move the crate and it stays in the same position. Calculate the work done by the student.

3 By considering the equation for work done, $E_W = Fd$, derive the equation for gravitational potential energy. What assumption have you made in this derivation?

4 Calculate the potential energy gained by lifting a mass of 1.6 kg through a height of 11 m.

5 A car of mass 1270 kg is driving up an incline as shown in the diagram below.

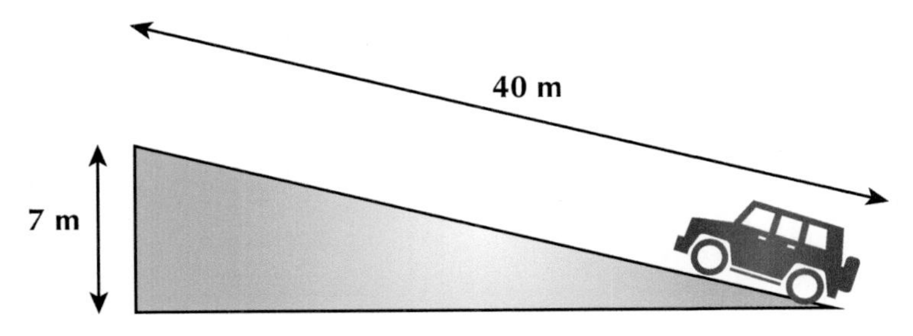

Fig 1.123

The incline is a length of 40 m and a height of 7 m. A constant frictional force of 250 N acts on the car.

a) Calculate the work done against friction climbing the incline.

b) What was the gain in potential energy of the car in climbing the incline?

c) Find the total energy required for the car to climb the incline.

6 An object of mass 24 kg is moving with a velocity of 9 m s^{-1}. Calculate the kinetic energy of this object.

7 A German Shepherd dog has a mass of 40 kg. It is found to have a kinetic energy of 800 J. Find the velocity of the dog.

Fig 1.124

2.3.2 Power

Power is the rate of change of energy – the greater the power, the faster energy is transformed from one type to another. Power can be found using:

$$P = \frac{E}{t}$$

where E is the energy transferred (J) and t is the time taken in seconds. Power is measures in Watts (W) where, 1 Watt = 1 Joule per second.

Key point

Power is the rate of change of energy, measured in Watts:

$$P = \frac{E}{t}$$

Worked example

A bus of mass 9500 kg climbs a hill on a motorway. Frictional forces of 1200 N act on the bus throughout the climb, which has a length of 1 km. The bus gains a height of 250 m during the climb. If the bus takes 20 minutes to complete the climb, calculate the average power of the bus engine.

Fig 1.125

First of all, work out the total energy gained by the bus during the 20-minute climb.

This is the combination of the work done against friction:

$$E_W = Fd$$
$$E_W = (1200)(1000)$$
$$E_W = 1200000\ J$$

and the gravitational potential energy gained:

$$E_P = mgh$$
$$E_P = (9500)(9.8)(250)$$
$$E_P = 23275000\ J$$

The total energy is therefore:

$$E = 1200000 + 2327500$$
$$E = 24475000\ J$$

The power of the bus engine is then given by the equation for power:

$$P = \frac{E}{t}$$
$$P = \frac{24475000}{1200}$$
$$P = 20.4\ kW$$

Exercise 2.3.2 Power

1 Power is the rate of transfer of energy, measured in Watts where 1 Watt is equal to 1 Joule per second. Express this definition as an equation.

2 A lift is used to hoist a mass of 24 kg through a height of 8 m. If the lift must achieve this in a time of 12 s, calculate the minimum power of the lift.

3 A car's engine has a power, P, of 15 kW. The car has a mass of 900 kg and experiences constant frictional forces, F, of 5600 N.

a) By considering the work done against friction of the car, show that the maximum velocity the car will reach is given by,

$$v = \frac{P}{F}$$

b) Find the maximum speed of the car.

c) The car now begins to climb a slope. State and explain the effect this has on the maximum speed of the car.

4 A goods train is pulling a total mass of 6.6×10^6 kg of oil. This mass includes the mass of the engines pulling the train. The train is pulled up an incline of length 1.2 km and gains a height of 100 m, ignoring the effects of friction.

If the train is pulled up the incline in a time of 6 minutes, find:

a) The average velocity of the train.

b) The minimum power required by the train's engines.

5 A lorry of mass 44,000 kg is travelling at a constant velocity of 25 $m s^{-1}$. It experiences a frictional force of 11,000 N at this velocity. The lorry climbs an incline where it gains a height of 350 m over a distance of 2.5 km.

a) Find the time taken for the lorry to climb the incline.

b) Calculate the minimum power of the lorry's engine.

6 During take-off, an aircraft passes through two speeds before lifting off of the ground:

- V_1 – take-off decision speed, above which the plane will take off even in the event of an engine failure.
- V_2 – safe take-off speed, the speed at which the plane can lift off even with only one engine running.

A twin-jet is an aircraft with two engines. A particular aircraft has a mass of 95,000 kg. At V_2, the total friction acting on the plane is 170,000 N when just lifting off. Given that the plane must be able to reach this speed with only one engine, calculate the minimum power of a single engine on this aircraft.

Fig 1.126

2.3.3 Conservation of energy

Energy can be transformed from one form into another; for example, potential energy to kinetic energy when an object falls from a height towards the ground. Energy cannot be created or destroyed, only converted into different forms. This principle is known as energy conservation.

> **Key point**
>
> Energy cannot be created or destroyed. We can only transform energy from one form into another.

Falling objects – no friction

One of the classic applications of energy conservation is the falling object. An object of mass, m, lifted to a height, h, will have gravitational potential energy:

$$E_P = mgh$$

When this object falls to the ground, this potential energy is converted into kinetic energy:

$$E_K = \frac{1}{2}mv^2$$

In the absence of friction, no energy is 'lost' (i.e. no energy is converted into other forms such as heat or sound). The potential energy is equal to the kinetic energy:

$$\frac{1}{2}mv^2 = mgh$$

$$v = \sqrt{2gh}$$

This has an important consequence: the speed of the object when it hits the ground does not depend on its mass. In practice, this is not the case on Earth – different objects fall at different rates. However, this is because air resistance on Earth is different for different objects, and this introduces differing resistance forces. On the Moon, where there is no atmosphere to cause air resistance, two objects with different shape and mass will fall at the same rate.

The effects of friction

In the case of friction, some of the energy is converted into other forms. In the example above of the falling object, the gravitational potential energy will be transformed into heat energy (caused by air friction) as well as kinetic energy. Total energy is still conserved, however. So, for the ball falling to the ground, the gravitational potential energy:

$$E_P = mgh$$

has been transformed to kinetic energy:

$$E_K = \frac{1}{2}mv^2$$

and heat energy:

$$E_h = cm\Delta T$$

Hence:

$$mgh = \frac{1}{2}mv^2 + cm\Delta T$$

Worked example

A ball of mass 1.2 kg is dropped from a height of 2.0 m. It lands on the ground with a velocity of 5 m s^{-1}. All energy losses are due to air friction, which causes the temperature of the ball to rise slightly. Use the principle of energy conservation to calculate the rise in temperature of the ball due to air friction (specific heat capacity of the ball is 910 J kg^{-1} °C^{-1})

First of all, calculate the total energy that is the gravitational potential energy of the ball before it was dropped:

$$E_P = mgh$$
$$E_P = 1.2 \times 9.8 \times 2.0$$
$$E_P = 23.52\ J$$

Now calculate the kinetic energy of the ball when it reaches the ground:

$$E_K = \frac{1}{2}mv^2$$
$$E_K = \frac{1}{2} \times 5 \times 1.2^2$$
$$E_K = 15\ J$$

The missing energy has been converted to heat:

$$E_h = 23.52 - 15$$
$$E_h = 8.52\ J$$

Use the equation for specific heat capacity (covered at National 5 level) to find the rise in temperature:

$$E_h = cm\Delta T$$

$$8.52 = 910 \times 1.2 \times \Delta T$$

$$\Delta T = \frac{8.52}{1092}$$

$$\Delta T = 0.0078°C$$

Vehicle braking

Many road vehicles use disc brakes or drum brakes to slow down. Typical disc brakes are shown in Figure 1.127.

Typical drum brakes are shown in Figure 1.128 and brakes on a bike are also very similar in style (see Figure 1.129).

Despite different styles of brake, the underlying principle is the same. A friction material (such as the brake pad or brake shoe) makes contact with another material such as a brake disc, drum or wheel rim. This increases the friction and generates heat. The vehicle slows down as the kinetic energy is converted to heat energy in the brakes. Occasionally, some energy is converted to sound and you can hear the brakes squeak.

Energy conservation applies to this system too. The kinetic energy of the vehicle:

$$E_K = \frac{1}{2}mv^2$$

is converted to heat energy in the brakes:

$$E_h = cm\Delta T$$

So:

$$\frac{1}{2}mv^2 = cm\Delta T$$

Fig 1.127

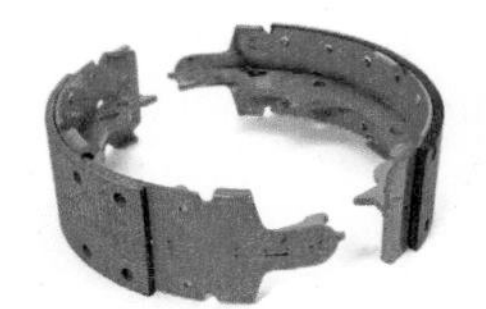

Fig 1.128

Fig 1.129

Worked example

A mountain bike and rider has a mass of 90 kg. The bike was on a downhill stretch of track and reached a velocity of 11 m s^{-1}. The front brakes, which have a mass of 3 kg and a specific heat capacity of 700 J kg^{-1} °C^{-1} are used to bring the bike to rest. Calculate the rise in temperature of the brakes.

Applying energy conservation, the kinetic energy of the bike is equal to the increase in heat energy:

$$\frac{1}{2}mv^2 = cm\Delta T$$

Substituting and solving, we get the following (remember, the mass of the bike and rider is different from the mass of the front brake):

$$\frac{1}{2} \times 90 \times 11^2 = 700 \times 3 \times \Delta T$$

$$5445 = 2100\Delta T$$

$$\Delta T = \frac{5445}{2100}$$

$$\Delta T = 2.59°C$$

Fig 1.130

Exercise 2.3.3 Conservation of energy

1 A ball of mass 1.7 kg is held at a height of 4.3 m above the ground. The ball is dropped. Calculate the speed of the ball when it hits the ground. You may ignore the effects of friction.

2 A football, of mass 600 g, is thrown upwards with an initial velocity of 7.2 ms^{-1}. Use the principle of conservation of energy to find the maximum height reached by the ball. You may ignore the effects of friction.

Fig 1.131

3 A car of mass 900 kg is parked at the top of an incline as shown in Figure 1.132 below.

The car is allowed to roll down the slope. The effects of friction may be ignored. Use the principle of energy conservation to calculate the velocity of the car at the bottom of the slope.

4 A metal ball of mass 1.6 kg is dropped from a height of 2.1 m and lands on the ground with a velocity of 5.5 $m s^{-1}$. Apply the principle of energy conservation to calculate the rise in temperature of the ball when it reaches the ground. You may assume that all of the frictional losses cause the temperature of the ball to increase and that the specific heat capacity of the ball is 910 J kg^{-1} $°C^{-1}$.

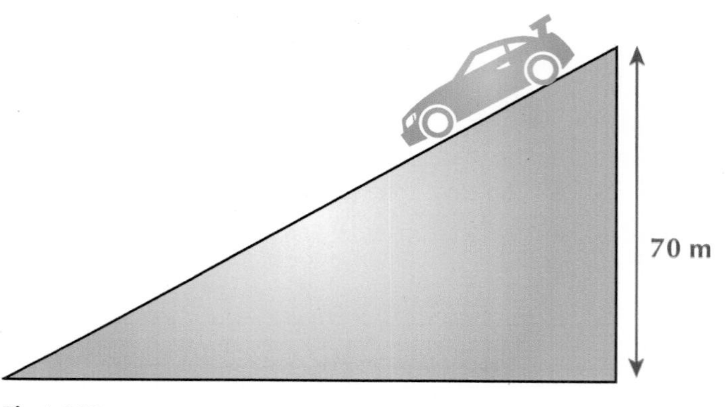

Fig 1.132

5 A lorry is travelling with a velocity of 20 $m s^{-1}$. The lorry has a mass of 44,000 kg. If the lorry is brought to rest by brakes that have a combined mass of 70 kg and specific heat capacity of 1100 J kg^{-1} $°C^{-1}$, calculate the increase in temperature of the brakes.

6 The actual speed of the car at the bottom of the slope is lower than calculated in question 3. Frictional forces of 2500 N act against the car as it rolls down the slope. This can be assumed to be constant.

a) If the length of the slope is 180 m, calculate the work done against friction by the car as it rolls down the slope.

b) Use the principle of energy conservation to calculate the velocity of the car at the bottom of the slope.

c) If the brakes of the car have a mass of 12 kg and specific heat capacity of 600 J kg^{-1} $°C^{-1}$, calculate the increase in temperature of the brakes bringing the car to rest again on the level ground. You may ignore frictional forces for this part of the question.

3 Collisions, explosions and impulse

You should already know (National 5)

- Newton's laws of motion:
  - Newton's first law – an object will remain at rest or continue with a constant velocity unless acted on by an unbalanced force.
  - Newton's second law – an unbalanced force will produce an acceleration ($F = ma$)
  - Newton's third law – every action force has an equal and opposite reaction force

Learning intentions

- Definition of momentum as the product of the mass and velocity of an object:
  - Momentum as a vector quantity
  - Calculation of the momentum of an object
  - Calculation of the total momentum of a system of objects moving in one dimension
- Principle of Conservation of Momentum applied to collisions:
  - Calculations involving the momentum before and after a collide-and-lock collision
  - Calculations involving the momentum before and after a collide-and-split collision
  - Calculations involving the momentum before and after an explosion
- Elastic and inelastic collisions
  - Calculations involving the kinetic energy before and after a collision to determine whether the collision is elastic or inelastic.
- Impulse and change in momentum
  - Define impulse as the average force acting multiplied by the time the force acts for
  - Link impulse to the change in momentum of an object
  - Carry out calculations involving average force, contact time and change in momentum of an object
  - Use the concept of impulse to explain how forces can be reduced in a car accident
  - Define impulse as the area under a force–time graph
  - Carry out calculations involving impulse and change in momentum using force–time graphs
  - Highlight the link between impulse, change in momentum and Newton's third law of motion

3.1 Momentum

Consider trying to catch a tennis ball and trying to catch a house brick. With one hand it is easy to catch a tennis ball and it should not injure your hand. A house brick travelling at the same velocity is more difficult to catch and it is likely to cause you a serious injury! This is due to the momentum of the object – a house brick has more momentum than a tennis ball travelling at the same speed, because it has a greater mass.

Fig 1.133

3.1.1 Defining momentum

The momentum, p, of an object depends on its mass (kg) and velocity ($m\,s^{-1}$):

$$p = mv$$

Momentum is a vector quantity measured in kilogram metres per second ($kg\,m\,s^{-1}$).

Worked example

Calculate the momentum of a car of mass 1200 kg travelling at 20 $m\,s^{-1}$.

$$p = mv$$

$$p = 1200 \times 20$$

$$p = 24{,}000\ kg\ ms^{-1}$$

3.1.2 Total momentum

In this section, many examples with more than one object moving are considered. In such cases, the total momentum is obtained by adding up the momentum of each of the objects:

$$p_{Total} = p_A + p_B + \ldots$$

Key point

The momentum of an object depends on both the mass and the velocity:

$$p = mv$$

Remember that momentum is a vector quantity, so direction is important! All problems considered here will be in one dimension, so it is possible to state that one direction is positive and the other is negative. For the two cars below, the car travelling to the right has positive momentum and the car travelling to the left has negative momentum:

Key point

Momentum is a vector quantity. In one dimension, the direction can be defined using positive and negative signs.

Positive momentum **Negative momentum**

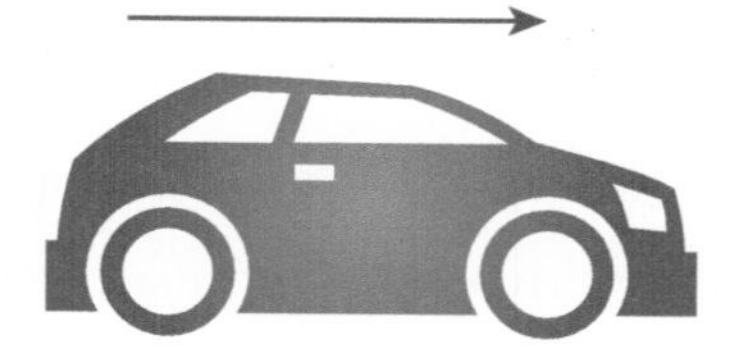

Fig 1.134

Worked example

Calculate the total momentum of the two cars shown in the diagram below. Car A has a mass of 1600 kg and is travelling to the right at 30 m s^{-1}. Car B has a mass of 1800 kg and is travelling to the left at 35 m s^{-1}.

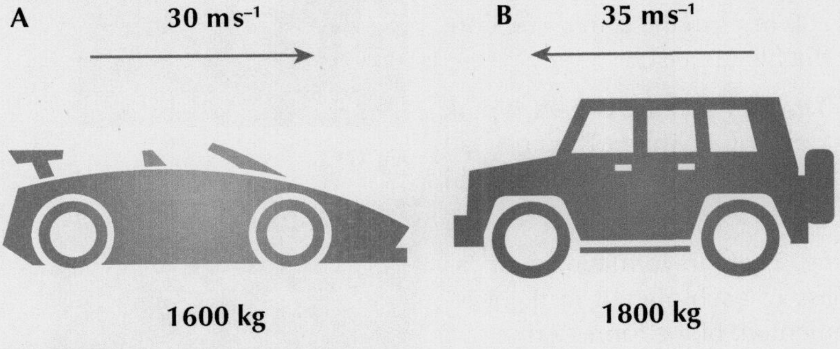

Fig 1.135

$p = mv$

$p = 1600 \times 30$

$p = 48{,}000\ kg\ ms^{-1}$

$p = mv$

$p = 1800 \times -35$

$p = -63{,}000\ kg\ ms^{-1}$

$p_{total} = 48{,}000 - 63{,}000$

$p_{total} = -15{,}000\ kg\ ms^{-1}$

Exercise 3.1.1 Defining momentum

1 Calculate the momentum of the following objects:

a) A house brick with a mass of 1 kg travelling at 20 m s^{-1}

b) A ship with mass of 10,000 kg travelling at 5 m s^{-1}

c) A dog of mass 20 kg running at a speed of 7 m s^{-1}

d) A tennis ball of mass 50 g travelling at 25 m s^{-1}

e) A car of mass 750 kg travelling at 100 km hr^{-1}

2 Two bricks, each of mass 1.7 kg are thrown into the air. Each brick has a velocity of 6.7 m s^{-1}. Find the total momentum of the bricks combined.

3 In a tennis match, the tennis ball of mass 0.05 kg is hit across the net with a velocity of 40 m s^{-1}. The player accidentally lets go of his racket of mass 0.5 kg, which also travels across the net with a velocity of 10 m s^{-1}. Find the total momentum of the racket and the ball.

Fig 1.136

(continued)

4 Two railway carriages are travelling in the same direction along a straight stretch of track. Each carriage has a mass of 8000 kg. One carriage has a velocity of 8 $m s^{-1}$ and the other a velocity of 12 $m s^{-1}$. Calculate the total momentum of the carriages.

5 Two Eurofighter fighter jets are flying in formation. Each jet has a mass of 11,000 kg. If one is flying with a velocity of 150 ms^{-1} and the other has a velocity of 200 $m s^{-1}$ in the same direction, what is the total momentum of the fighter jets?

6 A snooker ball has a mass of 160 g. After a shot is played, a pink ball travels forwards with a velocity of 11 $m s^{-1}$. The cue ball travels backwards with a velocity of 4 $m s^{-1}$. Calculate the total momentum in this case.

7 The Red Arrows team are flying 9 jets in formation with a velocity of 150 $m s^{-1}$. If the mass of each aircraft in flight is 9000 kg, calculate the total momentum of the formation.

Fig 1.137

Fig 1.138

3.2 Conservation of momentum

It has been shown that momentum is a vector quantity that depends on both the mass and the velocity of an object,

$$p = mv$$

During a collision, the total momentum is conserved. This is true in the absence of external forces and is a fact that can be used to predict velocities of objects both before and after a collision. So, in the absence of external forces:

Total Momentum Before a Collision = Total Momentum After a Collision

Written as an equation, where u is the initial velocity and v is the final velocity of an object, this is:

$$m_1 u_1 + m_2 u_2 = m_1 v_1 + m_2 v_2$$

This principle is known as the Principle of Conservation of Linear Momentum.

Applying conservation of momentum

The principle of conservation of momentum says that the total momentum before a collision is equal to the total momentum after it. This can be used to work out missing velocities or masses in a collision:

1. Calculate the total momentum before the collision: $p_{before} = m_1 u_1 + m_2 u_2$
2. Calculate the total momentum after the collision: $p_{after} = m_1 v_1 + m_2 v_2$
3. Set the momentum before equal to the momentum after and solve for the unknown.

It may help to draw a diagram to represent the objects and their velocities before and after the collision. This is shown in the worked example on page 70.

Experiment 3.2 Investigating collisions

This experiment investigates the momentum before and after a collision of objects using a linear air track.

Apparatus

- Linear air track or low-friction track
- Two carts for the track
- Velcro, Blu Tack or cork and needle
- Various masses
- Light gates
- Timers

Instructions

1 Set up the linear air track and light gates. One of the light gates will be used to measure the velocity of one cart before the collision and the other light gate will measure the velocity of the other cart. Care must be taken as to what light gates will measure the velocities of the carts after the collision, as it will depend on the direction the carts move off in.

2 Construct a table similar to the one below for noting your results:

Mass of vehicle A (kg)	Mass of vehicle B (kg)	Velocity of vehicle A before ($m s^{-1}$)	Velocity of vehicle B before ($m s^{-1}$)	Total momentum before ($kg m s^{-1}$)	Velocity of vehicle A after ($m s^{-1}$)	Velocity of vehicle B after ($m s^{-1}$)	Total momentum after ($kg m s^{-1}$)

Table 1.7

3 Keep cart B stationary, and push cart A towards it – let both carts collide and move off separately.

4 Use the light gates to measure the velocities of the carts and record them in the table.

5 Calculate the total momentum before and after the collision.

6 Compare the total momentum before the collision to the total momentum after the collision.

7 Repeat with different masses fitted to both carts.

8 Repeat with different types of collision: both carts moving before; use Blu Tack or velcro or cork and needle so that the carts lock together after the collision; start with both carts stationary (explosion).

3.2.1 Collide-and-lock collisions

A collide-and-lock collision is one where the objects are moving separately before the collision then stick together and move off as one after the collision. An example of this would be a car accident where both cars stick together and the wreckage moves off as one.

Fig 1.139

The total momentum after the collision will use the combined mass of both vehicles:

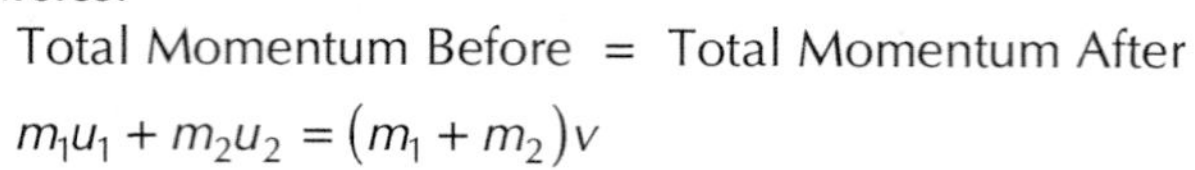

Total Momentum Before = Total Momentum After

$$m_1u_1 + m_2u_2 = (m_1 + m_2)v$$

Worked example 1

A train carriage of mass 8000 kg breaks free and runs along the track with a velocity of 12 $m s^{-1}$. It collides and locks with a carriage of mass 6000 kg. Calculate the velocity of the wreckage immediately after the collision.

Construct a diagram to show the masses and velocities before and after the collision:

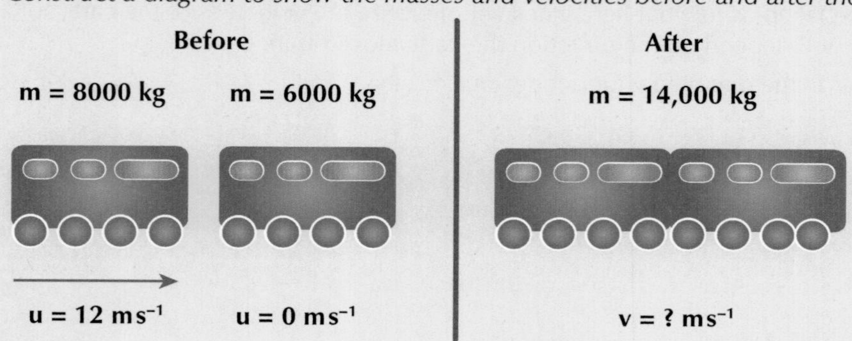

Fig 1.140

Use the principle of conservation of momentum to find the missing velocity:

Total Momentum Before = Total Momentum After

$$(8000 \times 12) + (6000 \times 0) = (14{,}000 \times v)$$

$$96{,}000 + 0 = 14{,}000v$$

$$v = \frac{96{,}000}{14{,}000}$$

$$v = 6.86\ ms^{-1}$$

The wreckage moves off with a velocity of 6.85 $m s^{-1}$ in the positive direction.

Exercise 3.2.1 Collide-and-lock collisions

1 In a lab experiment, collide-and-lock collisions are investigated with two cars using the experiment shown in Figure 1.141.

Cart A has a mass of 1.3 kg. Cart B has a mass of 2.1 kg.

Before the collision, Cart A is moving with a velocity of 4.5 m s^{-1}. Cart B is stationary. The carts collide, lock and move off together through the light gate.

a) Calculate the total momentum before the collision.

b) Find an expression for the total momentum after the collision.

c) Use conservation of momentum to find the speed of the carts after the collision.

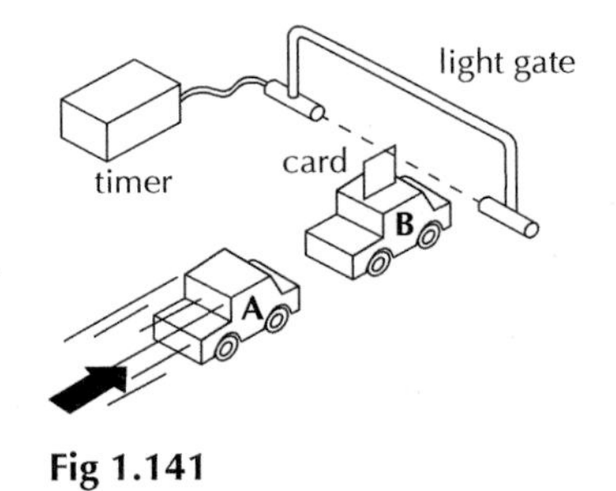

Fig 1.141

2 In a car accident, two cars are travelling towards each other. They collide and lock, and the wreckage moves off as one piece. Car A has a mass of 1200 kg and was travelling at 26 m s^{-1} before the collision. Car B has a mass of 1700 kg and was travelling in the opposite direction with a velocity of 19 m s^{-1}. Find the resultant velocity of the wreckage.

3 Police are investigating a car accident. The accident happened on a stretch of road where the speed limit is 14 m s^{-1}. Through analysing the cars and the tyre tracks on the road, they have found the following information:

- Mass of Car A = 700 kg
- Mass of Car B = 900 kg
- Velocity of both cars after collision = 8 m s^{-1}.

Car B is known to have been stationary before the collision. The cars locked together in the collision.

Calculate the total momentum after the collision took place.

Fig 1.142

Use the principle of conservation of momentum to work out the velocity of Car A before the collision. Was the car breaking the speed limit?

4 A bullet is fired from a gun and embeds itself in a box of sand. The box is suspended from a clampstand by a piece of string, the mass of which can be neglected. The bullet has a mass of 0.05 kg and is travelling with a velocity of 250 m s^{-1}. The box of sand has a mass of 3 kg.

a) Calculate the velocity that the bullet and box move off with after the collision.

b) Calculate the maximum height to which the bullet and box will swing.

5 In a car accident, two cars are travelling towards each other. They collide and lock together. The wreckage moves off together. Car A has a mass of 1200 kg and was travelling at 26 m s^{-1} before the collision. Car B has a mass of 1700 kg and was travelling in the opposite direction with a velocity of 19 m s^{-1}. Find the resultant velocity of the wreckage.

3.2.2 Collide-and-split collisions

A collide-and-split collision is a collision where both objects are moving individually before the collision and move off separately after the collision. An example of a typical collide-and-split collision would be in the game of snooker: the cue ball hits the object ball and both balls move off separately:

Total Momentum Before = Total Momentum After

$$m_1u_1 + m_2u_2 = m_1v_1 + m_2v_2$$

It is vital to pay close attention to the direction of the objects and ensure that you use the correct sign in the relationship.

Worked example

Snooker balls all have an equal mass of 0.16 kg. In a simple shot, the cue ball travels towards a stationary red ball at a speed of 8 m s^{-1}. The cue ball collides with the red ball and after the collision the red ball moves off with a speed of 6 m s^{-1}. Calculate the speed the cue ball moves off with.

Fig 1.143

First of all, calculate the total momentum before the collision:

$$p_{before} = m_1u_1 + m_2u_2$$
$$p_{before} = (0.16 \times 8) + (0.16 \times 0)$$
$$p_{before} = 1.28 + 0$$
$$p_{before} = 1.28\ kg\ ms^{-1}$$

Now calculate the total momentum after the collision:

$$p_{after} = m_1v_1 + m_2v_2$$
$$p_{after} = (0.16 \times v) + (0.16 \times 6)$$
$$p_{after} = 0.16v + 0.96$$

By conservation of momentum:

Total Momentum Before = Total Momentum After

$$1.28 = 0.16v + 0.96$$
$$0.32 = 0.16v$$
$$v = \frac{0.32}{0.16}$$
$$v = 2\ ms^{-1}$$

Exercise 3.2.2 Collide-and-split collisions

1 In a game of lawn bowls, the bowl with a mass of 2 kg collides with the stationary jack with a mass of 0.4 kg. The bowl is moving with a velocity of 12 m s^{-1} before the collision. After the collision, the jack moves off with a speed of 20 m s^{-1} in the same direction as the approaching bowl.

a) Calculate the total momentum of the bowl and the jack before the collision.

b) Find an expression for the total momentum of the bowl and the jack after the collision.

c) Use the principle of conservation of momentum to calculate the velocity of the bowl after the collision.

2 Two cars are involved in a collision. Car A, which has a mass of 1260 kg, is travelling with a speed of 12 $m s^{-1}$ when it collides with Car B, which is initially stationary. Car B has a mass of 2025 kg and moves off with a speed of 5 $m s^{-1}$ immediately after the collision.

   a) State the law of conservation of linear momentum.

   b) Calculate the speed of Car A immediately after the collision.

3 A snooker ball with a mass of 0.16 kg is travelling at 8 $m s^{-1}$. It collides with a stationary ball that has a mass of 0.16 kg, which moves off in the same direction with a speed of 3 $m s^{-1}$. Calculate the speed the second ball moves off with after the collision.

4 A train carriage with a mass of 4000 kg breaks free and runs down a hill and collides with a stationary engine of mass 9000 kg at the bottom. Just before the collision, the carriage is moving with a velocity of 40 $m s^{-1}$. After the collision, the carriage continues forward with a velocity of 10 $m s^{-1}$. What is the velocity of the train engine after the collision?

3.2.3 Explosions

The third type of collision starts with both objects stationary; after the 'collision' the objects move off separately. This type of collision is called an explosion. An example of an explosion would be a cannon firing a cannonball – the cannonball moves off in one direction and the cannon recoils in the opposite direction.

Worked example

A cannon has a mass of 1500 kg. It fires a cannonball of mass 10 kg with a velocity of 100 $m s^{-1}$. With what velocity does the cannon move after the cannonball is fired?

Fig 1.144

In this example, the total momentum before is zero because both the cannon and the cannonball are stationary:

$$p_{before} = 0 \text{ kg ms}^{-1}$$

Now calculate the total momentum after the collision:

$$p_{after} = m_1v_1 + m_2v_2$$
$$p_{after} = (100 \times 10) + (1500v)$$
$$p_{after} = 1000 + 1500v$$

By conservation of momentum,

Total Momentum Before = Total Momentum After

$$0 = 1000 + 1500v$$
$$-1000 = 1500v$$
$$v = \frac{-1000}{1500}$$
$$v = -0.67 \text{ ms}^{-1}$$

The velocity of the cannon is negative. This means that, as expected, the cannon is moving in the opposite direction to the cannonball. This velocity is known as the recoil velocity.

As with both previous collision types, the total momentum is conserved for explosions too. Before the explosion, the total momentum is zero. This is also the case after the collision. The momentum of one object is compensated for by the momentum of the other – this means that one velocity will be positive and the other will be negative.

$$\text{Total Momentum Before} = \text{Total Momentum After}$$
$$0 = m_1v_1 + m_2v_2$$

Exercise 3.2.3 Explosions

1 A cannon has a mass of 900 kg and it fires a cannonball with a mass of 2 kg. The cannonball is fired with a velocity of 150 m s^{-1}.

a) What is the total momentum before the explosion?

b) Write down an expression for the total momentum after the collision.

c) Use the principle of conservation of momentum to calculate the recoil velocity of the cannon.

Fig 1.145

2 A cannon with a mass of 100 kg is used to fire a cannonball with a mass of 8 kg. The recoil velocity of the cannon is found to be 3 m s^{-1}. Use the principle of conservation of momentum to find the velocity with which the cannonball is fired.

3.2.4 Elastic vs inelastic collisions

So far, the total momentum both before and after a collision has been investigated. In all cases, the total momentum is conserved and this makes it possible to find missing velocities and masses. However, this is not the case for kinetic energy. The total energy must always be conserved, because energy cannot be created or destroyed!

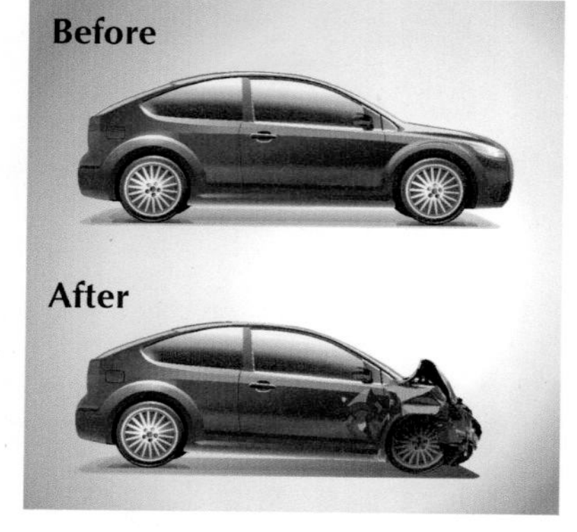

Fig 1.146

During a typical car crash, the velocity of the vehicles involved is changed. However, the vehicles themselves also sustain damage, as shown in Figure 1.146. It takes energy to cause this damage. This means that some of the original kinetic energy of the vehicles has been transferred into work done to change the shape of the car. You can also hear a car crash, so some of the kinetic energy has been transferred into sound energy.

If the kinetic energy is lost during the collision, we say that the collision is inelastic. There are some collisions, however, where kinetic energy is conserved; for example, collisions between atoms. If kinetic energy is conserved, the collision is called 'elastic'. There are also some collisions where kinetic energy increases. When explaining why a collision is inelastic, it is important to state clearly whether kinetic energy is being lost or gained as part of the collision; it is not sufficient to simply say that kinetic energy is not conserved.

3.3 Impulse

In contrast to what happens before and after a collision, during a collision there are forces acting for a given time that cause changes in momentum of the objects involved. In this section, the mechanics of the forces during a collision are considered with the concept of impulse.

3.3.1 Introducing impulse

When an unbalanced force acts on an object, it produces an acceleration according to Newton's second law of motion:

$$F = ma$$

Recall from above that acceleration can be found using:

$$a = \frac{v - u}{t}$$

Substituting this into Newton's second law gives:

$$F = m\left(\frac{v - u}{t}\right)$$

$$F = \frac{mv - mu}{t}$$

From the study of momentum above, it is known that the quantity $mv - mu$ is equal to the change in momentum, so:

$$F = \frac{\Delta p}{t}$$

$$Ft = \Delta p$$

where Δp is the change in momentum.

We define impulse as the average force, F, multiplied by the time for which the force is applied, t. Therefore we can state:

$$\text{Impulse} = Ft = \Delta p = mv - mu$$

The units for impulse are Newton seconds (N s), which is equivalent to the units for momentum, kg m s^{-1}.

Physically, this means that the greater the force applied, or the greater the time the force is applied for, the greater the impulse. Consider hitting a golf ball with a club. The harder you hit the ball (the greater the force), the faster the ball will move. Additionally, if you watch a golfer's swing, you will notice that they 'follow through' with the club. This increases the time of contact between the club and the ball, which increases the impulse and thus the change in momentum of the ball. As with conservation of momentum problems, the direction is key. You must be careful to represent direction with the correct sign. You must also ensure that you correctly substitute the initial velocity into 'u' and the final velocity into 'v' in the equation.

Key point

Impulse is the product of average applied force and contact time:

$$\text{Impulse} = \overline{F}t$$

The impulse is equal to the change in momentum:

$$Ft = mv - mu$$

Fig 1.147

Experiment 3.3.1 Investigating impulse

This experiment investigates the concept of impulse by considering the effects of force and contact time on the resultant speed of a golf ball.

Apparatus

- Golf club
- Golf ball
- Tin foil
- Wires
- 2 timing devices
- Light gate
- 'Tee' (a large pen cap works well)

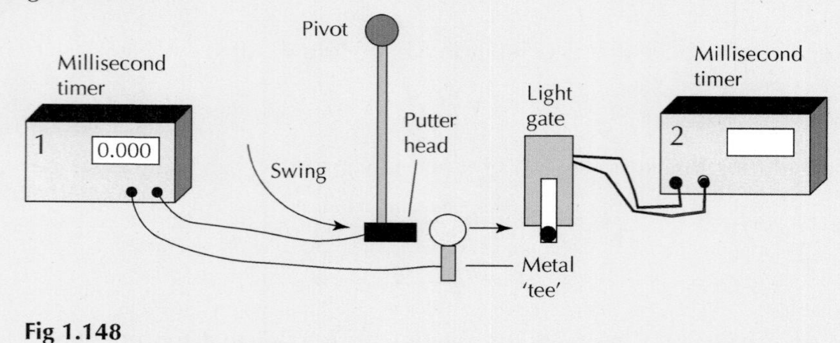

Fig 1.148

Instructions

1. Wrap tin foil around the golf ball, 'tee' (pen cap) and the end of the golf club.
2. Attach a wire to the tinfoil on the golf club and a second wire to the metal 'tee'. Connect both of these wires to your timing device to form a series circuit that is completed when the club comes into contact with the golf ball. This will be used to measure the contact time.
3. Connect the light gate to the second timing device. Ensure that the golf ball breaks the beam and this is recorded by the timing device. The time taken for the golf ball to pass through the beam will be used to calculate the speed of the ball as it leaves the club, using the equation:

$$v = \frac{d}{t}$$

   where d is the diameter of the golf ball.
4. Strike the golf ball with the club, using the first timer to measure the contact time. Calculate the speed the ball leaves the club with (as described in 2 and 3 above).
5. Repeat the experiment, trying to ensure the same force is applied to the golf ball but changing the contact time by rebounding the club or following through to a greater extent. Also, repeat the experiment but change the force applied to the club, trying to keep the contact time constant.
6. Write a short report that qualitatively describes the relationship observed between contact time, applied force and the speed the ball leaves the 'tee'.

Key point

In the event of a ball rebounding (for example, in the case of a tennis racket being used to return a serve), the initial velocity of the ball will be negative and the final velocity of the ball will be positive, as the ball has changed direction. The applied force will also be positive as it will be in the opposite direction to the initial velocity of the ball.

Fig 1.149

Worked example 1: initially stationary object

In a game of pool, the cue applies an average force of 5 N to the cue ball, which has a mass of 0.165 kg. The time of contact between the cue and the ball is 0.4 s. The cue ball was initially stationary.

Fig 1.150

Calculate:

a) The impulse between the cue and the cue ball.

b) The speed with which the cue ball moves off.

a) *To calculate the impulse, you need to use the average force and the time:*

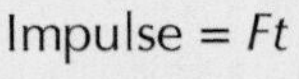

$$\text{Impulse} = Ft$$

$$\text{Impulse} = 5 \times 0.4$$

$$\text{Impulse} = 2Ns$$

b) *The speed after comes from the change in momentum, which is equal to the impulse:*

$$\Delta p = Ft = mv - mu$$

$$2 = mv - mu$$

$$2 = (0.165 \times v) - 0$$

$$0.165v = 2$$

$$v = \frac{2}{0.165}$$

$$v = 12.12\ ms^{-1}$$

Worked example 2: rebounding object

A tennis player returns a serve successfully. The ball, of mass 0.1 kg, hits the player's racket with a velocity of 40 m s^{-1}. She exerts an average force of 80 N for a time of 100 ms. Calculate the velocity with which the ball leaves her racket.

Fig 1.151

Here, the rebound velocity of the ball can be calculated by setting the impulse equal to the change in momentum:

$$Ft = mv - mu$$

Substitute what you know, assuming the initial velocity to be negative. If you assume this, the applied force will be positive, as it acts in the opposite direction

$$80 \times 100 \times 10^{-3} = 0.1v - (0.1 \times -40)$$

Solve for the rebound velocity:

$$8 = 0.1v - (-4)$$

$$8 = 0.1v + 4$$

$$0.1v = 4$$

$$v = 40\ ms^{-1}$$

The ball rebounds with a velocity of 40 m s^{-1} in the positive direction (opposite to the initial direction of the ball).

Exercise 3.3.1 Introducing impulse

1. A rugby player is taking a penalty kick. He exerts a force of 90 N for a time of 0.02 s on the rugby ball. If the ball has a mass of 0.450 kg, calculate the speed with which the ball leaves the player's boot.

Fig 1.152

2. A snooker cue exerts a force of 15 N on a ball of mass 0.16 kg for a time of 35 ms. Find the speed with which the ball leaves the cue, assuming the ball was initially stationary.
3. A golf ball of mass 0.160 kg is hit by a golf driver. The driver is in contact with the ball for a time of 40 ms, and the ball leaves the tee with a speed of 45 $m\,s^{-1}$. If the ball was initially stationary, find the average force exerted by the club on the ball.
4. During a game of football, a player headers the ball towards the goal. The ball, of mass 0.5 kg, was initially travelling towards the player's head with a velocity of 15 $m\,s^{-1}$. If the player exerts a force of 250 N for time of 70 ms, calculate the speed with which the ball leaves the player's head.
5. Students are investigating the principle of impulse in the lab. They set up an experiment that records the contact time between a club and a ball. Keeping the same average force, they vary the contact time and measure the speed with which the ball leaves the club. They notice that the greater the contact time, the greater the speed of the ball leaving the club. Use your knowledge of impulse and momentum to explain this observation.

Fig 1.153

6. A car collides head-on with a wall. The car, of mass 1750 kg, was initially travelling at 25 $m\,s^{-1}$ and was brought to rest in a time of 0.45 s. Calculate the average force exerted on the car.
7. A ball of mass 1.20 kg is dropped from a height of 2.00 m above the ground. It bounces off the ground and rebounds to a height of 1.75 m. The ball is in contact with the ground for a time of 270 ms.
   a) Calculate the speed with which the ball hits the ground.
   b) Calculate the speed with which the ball leaves the ground. You may assume air resistance is negligible.
   c) Find the average force exerted on the ball by the ground – state both the magnitude and direction of this force.
8. The fire service are called to deal with a fire in an old factory. To extinguish the fire, water is ejected from a fire hose at a rate of 45 kg every second. The water leaves the fire hose with a velocity of 50 $m\,s^{-1}$ and hits the wall of the building with the same velocity.
   a) Assuming the water does not rebound from the building, calculate the force exerted on the wall of the building each second.
   b) Use your knowledge of momentum and impulse to explain why the fireman feels a recoil force from the hose when it is ejecting water.

Fig 1.154

9. In a game of snooker, the player can control the speed the cue ball leaves the cue by changing two things – the force applied to the ball and the length of time the cue is in contact with the ball. Use your knowledge of physics to explain the effect of changing each of these variables on the resultant speed of the cue ball.

3.3.2 Impulse from force–time graphs

So far, we have considered impulse where the force is constant or is described as an 'average force'. During most collisions, the force applied is not constant. In this case, the force applied can be plotted as a graph against time. The impulse is then the area under the force–time graph.

Consider the example of a golf club striking a golf ball. The club exerts a force that is small at first as it comes into contact with the ball, increasing as the club is pushed through the ball and then decreasing as the ball leaves the club. This can be plotted on a graph as shown below:

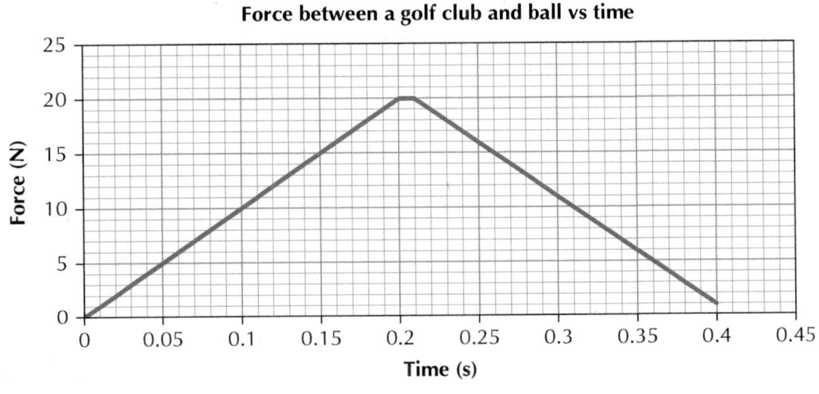

Fig 1.155

Calculating the area under the graph will give the impulse, which is then equal to the change in momentum of the golf ball. Here, the area is given by the area of a triangle:

$$A = \frac{1}{2}bh$$

Resultant velocity using force–time graphs

The impulse is given by finding the area under the force–time graph. This can then be used to find the velocity of an object after a collision:

1. Find the impulse by finding the area under a force–time graph. Remember that the area of a triangle is given by:

$$A = \frac{1}{2}bh$$

2. Calculate the velocity of the object after the collision using:

$$\text{Impulse} = mv - mu$$

Remember that velocity is a vector so direction, represented by positive or negative, is important!

Worked example

During a tennis serve, the ball of mass 0.1 kg is initially at rest. It is hit by the racket, which is in contact with the ball for 150 ms. The graph below shows how the force applied to the ball varies with time.

The impulse is given by the area under the graph:

$$A = \frac{1}{2}bh$$

$$A = \frac{1}{2} \times (150 \times 10^{-3}) \times 60$$

$$A = 4.5 Ns$$

This impulse is equal to the change in momentum:

$$\Delta p = mv - mu = 4.5$$

$$(0.1 \times v) - (0.1 \times 0) = 4.5$$

$$0.1v = 4.5$$

$$v = \frac{4.5}{0.1}$$

$$v = 45\ ms^{-1}$$

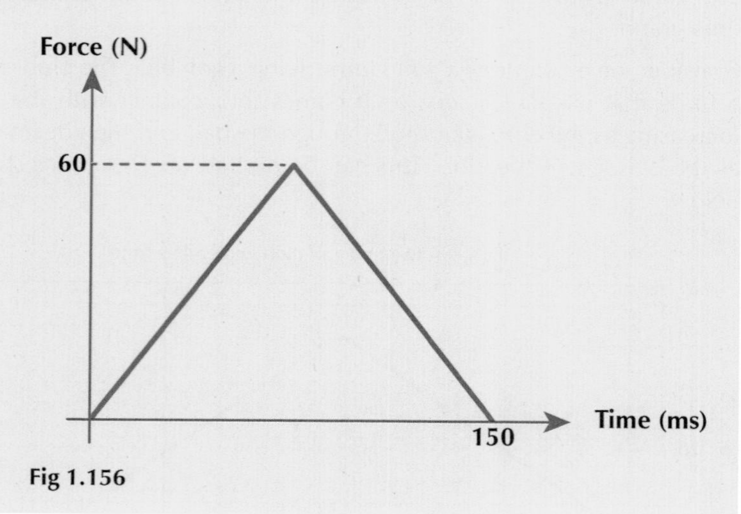

Fig 1.156

Fig 1.157

Applying impulse to car safety – the airbag

It has been shown that the average force acting during a collision is linked to the length of time of the collision and the change of momentum. Common sense dictates that the greater the change in momentum (the greater the speed before the crash), the greater the forces acting during the collision. Studies of impulse reveal that as you increase the time of the impact, the average force is reduced. This can be applied to car safety to make accidents survivable.

An airbag (also known as a *Supplementary Restraint System* or *SRS*) is used to increase the time of contact between the occupant of the car and solid objects of the interior.

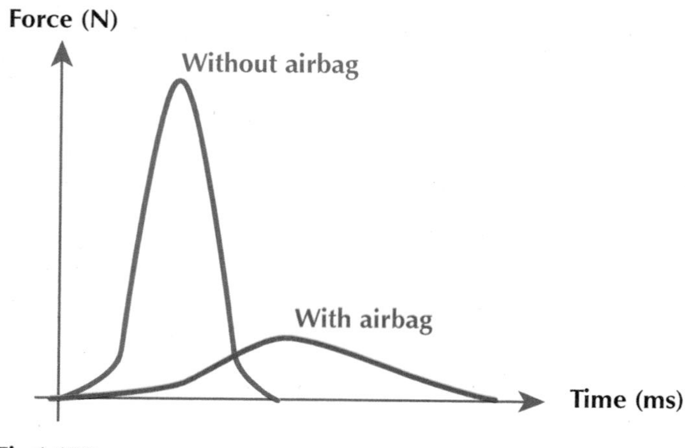

Fig 1.158

The force–time graph in Figure 1.158 shows the forces acting during a collision with and without an airbag.

The change in momentum is the same in both cases as the driver has the same mass and is being brought to rest from the same speed. Therefore, the impulse and the area under the force–time graph must be the same in both cases. However, if the contact time is increased, the peak force will be decreased; this will reduce the severity of the injuries received by the driver.

Momentum, impulse and Newton's third law

The link between the concept of momentum conservation and Newton's third law has been made. This is most obvious in terms of action-reaction from the recoil of the cannon when it fires a cannonball. Having studied impulse, the link can be completed. Consider an explosion, shown in Figure 1.159:

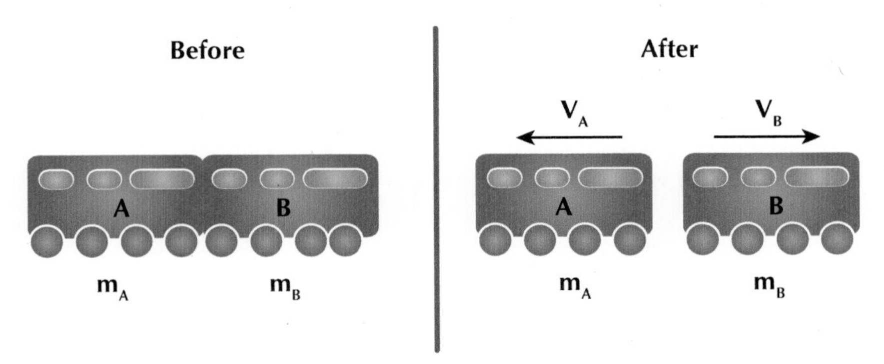

Fig 1.159

Before the explosion, both carts are stationary and the total momentum is 0:

$$m_A u_A + m_B u_B = 0$$

During the explosion, forces act on both carts: F_A acts on cart A and F_B acts on cart B. These forces cause a change in momentum of both carts and they move off in opposite directions with velocities v_A and v_B. The total momentum after the collision must be 0, giving:

$$m_A v_A + m_B v_B = 0$$

Rearranging this gives:

$$m_A v_A = -(m_B v_B)$$

The right-hand side and left-hand side of the above equation represent the changes in momentum of each cart, so:

$$\Delta p_A = -(\Delta p_B)$$

As shown before, the changes in momentum are equal in size but are in the opposite direction. During the collision, forces act for the same amount of time. Dividing both sides by this time:

$$\frac{\Delta p_A}{t} = \frac{-(\Delta p_B)}{t} \quad (*)$$

Using the definition impulse:

$$\text{Impulse} = Ft = \Delta p$$

and rearranging for force:

$$F = \frac{\Delta p}{t}$$

this can now be written (*) as:

$$p = mv$$

* This is a statement of Newton's third law of motion – every action force has an equal and opposite reaction force.

Exercise 3.3.2 Impulse from force–time graphs

1 In a game of shinty, a player hits the ball with a stick towards the goal. The force–time graph below shows the force he exerts on the ball with the stick. If the ball was initially stationary and has a mass of 0.25 kg, calculate the speed with which it leaves the stick.

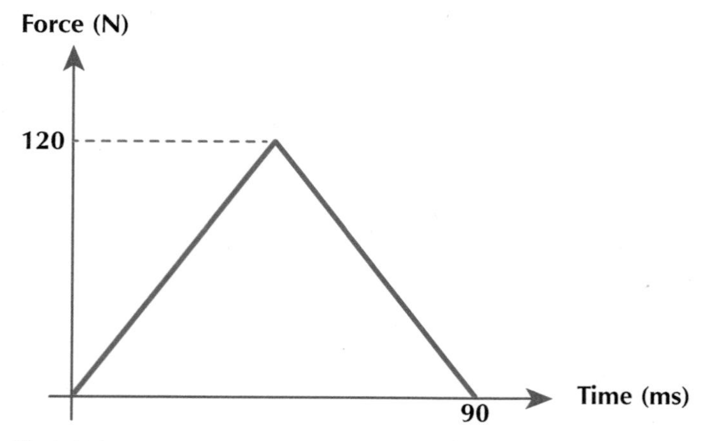

Fig 1.160

Fig 1.161

2 A van is involved in a collision with a wall. The force–time graph below shows the force exerted by the wall on the van as the van is brought to rest.

If the van had a mass of 2100 kg, calculate the speed at which the van was travelling before it made contact with the wall.

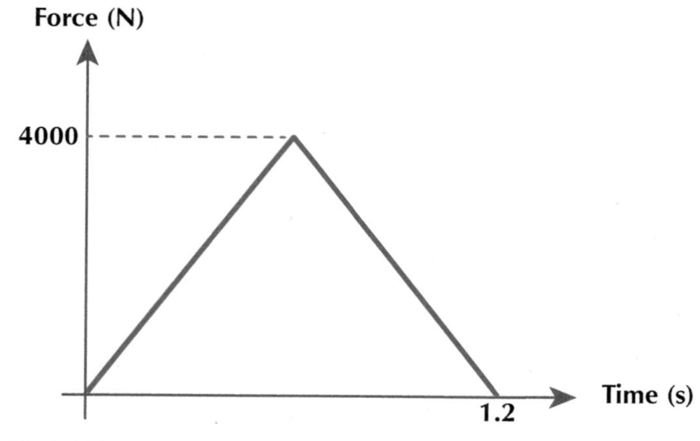

Fig 1.162

3 During a game of tennis, a player returns the serve. The ball hits her racket with a velocity of 45 $m s^{-1}$ and leaves her racket with a speed of 60 $m s^{-1}$ in the opposite direction. The mass of the ball is 0.2 kg. Calculate the missing value of peak force on the graph below, which represents the interaction between the ball and the racket.

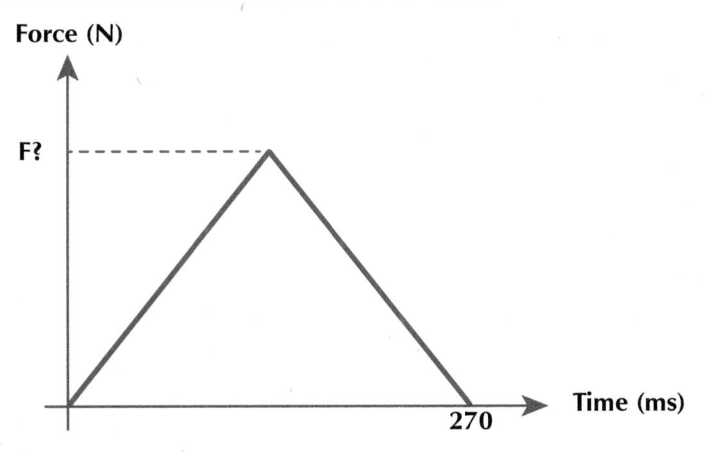

Fig 1.163

4 A ball is dropped onto a floor. The ball, of mass 0.8 kg, falls from a height of 3 m and rebounds.

a) Calculate the velocity of the ball just before it hits the floor.

b) The force–time graph to the right shows the force acting on the ball as it is in contact with the floor.

(i) Calculate the impulse.

(ii) Calculate the velocity of the ball immediately after the bounce.

(iii) Assuming air resistance is negligible, find the height to which the ball rebounds.

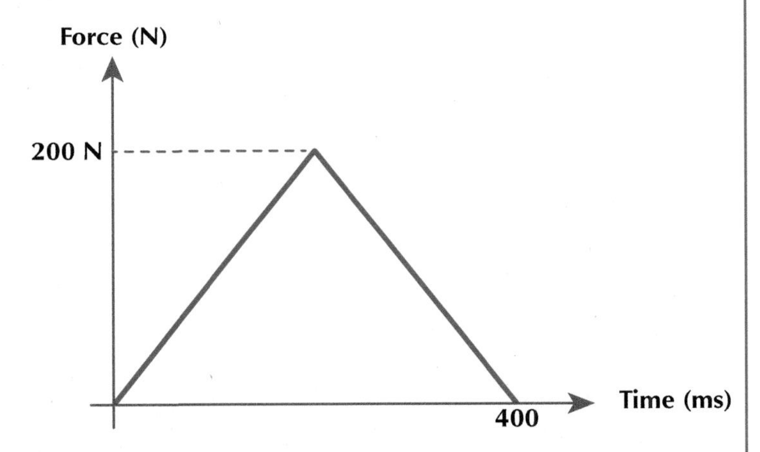

Fig 1.164

c) A second ball made with a softer material is dropped from the same height. This ball rebounds to the same height as the first ball. Copy the force–time graph for the first ball above and on the same graph, sketch what the force–time graph would look like for the second ball. Numerical values are not required.

5 Use your knowledge of impulse and momentum to explain how a crumple zone in a car helps to reduce the forces acting during a collision.

6 Students are discussing how a rocket engine is used to propel a rocket into space. They use Newton's third law of motion to explain how the rocket engine produces the force required to lift the rocket off the surface of Earth. Using your knowledge of impulse and momentum, and that a rocket engine works by ejecting a given mass of gas in a given time, explain how a rocket is propelled upwards from the surface of a planet. In your description, show the link between impulse, momentum and Newton's third law.

4 Gravitation

You should already know (National 5):

- The motion of a projectile can be split into horizontal and vertical components
- The horizontal motion of a projectile is constant
- The vertical motion of a projectile is subject to constant acceleration due to gravity
- A horizontally launched projectile starts at its maximum height and follows a curved path (called a trajectory) to the ground
- The range of a projectile can be found using the time of flight and horizontal (launch) velocity
- The height a projectile falls through can be found by plotting a velocity–time graph of the motion and finding the area under the graph
- Acceleration due to gravity on Earth is directed downwards and is constant (= 9.8 $m s^{-2}$)
- The gravitational field on Earth gives rise to a downwards force experienced by all objects with mass – this force is called **weight**.

Learning intentions

- The motion of a projectile can be split into horizontal and vertical components
- The horizontal motion of a projectile is constant and the equation $d = vt$ can be used to analyse it
- The vertical motion of a projectile is subject to constant acceleration due to gravity so can be analysed using the equations of motion
- The initial horizontal and vertical components of the motion of a projectile can be found using SOH-CAH-TOA applied to the vector diagram for the velocity
- The range of a projectile is calculated using the horizontal motion
- The maximum height of a projectile is calculated using the vertical motion
- The time of flight can be found using either horizontal or vertical motion dependent on the information supplied
- Any object with a mass produces a gravitational field
- The gravitational field strength from a point mass, such as a planet, decreases with distance squared away from the object (inverse-square law) – Newton's universal law of gravitation
- The force of gravity is responsible for keeping planets in orbit around stars, and also for the formation of stars and planets

4.1 Horizontally launched projectiles

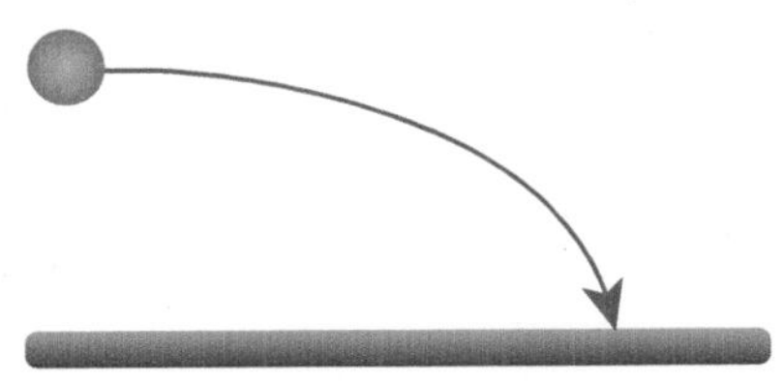
Fig 1.165

We have seen horizontally launched projectiles at National 5 level. The velocity of a projectile can be split into two rectangular components: a horizontal component where the velocity is constant; and a vertical component where the projectile is subject to a constant acceleration due to gravity. The combination of these two components leads to a curved path as shown in Figure 1.165:

Horizontal component

The horizontal velocity of a projectile is constant. This means we can apply the simple equation of motion that links velocity, distance and time:

$$d = \bar{v}t$$

This equation will govern the range of the projectile.

Vertical component

The vertical velocity of a projectile is subject to a constant acceleration due to gravity, $a = -9.8\ \text{m}\,\text{s}^{-2}$. This means we can apply our equations of motion to the vertical component of the motion:

$$v = u + at$$

$$s = ut + \frac{1}{2}at^2$$

$$v^2 = u^2 + 2as$$

For more information on the use of these equations, refer to Chapter 1.

4.2 Oblique projectiles

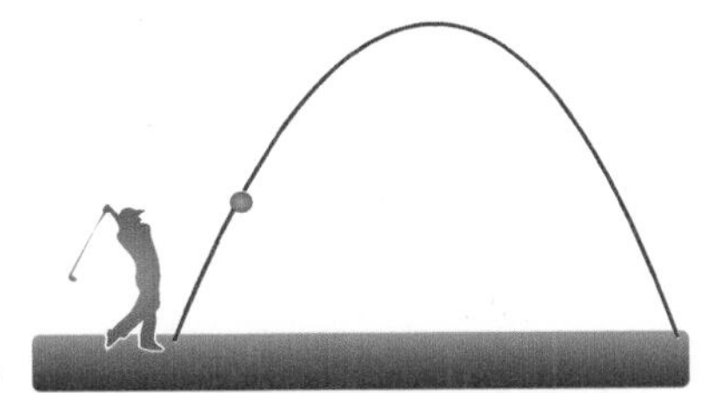
Fig 1.166

So far we have looked at projectiles that are launched **horizontally** – that is, projectiles where the initial vertical velocity is zero. You can also launch projectiles either upwards or downwards, which gives them an initial vertical velocity. An example of this would be a golf ball, which is struck from the 'tee' – it follows a curved path similar to that shown in Figure 1.166.

4.2.1 Analysing oblique projectiles

The analysis of an oblique projectile is exactly the same as the horizontal projectiles from National 5. The motion is split into horizontal and vertical components as before.

Combining the horizontal and vertical components gives the motion of the projectile shown here.

Key point

The vertical component of the velocity of a projectile is zero when the projectile is at its maximum height.

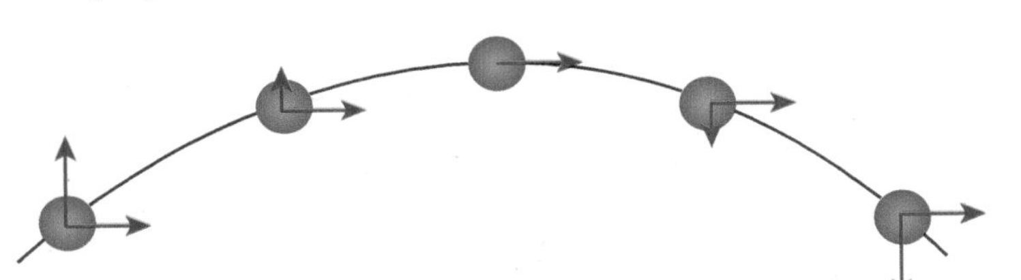
Fig 1.167

The horizontal component remains constant throughout the motion. The vertical component is always accelerating towards the ground – for a projectile launched upwards, this means the vertical component slows down to zero at the maximum height and then begins to increase in the downward direction as shown in Figure 1.167.

Horizontal and vertical components

We have seen previously that velocity is a vector quantity and can be split into two rectangular components: horizontal and vertical. These velocities can then be analysed separately.

Let us consider a 'typical projectile', which is launched at a certain angle, θ, above the horizontal as shown in Figure 1.168.

The initial velocity, *u*, can be separated into two separate components:

- Horizontal component: using trigonometry, we find the horizontal component to be:

$$u_h = u\cos\theta$$

- Vertical component: using trigonometry, we find the vertical component to be:

$$u_v = u\sin\theta$$

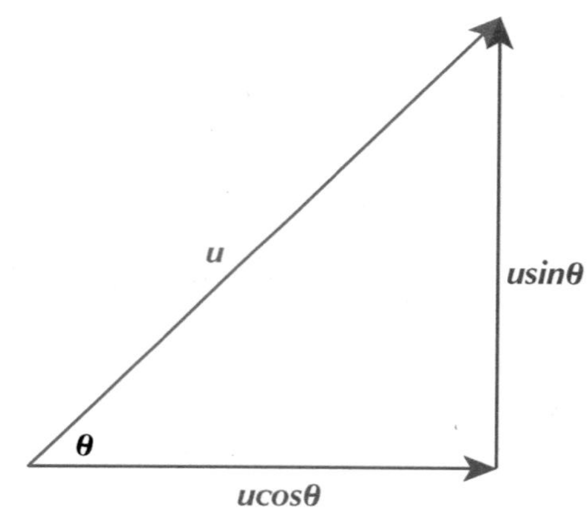

Fig 1.168

From our previous examples, we know that the horizontal velocity of a projectile is **constant**, and obeys the following equation:

$$s = vt$$

The vertical velocity of a projectile is subject to acceleration under gravity of $g = -9.8\ m\,s^{-2}$. As this is a constant acceleration, the equations of motion apply:

$$v = u + at$$

$$s = ut + \frac{1}{2}at^2$$

$$v^2 = u^2 + 2as$$

Worked example

A golf ball is launched with an initial velocity of 50 m s^{-1} at an angle of 30° above the horizontal. Calculate the initial horizontal and vertical components of the motion.

The horizontal component can be found using:

$$u_h = u\cos\theta$$

$$u_h = 50\cos 30$$

$$u_h = 43.3\ ms^{-1}$$

The vertical component can be found using:

$$u_v = u\sin\theta$$

$$u_v = 50\sin 30$$

$$u_v = 25\ ms^{-1}$$

Calculations involving the maximum height

At the maximum height of a projectile, the vertical velocity is zero. Knowing this allows us to work out the time taken to reach the maximum height, or what the maximum height will be for a given launch velocity.

Time taken

We use the initial vertical velocity, u_v, to find the time taken to reach the maximum height. The initial vertical velocity comes from the vertical component of the launch velocity:

$$u_v = u\sin\theta$$

At the maximum height, the vertical velocity is zero:

$$u_v = 0\ ms^{-1}$$

We are dealing with vertical motion, so the equations of motion for constant acceleration are used. Write down the '*suvat*' table:

s	**u**	**v**	**a**	**t**
X	$u\sin\theta$	0	–9.8	?

Table 1.8

The relevant equation here for time of flight to maximum height is:

$$v = u + at$$

Maximum height

The initial velocity can also be used to work out the maximum height. As above, we use the vertical component of the launch velocity:

$$u_v = u\sin\theta$$

As before, at the maximum height, the vertical velocity is zero:

$$u_v = 0\ ms^{-1}$$

This is again vertical motion, so we write down the '*suvat*' table:

s	**u**	**v**	**a**	**t**
?	$u\sin\theta$	0	–9.8	X

Table 1.9

The relevant equation here for working out the maximum height reached, s, is:

$$v^2 = u^2 + 2as$$

Calculations involving range

The range, R, of a projectile is the horizontal distance travelled between the starting point and the finishing point. It depends on the horizontal velocity and the time of flight:

$$R = v_h t$$

The horizontal velocity is constant, and is worked out using trigonometry, to be:

$$v_h = u \cos \theta$$

The time of flight can be worked out by considering the vertical velocity. For a projectile that takes off and lands at the same height, the time of flight is double the time taken to reach the maximum height. Such a projectile has a symmetrical path.

The equations of motion can also be used to work out the time of flight. This will work for situations where the projectile does not land at the same height as it was launched from. As we are dealing with vertical motion, we can write down the '*suvat*' table and fill in what we know and what we are trying to find out:

s	**u**	**v**	**a**	**t**
h	$u \sin \theta$	X	−9.8	?

Table 1.10

We assume the projectile lands at a height, h, either above or below its starting point. If it lands below its starting point, h will be negative. One relevant equation here is:

$$s = ut + \frac{1}{2}at^2$$

This equation is a quadratic in time, so to solve it requires rearranging into the following:

$$\frac{1}{2}at^2 + ut - s = 0$$

If the height is zero (symmetric projectile), then time will be a common factor. If not, then the quadratic formula is required to solve for t:

$$t = \frac{-b \pm \sqrt{b^2 - 4ac}}{2a}$$

This problem can also be solved in two stages by finding the velocity when the projectile lands using the equation,

$$v^2 = u^2 + 2as$$

and then finding the time of flight using the equation,

$$v = u + at$$

This technique means that solving a quadratic equation is not required.

Worked example 1

A javelin thrower launches her javelin with an initial velocity of 20 m s^{-1} at an angle of 40° above the horizontal on level ground. The javelin is assumed to land at the same height it was launched from.

Fig 1.169

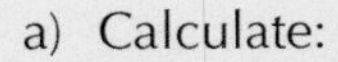

a) Calculate:

   (i) The initial horizontal velocity.

   (ii) The initial vertical velocity.

b) What is the maximum height the javelin will reach?

c) What is the range of the javelin, assuming it lands at the same height from which it was thrown?

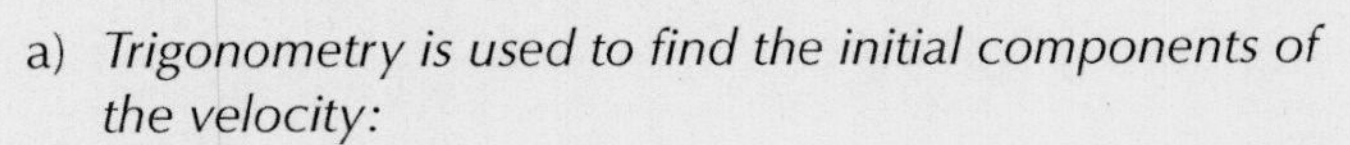

a) *Trigonometry is used to find the initial components of the velocity:*

Horizontal:

$$u_h = u\cos\theta$$
$$u_h = 20\cos 40$$
$$u_h = 15.3\ ms^{-1}$$

Vertical:

$$u_v = u\sin\theta$$
$$u_v = 20\sin 40$$
$$u_v = 12.9\ ms^{-1}$$

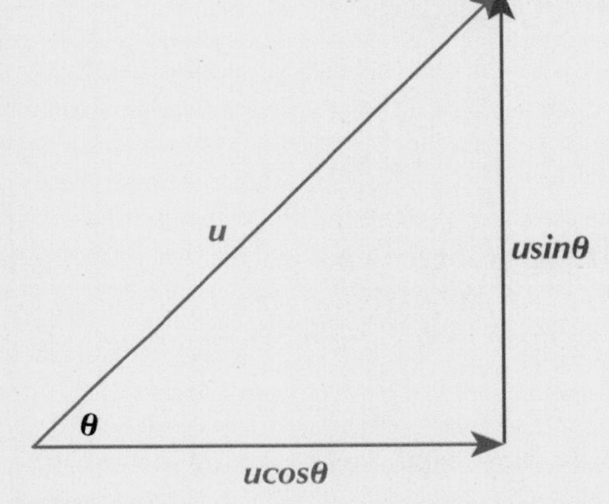

Fig 1.170

b) *The maximum height of the javelin depends on the vertical velocity, so we use the equations of motion. Write down the 'suvat' table:*

s	u	v	a	t
?	12.9	0	–9.8	X

Table 1.11

The relevant equation here is:

$$v^2 = u^2 + 2as$$

Substitute what we know and solve for the maximum height, s:

$$0^2 = 12.9^2 + 2(-9.8)s$$
$$0 = 166.4 - 19.6s$$
$$19.6s = 166.4$$
$$s = 8.5\ m$$

(continued)

c) *The range of a projectile will depend on the time of flight. The time of flight is controlled by the vertical velocity, so we use the equations of motion. Write down the 'suvat' table:*

s	u	v	a	t
0	12.9	X	−9.8	?

Table 1.12

The relevant equation here is:

$$s = ut + \frac{1}{2}at^2$$

Substitute what we know and solve for t. This involves solving a quadratic equation, where t is a common factor:

$$0 = 12.9t + \frac{1}{2}(-9.8)t^2$$
$$12.9t - 4.9t^2 = 0$$
$$t(12.9 - 4.9t) = 0$$
$$t = 0 \;\; ; \;\; 4.9t = 12.9$$
$$t = 0 \;\; ; \;\; t = 2.6s$$

The relevant time here is t = 2.6 s. The time of flight can also be found by working out the time taken to reach the maximum height and then doubling this value because the projectile is symmetric. Now, use the equation for horizontal motion to work out the range:

$$R = v_h t$$

Substitute what we know and solve for the range:

$$R = 15.3 \times 2.6$$
$$R = 39.8\ m$$

Worked example 2

A basketball player is taking a penalty shot. He launches the ball from a height of 2.1 m above the ground with a velocity of 10 $m s^{-1}$ at 40° above the horizontal as shown in Figure 1.172. The basketball hoop is a height of 3.5 m off the ground and a distance of 4 m away.

Does the player score his penalty? (Does the ball go through the hoop, without rebounding off the backboard?)

Answering this is tricky as it first requires you to understand exactly what the question is asking and how to apply knowledge of projectiles to it. If the ball successfully goes through the hoop to score the goal, then the height of the ball at the position of the hoop must be equal to the height off the ground – this is the key! We need to find out the height of the ball above the ground when it has travelled a horizontal distance of 4 m away from the player. If it is equal to the height of the basket, a goal was scored.

Fig 1.171

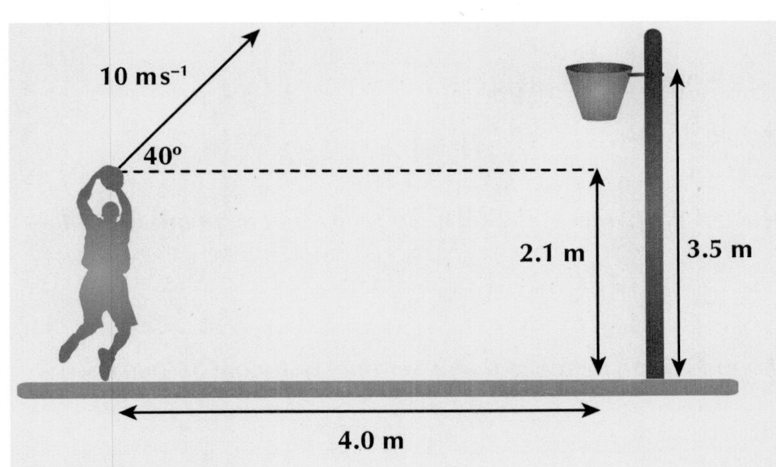

Fig 1.172

First of all, we need to work out the time taken for the ball to reach the hoop. This can be done using the horizontal motion of the projectile, which is constant, and the equation for horizontal motion:

$$d = vt$$

The horizontal velocity is given using trigonometry:

$$v_h = v\cos\theta$$
$$v_h = 10\cos 40^\circ$$
$$v_h = 7.66\ ms^{-1}$$

Now work out the time to reach the hoop:

$$d = vt$$
$$4.0 = 7.66\ t$$
$$t = \frac{4.0}{7.66}$$
$$t = 0.52\ s$$

Having worked out the time taken to reach the hoop, the height of the projectile above its starting point can be found by using the vertical motion. The vertical velocity of the ball is given by trigonometry:

$$v_v = v\sin\theta$$
$$v_v = 10\sin 40^\circ$$
$$v_v = 6.43\ ms^{-1}$$

Vertical motion uses the equations of motion, so write down the 'suvat' table and fill in what is known:

s	*u*	*v*	*a*	*t*
?	6.43	X	–9.8	0.52

Table 1.13

The relevant equation here is:

$$s = ut + \frac{1}{2}at^2$$

(*continued*)

Substitute what we know and solve for s:

$$s = (6.43)(0.52) + \frac{1}{2}(-9.8)(0.52)^2$$
$$s = 3.34 - 1.32$$
$$s = 2.02\ m$$

This means that at the time the ball reaches the basket, it is 2.02 m above its starting point. The ball is therefore at a total height of

$$s = 2.1 + 2.02$$
$$s = 4.12\ m$$

This is greater than the height of the hoop (3.5 m), therefore the player does not score his penalty.

Exercise 4.2.1 Projectile motion

1 A stuntman is performing a stunt where he drives his car over a ravine. The car leaves one bank of the ravine and lands on the bank on the opposite side, as shown in the diagram below.

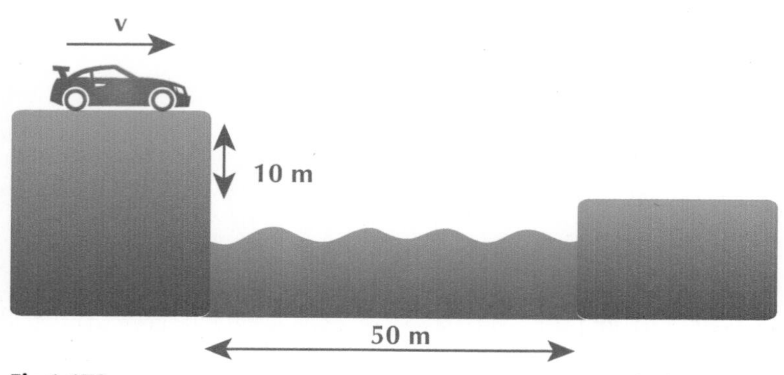

Fig 1.173

The ravine is 50 m wide and the difference in height is 10 m as shown.

a) Calculate the minimum initial velocity required for the stuntman to successfully clear the ravine and land safely on the other side.

b) Assuming the car's initial velocity is that calculated in a) above, find the resultant velocity (magnitude and direction) of the car when it lands on the other side of the ravine.

2 A lorry of mass 12,000 kg is parked at the top of a ramp next to a cliff as shown in the diagram. The ramp has a height of 20 m and the cliff a height of 15 m above a river.

The lorry's handbrake fails and it rolls freely down the ramp. At the bottom of the ramp, the lorry rolls freely along a horizontal stretch before falling over the edge of the cliff. The effects of friction can be ignored.

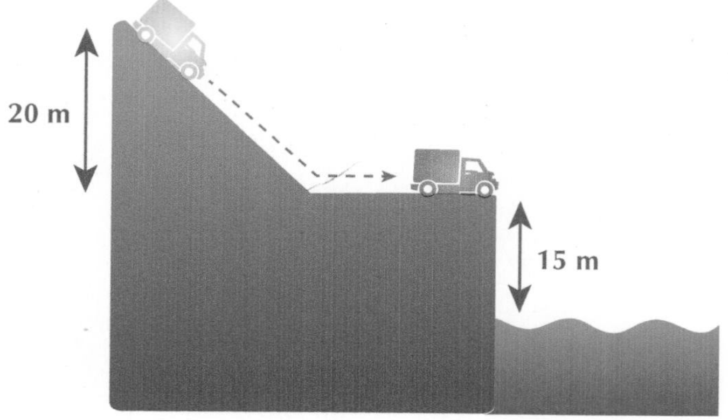

Fig 1.174

a) Calculate the speed of the lorry at the bottom of the ramp.

b) Assuming the lorry goes over the edge of the cliff at the speed calculated in part a), find:

(i) The time taken to reach the water.

(ii) The distance away from the edge of the cliff that the lorry lands.

(iii) The magnitude of the resultant velocity of the lorry as it hits the water.

c) Calculate the resultant velocity of the lorry as it hits the water using **energy conservation**. Comment on your answer compared to b) (iii) above.

3 A golf ball is hit on level ground during a driving shot. The golfer hits the ball with an initial velocity of 35 m s^{-1} at an angle of 35° above the ground.

a) Calculate:

(i) The initial horizontal component of the velocity of the ball.

(ii) The initial vertical component of the velocity of the ball.

b) Calculate the time taken for the ball to reach its maximum height.

c) How far from the initial position does the ball land?

4 A cannon is used to fire at a target that is on level ground (same height as the cannon). The cannonball is launched with an initial velocity of 150 m s^{-1} at an angle of 15° above the horizontal.

a) Calculate:

(i) The initial horizontal component of the velocity of the ball.

(ii) The initial vertical component of the velocity of the ball.

b) Calculate the maximum height reached by the cannonball during its flight.

c) By calculating the total time of flight, find the horizontal distance between the target and the cannon.

5 In a game of tiddlywinks, a small disc is 'pinged' into a cup. The cup is placed on a table above the launch point of the disc as shown in the diagram. The cup is a distance of 0.4 m away and the disc has a launch velocity of 8 m s^{-1} at an angle of 50° above the horizontal.

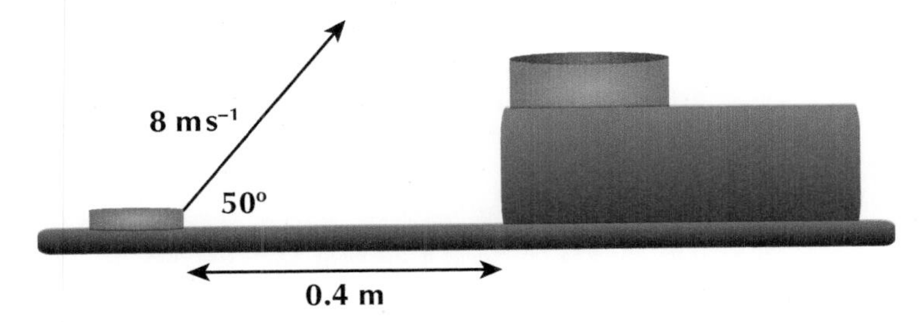

Fig 1.175

a) Calculate the horizontal and vertical components of the launch velocity.

b) Calculate the length of time taken to reach the cup.

c) If the disc lands in the cup, calculate the height of the top of the cup above the table.

(continued)

6 A ball rolls down a ramp and off the end of a table as shown in Figure 1.176. The table is a height of 1.2 m above the ground and the ball leaves the end of the ramp with a velocity of 9 m s^{-1}. The ramp is elevated at an angle of 25° above the horizontal.

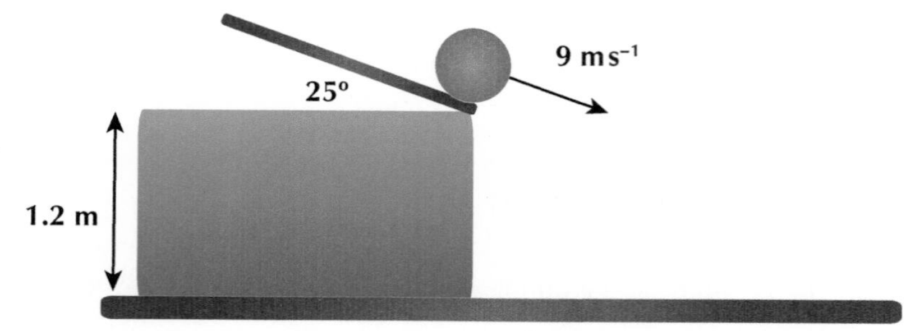

Fig 1.176

a) Calculate the length of time taken for the ball to reach the floor.

b) How far from the end of the table does the ball land?

c) The ramp is now raised to a greater angle above the table. The speed of the ball leaving the ramp is kept constant. How does the horizontal distance away from the table compare to that calculated in part b)? Explain your answer.

7 A rugby player is taking a penalty kick. He kicks the ball with an initial velocity of 14 m s^{-1} at an angle of 35° above the horizontal. A set of goal posts are 11 m away from where the ball is kicked. The goal posts are 7 m high. Does the football go over the goal posts? You must justify your answer by calculation.

Fig 1.177

8 A cannon is used to fire at an enemy ship. The cannon is placed on top of a cliff, 150 m above sea level. The ship is 300 m away from the cannon as shown in Figure 1.178. If the cannon launches the cannonball at an angle of 30° above the horizontal, calculate the speed with which the cannonball was launched.

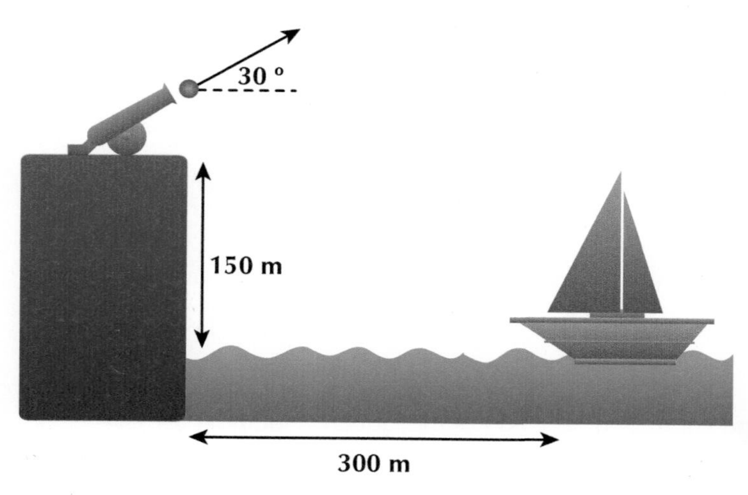

Fig 1.178

9 When firing at a target, an archer must ensure that he points the arrow upwards slightly rather than directly at the target. Use your knowledge of projectile motion to explain why the archer must do this in order to strike the target.

Fig 1.179

4.2.2 Satellite motion and Newton's thought experiment

We know that a projectile that is launched horizontally from a given height (for example, a ball rolling off a table) will follow a curved path until it hits the ground. The object continues horizontally with a constant speed until gravity pulls it to the ground.

If we change the launch velocity of the object, we will change the range of projectile – the greater the launch velocity, the greater the range of the projectile – this is illustrated in the diagram below:

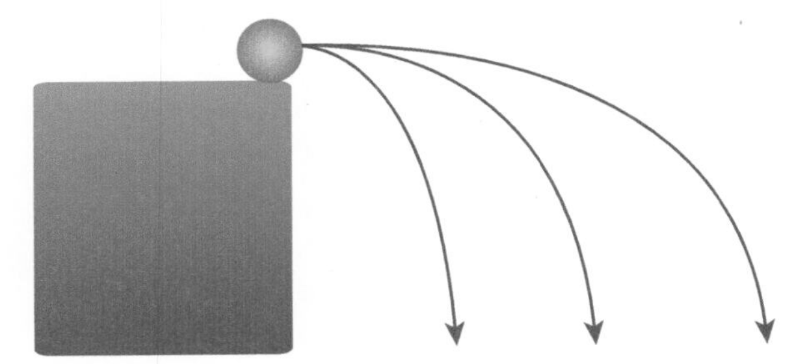

Fig 1.180

Fig 1.181

Fig 1.182

Additionally, if we change the launch height of the projectile it will also affect the range – the greater the launch height, the greater the range. This is because the time of flight of the projectile is greater.

The above ideas can be extended to explain satellite motion. If a cannonball is launched from the top of the Earth, it will follow a projectile path as shown in Figure 1.181. The gravity of the Earth will pull the cannonball down to the Earth's surface.

If the cannon is raised to a greater height, then the cannonball will travel further before gravity pulls it to the surface of the Earth. The range of the projectile is greater, as illustrated in Figure 1.182.

If the height of the cannon is great enough, then the cannonball is never accelerated to the surface of the Earth by gravity – it keeps missing and it makes a complete orbit of the Earth as shown in Figure 1.183. This is satellite motion. Gravity continually accelerates the satellite to the surface of Earth, but its horizontal speed is great enough that it keeps missing and stays in orbit.

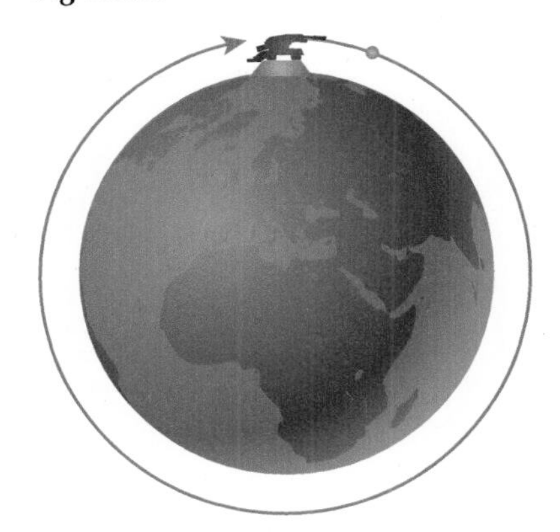

Fig 1.183

4.3 Newton's universal law of gravitation

Newton's universal law of gravitation can be used to explain why objects in the universe follow the paths that they do; for example, why the Earth stays in orbit around the Sun and why the Moon stays in orbit around the Earth. Gravity is also instrumental in the formation of stars, planets and solar systems, and is certainly responsible for keeping these objects together.

Fig 1.184

Key point

A gravitational field is a region in space where an object will feel a force due to gravity. The Earth produces a gravitational field due to its mass, and it pulls other objects with mass down towards the Earth.

4.3.1 Introducing the gravitational field

Think about the force that is responsible for keeping you on the surface of the Earth. It is the force due to gravity, which is called your weight, that is often thought of as a force that pulls everything towards the ground. Many people talk of the 'discovery' of gravity by Isaac Newton when an apple fell on his head. While it is debated that the apple fell on Newton's head, it is said that watching the apple fall to the ground led him to describe the phenomenon known as gravity.

Newton's own laws of motion, described in detail in Chapter 2 and also at National 5 level, state that an object will remain at rest unless acted on by an unbalanced force. The effect of the unbalanced force would be to cause an object to accelerate. Clearly, the apple is being acted on by an unbalanced force as it is accelerating to the ground. However, there is no clear indication of what this force is – there is no other object physically touching the apple, it appears to be falling on its own!

It is now known that the apple falls to the ground due to the gravitational field produced by the Earth. Any object with a mass will 'feel' a force due to the gravitational field; the apple therefore experiences a force directed towards the ground. This force acts at a distance. The concept of force fields will also be explored with electric fields.

4.3.2 The Earth's gravitational field: close to the surface

The Earth produces a gravitational field, due to its mass. Close to the surface of the Earth, the strength of this field is approximately constant. In earlier studies of physics, the strength of this field is given the symbol *g* and the gravitational field strength is stated as:

$$g = 9.8\ Nkg^{-1}$$

Fig 1.185

This means that any object on Earth will feel a force with a magnitude of 9.8 N for every kilogram of mass that it has. For example, a crate that has a mass of 10 kg will feel a force of 98 N downwards due to its mass being in the Earth's gravitational field.

The downward force felt by an object due to gravity has been named 'weight'. The weight of an object, *W*, is a force measured in Newtons and is given by:

$$W = mg$$

where *m* is mass (in kg) and *g* is the strength of the gravitational field (in N kg^{-1}).

4.3.3 Gravitational field strength: large distances

It has already been shown that the gravitational field strength of the Earth is approximately constant near the surface of the Earth. However, this is clearly not the case at large distances. If you were on the Moon, you would not feel acceleration due to the Earth's gravitational field equal to $-9.8\ m s^{-2}$!

By studying the field lines of the Earth's gravitational field, it is possible to understand how the strength of the field (and thus the magnitude of the force due to gravity) changes. Any field can be represented by field lines – the lines point in the direction of the force produced by the field and the closer together the lines, the stronger the field. The gravitational field lines for the Earth are as shown in Figure 1.186.

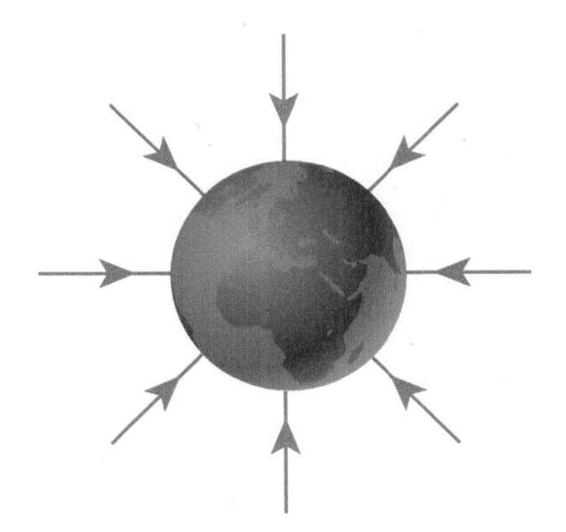

Fig 1.186

The gravitational field of any planet looks like the field in the diagram. You can see that the further away from the Earth you move, the further apart the field lines are – this means that the strength of the gravitational field decreases as you move further away from the planet. If you 'zoom in' very close to the surface of the Earth, the field lines will look like they are equal distances apart. This is why the strength of the gravitational field is approximately constant near the surface.

Key point

A force field can be represented on a diagram using field lines. These field lines point in the direction that a particle will move due to the field. The closer together the field lines, the stronger the field.

4.3.4 Newton's universal law of gravitation

Looking at the gravitational field of the Earth at large distances shows that the force due to gravity changes with distance from a planet. This is described by Newton's universal law of gravitation, which states that the force of gravity depends on:

- The mass of the planet, M
- The mass of the object, m
- The distance between the centre of the object and the centre of the planet, r

The force is given by:

$$F = \frac{GMm}{r^2}$$

where G is the gravitational constant:

$$G = 6.67 \times 10^{-11} N\, m^2\, kg^{-2}$$

This relationship shows that the force is directly proportional to the product (multiplication) of the two masses and inversely proportional to the square of the distance between the masses.

It makes sense that the greater the masses involved, the greater the force experienced. To understand the inverse square dependence on distance, we consider the Earth as an infinitesimally small point (a point mass – all of the mass of the Earth concentrated at an infinitesimally small point). The field lines spread out in all directions (in three dimensions) from this point, covering the surface area of a sphere. This means the force can be thought of as being 'spread out' over the surface area of a sphere, which is given by $A = 4\pi r^2$.

We assume all masses are point masses when applying this relationship. This means that for relatively large objects such as a planet and a moon, the relevant distance r is the distance between the centres of the planets, as shown in Figure 1.187. The force is an attractive force – the masses are attracted to each other:

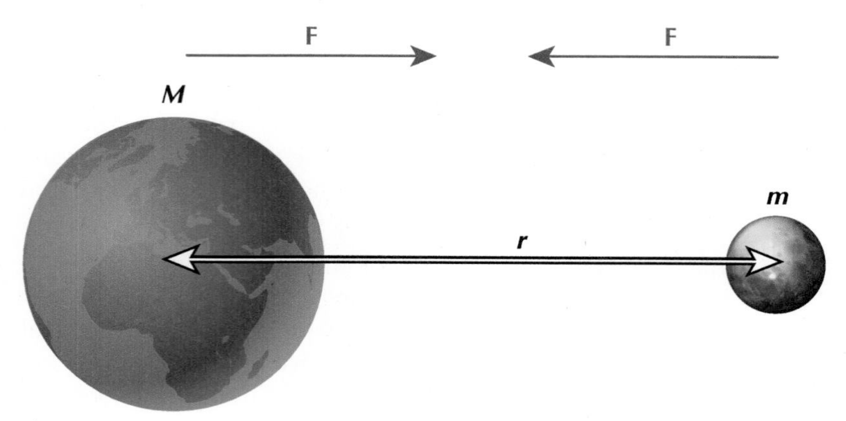

Fig 1.187

This equation has an important consequence – it reveals that any two objects that have mass will feel an attractive force. In other words, you are attracted by gravity to the person sitting next to you! However, you don't feel this force. If you substitute into the relationship the masses of two typical people and assume them to be sitting one metre apart, you will find the magnitude of the force of attraction is tiny – far too small to feel. It is due to the relatively large mass of the Earth that we feel the gravitational force due to it.

Using Newton's universal law of gravitation

To apply Newton's universal law of gravitation, identify the masses and the distance between the centres of these masses. These can then be substituted into the relationship to find the force of attraction between the masses. Force is a vector quantity – the direction of the force is towards the centre of the other mass as shown in Figure 1.188.

The universal constant of gravitation, $G = 6.67 \times 10^{-11} N\, m^2\, kg^{-2}$ is given on the data sheet.

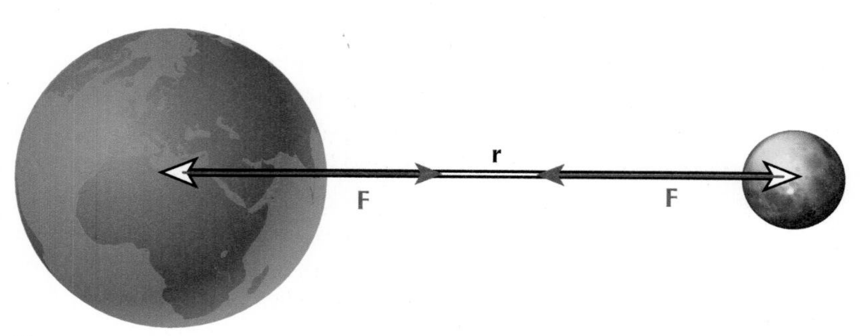

Fig 1.188

Worked example

At one point during the Moon's orbit, its centre is 384,000 km away from the centre of the Earth. If the Moon has a mass of 7.35×10^{22} kg and the Earth has a mass of 5.97×10^{24} kg, calculate the magnitude of the force of attraction due to gravity between the Earth and the Moon.

Fig 1.189

Here, apply the universal law of gravitation:

$$F = \frac{GMm}{r^2}$$

The distance quoted is between the centres of the planet and moon so you can use this distance directly. Substitute and solve for force:

$$F = \frac{(6.67 \times 10^{-11})(7.35 \times 10^{22})(5.97 \times 10^{24})}{(384{,}000 \times 10^3)^2}$$

$$F = \frac{2.93 \times 10^{37}}{1.47 \times 10^{17}}$$

$$F = 1.99 \times 10^{20} N$$

The effects of the gravitational attraction on the Earth by the Moon can be noticed on Earth. This is the force that is responsible for the tides. Water is a fluid and so can be 'pulled' by the gravitational attraction of the Moon. The Sun also affects the tides here on Earth. The effects of the Sun and Moon are illustrated in the diagram below.

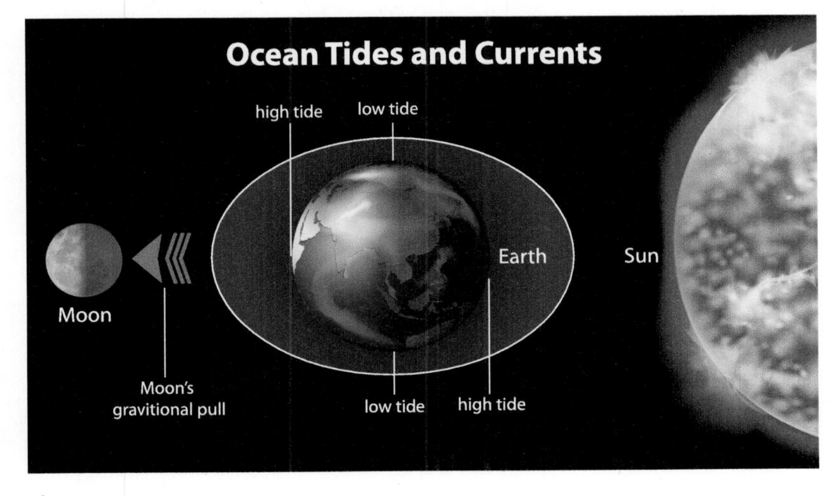

Fig 1.190

Exercise 4.3.4 Newton's universal law of gravitation

1 Using your knowledge of physics, explain what Newton's universal law of gravitation states about the magnitude of the force due to gravity from a planet, and how this changes with distance away from the planet. Include in your description any assumption made about the masses.

2 The centre of the planet Jupiter is a distance of 778×10^6 km from the centre of the Sun. If the mass of Jupiter is 1.90×10^{27} kg and the mass of the Sun is 2.00×10^{30} kg, calculate the magnitude of the force of attraction due to gravity between Jupiter and the Sun.

3 Jupiter, which has a mass of 1.90×10^{27} kg, has several moons orbiting it. One of these moons is called Ganymede and it has a mass of 1.48×10^{23} kg. Its centre is approximately 1.07×10^6 km away from the centre of Jupiter.

a) Explain, using your knowledge of gravity, how Ganymede remains in orbit around Jupiter.

b) Calculate the magnitude of the force exerted on Ganymede by Jupiter due to gravity.

Fig 1.191

4 Shown below are key masses and distances.

- Mass of Earth = 5.97×10^{24} kg
- Mass of Moon = 7.35×10^{22} kg
- Mass of Sun = 2.00×10^{30} kg
- Distance from Earth to Sun (centre to centre) = 1.50×10^8 m
- Distance from Earth to Moon (centre to centre) = 0.38×10^6 m

a) Show by calculation which object, the Sun or the Moon, produces the greatest gravitational attraction for the Earth.

b) Explain, using your knowledge of gravity, how the Sun and the Moon produce tides on Earth.

5 Two students are sitting 1.5 m apart. One student has a mass of 49 kg and the other student has a mass of 61 kg.

a) Calculate the magnitude of the gravitational attraction force between the two students.

b) Explain, based on your answer to part a), why gravity is called the weak force.

6 Two asteroids pass each other in space and experience a slight deflection due to their gravitational fields. One has a mass of 259,000 kg and the other has a mass of 190,000 kg. If the force of attraction between these is 4000 N, calculate the distance between the asteroids.

7 A binary star is a system of two stars that orbit a common centre of mass. At one point during the orbit, the stars are a distance of 2.59×10^6 km away from each other (centre to centre). The force of attraction between the stars is 2.24×10^{25} N. If the mass of one of the stars is 3.45×10^{30} kg, calculate the mass of the second star.

Fig 1.192

4.4 Gravitation and the universe

Gravity plays a big role in many aspects of the known universe. Gravity is the reason the Earth stays in orbit around the Sun, and the Moon stays in orbit around the Earth. Gravity is also responsible for planet and star formation; lensing effects of light travelling through space; and for keeping objects on planets. In fact, simulations of the Big Bang where the universal constant of gravity has been changed show how important gravity is: if the value was lower, the universe would simply fly apart and no stars would form; if it was higher, the universe would end in a series of black holes.

4.4.1 Gravity and the solar system

Fig 1.193

The studies of Newton's universal law of gravitation above showed that any two objects that have a mass will be attracted to each other due to gravity. The following equation gives the magnitude of this force:

$$F = \frac{GMm}{r^2}$$

As the Figure 1.194 shows, this force acts on both of the masses. This is also Newton's third law in action: if the Earth exerts a force on the Moon, the Moon exerts an equal force in the opposite direction on the Earth.

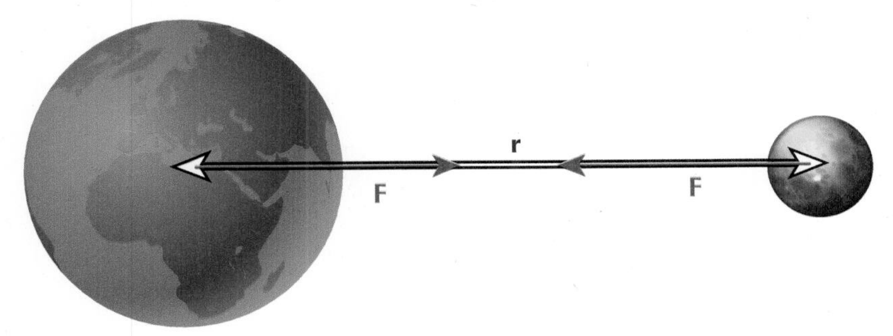

Fig 1.194

In the section above, the magnitude of the force of attraction between the Earth and the Moon was calculated to be:

$$F = 1.99 \times 10^{20} N$$

Compared to the mass of the Moon, the magnitude of this force is significant (100 times smaller than the mass). Therefore, it affects the motion of the Moon and causes it to stay in orbit around the Earth. Compared to the mass of the Earth, this force is less significant, so the Earth itself does not change its motion by a notable amount due to the gravitational pull of the Moon.

The same can be said for the Sun and the planets orbiting it. Each planet orbits the sun due to the force of gravity. Compared to the mass of the planet, the force is significant, so the planet's direction is affected in a measurable way – the planet stays in orbit. The size of the force is negligible compared to the mass of the Sun, so the Sun's direction remains unchanged to a good approximation by the planets of the solar system.

4.4.2 Gravity and star formation

Newton's universal law of gravitation shows that any particle with mass will be attracted to any other particle with mass. The greater the mass, the greater the force of attraction due to gravity. Following the Big Bang (more details in Area 1, Chapter 9), particles of matter in space were attracted to each other due to gravity. A nebula is a cloud of dust particles in space – these are the starting points of new stars, making nebulae the birth place of stars.

Fig 1.195

Particles join together, attracted to each other by gravity to form bigger and bigger particles. The particles keep growing in size and gravity keeps increasing. At its extreme, the force of gravity is so great that the speed of the particles becomes so fast they smash into each other with huge amounts of energy. This kinetic energy is linked to thermal energy, so the system is very hot – eventually, the particles smash into each other with enough energy to kick-start nuclear fusion and the star 'ignites'. When this happens, a disc of dust particles is sent out into space, as shown in Figure 1.195.

Dust from this disc coalesces under the influence of gravity to form planets, moons and asteroids – in other words, a solar system – around the star.

5 Special relativity

You should already know (National 5):

- Relative motion refers to the motion of an object relative to another object; for example, the motion of a person walking down a train carriage relative to either someone standing on the station platform or sitting on the same carriage

Learning intentions

- A frame of reference is a coordinate system, inside which the motion of an object can be described
- Newtonian relativity states that all motion is relative
- The speed of light postulate states that the speed of light is constant for all observers
- The measurement of time depends on the relative motion of the observer to the event – time dilation
- The measurement of length depends on the relative motion of the observer to the length being measured – length contraction

5.1 Galilean invariance and Newtonian relativity

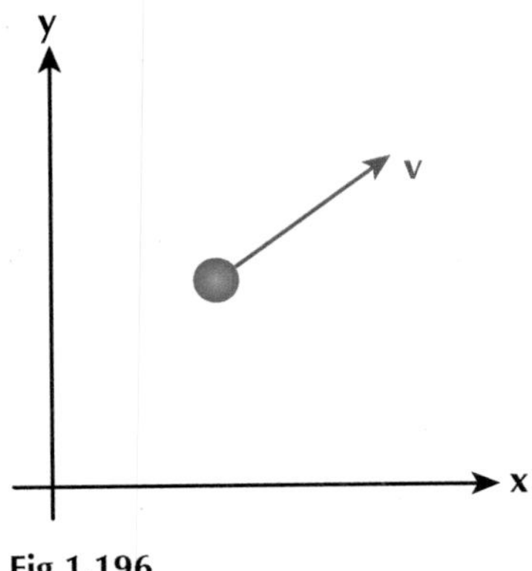

Fig 1.196

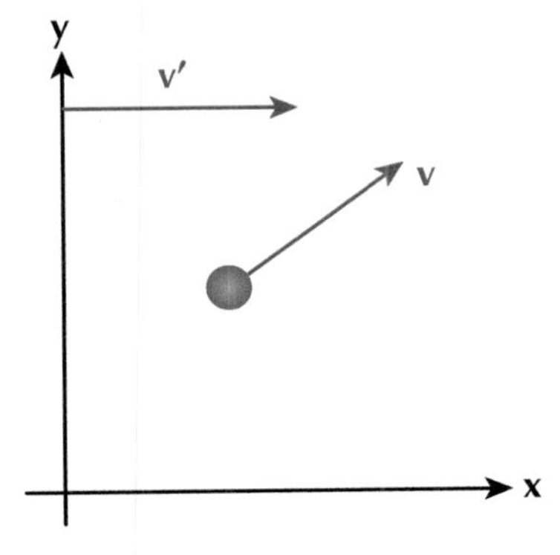

Fig 1.197

So far, it has been assumed that Newton's laws of motion are obeyed at all times. The motion of an object is studied relative to a fixed point in space. For example, if a car is travelling at 60 mph, it means the car is travelling at 60 mph relative to the Earth, which is assumed to be stationary.

In order to understand the concept of relativity, it is necessary to introduce the idea of a frame of reference. A frame of reference can be thought of as a coordinate system, where the position and motion of an object is relative to points within the frame. Such a frame is shown in Figure 1.196, with an object at a certain position.

Say the object is moving with a velocity, v. This velocity is relative to the frame of reference. Here it is assumed the frame is stationary. However, this frame can also be moving – Figure 1.197 shows the same object moving with the same velocity, but this time in a frame of reference moving with a speed v'.

The velocity of this object relative to the above frame of reference is still the same! It does not matter what velocity the frame of reference has, the velocity of the object relative to the frame will always be the same. However, the velocity of the object observed from another frame of reference will be different. In other words, the velocity of the object depends on the frame of reference you are measuring it relative to.

Fig 1.198

Key point

The laws of physics are the same in any frame of reference. That is, Newton's laws of motion will apply to any frame of reference.

However, the value of quantities that you measure (such as velocity) will depend on the frame of reference in which you measure them.

This concept can be illustrated in practice. Consider travelling on a train moving with a velocity of 50 m s^{-1}. The train is a moving frame of reference. If you are sitting still on a seat on the train, then you are stationary relative to the train. Your velocity is zero measured in the frame of reference that is the train. If you throw a ball at 5 m s^{-1} along the train carriage, then the ball will have a velocity of 5 m s^{-1} relative to the train. However, the velocity of the ball (and yourself) will be different in a different frame of reference. Consider a person standing on a station platform and the train travels through the station. The station is a second frame of reference. The velocity of the ball relative to the station will be 55 m s^{-1}, and your velocity would be 50 m s^{-1} relative to the station. This further highlights the important consequence that the velocity of the object depends on the frame of reference.

This example shows Newtonian relativity. Where the speeds of objects are low (relative to the speed of light), the velocities combine as expected to give the relative velocity – the addition of the train velocity and that of the ball. As we will see later in this chapter, however, this is not always the case!

Newton's laws of motion are applied to objects that are moving relative to a fixed point. As shown above, the velocity of an object depends on the frame of reference in which the velocity is measured. However, Galileo stated that laws of motion hold true in every frame of reference providing it is stationary or moving with a constant velocity. This concept is known as Galilean invariance. This means that Newton's laws of motion are the same in all frames of reference that are stationary or moving with a constant velocity – we call these inertial frames of reference.

Experiment 5.1 Galilean invariance thought experiment

This thought experiment explains Galilean invariance.

You will not need any equipment – this is a thought experiment.

Consider two people carrying out individual experiments to analyse Newton's second law of motion using hanging masses and dynamics carts – this experiment is shown below:

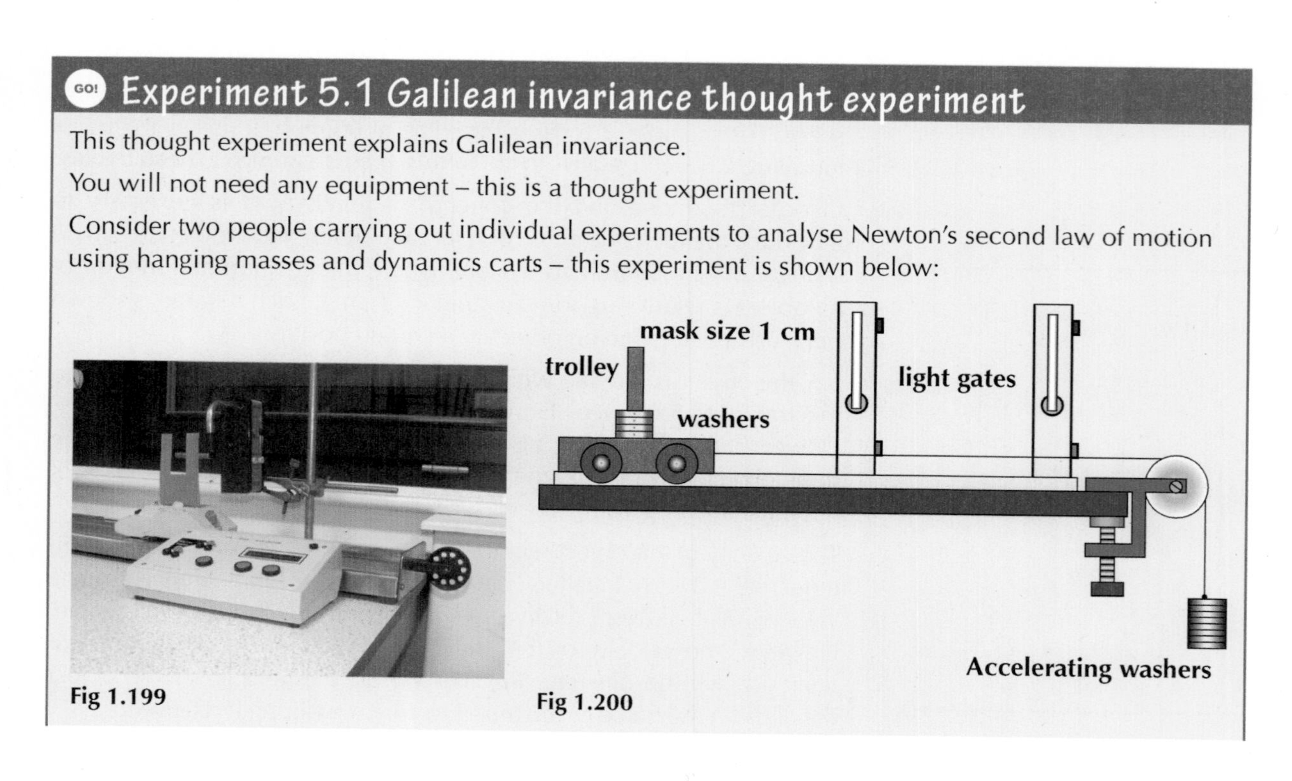

Fig 1.199

Fig 1.200

One person is carrying this experiment out in a lab beside a train track. They are carrying out the experiment in a stationary frame of reference. The second person is carrying out the experiment in a moving train; they are carrying out the experiment in a moving frame of reference. However, they are working in a carriage with no windows.

If the train is moving at a constant speed, can the person carrying out the experiment in the train carriage tell that they are moving based on the results of their experiment? Will the results of their experiment be different from those obtained by the observer in the stationary lab?

Galileo postulated that the person in the moving train carriage would not be able to tell from the results of their experiment whether or not they were moving. They would obtain the same results as the observer in the stationary laboratory. As the results from the Newton's laws experiment are the same (Galilean invariance), the moving person cannot tell they are moving from the results of their experiment.

Exercise 5.1 Newtonian relativity

1 It can be said that the velocity of the object has to be measured relative to something, such as a point in the laboratory or the seat of a train carriage.

a) Explain what is meant by the term 'frame of reference', giving two examples of frames of reference.

b) For the two examples of frames of reference given above, explain whether or not these frames are moving.

c) Explain why the velocity of an object can change depending on the frame of reference in which the measurement is made.

2 You are travelling on a train to London. The train is moving with a velocity of 60 $m\,s^{-1}$ in a southerly direction. You walk to the buffet car at a velocity of 2 $m\,s^{-1}$, also in a southerly direction.

a) In the frame of reference of the train, what is your velocity?

b) The train passes through a station while you are walking to the buffet car. What is your velocity as observed by a person standing on the station platform (the frame of reference of the station)?

Fig 1.201

3 Explain what is meant by the term Galilean invariance.

4 Two pupils are carrying out an experiment to measure Newton's second law of motion. One pupil is carrying out the experiment on a beach and the other on a boat that is moving with a constant speed. The boat has no windows so the pupil cannot see out.

a) The pupil on the beach correctly derives Newton's second law using their apparatus. If the pupil on the boat has identical apparatus and conducts the experiment in the same way, would they also derive Newton's second law correctly?

b) Can the pupil on the boat tell that she is moving from the results of her experiment?

5.2 Special relativity

Fig 1.202

In physics, events such as a flash of light from a bulb or the collision of two cars are measured. Measurements are concerned with where the event happens (position) and when the event happens (time). If two events happen, such as two flashes of light, then the time difference between these events may also be of interest. As discussed above, the physical laws that govern these events are the same, regardless of the frame of reference for measuring the event; Newton's laws of motion hold in any frame of reference so long as it is either stationary or moving at a constant speed.

However, work on electromagnetic waves revealed a key concept that threw Newtonian relativity into doubt! Albert Einstein developed his theory of 'special relativity'.

It is called *special* relativity to show that it deals only with the special case of inertial reference frames that are either stationary or moving at a constant velocity. This theory stunned the scientific world at the time, as it had taken Newtonian relativity to be common sense. The problem was that this belief was based on objects moving at comparatively low speeds. Einstein's theory of special relativity holds for objects moving at any speed, in an inertial frame of reference.

5.2.1 Speed of light postulate

James Clark Maxwell was a Scottish physicist who investigated the properties of electromagnetic waves (of which visible light is an example). A key consequence of Maxwell's work was that the speed of light, c, in a vacuum was fixed and depended on two constants:

- The permittivity of free space, $\varepsilon_0 = 8.85 \times 10^{-12}$ F/m
- The permeability of free space, $\mu_0 = 4\pi \times 10^{-7}$ H/m

These constants relate to the electric and magnetic fields that make up an electromagnetic wave. The speed of light in a vacuum is given by:

$$c = \frac{1}{\sqrt{\varepsilon_0 \mu_0}}$$

Substituting the values for the constants gives:

$$c = \frac{1}{\sqrt{(8.85 \times 10^{-12})(4\pi \times 10^{-7})}}$$

$$c = 299939418\ ms^{-1}$$

This is approximated to:

$$c = 3.00 \times 10^{8}\ ms^{-1}$$

which is the value that has long been used for the speed of light. Indeed, it has always been assumed that the speed of light is constant. However, this postulate introduces issues to Newtonian relativity that the thought experiment below highlights.

Experiment 5.2.1 Light speed thought experiment

This thought experiment highlights how Newtonian relativity breaks down when large velocities are considered and the speed of light postulate is taken into account.

Consider standing stationary on the surface of the Earth and observing a large rocket. To begin with, the rocket hovers high above you, but is stationary with respect to the Earth. If there is a bulb on the front of the rocket, how fast does the light travel from it to you? In other words, what is the speed of the light waves relative to you? That's easy! It's the speed of light: c.

Now consider the rocket moving towards you at a velocity equal to one tenth the speed of light, $0.1c$. In the rocket's frame of reference, the light emitted from the bulb is still leaving the rocket at a speed c. In other words, the light from the rocket is travelling at a speed c in the frame of reference of the rocket. However, how fast is the light moving in *your* frame of reference according to Newtonian relativity? According to Newtonian relativity, the light will be moving at a speed $c + 0.1c$ (faster than the speed of light!). According to Maxwell, this is not possible. Another theory of relativity is needed to account for this.

The ultimate speed

Maxwell's postulate that the speed of light is the same in all reference frames is at odds with Newtonian relativity, which states that the velocity of an object depends on the frame of reference in which it is measured. The postulate can be rephrased to say that there exists an ultimate speed, and no object can travel faster than this. The existence of this speed limit was demonstrated in 1964 by William Bertozzi. He accelerated electrons to known speeds using a potential difference and measured the speed of the electrons directly. The kinetic energy of the electrons was found using the potential difference – a method independent of that used to find the speed. Bertozzi found that as the kinetic energy of the electrons increased, the speed increased as expected. This only happened up to a point, however, after which increasing the kinetic energy saw smaller changes in velocity until a maximum velocity was reached. These results are shown in the graph below:

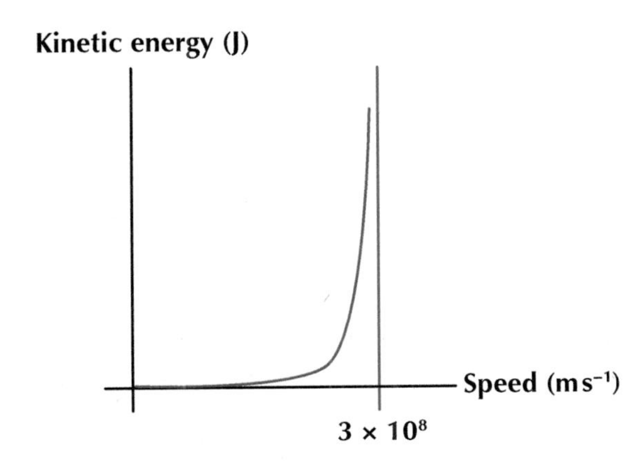

Fig 1.203

> **Key point**
>
> The ultimate speed in nature is approximately 3×10^8 m s^{-1}. No object can travel faster than this ultimate speed.

Testing the speed of light postulate

The speed of light postulate says that the speed of a light wave is the same in any inertial reference frame, regardless of the frame's speed.

In other words, the speed of light emitted by a source that is moving should be the same as the speed of light emitted by a source that is stationary. This postulate was tested in 1964 at CERN, near Geneva in Switzerland. Physicists tested the postulate using gamma rays emitted by subatomic particles called pions. As gamma rays are a member of the electromagnetic spectrum, they also travel at the speed of light. The pions in the experiment were accelerated to a very high velocity of $0.99975c$. It was found that the gamma rays emitted by these fast-moving pions were travelling at the speed of light – they were travelling at the same speed as gamma rays emitted by stationary pions. The speed of light was the same in both reference frames!

5.2.2 Time dilation

One of the consequences of Einstein's theory of special relativity is that time does not pass at a fixed rate – the length of time taken between two events changes depending on the frame of reference. In other words, in some frames of reference, time can pass more slowly. This concept is known as time dilation.

GO! Experiment 5.2.2 Time dilation thought experiment

This thought experiment investigates how measured time varies between different frames of reference. The experiment involves two people measuring an event. In this case, the event is a pulse of light leaving a source, S, travelling up to a mirror and reflecting off it back down to the source as shown below.

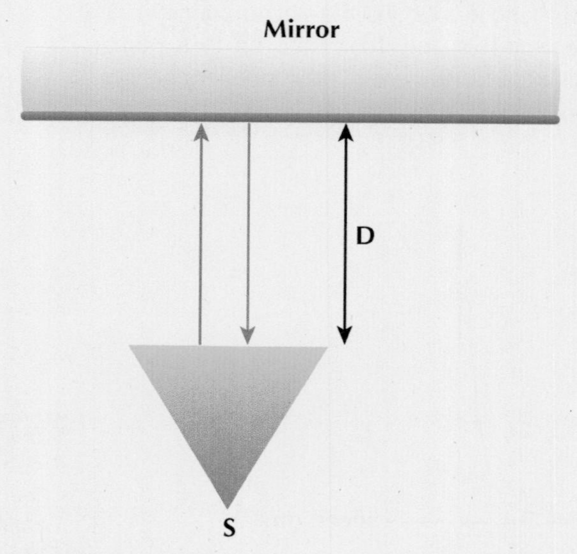

Fig 1.204

This experiment is carried out in a spacecraft that is moving with a velocity, v.

Two pupils, Jack and Lewis, measure the time taken for the light to travel to and from the mirror. Jack is on the spacecraft, while Lewis is standing on Earth watching the experiment.

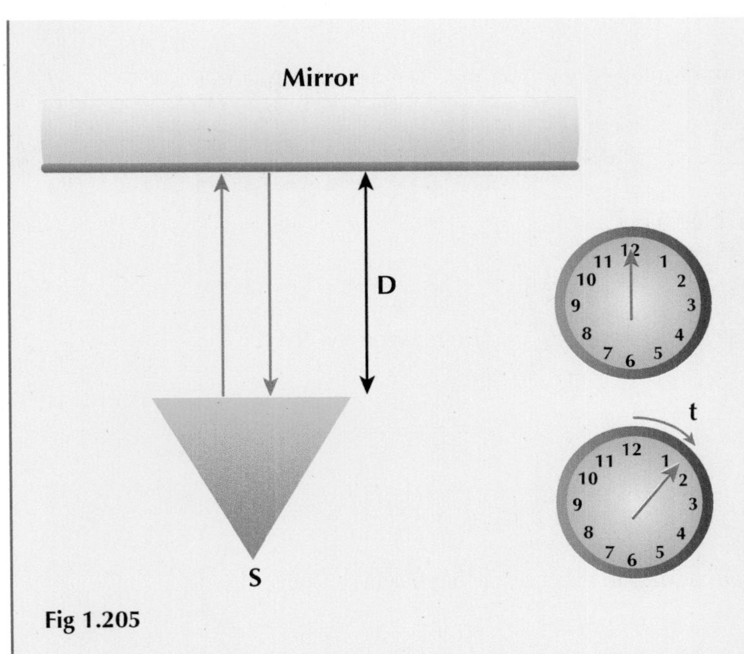

Fig 1.205

Jack is moving in the spacecraft with the experiment. Jack is therefore in the spacecraft's frame of reference, so the experiment appears to be still. He measures the time taken, t, for the light to travel to the mirror and back again, a distance of $2D$ using the single clock shown in Figure 1.205.

The time measured by Jack would be given by the classic distance, speed, time equation:

$$t = \frac{2D}{c}$$

Lewis is in a different frame of reference as he is outside the spacecraft. To him, the experiment appears to be moving at a speed, v, horizontally. This means that the distance the light travels is different, and the two events (light emitted and light received) happen at different locations. The distance the light travels in Lewis' frame of reference is $2L$.

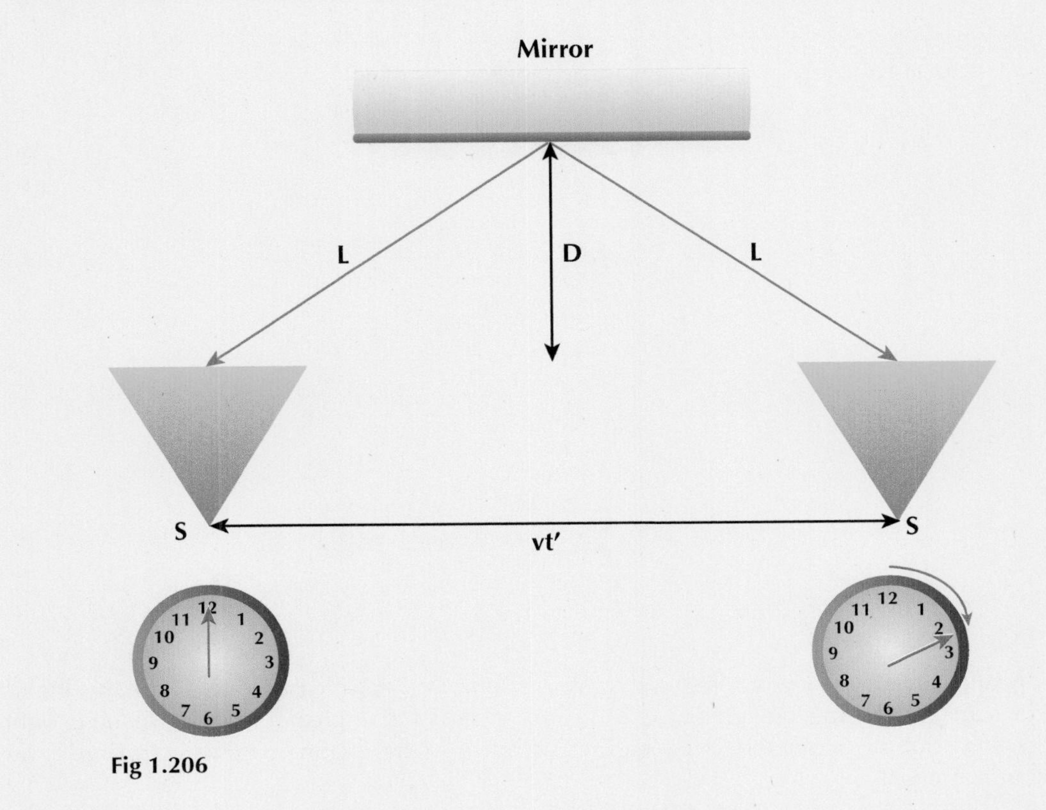

Fig 1.206

In the time taken for the light to travel to and from the mirror, the source has moved a distance:

$$d = vt'$$

(continued)

in Lewis' frame of reference, where t is the time interval between the light being emitted and received, he measures this time interval to be:

$$t' = \frac{2L}{c}$$

Using Pythagoras' theorem, L can be written as:

$$L^2 = D^2 + \left(\frac{vt'}{2}\right)^2$$

Using Jack's time measurement, D in the above equation can be replaced with:

$$D = \frac{ct}{2}$$

This gives:

$$L^2 = \left(\frac{ct}{2}\right)^2 + \left(\frac{vt'}{2}\right)^2$$

From Lewis' measurement, L in the above equation can be replaced with:

$$L = \frac{ct'}{2}$$

Hence:

$$\left(\frac{ct'}{2}\right)^2 = \left(\frac{ct}{2}\right)^2 + \left(\frac{vt'}{2}\right)^2$$

$$\frac{c^2t'^2}{4} = \frac{c^2t^2}{4} + \frac{v^2t'^2}{4}$$

$$c^2t'^2 = c^2t^2 + v^2t'^2$$

$$t'^2 = t^2 + \frac{v^2t'^2}{c^2}$$

$$t'^2 - \frac{v^2t'^2}{c^2} = t^2$$

$$t'^2\left(1 - \frac{v^2}{c^2}\right) = t^2$$

$$t'^2 = \frac{t^2}{\left(1 - \frac{v^2}{c^2}\right)}$$

$$t' = \frac{t}{\sqrt{\left(1 - \frac{v^2}{c^2}\right)}}$$

This end result shows how Lewis' measured time compares to Jack's measured time. Clearly, the times are not the same. As the speed of an object can never be equal to or greater than the speed of light, what does this result say about the time measured by Lewis in the stationary frame compared to Jack in the moving frame?

It demonstrates that the time measured in the stationary frame is longer. In other words, relative motion is affecting the time between two events happening. This is time dilation.

Proper time and Lorentz factor

The thought experiment above shows that the time measured between two events depends on the relative motion of the frame of reference.

Proper time is a quantity defined as t. When two events happen at the same location in the same inertial reference frame, the time measured between these events is called the proper time. Hence, the time measured by Jack in the thought experiment above is the proper time. Any measurements of the time interval, t, in any other inertial reference frame will always be larger:

$$t' = \frac{t}{\sqrt{\left(1 - \frac{v^2}{c^2}\right)}}$$

Here, c is the speed of light and interval v is the speed of the frame of reference. This means the denominator of the fraction will always be less than one, increasing the measured time.

Consequences of time dilation

Time dilation – predicted by Einstein's theory of special relativity and later proved by scientific observation – revealed that time does not tick by at a constant rate. This cast common sense and accepted knowledge into doubt. When a new scientific idea is tested, proven and accepted, it forces us to view the world (and the wider cosmos) in a different way.

However, consider more closely the equation above. There is a correction factor, called the Lorentz factor, given by:

$$\gamma = \frac{1}{\sqrt{\left(1 - \frac{v^2}{c^2}\right)}}$$

If you plot the Lorentz factor, γ, against the speed of the frame of reference, v, you see the following:

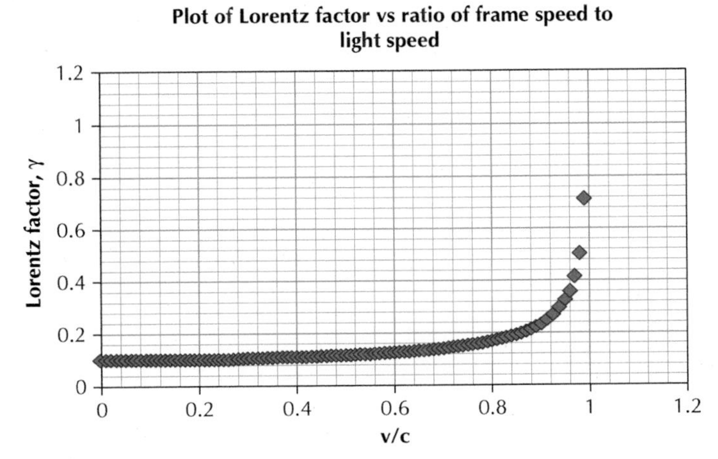

Fig 1.207

At low speeds, where $v < c$ and the ratio v/c is very small, the Lorentz factor is very nearly equal to 1. In other words, at low speeds there is no noticeable difference between the measured time and proper time.

Key point

The time measured between two events depends on the velocity of the frame of reference. If the time measurements are made in the same frame of reference as the events and the events occur at the same point, the time interval measured between these events is called the proper time. Another measurement of the same time interval in a different frame of reference will be longer by a scaling factor equal to the interval Lorentz factor,

$$t' = \frac{t}{\sqrt{\left(1 - \frac{v^2}{c^2}\right)}} = \gamma t$$

What was known about time from before the introduction of special relativity still holds true so long as the speeds are kept low. Indeed, the correction factor only really comes into play when the speed of the object is greater than 10% of the speed of light.

In other words, the correction factor is only noticeable at very high speeds, which is why it was not previously observed and why Newtonian relativity held as common sense for so long.

Using the time dilation equation

In order to properly apply the time dilation equation, you need to know the variables involved, specifically:

- what corresponds to proper time
- what corresponds to a measured time that is being corrected by the Lorentz factor
- the relative velocity.

Always write down which time is the proper time and then build what you know from there!

The proper time is the time in the frame of reference where the experiment is stationary. For example, if an experiment is carried out on a spacecraft that is moving at a given constant speed, the proper time is the time measured on the spacecraft, because the experiment is stationary within this frame. Any other time measured in any other frame will be subject to time dilation.

Worked example

An experiment is carried out on a space shuttle that is moving through space with a velocity of 1.5×10^8 m s^{-1}. On the shuttle, the time between two events occurring at the same point is measured to be 15 ns. Calculate the time measured by a stationary observer on Earth.

Here, the proper time can be identified as the time measured on the space shuttle:

$$t = 15\ ns$$

This is because the experiment is stationary as the space shuttle is the frame of reference. The shuttle is moving at:

$$v = 1.5 \times 10^8\ ms^{-1}$$

This is the speed of the frame of reference of the Earth relative to the experiment. Hence, the measured time on Earth can be found using:

$$t' = \frac{t}{\sqrt{\left(1 - \frac{v^2}{c^2}\right)}}$$

Substitute what you know and solve for the measured time:

$$t' = \frac{15}{\sqrt{\left(1 - \frac{(1.5 \times 10^8)^2}{(3.0 \times 10^8)^2}\right)}}$$

$$t' = \frac{15}{\sqrt{1 - \frac{2.25 \times 10^{16}}{9.0 \times 10^{16}}}}$$

$$t' = \frac{15}{\sqrt{1 - 0.25}}$$

$$t' = \frac{15}{0.87}$$

$$t' = 17.2\ ns$$

Exercise 5.2.2 Time dilation

1 Explain what the speed of light postulate says, making reference to the concept of an ultimate speed in nature.

2 Explain what the concept of time dilation says about the time interval measured in inertial reference frames. In your answer, make reference to the concept of proper time.

3 Two students, *A* and *B*, measure the time taken for light to travel from a source, reflect off a mirror and then return to the source. This experiment, shown in Figure 1.208, is taking place on a spacecraft that is moving at 2.0×10^8 m s^{-1}.

a) Student A is on board the spacecraft with the experiment. If the distance between the mirror and the light source is 2 m, calculate the measured time between the light leaving the source and returning to it.

b) Student B is watching the experiment from a second spacecraft that is stationary. Calculate the time between the light leaving the source and returning to it.

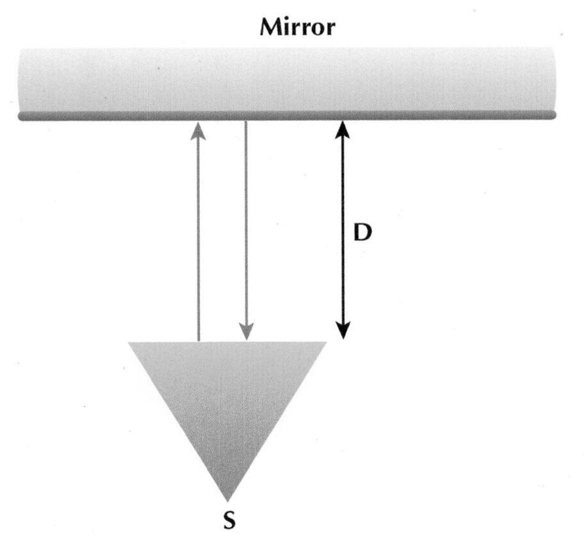

Fig 1.208

4 Two physicists are measuring the results of an experiment in which a pulse of light travels out from a source and returns to it after reflecting off a mirror. One of the physicists is moving relative to the experiment; the other is stationary relative to the experiment.

a) The physicist who is stationary relative to the experiment measures the time between sending and receiving the pulse of light to be 2.5 μs. Calculate the distance light travels in the experiment.

b) The other physicist, who is moving relative to the experiment, measures the time between the light being sent and received to be 2.9 μs. Calculate the speed he is moving relative to the experiment.

(continued)

5 The concept of time dilation is a consequence of special relativity. Explain why the relativistic effect of time dilation is not observed in day-to-day life by considering the equation for the Lorentz factor,

$$\gamma = \frac{1}{\sqrt{\left(1 - \frac{v^2}{c^2}\right)}}$$

Under what conditions can Newtonian relativity be successfully applied?

6 You are being sent on a mission to investigate a distant planet called EX7611. Your spacecraft leaves the Earth on 1 June 2014 and travels to this planet at a constant speed of 2.5×10^8 m s^{-1}. It takes a time of 1 year (measured by you) to reach the distant planet. After collecting samples from the planet, you fly home at the same constant speed, which takes a further year as measured by you. What date do you land on Earth?

5.2.3 Length contraction

As shown above, the time measured between two events depends on the frame of reference in which the measurement is made. This also applies to *distances* that are measured. Consider trying to measure the length of a fast-moving object. You are standing still on the surface of the Earth and the object is moving quickly past you. To measure its length, you need to measure the position of the front of the object and the back of the object at the same time. If there is a very small difference in the time between the measurements, the movement of the object will make your measurement inaccurate. Alas, making simultaneous measurements is very difficult and, as noted above, time is affected by the relative motion of the object. Thus, the length you measure is also going to be affected by the motion of the object relative to you.

Key point

The length measured between two events depends on the velocity of the frame of reference. The length of the object measured in the frame of reference where the object is at rest is called the proper length, L_0. Another measurement of the same length in a different frame of reference will be shorter by a scaling factor equal to the Lorentz factor:

$$L = L_0\sqrt{\left(1 - \frac{v^2}{c^2}\right)}$$

Using the length contraction equation

As with the time dilation equation, the key to successfully applying the length contraction equation is correctly identifying the proper length in the question. Once this has been identified, write down the proper length and other variables that you know:

- Proper length, L_0 = ______ m
- Measured length, L = ______ m
- Relative velocity, v = ______ m s^{-1}

Remember that the proper length is the length measured in the frame of reference where the object is stationary.

Worked example

Blair and Connor are measuring the length of an airport runway. Blair is on the ground and measures the length of the runway to be 2500 m. Connor measures the length of the runway from a very fast aircraft that is flying horizontally over the runway at 1×10^8 m s^{-1}. Calculate the length of the runway as measured by Connor.

First of all, identify the proper length in the question – this is the length of the runway measured by Blair because the runway is stationary in Blair's frame of reference. So:

$$L_0 = 2500\ m$$

Use the length contraction equation to find the length of the runway measured by Connor:

$$L = L_0\sqrt{\left(1-\frac{v^2}{c^2}\right)}$$

$$L = 2500 \times \sqrt{\left(1-\frac{(1\times10^8)^2}{(3\times10^8)^2}\right)}$$

$$L = 2500 \times \sqrt{\left(1-\frac{1\times10^{16}}{9\times10^{16}}\right)}$$

$$L = 2500 \times \sqrt{1-0.11}$$

$$L = 2500 \times 0.94$$

$$L = 2357\ m$$

Exercise 5.2.3 Length contraction

Fig 1.209

1. Explain what the concept of length contraction says about the length of an object measured in different inertial reference frames. In your answer, make reference to the concept of proper length.
2. A new space station has been built on the Moon. An astronaut on the Moon measures the length of the space station to be 450 m. Calculate the length measured by an astronaut who flies past the space station at a velocity of 1.8×10^8 m s^{-1}.
3. A stretch of road is 1400 m long. A passenger in a very high-speed plane measures the length of the road to be 1350 m. Calculate the speed the plane was flying at relative to the road when the measurement was made.
4. Two astronomers are measuring the length of the space station. One of the astronomers is in a spacecraft moving at 1.6×10^8 m s^{-1} relative to the space station. She measures the length of the space station to be 210 m. What is the proper length of the space station, as measured by the second astronomer, who is at rest relative to the station?
5. By considering the Lorentz factor and the following equation for the measured length, explain why the relativistic effect of length contraction is not observed in everyday life.

$$L = L_0\sqrt{1-\left(\frac{v}{c}\right)^2}$$

6 The expanding universe

You should already know (National 5):

- The wavelength of a wave is measured between two identical points, e.g. from peak to peak
- The frequency of the wave is the number of complete waves in one second
- The frequency of a sound wave is linked to the pitch of the sound
- The frequency of a light wave is linked to the colour of the light
- Wavelength, frequency and velocity of a wave are linked by the equation $v = f\lambda$
- The universe started with the Big Bang
- The universe is expanding and cooling following the Big Bang

Learning intentions

- The Doppler effect is observed for both sound and light
- The Doppler effect links the change in observed frequency and wavelength of a wave to the relative motion of the source and observer
- Sound from an object moving towards/away from an observer is increased/ decreased in frequency
- Light from an object moving towards/away from an observer is blue/red shifted
- Red shift of a celestial object refers to the shift in apparent wavelength emitted due to the motion of the object away from Earth
- Hubble's law links the velocity of a celestial body moving away from Earth to the distance away from Earth
- Hubble's law can be used to estimate the age of the universe
- Expansion of the universe – evidence based on Hubble's law
- Evidence for dark matter and dark energy linked to the velocity of stars orbiting in galaxies.
- The red-shift ratio is the ration of the velocity of the object to the velocity of light (for slow moving objects)
- The temperature of an object is linked to the wavelengths being emitted by the object
- Wien's law can be used to estimate the temperature of a stellar object if the peak wavelength emitted by the object is known
- The life-cycle of a star and the Hertzsprung–Russell diagram
- Evidence for the Big Bang: cosmic microwave background radiation; Olbers' paradox; light element abundance

6.1 The Doppler effect

The Doppler effect is part of day-to-day life. Imagine standing on a train station platform. When a train comes through the station without stopping, the noise you hear from it changes depending on whether the train is coming towards you or going away from you. As the train comes towards you, the sound coming from its engine seems to be higher than when the train is moving away from you. This is an effect caused by the motion of the train relative to you, called the Doppler effect.

6.1.1 Doppler effect for sound

The pitch of the sound heard by an observer depends on the relative motion of the object emitting the sound. Either the object or the observer can be moving (or both).

Consider a loudspeaker emitting sound waves with a single frequency, as shown in the diagram.

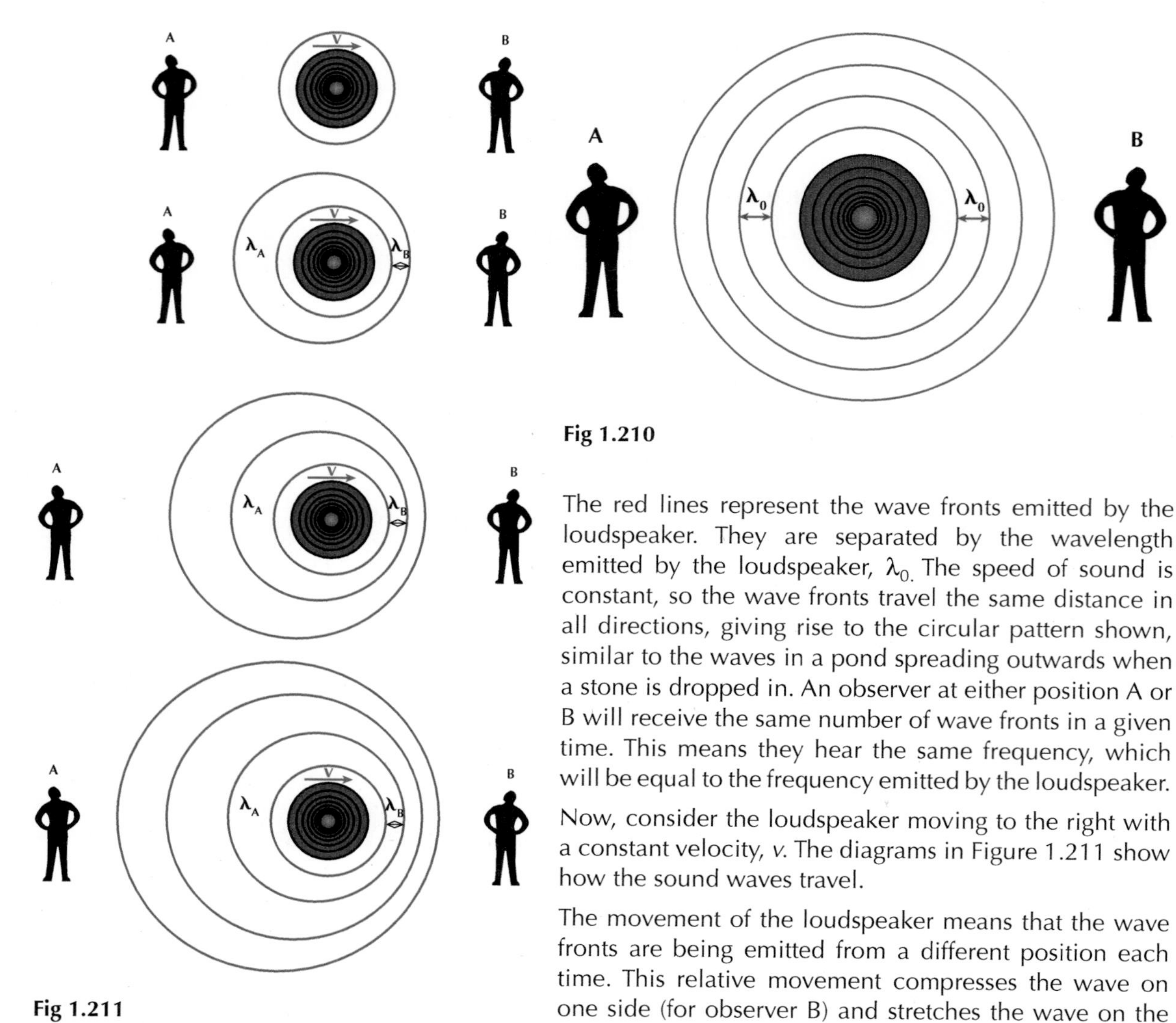

Fig 1.210

The red lines represent the wave fronts emitted by the loudspeaker. They are separated by the wavelength emitted by the loudspeaker, λ_0. The speed of sound is constant, so the wave fronts travel the same distance in all directions, giving rise to the circular pattern shown, similar to the waves in a pond spreading outwards when a stone is dropped in. An observer at either position A or B will receive the same number of wave fronts in a given time. This means they hear the same frequency, which will be equal to the frequency emitted by the loudspeaker.

Now, consider the loudspeaker moving to the right with a constant velocity, v. The diagrams in Figure 1.211 show how the sound waves travel.

The movement of the loudspeaker means that the wave fronts are being emitted from a different position each time. This relative movement compresses the wave on one side (for observer B) and stretches the wave on the

Fig 1.211

other (for observer A). The difference here means that observer B will receive more wavefronts per second; this means observer B hears a higher frequency. Observer A will receive fewer wavefronts per second, meaning a lower observed frequency. It is important to remember the emitted frequency is unchanged!

The frequency observed by the observer (f_0) will be given by the equation:

$$f_0 = f_S\left(\frac{v}{v \pm v_S}\right)$$

where f_S is the frequency of the source, v is the velocity of sound in air and v_S is the velocity of the source. The ± refers to the source moving either towards the observer or away: plus for the source moving away; minus for the source moving towards the observer.

Key point

The frequency of sound observed depends on the velocity of the source relative to the observer. This is known as the Doppler effect. The observed frequency f_0 is given by:

$$f_0 = f_S\left(\frac{v}{v \pm v_S}\right)$$

where f_S is the frequency emitted by the source, v is the speed of sound and v_S is the speed of the source.

The observed frequency is higher when the source is moving towards you and lower when the source is moving away from you.

Worked example

A car is driving towards you at 25 m s^{-1} when the driver sounds her horn. If the frequency emitted by the horn is 256 Hz, calculate the frequency of the sound that you hear. Speed of sound is 340 ms^{-1}

To answer this question, use the equation for the Doppler effect. As the car is coming towards you, the frequency of the sound will be higher, so use:

$$f_0 = f_S\left(\frac{v}{v - v_S}\right)$$

Substitute what you know and solve for the observed frequency:

$$f_0 = 256\left(\frac{340}{340 - 25}\right)$$

$$f_0 = 256 \times 1.08$$

$$f_0 = 276\,Hz$$

Fig 1.212

Exercise 6.1.1 Doppler effect for sound

1 A fire engine siren emits two frequencies – one at 240 Hz and the other at 350 Hz. If the fire engine is moving away from you at a velocity of 35 m s^{-1}, calculate the frequencies that you hear.

Fig 1.213

2 A train is approaching a station platform at a constant speed. The engine of the train is emitting a sound with a frequency of 450 Hz. If the sound heard by a person on the platform is 480 Hz, calculate the speed of the train.

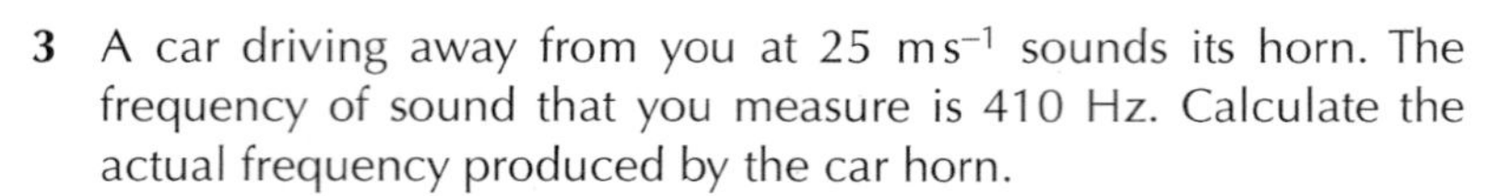

3 A car driving away from you at 25 m s^{-1} sounds its horn. The frequency of sound that you measure is 410 Hz. Calculate the actual frequency produced by the car horn.

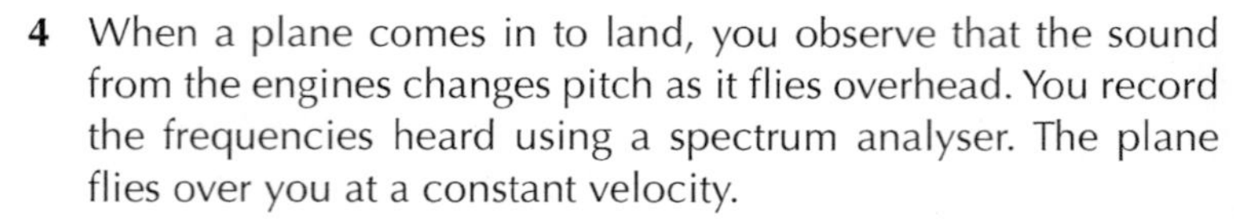

4 When a plane comes in to land, you observe that the sound from the engines changes pitch as it flies overhead. You record the frequencies heard using a spectrum analyser. The plane flies over you at a constant velocity.

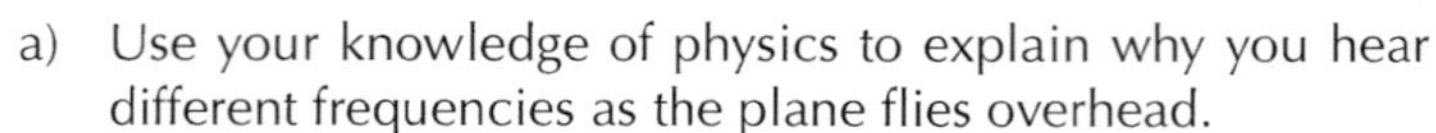

a) Use your knowledge of physics to explain why you hear different frequencies as the plane flies overhead.

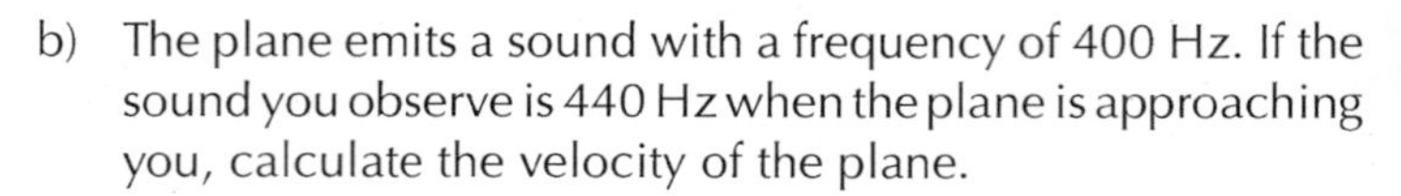

b) The plane emits a sound with a frequency of 400 Hz. If the sound you observe is 440 Hz when the plane is approaching you, calculate the velocity of the plane.

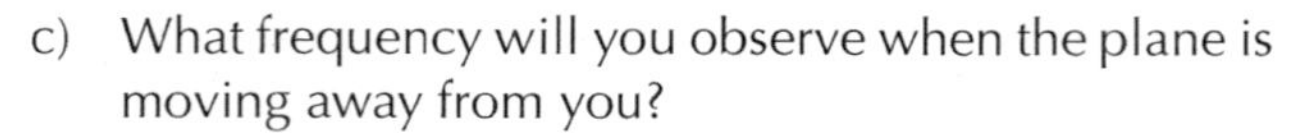

c) What frequency will you observe when the plane is moving away from you?

Fig 1.214

5 Students reading about Doppler shift see Figure 1.215 in a textbook.

Use this diagram, and your knowledge of physics, to explain the Doppler effect for sound. Make reference to the observed changed in frequency and wavelength for a moving source.

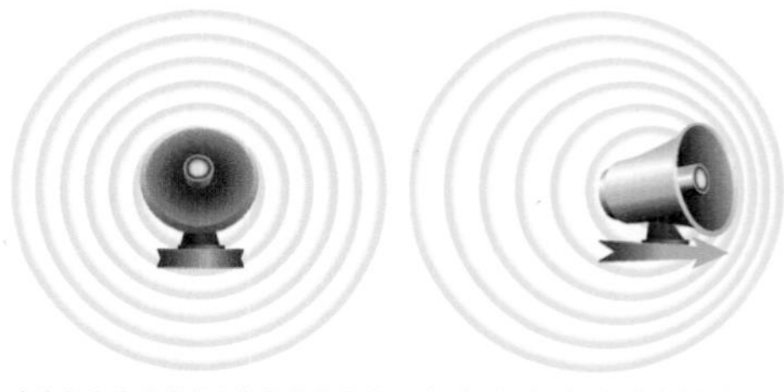

Fig 1.215

6.1.2 Doppler effect for light

The Doppler effect applies to any source of energy that travels as a wave. Therefore, the Doppler effect also applies to light. However, due to the speed of light being much greater than the speed of sound, the Doppler effect for light is only really observed when the relative velocity of the source and observer is very large.

It can be shown that the observed wavelength from a source of light that is travelling with a velocity v is given by

$$\lambda_0 = \lambda_S\left(1 \pm \frac{v_S}{c}\right)$$

This equation is not part of the Higher Physics course.

The equation shows the concept of red shift. When an object is moving away from you, the observed wavelength will be greater. Greater wavelengths correspond to colours towards the red end of the spectrum;

an object moving away from you will be observed to be emitting light that is red-shifted. We consider this in more detail in the next section.

This effect can be used in radar systems to measure the speed of objects. The greater the speed of the object, the greater the wavelength shift. Examples of the application of Doppler effect include radar speed guns used by the police and weather radar.

6.1.3 Red shift

As described above, the Doppler effect can be applied to light as well as sound. Astronomers measure the shift in observed frequencies from celestial bodies in order to determine their velocities, relative to Earth. Objects in the night sky that are moving towards Earth have shorter observed wavelengths and are shifted towards the blue end of the spectrum. Objects that are moving away from the observers on Earth appear to have longer wavelengths and are shifted towards the red end of the spectrum.

The red-shift parameter, z, is the ratio of the change in observed wavelength to the emitted wavelength, defined as:

$$z = \frac{\lambda_0 - \lambda_S}{\lambda_S}$$

where λ_0 is the observed wavelength and λ_s is the emitted wavelength. It is known that the observed wavelength can be either longer or shorter than the actual wavelength, so the red shift can be either positive or negative. The red-shift ratio only works for stars if the speed of the star is less than 0.1 times the speed of light.

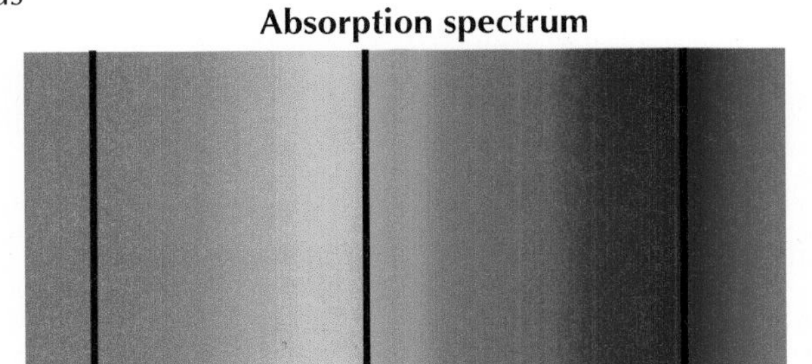

Fig 1.216

Consider an absorption spectrum for a distant star shown in Figure 1.216. Three Fraunhofer lines are shown with their corresponding wavelengths λ_A, λ_B and λ_C. These lines are produced by the presence of certain elements in the star's atmosphere. The relative motion of a star will cause the lines above to shift either up or down the spectrum by an amount, $\Delta\lambda$.

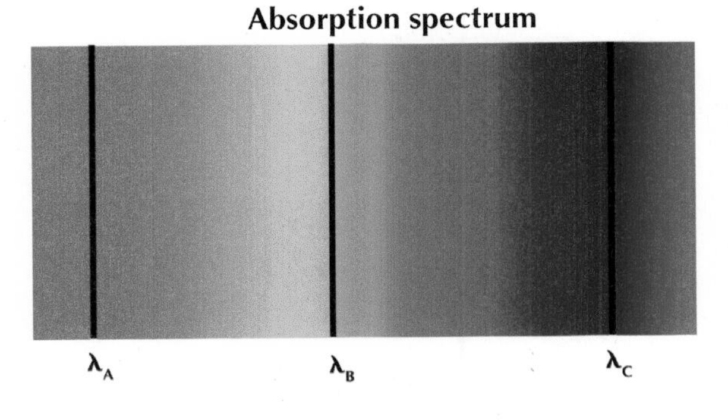

Positive red shift

If the star is moving away from Earth, then the observed lines will move towards the red end of the spectrum – they are red-shifted. This is shown in Figure 1.217.

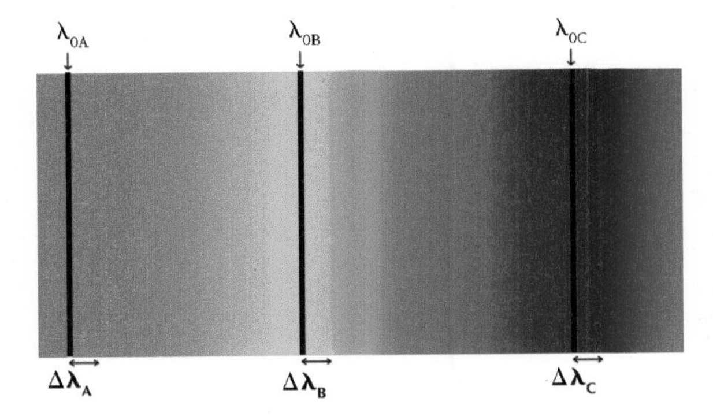

Fig 1.217

The size of the change in wavelength depends on the speed of the star according to the Doppler effect. The observed wavelength is given by:

$$\lambda_0 = \lambda_S\left(1 + \frac{v_S}{c}\right)$$

The wavelength shift is then given by:

$$\lambda_0 - \lambda_S$$

This wavelength shift is positive because the observed wavelength is greater than the emitted wavelength. This leads to a positive red-shift ratio.

Negative red shift (blue shift)

If the star is moving towards the Earth, then the observed lines will move towards the blue end of the spectrum – we say that they are blue-shifted. This is shown in Figure 1.218.

The observed wavelength is given by:

$$\lambda_0 = \lambda_S\left(1 - \frac{v_S}{c}\right)$$

The wavelength shift is then given by:

$$\lambda_0 - \lambda_S$$

This wavelength shift is negative because the observed wavelength is less than the emitted wavelength. This leads to a negative red-shift ratio.

Absorption spectrum

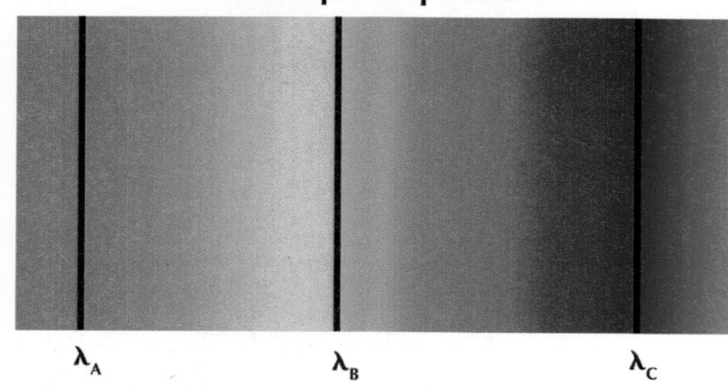

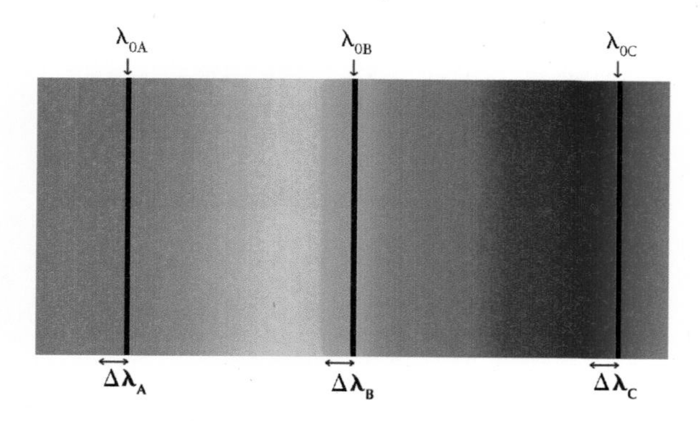

Fig 1.218

Red-shift ratio for non-relativistic objects

The red-shift ratio described above can be used to work out the velocity of a celestial body such as a star or galaxy. It can be applied so long as the speed of the object is non-relativistic – that is, the speed of the object is less than 10% of the speed of light:

$$v \leq 3 \times 10^7 ms^{-1}$$

Consider the equation for red shift:

$$z = \frac{\lambda_0 - \lambda_S}{\lambda_S}$$

Substituting for the observed wavelength as:

$$\lambda_0 = \lambda_S\left(1 \pm \frac{v_S}{c}\right)$$

obtains:

$$z = \frac{\lambda_S\left(1 \pm \frac{v_S}{c}\right) - \lambda_S}{\lambda_S}$$

$$z = \frac{\lambda_S \pm \frac{\lambda_S v_S}{c} - \lambda_S}{\lambda_S}$$

$$z = \frac{\left(\frac{\lambda_S v_S}{c}\right)}{\lambda_S}$$

$$z = \frac{\lambda_S v_S}{c\lambda_S}$$

$$z = \frac{v_S}{c}$$

The derivation of this equation is not part of the Higher course. However, the application of the red-shift ratio equation is a key part of the course.

Key point

The red-shift ratio can be used to work out the speed of a non-relativistic celestial body:

$$z = \frac{v_S}{c}$$

where v_S is the velocity of the source and c is the speed of light.

Key point

The red-shift ratio describes the size of the wavelength shift caused by relative motion of a celestial body:

$$z = \frac{\lambda_{observed} - \lambda_{rest}}{\lambda_{rest}}$$

The red-shift ratio can be either positive or negative. A positive red-shift ratio indicates that the body is moving towards the Earth and a negative ratio indicates that the body is moving away from the Earth.

The red-shift ratio only works if the speed of the star is less than 0.1c, where c is the speed of light.

Worked example

The red absorption line for hydrogen has a wavelength of 656 nm. The absorption line is observed to have a wavelength of 669 nm on Earth for a distant galaxy.

a) Calculate the red-shift ratio for this galaxy.

b) Find the velocity of the galaxy relative to Earth.

c) In what direction is this galaxy moving relative to Earth?

Fig 1.219

a) *The red-shift ratio is the ratio of the wavelength shift ($\lambda_0 - \lambda_S$) to the emitted wavelength:*

$$z = \frac{\lambda_{observed} - \lambda_{rest}}{\lambda_{rest}}$$

Substitute the source and observed wavelengths from above and solve for the red-shift ratio:

$$z = \frac{669 - 656}{656}$$

$$z = 0.020$$

Note that z is unitless because it is a ratio.

b) *The red-shift ratio can be used directly to find the velocity that the galaxy is moving from:*

$$z = \frac{v_S}{c}$$

Substitute the red-shift ratio from above and the speed of light that is known to get the velocity of the galaxy:

$$0.020 = \frac{v_S}{3 \times 10^8}$$

$$v_S = 0.020 \times 3 \times 10^8$$

$$v_S = 6 \times 10^6 ms^{-1}$$

c) *The red-shift ratio is positive so the galaxy is moving away from the Earth. This can be seen by thinking about the wavelength observed: it is stretched out compared to the source wavelength, so the galaxy must be moving away from Earth.*

Exercise 6.1.3 Red-shift ratio

1 Astronomers are studying a distant star in the galaxy. They calculate the red-shift ratio for the star to be 0.085. Find the velocity of the star relative to Earth and state whether it is moving towards or away from Earth.

2 Light emitted from the tail of a comet is being studied on Earth. It is known that an element present on the comet has an absorption wavelength of 550 nm.

If the comet is moving with a velocity of 2.5×10^6 m s^{-1} towards the Earth, calculate:

a) The red-shift ratio for the comet.

b) The absorption wavelength detected on Earth.

3 Aliens on a distant planet are studying light coming from our sun. One of the Fraunhofer lines from our sun has a wavelength of 653 nm.

If the alien planet is moving away from our sun with a velocity of 4.8×10^6 m s^{-1}, find:

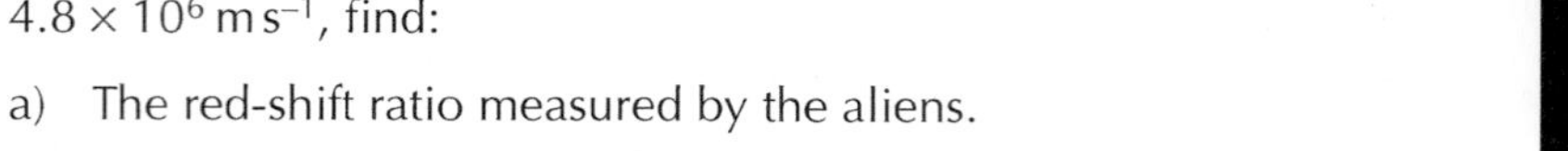

a) The red-shift ratio measured by the aliens.

b) The wavelength of the Fraunhofer line detected by the aliens on their home planet.

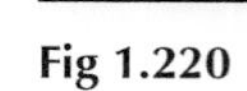

Fig 1.220

4 Astronomers are studying light from a distant galaxy. They study the hydrogen absorption line, which has a wavelength of 658.5 nm. The hydrogen absorption line detected by the distant planet is found to have a wavelength of 677.9 nm. Calculate the red-shift ratio for this galaxy and find the velocity that this galaxy is moving with, relative to the Earth. Include in your answer whether the galaxy is moving towards or away from Earth.

Fig 1.221

5 Galactic red-shift allows observers to measure the velocity of a galaxy either towards or away from Earth by measuring the wavelength shift of known spectral lines. Use your knowledge of the Doppler effect and the red-shift ratio to explain how measurement of the shift in wavelength allows the velocity of a galaxy relative to Earth to be found.

6 Starting with the equation for the wavelength shift due to the Doppler effect:

$$\lambda_0 = \lambda_S \left(1 \pm \frac{v_S}{c}\right)$$

derive an expression for the red-shift ratio and a further expression for the velocity of the object in terms of the speed of light and the red-shift ratio.

6.2 Hubble's law

Edwin Hubble studied the recession velocity of galaxies and their estimated distance from Earth. He found a crucial link between these, which is powerful evidence for the expansion of the universe.

In 1929, Edwin Hubble announced his discovery that the recession velocity, v, of a star was directly proportional to its distance, d, away from us:

$$v = H_0 d$$

where H_0 is Hubble's constant. At the time of Hubble's discovery, the Hubble constant was taken to be:

$$H_0 = 1.62 \times 10^{-17} s^{-1}$$

Very recent work using much more sensitive telescopes has revealed the now accepted value for Hubble's constant to be:

$$H_0 = 2.34 \times 10^{-18} s^{-1}$$

The changing values for Hubble's constant reflect the difficulties experienced with measurements in astronomy. Big advancements in technology have allowed astronomers to study celestial objects in greater detail and look deeper into the universe, and this has led to more accurate calculations of Hubble's constant.

Hubble's law relies on the measurement of the distance between Earth and a star, and also the velocity of the star. The velocity of a star can be found using Doppler red shift as described above. Estimating the distance between the Earth and a star can be done using either direct techniques or indirect techniques. Direct techniques, such as parallax, can be used for celestial objects that are close to the Earth. Indirect techniques, using Cepheid variables, need to be used for greater distances.

Fig 1.222

Key point

Hubble's law states that the recession velocity of a celestial object such as a star is directly proportional to the distance the star is away from Earth:

$$v = H_0 d$$

Parallax

Parallax refers to the different apparent position of an object depending on where it is viewed.

Consider viewing an object from two different viewpoints as shown in Figure 1.223.

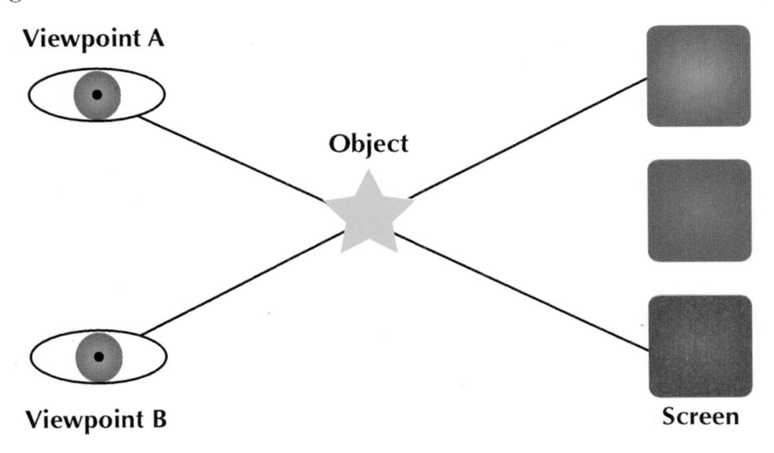

Viewpoint A

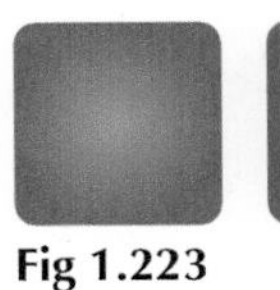

Viewpoint B

Fig 1.223

When standing at viewpoint A, the object appears to be in one position. When you move to viewpoint B, the object appears to be in a different position (having seemingly moved in the opposite direction to you). This effect is known as parallax. You can see this effect when you look out of the window of a moving object such as a train – objects that are nearby appear to be moving faster than objects far away. As the parallax effect varies with distance, it can be used to measure distances.

A nearby star's position in the sky changes as the Earth goes around the Sun, as shown in Figure 1.224:

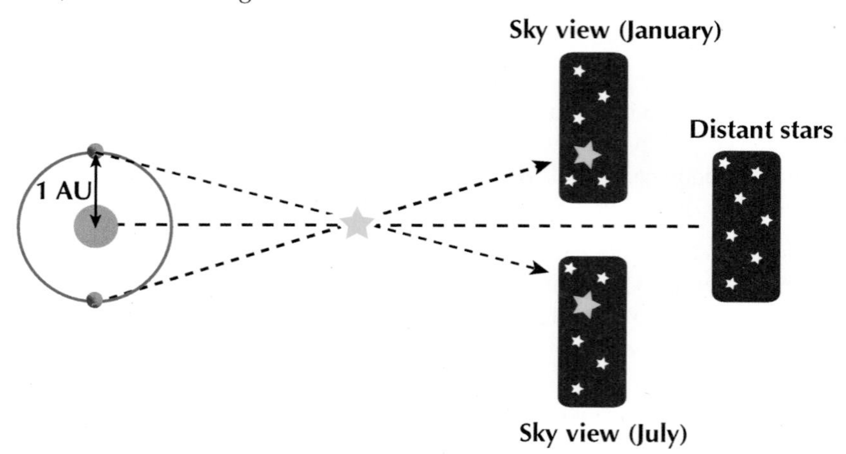

Fig 1.224

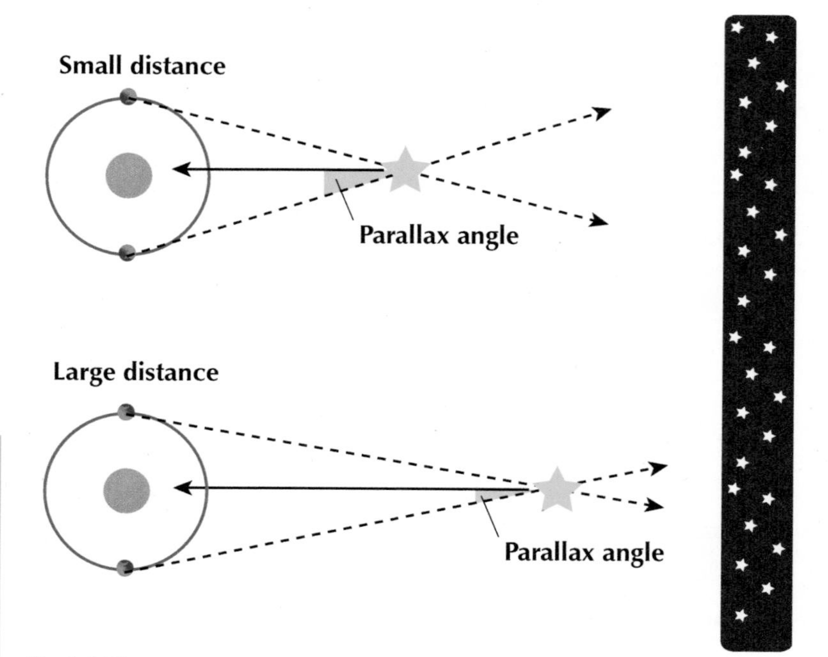

Fig 1.225

If the star is close enough to our sun, it will have a parallax that can be measured.

Stars that are a large distance away cannot be measured using parallax because the angle will be too small.

Key point

Parallax is the difference in apparent position of an object when viewed from different places. It can be used to measure the distance an object is away from us as the parallax depends on the distance. Parallax can be used to measure distances between our Sun and other stars, providing the star is close enough.

Cepheid variables

The technique of parallax can only be used to measure the distance to stars that are fairly close to our Sun. Stars a greater distance away don't have a measurable parallax, so a different technique must be used to measure their distance.

Fig 1.226

One possible technique involves using Cepheid variables. A star's luminosity is a measure of the total amount of energy that it emits. Cepheid variable stars exhibit a brightness that varies periodically with time in a predictable way. In 1929, Henrietta Leavitt discovered that the period of oscillation was linked to the luminosity of the Cepheid star. This allowed these stars to be calibrated for distance measurement. Such stars are called standard candles.

Once the luminosity of the star is found (using the luminosity-period relationship), its apparent brightness measured on Earth would allow a measure of the star's distance away. The star can be treated as a point source of light, so it obeys the inverse square law for intensity (brightness), I:

$$I \propto \frac{1}{d^2}$$

Cepheid variables can be viewed and distances measured up to 20 million light years, significantly more than is possible with parallax measurements (just over 300 light years).

Towards Hubble's law

Using both methods described above to measure distance, and red shift to measure velocity, Hubble's law can be established:

$$v = H_0 d$$

Recent measurements from modern, highly sensitive telescopes have allowed determination of Hubble's constant to be:

$$H_0 = 2.34 \times 10^{-18} s^{-1}$$

This means that through calculating the velocity of a star or galaxy using red shift, its distance from Earth can be calculated.

Worked example

The red shift calculation for a distant galaxy shows that it is moving with a velocity of 2.1 × 107 $m s^{-1}$ relative to Earth. Calculate the distance between Earth and this galaxy.

The distance between Earth and the galaxy can be found using Hubble's law:

$$v = H_0 d$$

Fig 1.227

Substitute what is known and solve for distance:

$$2.1 \times 10^7 = 2.34 \times 10^{-18} d$$

$$d = \frac{2.1 \times 10^7}{2.34 \times 10^{-18}}$$

$$d = 8.97 \times 10^{24} m$$

This can be converted to light years:

$$d = \frac{8.97 \times 10^{24}}{9.46 \times 10^{15}}$$

$$d = 9.47 \times 10^8 Ly$$

Exercise 6.2 Hubble's law

1 Hubble's law links the recession velocity of a star to the distance the star is from Earth.

a) State the relationship that describes Hubble's law.

b) Explain how the velocity of a distant star can be measured.

c) Explain two methods that can be used to measure the distance to a distant star.

2 The red shift for a distant star shows that it is moving away from Earth with a velocity of 2.6×10^6 m s^{-1}. Use Hubble's law to determine the distance between Earth and the star.

3 A distant galaxy is a distance of 25 Ly away from Earth. If 1 Ly = 9×10^{15} m, use Hubble's law to determine the velocity of the galaxy as it moves away from Earth.

4 Since 1929, when Hubble discovered the relationship between recession velocity and distance, the constant (Hubble's constant) has been updated regularly. Explain why the accuracy of Hubble's constant has been improved since 1929 to the present day.

Age of the universe

An important consequence of Hubble's law is that it shows the universe is expanding. If the rate of expansion is assumed to be constant, then Hubble's law can be used to estimate the age of the universe. Consider Hubble's law:

$$v = H_0 d$$

Applying the classic speed, distance, time equation:

$$d = vt$$

obtains:

$$v = H_0 vt$$

$$t = \frac{1}{H_0}$$

This time tells how long a celestial body has been moving – in other words, it tells the age of the universe!

In 1929, Hubble's constant was taken to be:

$$H_0 = 1.62 \times 10^{-17}\,s^{-1}$$

Substituting this into the equation above for the age of the universe gives:

$$v = H_0 vt$$

$$t = \frac{1}{H_0}$$

$$t = \frac{1}{1.62 \times 10^{-17}}$$

$$t = 6.2 \times 10^{16}\,s$$

$$t = 1.9 \times 10^{9}\ years$$

This value raised problems at the time. Radioactive dating had placed the age of the Earth to be around 3 billion years, so the universe could not possibly be younger than this! Higher-sensitivity telescopes revised the value for Hubble's constant to the value now accepted, which is:

$$H_0 = 2.34 \times 10^{-18} s^{-1}$$

This gives the age of the universe as:

$$t = \frac{1}{H_0}$$

$$t = \frac{1}{2.34 \times 10^{-18}}$$

$$t = 4.3 \times 10^{17} s$$

$$t = 1.4 \times 10^{10} years$$

6.3 Expansion of the universe

The most commonly accepted theory for the start of the universe is the Big Bang (discussed in more detail in the next section). Following the Big Bang, the universe has been continually expanding. Hubble's law can be used to show that the universe is expanding, and also to find the age of the universe, assuming a constant rate of expansion.

The age of the universe was estimated using Hubble's law by assuming that the universe was constantly expanding. Hubble's law can be used to demonstrate that the universe is expanding, as it shows that objects that are further away are receding with a greater velocity. Working backwards, this must mean that the universe started at a single point and then following this all of the matter expanded outwards and is continuing to expand.

GO! Experiment 6.3 Expansion of the universe

This experiment demonstrates the link between Hubble's law and the expansion of the universe.

Apparatus

- A balloon
- A marker pen

Instructions

1. Blow up the balloon to a very small size. Mark an X on the balloon to represent the Earth. Mark dots on the balloon at different distances away from the X to represent near and far celestial bodies.
2. Blow up the balloon slowly, watching how the near and far dots move relative to the X. Which dots are moving more quickly away from Earth?

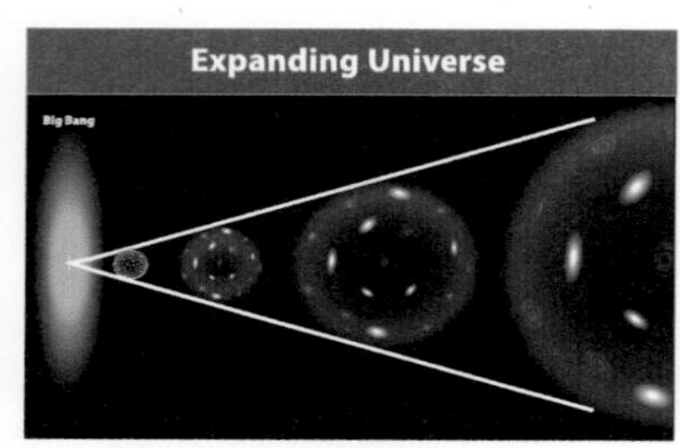

Fig 1.228

The experiment above demonstrates that objects near to Earth appear to be moving away more slowly than objects that are far away. This is in agreement with Hubble's law, which states that the recession velocity increases with distance away from Earth. Figure 1.228 illustrates the expansion of the universe demonstrated by the experiment above. Notice how the galaxies are moving further and further apart. The further away a galaxy is, the faster it appears to be receding.

The ultimate fate of the universe will depend on its mass (all of its constituent parts) and how this mass is distributed. The distribution of the mass of an object is known as the density. Scientists have long studied the universe to find out its ultimate fate, and in so doing they have made many startling discoveries. Two of these are evidence for matter that cannot be seen, known as dark matter, and energy that cannot be accounted for, known as dark energy.

6.3.1 Dark matter

Fig 1.229

There is matter that exists in the galaxy that cannot be seen. When large stars die, they can form a black hole. The gravity for a black hole is so intense that even light cannot escape – this means that a black hole cannot be seen, but there is still matter to produce the gravitational field to trap the light.

Newton's universal law of gravitation (Area 1, Chapter 4) shows how the force of gravity changes with distance:

$$F = \frac{GMm}{r^2}$$

In our solar system, this causes each planet to follow a circular path. The force of gravity acts towards the centre of the solar system, where you find the Sun. The greater the distance between the planet and the Sun, the smaller the force of gravity. The force gives rise to a circular motion that can be described by:

$$F = \frac{mv^2}{r}$$

This equation is not required as part of the Higher Physics course (it is covered at Advanced Higher), but it is used here to highlight how the velocity of a planet depends on its distance from the Sun. The force of gravity is set equal to the gravitational force:

$$\frac{mv^2}{r} = \frac{GMm}{r^2}$$

This gives the following for the velocity of the planet:

$$v = \sqrt{\frac{GM}{r}}$$

where M is the mass of the Sun and r is the distance from the Sun to the planet. As you move to the outer planets of the solar system, the speed of the planet decreases. A plot of the velocity of the planets vs the distance away from the Sun will give a graph as shown below:

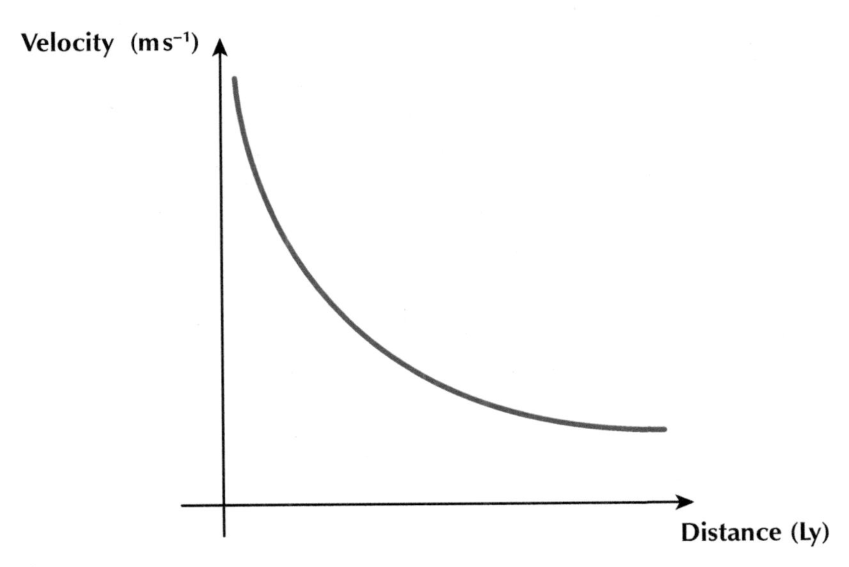

Fig 1.230

This graph is known as a rotation curve and it demonstrates how velocity depends on distance away from the body being orbited. It is governed by Newton's law of gravitation.

The same should apply to a galaxy. In a spiral galaxy, stars orbit around a super-massive black hole at the centre. The orbital speed of stars allows a measure of the mass of the galaxy. This black hole gives the force of gravity required for all stars in the galaxy to orbit. By Newton's law of gravitation, the stars on the outside of the galaxy should be moving more slowly than the stars at the centre. However, this is not what was observed by Vera Rubin in the 1970s when she studied our nearest neighbour galaxy, Andromeda.

Fig 1.231

Rubin studied the velocity of stars on the outside of the galaxy Andromeda. She found that the velocity of the stars remained constant despite changing the distance away from the centre of the gravity. This led to a velocity–distance graph similar to the following:

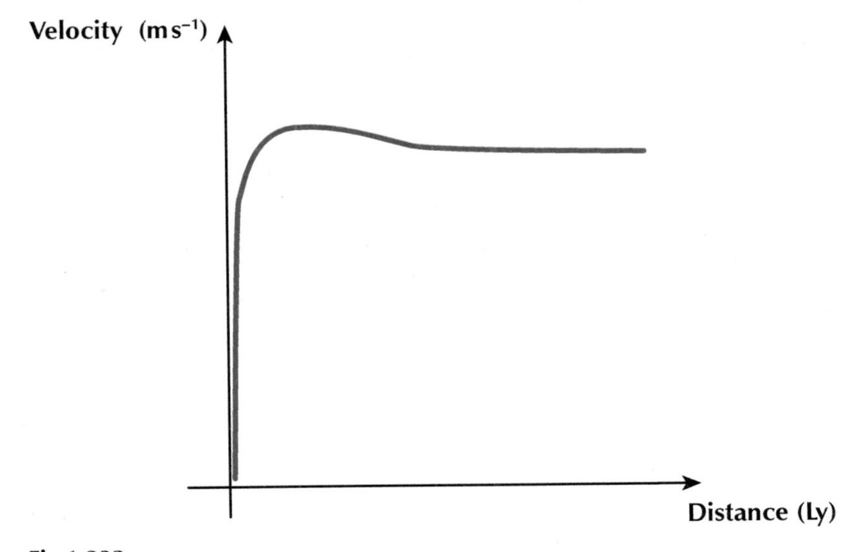

Fig 1.232

This curve shows that the stars on the outside of the galaxy were moving just as fast as the ones near the centre. The stars appeared to defy gravity. One explanation for this observation was the presence of additional mass in the galaxy. This additional mass provided the extra gravity required for the velocity of the stars to remain high as you move away from the centre of the galaxy. This extra mass takes the form of dark matter because it cannot be observed on Earth. Only the effects can be seen.

6.3.2 Dark energy

The above discussions for the age of the universe assume that the rate of expansion of the universe is constant. The expansion of the universe has fascinated scientists for many years. In theory, gravity between the different masses in the universe should act to slow down its expansion. However, scientists studying supernovae explosions in distant galaxies uncovered results that showed this not to be true. Instead, they found the rate of expansion of the universe was actually increasing!

The image here shows the current picture of how it is thought the universe has evolved since its inception at the Big Bang.

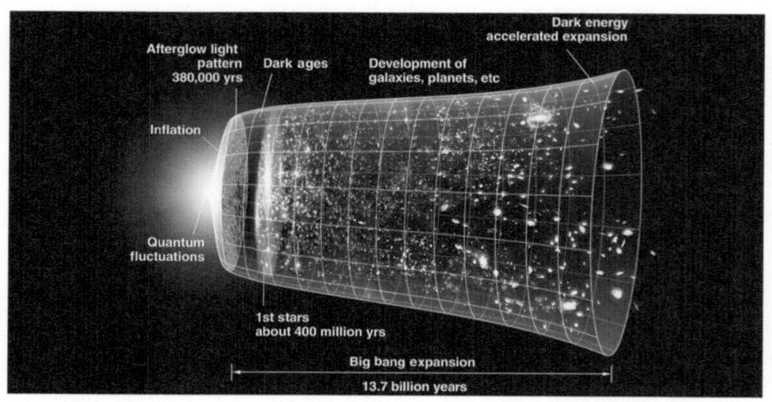

Fig 1.233

The universe started with the Big Bang; following this there was a period of very rapid expansion referred to as 'inflation'. No stars were present at this time; in fact, matter as we know it was not even present then. Once matter had formed, its mass resulted in gravitational attraction, which acted to slow the rapid rate of expansion. Stars, planets and galaxies formed. At the present day, the rate of expansion of the universe is now found to be speeding up – we are in a period of accelerated expansion. It is currently thought that dark energy is responsible for the accelerated expansion.

6.3.3 Blackbody radiation

Any object will emit radiation. Quite often this radiation is felt as heat; for example, when standing next to a building that has been illuminated by the sun for a period of time, the building radiates heat that can be felt. A substance that is heated will emit visible light when it gets hot enough. A prime example of this is a light bulb, where the filament glows white hot and emits light that illuminates a room. The link between the thermal energy of an object and the radiation it emits is an important one, and one that has been used in the study of stellar physics.

Fig 1.234

An object that absorbs all of the radiation that is incident on it (reflects no radiation) and radiates energy that is dependent only on the object itself is called a blackbody. The only radiation emitted by a blackbody is due to the object itself, and is linked to the temperature of the object. An example of a blackbody would be a furnace with walls that absorb all incident radiation but that has a very small opening for radiation from inside the furnace to escape. The walls of the furnace absorb all of the energy that is incident on them, so no radiation is reflected. The only radiation emitted by the furnace comes from inside the furnace itself.

The radiation emitted by an object depends on the temperature of the object. At room temperature, the object does not emit any visible radiation and so it appears black. At higher temperatures, the object begins to emit radiation in the visible part of the spectrum so it starts to glow. This can be seen in practice when steel is heated. At low temperatures, the steel simply appears grey, but as it gets hotter it begins to glow. The steel glows red at first, and then yellow as it gets hotter. Figure 1.235 shows steel being heated; notice the hottest region of the steel at the centre, where it glows yellow, and the cooler region around the outside, where it glows red.

Fig 1.235

In the early 20th century, Max Planck used a similar simple blackbody to study blackbody radiation. Plank demonstrated by experiment the same effect as seen by the steel in Figure 1.235 – for higher temperatures, the wavelength of the radiation emitted moves from the red end of the spectrum towards the yellow and ultimately towards the blue. He also found that the total power emitted increased with temperature. This is summarised in Figure 1.236.

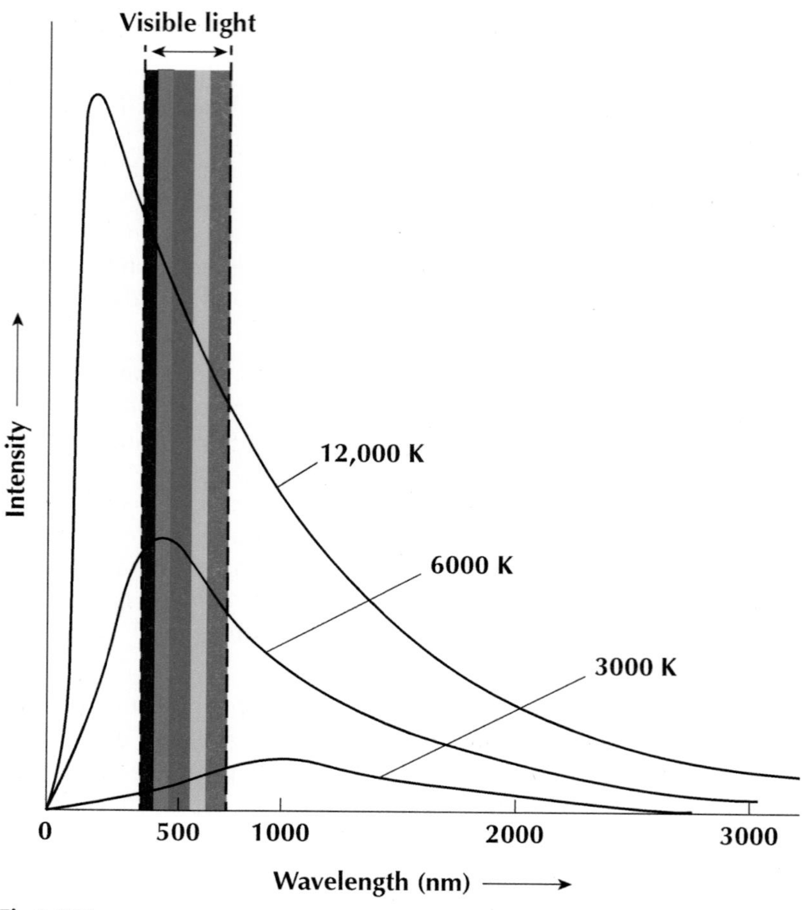

Fig 1.236

A light bulb filament has a surface temperature of around 3000 Kelvin (K), and this is why its light emissions are visible. The peak wavelength, λ_{peak} is in the infrared. Our sun has a surface temperature of 6000 K and thus emits in the visible spectrum (and infrared and ultraviolet). The greater temperature results in a shorter peak wavelength.

6.3.4 Stellar temperatures

If you look at the night sky on a clear night, you can see many stars. To the naked eye, these stars all appear to be white specks of light in the sky, with little or no difference in the colour. However, if you look more closely at the stars using powerful telescopes, you will find that the stars seem to glow with different colours. This is shown in Figure 1.237, which shows three stars: Sirius (bottom), Betelgeuse (top right) and Procyon (top left).

Sirius appears to be glowing a blue-white colour compared to Betelgeuse, which is a red colour. The difference in observed colour is due to the difference in surface temperature of the stars.

Key point

A blackbody emitter is a source that only emits radiation due to its own properties – it does not reflect radiation from its surroundings.

Blackbody radiation follows the same basic shape for all temperatures, with two key points:

- The greater the temperature, the shorter the peak wavelength emitted.
- The greater the temperature, the greater the power emitted.

Peak wavelength and temperature

Although a star is not a perfect blackbody, it approximates closely enough that the theory of blackbody radiation can be applied. Measuring the peak wavelength of radiation emitted by a star allows the surface temperature of the star to be estimated. Planck showed that the greater the temperature of a blackbody emitter, the shorter the emitted peak wavelength (see above for more details).

Wilhelm Wien quantified the link between the temperature of a blackbody and the peak wavelength with the following equation:

$$\lambda_{peak}T = 2.898\times10^{-3}$$

Hence, if the peak wavelength emitted by a star can be measured, its surface temperature can be found. The blackbody radiation acts as a stellar thermometer. The peak wavelength of our Sun is approximately 500 nm, so the surface temperature can be estimated to be:

$$\lambda_{peak}T = 2.898\times10^{-3}$$

$$500\times10^{-9}T = 2.898\times10^{-3}$$

$$T = \frac{2.898\times10^{-3}}{500\times10^{-9}}$$

$$T = 5796K$$

The hotter the star, the shorter the peak wavelength. Hence, Sirius must have a greater peak wavelength than Betelgeuse because it emits peak wavelengths in the blue region of the spectrum compared to the red region.

Fig 1.237

Key point

When viewed using powerful telescopes, stars are found to be glowing different colours. This is due to the difference in surface temperature of the star. Assuming a star to be a perfect blackbody, the shorter the peak wavelength emitted, the greater the surface temperature of the star.

Key point

The peak wavelength emitted by a blackbody and the temperature of the blackbody are linked by the following equation:

$$\lambda_{peak}T = 2.898\times10^{-3}mk$$

This can be used to find the temperature of a blackbody.

Worked example

Sirius is the brightest star that can be seen from the northern hemisphere. It emits a peak wavelength of 72 nm. Use Wien's law to find the surface temperature of Sirius.

Here, use Wien's law to find the surface temperature of Sirius:

$$\lambda_{peak}T = 2.898\times10^{-3}mK$$

Substitute what you know and solve for the temperature:

$$(72\times10^{-9})T = 2.898\times10^{-3}$$

$$T = \frac{2.898\times10^{-3}}{72\times10^{-9}}$$

$$T = 40250K$$

Fig 1.238

Luminosity and temperature

As well as the peak wavelength changing with surface temperature, the power radiated by the star also changes with surface temperature. Studying Figure 1.236 for the blackbody emitters, the area under the curves gives a measurement of the amount of energy radiated by the blackbody. The hotter the body, the more energy is radiated. Hence, hotter stars glow more brightly in the sky. The power radiated by an object also depends on its surface area; the greater the surface area, the greater the radiated power. Hence, larger stars glow more brightly in the sky than smaller ones.

Exercise 6.3.4 Stellar temperature

1 It is possible to approximate a star by a blackbody.

a) Explain what is meant by the term blackbody radiation.

b) Use the concept of blackbody radiation to explain why a metal bar glows red when it is heated to high temperatures, and then glows yellow when it is heated to higher temperatures.

c) How does the power emitted by a blackbody depend on its temperature?

2 Students studying the night sky have noticed that stars appear to be different colours when viewed through a telescope. Some of the stars are glowing red, while others appear to be glowing blue, as shown in Figure 1.239

a) Explain why stars emit different wavelengths of light.

b) What colour of light do the hottest stars in Fig 1.239 emit?

c) What colour of light do the coolest stars in Fig 1.239 emit?

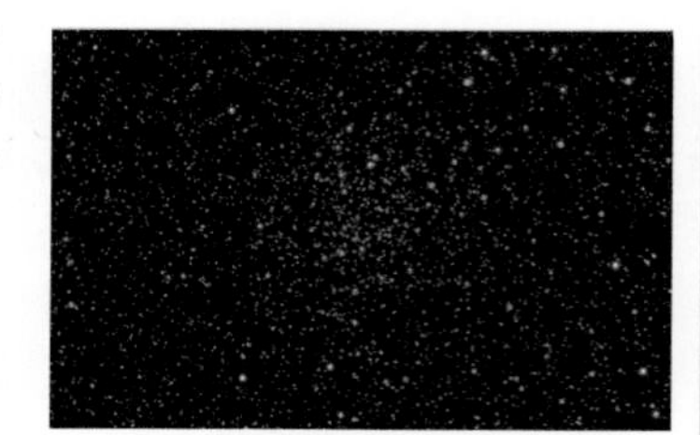

Fig 1.239

3 The red giant Betelgeuse is a star in the Orion nebula. It emits a reddish glow with a peak wavelength of 970 nm. Use Wien's law to find the surface temperature of Betelgeuse.

4 A lightning bolt causes the air around it to rise in temperature to approximately 10,000 K. Use Wien's law to estimate the peak wavelength of emission of photons produced by this thermal energy. What part of the electromagnetic spectrum do these photons belong to?

5 The radiated power of a star depends on its surface temperature.

a) Explain why hotter stars glow more brightly in the sky.

b) Use your knowledge of the inverse square law to explain why astronomers introduced the quantity called absolute magnitude to compare different stars in the sky.

Fig 1.240

6.3.5 Evidence for the Big Bang

In 1927, Georges Lemaître wrote a paper where he proposed a relationship between a galaxy's distance from Earth and its observed Doppler red shift – the further away the galaxy, the greater the red shift. Many scientists of the time had not read this paper, and two years later Edwin Hubble published the same relationship – the relationship was labelled as Hubble's law, discussed above, and it is a key concept to the Big Bang. However, as it became clear that Lemaître first proposed that the universe is expanding, he is considered to be the 'Father of the Big Bang'.

The Big Bang

The fact that the universe is expanding suggests that a long time ago the universe started comparatively small, with an event that triggered the expansion. This event is known as as the Big Bang. In order to justify the theory, astronomers have searched for evidence of the Big Bang, some of which is detailed below.

It is important not to think of the Big Bang as something that you could watch happen. It would not have been like standing next to a firework and watching it explode, with the parts expanding. Before the Big Bang, there would have been nowhere to stand and watch from, because it represents the start of time itself. This means that there was no such thing as *before the Big Bang*. Time started with the Big Bang, so the idea of something before it has no meaning. Also, as the universe is still expanding, it can be argued that the Big Bang is still happening.

The Big Bang is represented by Figure 1.241.

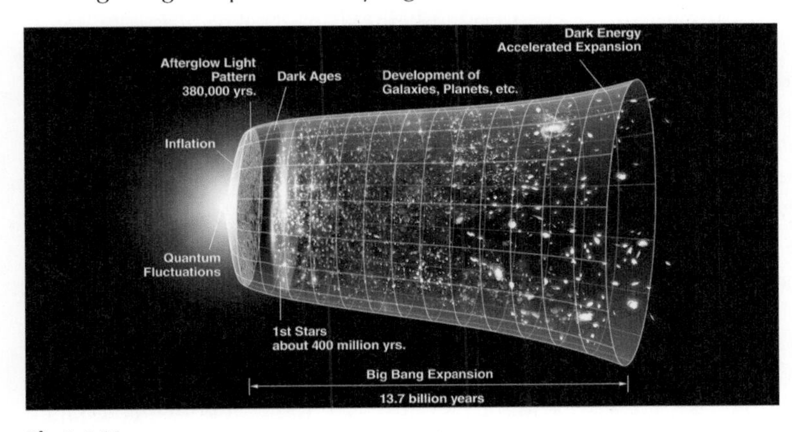

Fig 1.241

The actual processes during the rapid inflation are not well known, but currently it is thought that the universe started smaller than a proton and with an initial temperature of around 10^{34} K. Immediately following this, in a period of 10^{-9} s, the universe underwent a period of rapid expansion and increased in size by a factor of 10^{30}. It was now a dense sea of quarks, leptons and photons (see Area 2, Chapter 2, The standard model) and cooling rapidly.

The quarks then combined at lower temperatures to form protons and neutrons (and their antiparticles) and a battle between matter and antimatter began. Matter and antimatter collided and annihilated each other. A small excess of matter meant that there was matter that did not find any antimatter to annihilate, and this formed the matter that exists today.

Approximately 1 minute after the start of the rapid expansion, the universe cooled sufficiently to allow the protons and neutrons to combine and form nuclei such as helium and hydrogen. These positively charged nuclei interacted with photons; this meant that photons – the light-carrying particles – did not travel far after being emitted and so light could not escape. The universe at this stage appears to have been opaque.

Approximately 300,000 years after the start of the expansion, the universe had cooled to roughly 10,000 K and at this stage electrons could begin to stick to nuclei to form atoms. Photons of light no longer interacted with the nuclei and could now propagate freely through the universe, forming cosmic background radiation (discussed below). The atoms now began to clump together to form stars and galaxies, and the first stars shone light across the universe to end the dark ages.

Cosmic microwave background radiation (CMBR)

As discussed above, when the universe had cooled sufficiently to form atoms, photons of radiation were able to travel great distances. With nothing to stop this radiation, it was able to propagate across the entire universe and therefore be detectable on Earth.

Fig 1.242

Cosmic microwave background radiation (CMBR) was first detected by accident in 1965 by Arno Penzias and Robert Wilson (Figure 1.242). They were conducting research into telecommunication technology and were testing a sensitive microwave receiver. A faint noise ('hiss') was discovered, which remained constant regardless of the direction of the antenna. This shows that the radiation fills the universe and was a key piece of evidence to support the Big Bang theory.

The Big Bang theory suggested that in the time before atoms formed, a dense sea of quarks and latterly nuclei at very high temperatures existed. This 'sea' produced a hot glow that, once atoms formed, was free to travel across the universe. The existence of this glow was predicted in 1948 by Ralph Alpher, Robert Herman and George Gamow, who were working on the Big Bang theory. Its chance discovery in 1965 supported the Big Bang theory. Such was the significance of their discovery, Penzias and Wilson received the Nobel Prize for their work in 1978.

The CMBR has been shown (with increasing accuracy since its discovery) to have the form of blackbody radiation, with a peak wavelength that corresponds to a temperature of 2.725 K – this is the current temperature of the universe. The resulting peak wavelength of the radiation is in the millimetre wave region of the spectrum and is therefore invisible to the human eye – hence we don't see the background radiation. However, a millimetre wave detector is capable of picking it up.

Although the initial discovery of the CMBR added great weight to the theory of the Big Bang, there was one key problem – the fact that the radiation appeared to be constant and smooth. This suggested that all of the matter in the early universe was uniformly distributed. This is troubling, because following this period stars and galaxies formed – clumps of matter. Further evidence of the Big Bang came from seeing regions of light and dark in the CMBR, which would have been caused by clumps of matter.

In 1992, the Cosmic Background Explorer (COBE) satellite carried out measurements that showed that the CMBR was not uniform across all of space. In 2001, NASA launched the Wilkinson Microwave Anisotropy Probe (WMAP), which refined the results from COBE and began to show the fine detail in the background radiation. In 2013, a further satellite, called Planck, was launched, which again refined the detail of the background radiation to show the light and dark regions that point to the initial clumps of matter in the universe. The differing detail observed by the different satellites is shown in Figure 1.243.

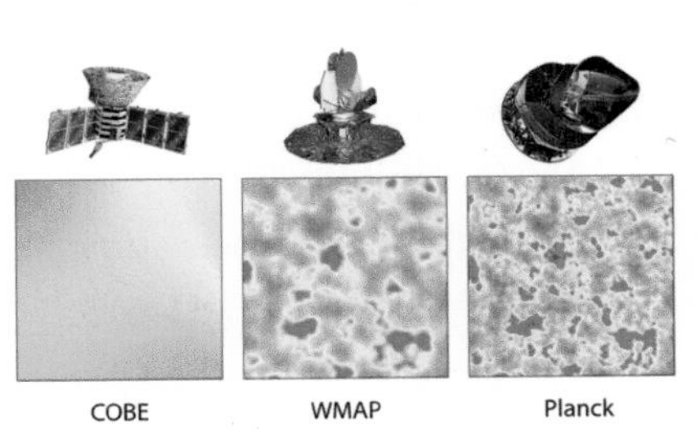

Fig 1.243

Light element abundance

During the early stages of the rapid expansion, the quarks began to combine to form the nuclei of hydrogen and helium. Further cooling of the universe allowed electrons to become bound to these nuclei to form atoms. These atoms were all light elements such as hydrogen and helium and they were produced in an abundance.

Initially, astronomers were unable to explain the large amount of helium in the universe. This became known as the 'Helium Problem'. There was too much helium to be accounted for by star formation, for example. An explanation for the large abundance of helium is that it was produced by the Big Bang, and for this reason the large abundance of helium is considered to be evidence for the Big Bang.

Olbers' paradox

Consider Figure 1.244, which is what the night sky looks like.

Fig 1.244

There are regions of light (stars) and dark (nothing). Olbers' paradox considers this image as further justification of the expansion of the universe (and thus the Big Bang). If the universe were not expanding, then everywhere you looked in the sky there would be a star and the sky should be an even brightness. This is clearly not the case – there are many dark regions. This can be explained by the universe expanding, where distant stars are moving so quickly that their light is red-shifted out of the visible spectrum and is thus not seen. This is in line with the idea of the Big Bang and expansion of the universe, that distant stars are moving very quickly.

Exercise 6.3.5 Evidence for the Big Bang

1 Explain why Georges Lemaître is considered to be the 'Father of the Big Bang'. Include in your description Lemaître's significant discovery and how this leads to the idea of the universe starting with a Big Bang.

2 The Big Bang can be said to mark the beginning of space and time itself. Explain the significance of this statement, making reference to our inability to stand and observe the Big Bang happening from another position in space.

3 During the first 300,000 years of expansion, the temperature of the universe was too great for atoms to form – electrons could not bind to nuclei. Explain why this meant that the universe was opaque during this time.

4 The existence of the cosmic microwave background radiation (CMBR) was predicted in 1948, and its existence was discovered by chance 17 years later, in 1965.

   a) Describe what the CMBR is.

   b) Explain why CMBR is not visible to the naked eye.

   c) The initial discovery of the CMBR in 1965 had a key problem for evidence of the Big Bang. What was this problem?

   d) What did the satellites COBE and WMAP discover that gave further weight to CMBR as evidence for the Big Bang?

5 Explain why the abundance of helium is evidence for the Big Bang.

6 What is Olbers' paradox, and how is it evidence for the Big Bang?

1. **Forces on charged particles**
    - Force fields
    - Electric fields
    - Moving charges and magnetic fields
    - Particle accelerators
2. **The standard model**
    - Orders of magnitude
    - The standard model
    - Particle facts
    - The neutrino
    - Hadrons
3. **Nuclear reactions**
    - Model of the atom
    - Nuclear fission
    - Nuclear fusion
4. **Inverse square law**
    - Irradiance
    - Inverse square law
5. **Wave–particle duality**
    - Quantum theory of light
    - The photoelectric effect
6. **Interference**
    - Review of waves
    - Interference of waves
7. **Refraction of light**
    - Emission spectra
    - Absorption spectra
    - The model of the atom
    - Bohr model of the atom
    - Photon emission and absorption
8. **Refraction**
    - Snell's law
    - Total internal reflection

AREA 2
Particles and Waves

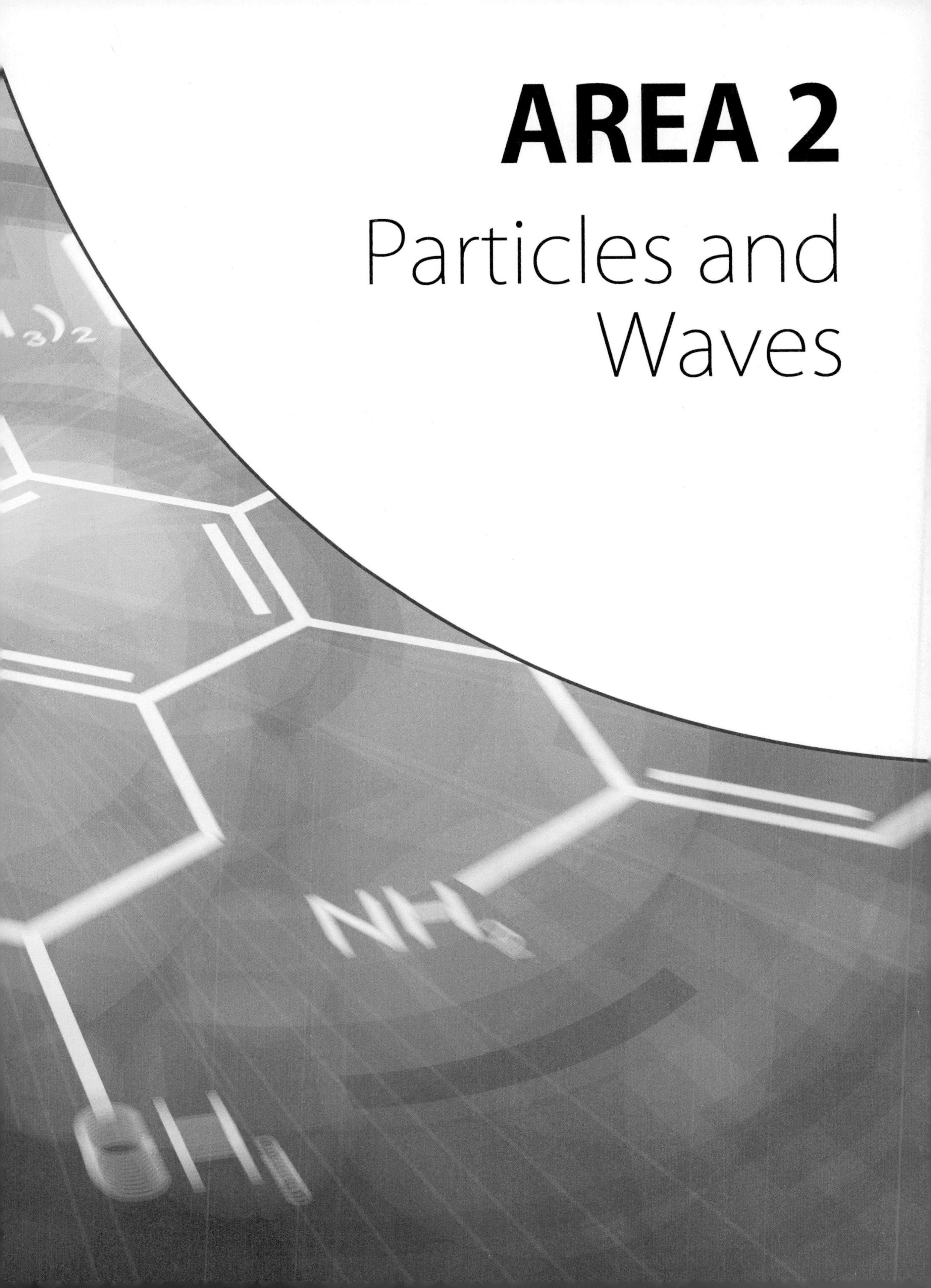

1 Forces on charged particles

What you should know (National 5)

- Effect of electric fields on charged particles
  - Charged particles move in electric fields
  - Opposite charges attract; like charges repel
- Basic electric field patterns

Learning intentions

- Nature of electric field around point charge, between two point charges (like and different) and between parallel plates
- Movement of charged particles in electric fields
- Definition of the volt
- Energy required to move a charge through an electric field
- Calculation of speed of a particle moving through a given potential difference
- Right hand grip rule (moving charge produces a magnetic field)
- How to find the direction a force acts on a charge moving through a magnetic field (left and right hand rules)
- Basic operation of particle accelerators

Fig 2.1

Key point

A region in space where an object feels a force acting at a distance is called a field.

1.1 Force fields

If you jump into the air, you will be pulled back down to the ground. As shown in Area 1, this is due to the force of gravity, which acts downwards, and produces a force called weight (W):

$$W = mg$$

There is no object that physically touches you to pull you back to the ground. However, you are accelerating downwards, so there must be a force acting on you. This force acts at a distance and is due to the mass of the Earth and the mass of you; you are in the Earth's gravitational field and you experience a force in this field because you have a mass.

A region in space where an object feels a force (without physically being touched) is known as a force field. An object in a gravitational field will feel a force due to its mass.

Similarly, an electrically charged particle will feel a force in an electric field. An example of this is the movement of a negatively charged electron towards a positively charged plate.

Field lines

In practice, the field is represented in space by field lines. These lines point in the direction that an object would move within the field. Like the contours of a map, the spacing of the field lines is linked to the strength of the field – the closer together the field lines, the stronger the field. The gravitational field lines of the Earth are shown in Figure 2.1 (see Area 1, Chapter 4 for more details).

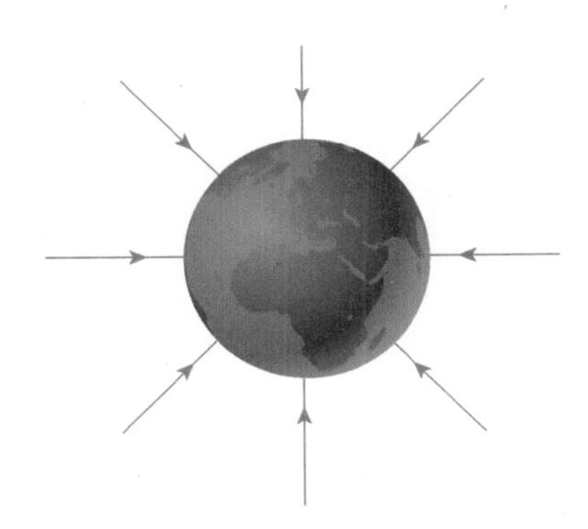

Fig 2.2

The field lines point in the direction of the surface of the Earth, which shows that an object with a mass will move towards the surface of the Earth. This is apparent every day – if you drop an object, it falls towards the Earth's surface. The closer you get to the Earth's surface, the closer together the field lines get. This shows that the strength of the Earth's gravitational field is increasing as you get closer to the Earth's surface. The value of gravity measured at sea level will be slightly greater than that measured at the top of Ben Nevis, for example.

1.2 Electric fields

Charged particles can move. An example of this is the movement of an electron through a wire as the flow of current – the negatively charged electron moves in the direction of the positive terminal of a battery. The electron must therefore feel a force, which is due to the presence of an electric field.

> **Key point**
>
> A force field is represented by field lines:
> - The field lines point in the direction an object would move in the field.
> - The closer together the field lines, the stronger the field.

1.2.1 Electric field patterns

At Higher Physics level, specific electric fields will be considered, namely the field caused by a point charge, a pair of point charges (called a 'dipole' if the charges are opposite) and parallel plates (like a capacitor). You will be expected to be able to sketch these field lines. It is important to remember that the field lines point in the direction a positive charge would move.

> **Key point**
>
> The field lines of electric fields always point in the direction a positive test charge would move.

Experiment 1.2.1 Electric field patterns

This experiment investigates the electric field produced by a point charge, point charges and parallel plates.

Apparatus

- EHT power supply*
- Castor oil
- Semolina seeds
- Electrodes with various shapes

* A Van de Graaff generator can be used instead of the EHT power supply to give the electric charge required.

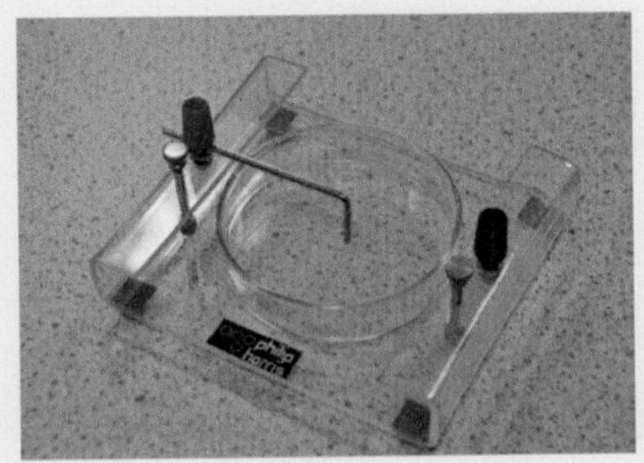

Fig 2.3

(continued)

Instructions

1. Set up the experiment as shown in Figure 2.3, taking great care when connecting the point charged rod to the EHT supply (not shown).
2. Fill the dish with a layer of castor oil that is approximately 0.5–1.0 cm deep. Sprinkle a layer of semolina grains onto the surface of the oil.
3. Connect the electrode with the shape of a point charge to the power supply and switch on, raising the voltage to between 3 kV and 5 kV. Take great care, very high voltages are used in this experiment!
4. Carefully examine the field lines. Copy a diagram of the field lines you see into your jotter.
5. Repeat with an electric dipole and two parallel plates, shown in Figures 2.4 and 2.5.

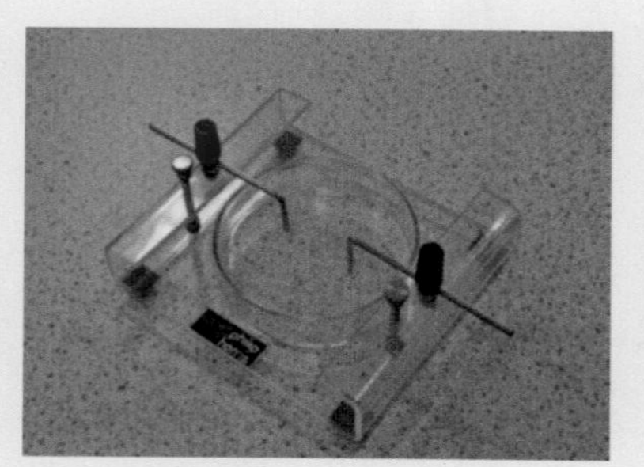

Fig 2.4

Fig 2.5

Point charge

A point charge is an object in space where all of the charge can be considered to be at one single point. When gravitation is considered, it is assumed that planets and stars are point masses. The electric field for a point charge is very similar to the gravitational field for a point mass. The field pattern from a point charge is shown in Figure 2.6.

When an electric field is sketched on a diagram, the arrows always point in the direction that a positive test charge will move. In other words, the field lines always point towards a negative charge. On a diagram, the electric field lines can be sketched from a point charge as shown below:

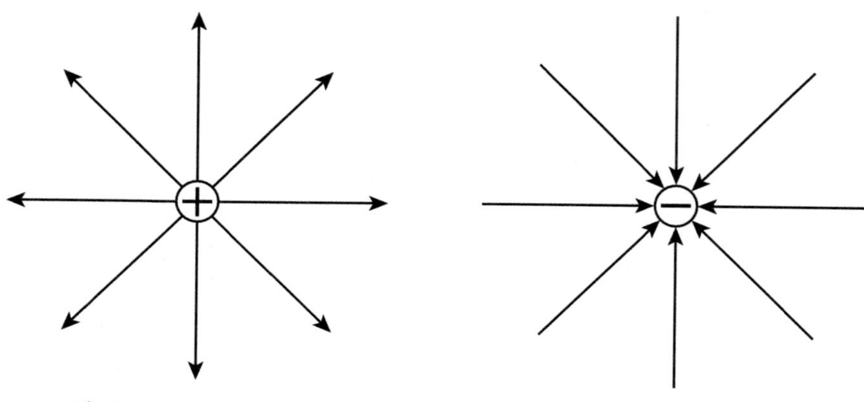

Fig 2.6

Two like charges

Two point charges that are either both positive or both negative produce an electric field similar to the one shown in Figure 2.7.

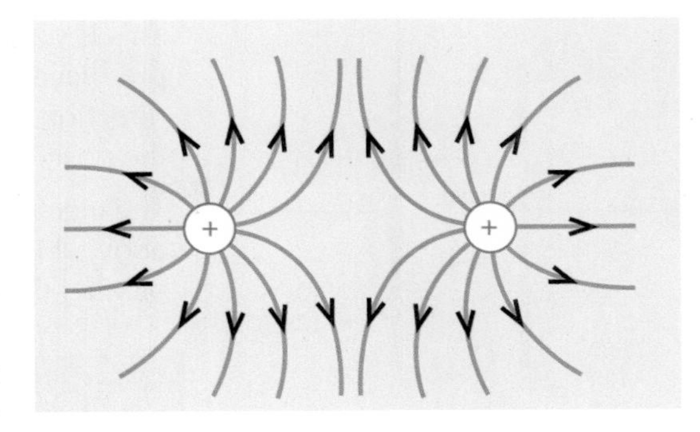

Fig 2.7

This diagram highlights an important effect of two like charges. There is a region in space where there are no field lines, i.e. where the electric field produced by the two charges is zero. The field from one charge cancels out the field from the other. If you consider the movement of a positive test charge, this will make sense; a positive test charge placed midway between the point charge above will be repelled by both charges by the same amount but in opposite directions, resulting in the effects cancelling each other out.

Fig 2.8

The electric dipole

An electric dipole consists of a positive point charge and a negative point charge separated by a set distance in space as shown in Figure 2.8. Field lines for a dipole are shown in Figure 2.9, which assumes that the charge of the positive and negative charges has the same magnitude.

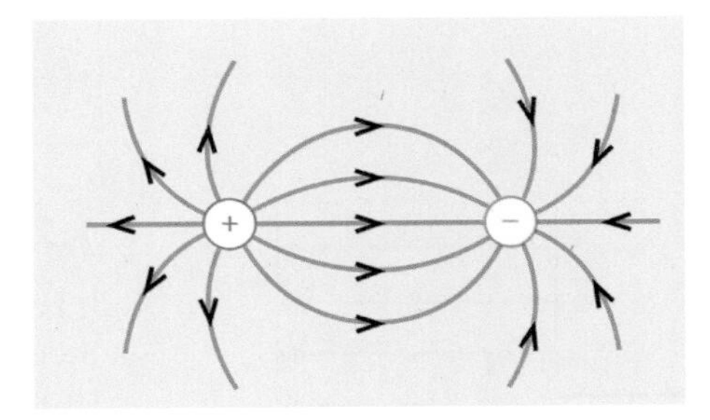

Fig 2.9

Parallel plates

In Area 3 on electricity, a parallel plate capacitor will be investigated. This sets up an electric field between two flat parallel plates by storing charge on the plates, which results in energy being stored. The diagram below shows a simple schematic of the parallel plate set-up.

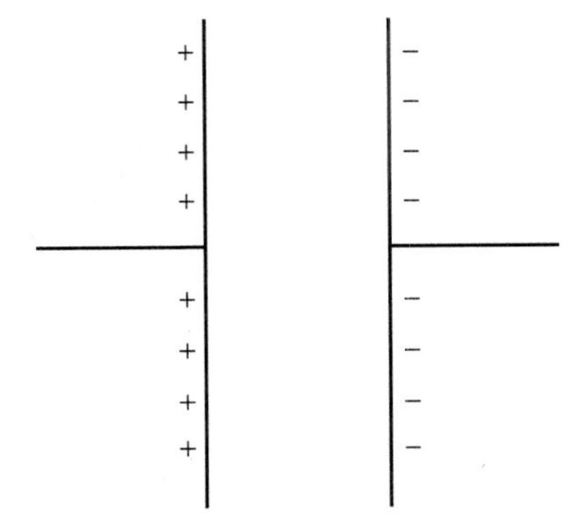

When parallel plate electrodes are used to produce an electric field, with grains in castor oil to view the field produced, the effect shown in Figure 2.10 is produced.

The grains show the electric field generated by two parallel plates. Two interesting effects can be seen for the electric field from parallel plates:

- The field is uniform between the two plates: equally spaced field lines that are parallel to each other.
- Around the edges of the plates there is a fringing field where the field lines are curved.

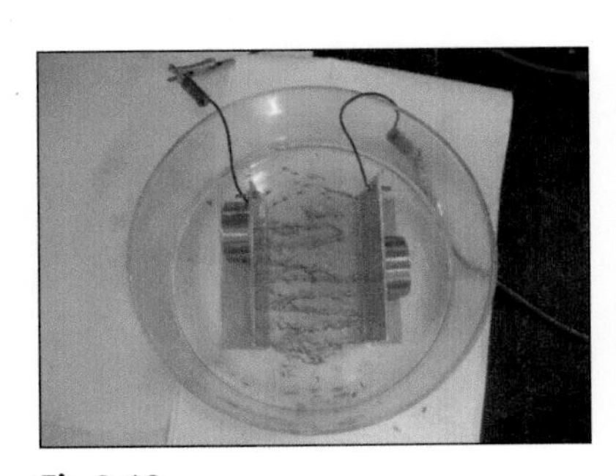

Fig 2.10

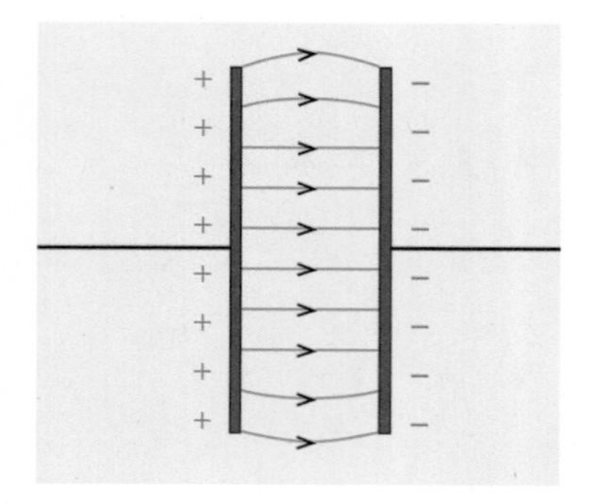

Fig 2.11

This is summarised in Figure 2.11, which shows the electric field produced by parallel plates. Notice that the field lines point in the direction that a positive test charge would move (i.e. away from the positive plate and towards the negative plate).

The electric field lines are equally spaced, which, as described above, highlights that the strength of the field is uniform between parallel metal plates.

Key point

The electric fields from various charge arrangements are shown in Figure 2.12. Remember that the closer together the field lines, the stronger the electric field.

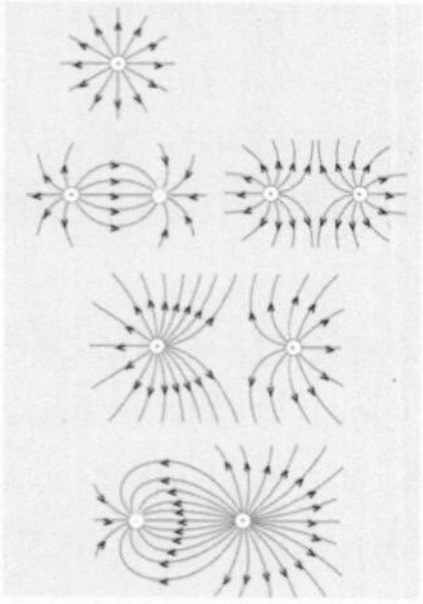

Fig 2.12

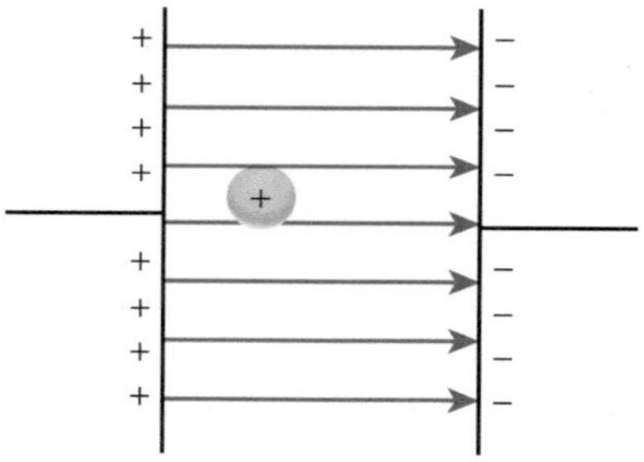

Fig 2.13

1.2.2 Electric field strength

It has been shown above that the strength of a field is represented by the proximity of the field lines. The closer together the field lines, the stronger the field. This demonstrates that for a point charge, the strength of the electric field reduces with distance away from the charge. For parallel plates, the strength of the electric field is uniform between the plates.

An electric field causes a charged particle to move – a positive charge will be attracted by a negative charge, for example. Consider placing a positive charge between two plates as shown in Figure 2.13. The positive charge is repelled by the positive plate and attracted to the negative plate, and will move in the direction of the field lines. A negative charge will move in the opposite direction to the field lines. The movement of a charge is due to a force acting on the charge. Like gravity, this force acts at a distance due to the presence of the electric field. The strength of the electric field will affect the force felt by the charge – the stronger the electric field, the greater the force. This leads to the definition of electric field strength, E as:

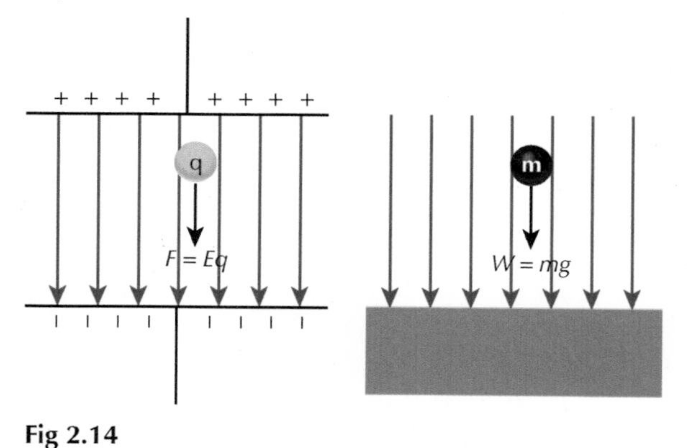

Fig 2.14

$$E = \frac{F}{q}$$

The electric field strength is the force, F, per unit charge, q, measured in NC^{-1}. Rearranging the above equation gives the force felt by a charged particle:

$$F = Eq$$

This equation is very similar to the equation for the force produced by a gravitational field:

$$F = W = mg$$

For a gravitational field, the size of the force depends on the mass of the particle and the gravitational field strength. For an electric field, the size of the force depends on the charge of the particle and the electric field strength.

Millikan's oil drop experiment

Robert Millikan devised an experiment in 1917 to measure the charge on an electron. The experiment involved balancing the force of the gravitational field acting on a droplet of oil with the force of an electric field. The arrangement shown in Figure 2.15 was used by Millikan to carry out this experiment.

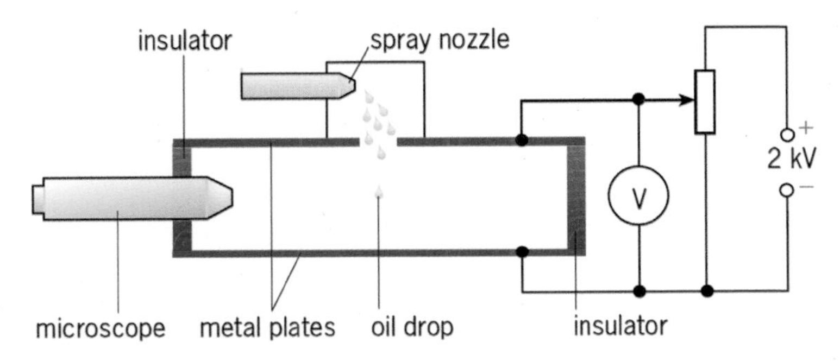

Fig 2.15

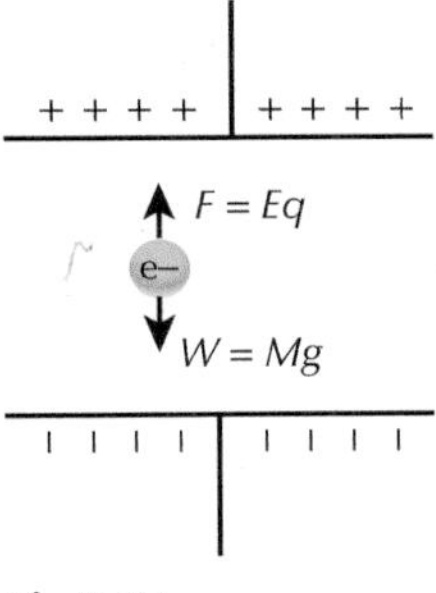

Fig 2.16

The density of the oil was known, so by measuring the size of the oil drop using a microscope, Millikan could estimate its mass and accordingly, its weight. A voltage across the plates produced an electric field, E. The voltage was increased until the oil drop was held stationary, at which point the force of gravity acting downwards was balanced by the electric field acting upwards:

This is:

$$qE = mg$$

Knowing the mass of the oil drop, the value of g on Earth and the electric field strength produced by the voltage across the plate, the charge of the oil drop could be calculated using:

$$q = \frac{mg}{E}$$

Millikan found that the charge was always an integer multiple of 1.6×10^{-19} C, which was concluded to be the charge on an electron.

Worked example

A point charge of +5 nC is placed in an electric field of strength 400 NC^{-1}. Calculate the force acting on the charge.

The force acting is given from the definition of electric field strength:

$$E = \frac{F}{q}$$

Re-arrange for force, and substitute the values given to find the magnitude of the force acting:

$$F = Eq$$
$$F = 5 \times 10^{-9} \times 400$$
$$F = 2 \times 10^{-6}\, N$$

Worked example

An oil droplet has a charge of -6.4×10^{-19} C and is suspended between two electric plates by an electric field strength of 250 NC^{-1}. The oil droplet is in a gravitational field.

Calculate the mass of the oil droplet.

The oil droplet is held stationary by balanced forces – the gravitation force balanced the electric field force, hence:

$$Eq = mg$$

Rearranging this for mass and substituting what you know, you can solve for the mass of the oil droplet:

$$m = \frac{Eq}{g}$$
$$m = \frac{(250)(6.4 \times 10^{-19})}{9.8}$$
$$m = 1.63 \times 10^{-17}\, kg$$

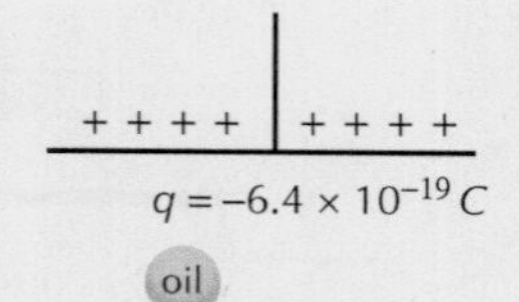

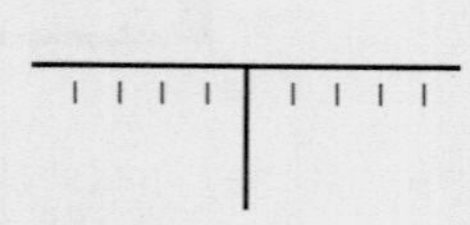

Fig 2.17

Exercise 1.2.2 Electric field strength

1. The force produced by an electric field acting on a charged particle is said to be a force acting at a distance. Explain what is meant by this statement.
2. A charged particle has a charge of 6 μC. Calculate the strength of the electric field required to produce a force of 2 mN on the particle.
3. Calculate the charge of a particle that experiences a force of 3 N in an electric field of 500 kNC^{-1}.
4. A charged particle has a charge of 15 nC. It is placed in an electric field with a strength of 1000 NC^{-1}. Calculate the force acting on the particle.

5 In 1917, Robert Millikan conducted an experiment to investigate the charge of an electron. He set up the experiment shown in the diagram below.

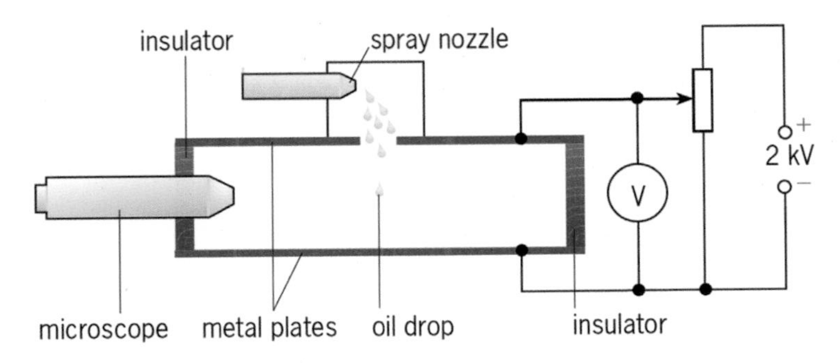

Fig 2.18

a) State the two forces that are acting on the oil drop in the above experiment.

b) By considering the forces acting on an oil drop that is held stationary, prove that the charge of a particle is given by the following equation:

$$q = \frac{mg}{E}$$

c) Millikan found that the magnitude of the charge of each oil drop was an integer multiple of 1.6×10^{-19} C. Explain the significance of this result.

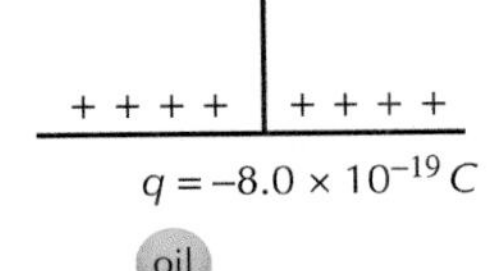

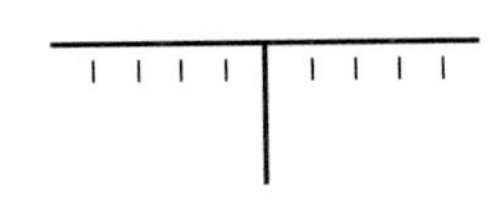

Fig 2.19

6 An oil particle has a mass of 1.9×10^{-12} kg and a charge of 8.0×10^{-19} C. It is held stationary in an electric field as shown in Figure 2.19.

Calculate the strength of the electric field required to hold the charged particle stationary between the plates.

7 A charged particle has a mass of 3 μg and a charge of 25 nC. It is held stationary in an electric field of strength 40 NC^{-1} as shown in Figure 2.20.

a) Calculate the force required to hold the particle stationary.

b) If the particle is released, calculate the initial acceleration of the particle towards the negative plate.

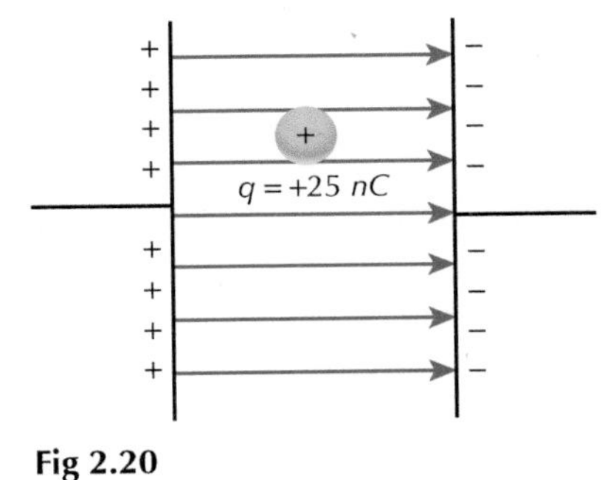

Fig 2.20

1.2.3 Movement in an electric field

In the section above, it was shown that charged particles experience a force in an electric field. The magnitude of this force is given by the equation:

$$F = Eq$$

where E is the electric field strength and q is the magnitude of the charge on the particle. This force will cause a charged particle to move in an electric field.

Experiment 1.2.3 The Teltron tube

This experiment investigates the motion of electrons in an electric field using a Teltron tube. The path of electrons released by the cathode can be seen on the fluorescent screen of the Teltron tube. An electric field can be applied to the region in space where the screen sits to change the path of the electrons.

Apparatus

- A high voltage (EHT) power supply
- Teltron tube
- (Optional Helmholtz coils)

!! Very high voltages are used in this experiment. Great care must be taken when setting up and operating the equipment used !!

Instructions

1 Connect the Teltron tube as shown in Figure 2.21, ensuring that the cathode is connected to a low voltage supply and that the plates are connected to the high voltage supply. A high voltage supply should also be connected to the anode to ensure that the electrons released are accelerated towards the screen region.

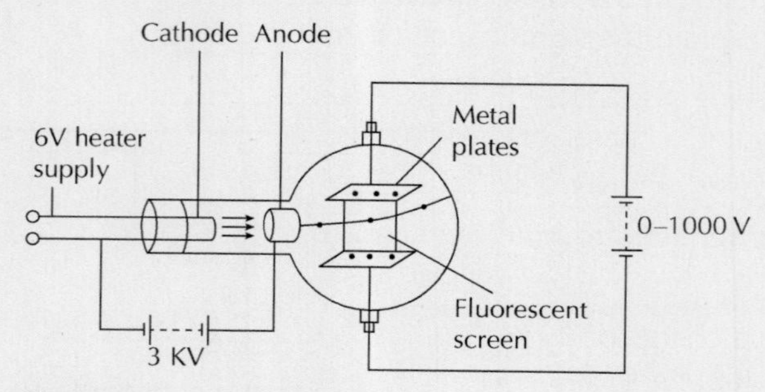

Fig 2.21

2 Carefully switch on the heater supply. This will begin the release of electrons from the cathode.

3 Switch on the high voltage supply to the anode to accelerate the electrons towards the screen. You should see a straight line going across the screen, which shows the path taken by the electrons as they move across the screen region. There is no electric field in the region of the screen so the electrons should pass undeflected.

4 Switch on the power supply to the metal plates. Increase the voltage slowly. Describe what happens to the path followed by the electrons between the metal plates. Can you explain this in terms of the electric field and the force acting on the charged particle?

Work done in an electric field

A charged particle in an electric field will feel a force due to the field, given by:

$$F = Eq$$

If a charged particle is moved against the electric field, then work must be done. It has already been shown that applying a force, F, over a distance, d, corresponds to work:

$$E_W = Fd$$

This equation applies if the strength of the field is constant, meaning the force is constant. If a particle moves through a field, attracted to its opposite charge, then work is not done to the particle. Instead, it loses potential energy and gains kinetic energy as it is accelerated through the field.

The work done in an electric field is analogous to the work done in a gravitational field, as the comparison below highlights. Consider moving a charge, q, through an electric field across a potential difference, V. Also consider moving a mass, m, through a gravitational field to a height, h.

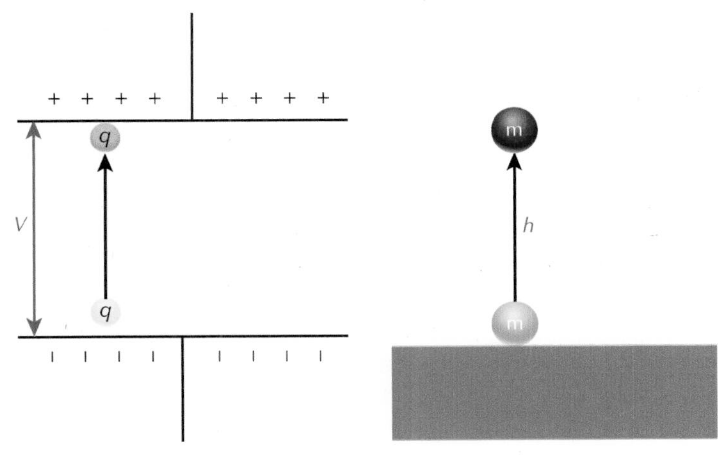

Fig 2.22

The work done lifting the mass to a height, h, is given by:

$$E_W = Fd$$
$$E_W = mgh$$

This is equivalent to the gravitational potential energy gained by the mass when it is lifted:

$$E_P = mgh$$

The work done moving the charge through the potential difference is given by:

$$E_W = QV$$

where Q is the charge of the particle and V is the potential difference it is moved through. This is analogous to the gravitational field example, where charge is equivalent to mass.

Defining the volt

Potential difference is measured in volts. Often, 'potential' is given to mean gravitational potential energy. When an object moves from one height to another, its gravitational potential energy changes – it effectively moves across a potential difference. In electricity, when a charged particle moves from one place in an electric field to another,

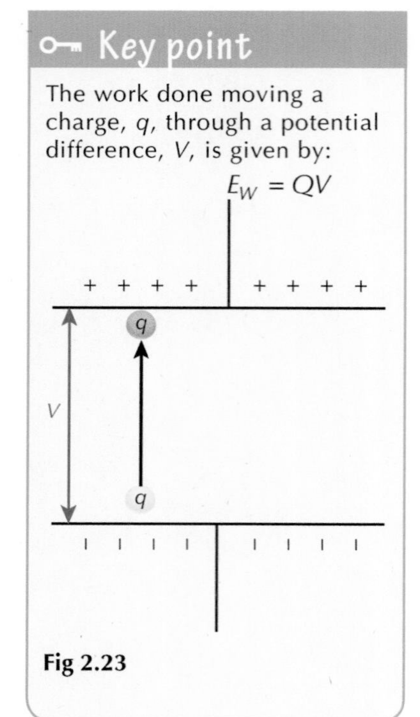

Fig 2.23

it will also move across a potential difference, which is measured in volts. The volt therefore refers to a change in energy.

The volt can be defined using the equation below:

$$E_W = QV$$

Rearranging this for voltage gives the following:

$$V = \frac{E_W}{Q}$$

Here, the link of voltage to energy becomes clear. The volt is defined as the energy per unit charge (JC^{-1}). In the context of an electrical circuit, a voltage across a bulb of, say, 12 V would mean that for every coulomb of charge passing through the bulb, 12 J of energy are transferred in the bulb. In the context of a charge moving, the volt is defined as the energy gained or lost by each coulomb of charge moving across the potential difference. The volt was named after Italian physicist Alessandro Volta (Figure 2.24) who is credited with inventing the voltaic pile, which was the first battery that could supply a constant current to a circuit.

Fig 2.24

Key point

The volt can be defined as the energy per unit charge,

$$V = \frac{E_W}{Q}$$

Converting work to kinetic energy: moving particles

A particle can be accelerated by an electric field as it is attracted to its opposite charge. In this case, the particle gains kinetic energy when moving across a potential difference, V. The amount of kinetic energy gained by the charged particle is equal to the work done by the field on the particle, given by:

$$E_W = QV$$

Setting this equal to the equation for kinetic energy:

$$E_K = \frac{1}{2}mv^2$$

an equation for the speed of a particle can be derived as follows:

$$\frac{1}{2}mv^2 = QV$$

$$v^2 = \frac{2QV}{m}$$

$$v = \sqrt{\frac{2QV}{m}}$$

It is important to note here that the final velocity of the charged particle depends only on the mass and charge of the particle and the strength of the electric field. It does not depend on the separation of the plates.

Consider an electron being accelerated between two plates as shown in Figure 2.25. The electron is attracted to the positive plate, moving in the direction of the red arrow. It will cross a potential difference, V, between the plates. The electron therefore gains energy:

$$E = E_W = eV$$

where e is the charge on an electron. This energy is converted to kinetic energy, so the speed of the electron when it hits the positive plate is given by applying energy conservation:

$$eV = \frac{1}{2}mv^2$$

This can be rearranged to give:

$$v = \sqrt{\frac{2eV}{m}}$$

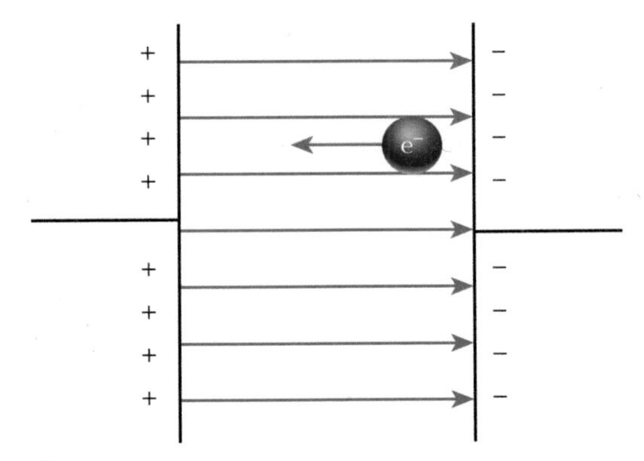

Fig 2.25

Key point

When a charged particle is accelerated through a potential difference, the energy gained is converted to kinetic energy. Energy is conserved, so the speed of the charged particle is given by energy conservation to be:

$$\frac{1}{2}mv^2 = QV$$

$$v^2 = \frac{2QV}{m}$$

$$v = \sqrt{\frac{2QV}{m}}$$

Worked example

A large charged particle has a mass of 1.7×10^{-5} kg and a charge of + 5 mC. The particle is accelerated across a potential difference of 2 kV. Calculate the resultant speed of the particle.

The work done to the particle moving through the potential difference is found from:

$$E_W = QV$$

Here you have:

$$E_W = (5 \times 10^{-3})(2 \times 10^3)$$

$$E_W = 10\ J$$

This energy is then converted to kinetic energy by energy conservation, giving:

$$10 = \frac{1}{2}mv^2$$

Substituting the mass of the particle, you can now find the velocity:

$$10 = \frac{1}{2}(1.7 \times 10^{-5})v^2$$

$$v^2 = \frac{2 \times 10}{(1.7 \times 10^{-5})}$$

$$v^2 = 1{,}176{,}471$$

$$v = 1085\ ms^{-1}$$

Exercise 1.2.3 Moving charges in an electric field

1 A woman places her hands on the dome of a Van de Graaff generator while the dome is charged. Her hair stands on end as shown in Figure 2.26.

Explain this effect in terms of the electric charge on the dome of the generator and on the strands of hair on the woman's head.

Fig 2.26

2 In an experiment to demonstrate the motion of charges in an electric field, a Teltron tube is connected to a high voltage supply to accelerate electrons. The path of the electrons is shown on a phosphorescent screen. When an electric field is applied across the screen, the path of the electrons is found to follow the path shown in Figure 2.27.

a) State which plate is positively charged. Explain your answer.

b) Sketch the parallel plates and the screen, and include the electric field lines that produce the motion of the electron shown in Figure 2.27.

c) Explain the motion of the electrons observed in Figure 2.27.

Fig 2.27

3 A positive charge of +2 nC is to be moved towards a positive plate in electric field as shown in Figure 2.28. The potential difference between the plates is held constant by a power supply.

a) Explain why work must be done to move the charge towards the positive plate.

b) The above example is equivalent to lifting a mass up to a given height. Explain this equivalence. What is mass equivalent to?

c) The charge is moved across a potential difference of 25 kV. Calculate the work done moving the charge across this potential difference.

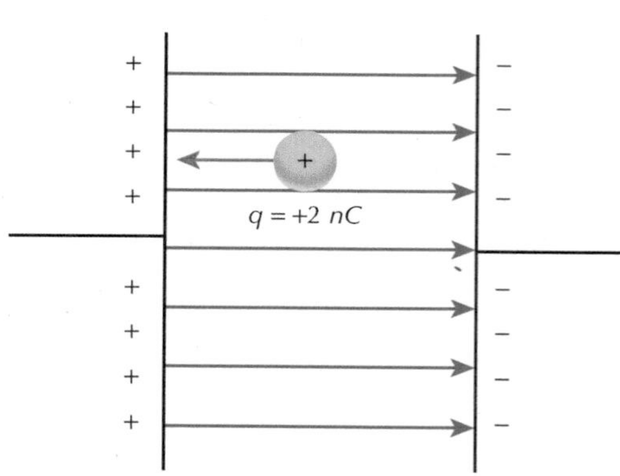

Fig 2.28

4 Define the volt.

5 State what is meant by a *potential difference of 200 V*.

6 In an experiment to investigate the effects of electric fields on charged particles, alpha particles (which have a positive charge of 3.2×10^{-19} C and a mass of 6.64×10^{-27} kg) are accelerated from rest across a potential difference of 5 kV.

a) Calculate the kinetic energy gained by the alpha particle moving across the potential difference.

b) Find the speed of the alpha particle after it is accelerated across the potential difference.

c) The experiment is repeated using a smaller potential difference. Describe and explain the effect this change has on the speed of the particle after being accelerated.

7 An electric field is used to accelerate an electron from rest to a speed of 2.5×10^6 m s^{-1}. Given the mass of an electron is 9.11×10^{-31} kg and it has a charge of 1.6×10^{-19} C, calculate the voltage required to accelerate the electron.

1.2.4 Applications of electric fields

As shown above, electric fields can be used to change the path of charged particles, and either accelerate or decelerate particles. Here, two main applications of electric fields are described.

Cathode ray tube (CRT)

A cathode ray tube contains one or more electron guns used to accelerate the electrons to a high speed towards a fluorescent screen. They were used in televisions and computer monitors (CRT monitors) before the introduction of plasma and LCD screens. A CRT monitor or television can be recognised by its shape – rather than being flat, they have a greater depth. Contained within is the cathode ray tube.

Fig 2.29

Figure 2.30 shows the inside of a typical CRT monitor.

Electrons from the cathode at the back of the tube are accelerated towards the anode by a high voltage. This accelerates the electrons to a very high speed given by:

$$v = \sqrt{\frac{2eV}{m}}$$

The electrons are then deflected towards the correct part of the fluorescent screen by electric and magnetic fields.

Fig 2.30

Worked example

A cathode ray tube is used to accelerate electrons to a high velocity. Electrons produced at the cathode are accelerated towards the anode by a potential difference, V, which produces a uniform electric field. In one cathode ray tube, the distance between the cathode and the anode is 25 cm and the accelerating voltage is 1.5 kV.

Given the mass of an electron is 9.11×10^{-31} kg and it has a charge of 1.6×10^{-19} C, calculate:

a) The speed of the electron when it reaches the anode.

b) The acceleration of the electron between the cathode and the anode (assume constant and u = m s^{-1}).

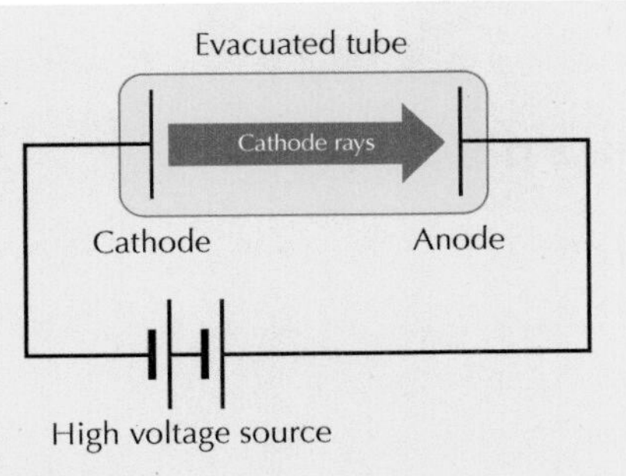

Fig 2.31

a) *First of all, apply energy conservation to find the velocity of the electron when it reaches the anode:*

$$QV = \frac{1}{2}mv^2$$

$$v = \sqrt{\frac{2QV}{m}}$$

$$v = \sqrt{\frac{2 \times (1.6 \times 10^{-19})(1500)}{(9.11 \times 10^{-31})}}$$

$$v = \sqrt{5.27 \times 10^{14}}$$

$$v = 2.30 \times 10^{7}\,ms^{-1}$$

b) *The acceleration of the electron between the cathode and the anode is given using the equations of motion because the acceleration is constant. Write down the 'suvat' table:*

s	u	v	a	t
0.25	0	2.30×10^7	?	X

Table 2.1

Choose the equation of motion that does not contain time:

$$v^2 = u^2 + 2as$$

Substitute and solve for the acceleration of the electron:

$$(2.30 \times 10^7)^2 = 0^2 + 2a(0.25)$$

$$(2.30 \times 10^7)^2 = 0.5a$$

$$a = \frac{(2.30 \times 10^7)^2}{0.5}$$

$$a = 1.05 \times 10^{15} ms^{-2}$$

Fig 2.32

The inkjet printer

An inkjet printer works by spraying tiny droplets of ink (volumes measured in pico-litres) at the page from a print head, which is moved side-to-side as the paper moves through. The ink droplets have to hit the paper in a very precise place to accurately represent what is being printed. Electric fields can be used to control the direction of the ink droplet so that it reaches the correct place on the page.

Inside the printhead, each tiny droplet of ink is given an electric charge before entering a region of an electric field set up by two parallel plates as shown in Figure 2.32.

The magnitude and direction of the electric field can be changed to change the direction in which the ink droplet moves. In practice, the electric field is usually held constant and the charge added to the ink droplet is changed. This also changes the force experienced by the ink droplet in the region of the field according to the equation,

$$F = Eq$$

A change in force results in a change in vertical acceleration and hence adjusts the path followed by the ink droplet.

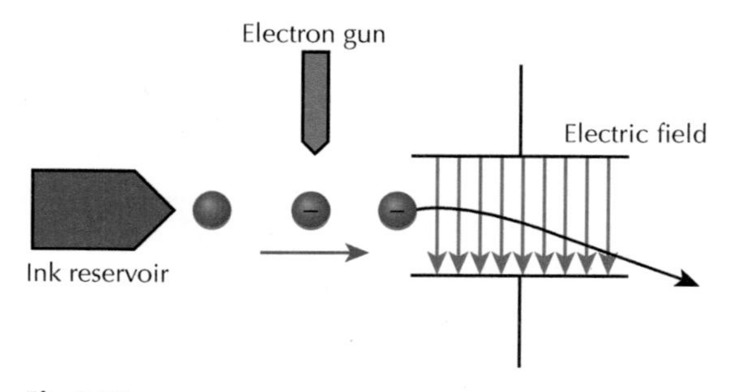

Fig 2.33

Exercise 1.2.4 Applications of electric fields

1 An electron is fired between two plates and follows the path shown in Figure 2.34. For each diagram, copy the plates and the path followed. Include the electric field lines in your sketch, including arrows to show their direction.

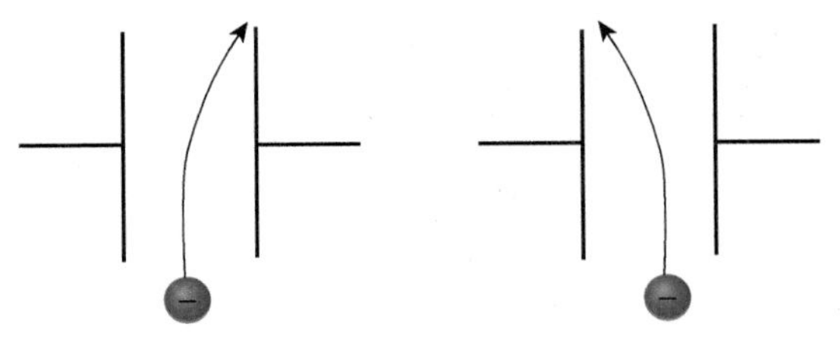

Fig 2.34

2 In a CRT monitor, electrons are accelerated through a potential difference of 750 V between the cathode and the anode. They then enter a region of a constant electric field as shown in Figure 2.35.

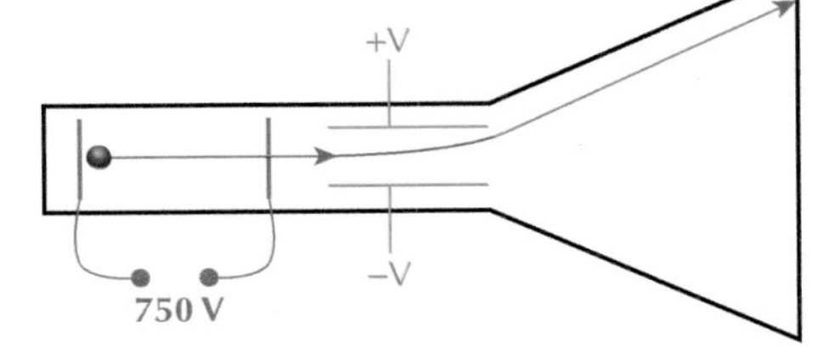

Fig 2.35

a) Assuming the electrons start from rest, calculate the speed of the electrons after being accelerated through the 750 V potential difference.

b) If the distance between the cathode and the anode is 40 mm, calculate the acceleration of the electrons.

c) Describe and explain the motion of the electrons between the parallel plates after the anode.

d) How could the size of the deflection be changed so that the electrons strike the screen closer to the centre?

3 In an old television, the picture is produced using a cathode ray tube and a phosphorescent screen. Electrons are accelerated across a potential difference of 600 V and then deflected by two parallel plates as shown in Figure 2.36.

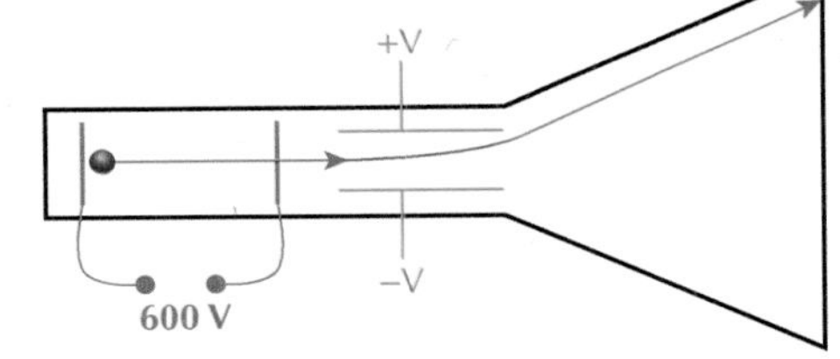

Fig 2.36

a) Calculate the speed of the electrons after being accelerated across the potential difference.

b) If the length of the parallel deflecting plates is 20 cm, calculate the time taken for the electrons to cross the region of the plates.

4 The diagram below shows how an inkjet printer shoots ink at a page during the printing process.

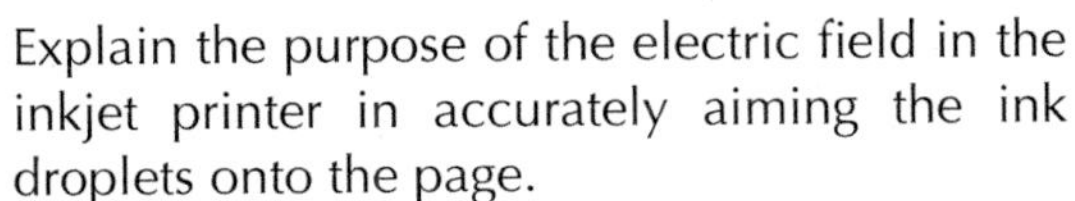

Explain the purpose of the electric field in the inkjet printer in accurately aiming the ink droplets onto the page.

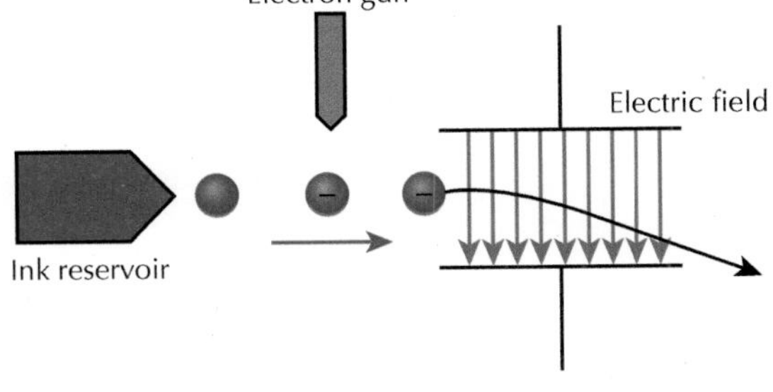

Fig 2.37

1.3 Moving charges and magnetic fields

Charged particles can be accelerated by an electric field, causing them to either start moving or change their velocity if they are already moving. The movement of a charged particle can also produce

Fig 2.38

a magnetic field. These magnetic fields can interact with other magnetic fields to produce interesting and very useful effects.

1.3.1 Magnetic fields

In the same way that electric and gravitational fields can produce a force at a distance, a magnetic field will produce a force at a distance on a magnetic particle. For example, a magnet can be used to pick up and hold drawing pins as shown in Figure 2.38. The Earth's gravitational field produces a force on the pins (due to their mass) that acts to pull them to the ground. The magnet produces a force that attracts the pins upwards towards the magnet.

Just like gravitational and electric fields, magnetic fields have field lines to show the strength and shape of the field.

Experiment 1.3.1 Magnetic fields

This experiment investigates different magnetic field patterns using iron filings and bar magnets.

Apparatus

- Iron filings
- Bar magnets
- Sheet of paper

Instructions

1. Place a single bar magnet underneath a sheet of paper.
2. Sprinkle iron filings onto the sheet of paper with the bar magnet underneath and observe the pattern. The iron filings line up to show the magnetic field lines.
3. Repeat with multiple magnets and combinations to see the different field lines produced. Sketch these field lines. Remember to clearly mark the nature of the magnetic poles on each magnet.

A bar magnet will produce a magnetic field that iron filings can be used to view. Notice that the magnetic field lines point from north to south (always away from north and towards south).

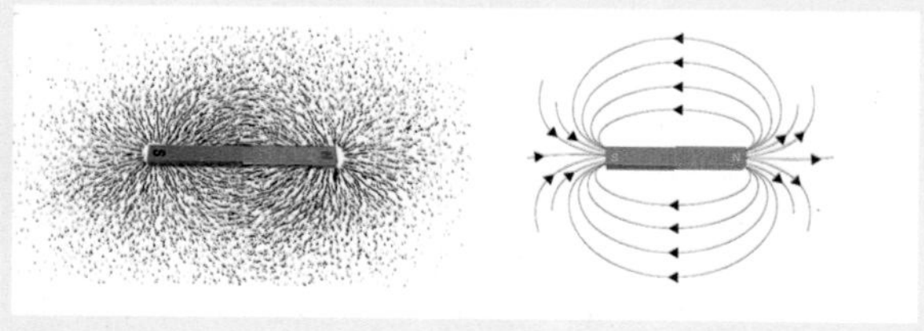
Fig 2.39

The field lines for two bar magnets are shown in Figure 2.40. The picture highlights that two like poles repel. For magnets, it is a case of opposites attract, just like charges in electrostatics.

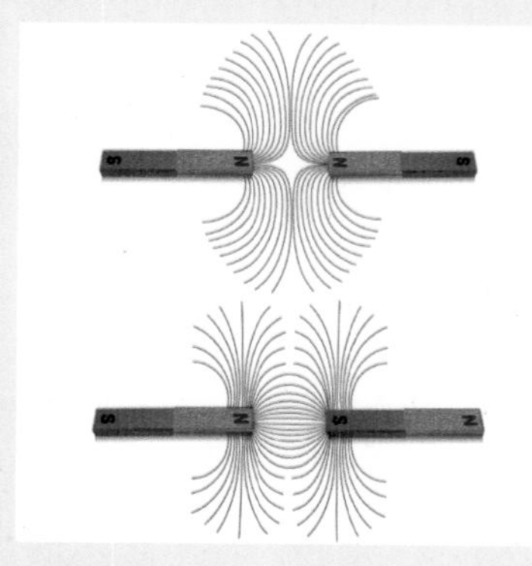
Fig 2.40

1.3.2 Producing a magnetic field (flow of charges)

A current-carrying wire produces a magnetic field around it. This discovery shows that there is a link between electricity and magnetism. This link is now utilised in devices such as the electric motor.

Key point

A magnetic field produces a force on magnetic particles, that acts at a distance. The field is represented by field lines that always point from north to south.

For magnetic fields, opposite poles attract.

Experiment 1.3.2 Current-carrying wire

This experiment demonstrates the magnetic field produced around a current-carrying wire using compasses.

Apparatus

- Wire (not insulated)
- DC power supply
- Compasses
- Sheet of card

Fig 2.41

Instructions

1. Use the wire to carefully pierce a hole in the middle of the card and slide the wire through the hole so that half is on one side and half is on the other, as shown Figure 2.41.
2. Using Figure 2.41 as a guide, position small compasses around the wire on the cardboard.
3. Connect the wire to a DC power supply and switch it on, so that a current flows through the wire. Observe what happens to the compass needles when the current is flowing through the wire.
4. Switch the direction of the current by switching the connections on the power supply – what happens to the compass needles?
5. The compasses can be substituted for iron filings – how do they appear around the wire when a current flows?

The magnetic field forms in a circle around the wire, circling clockwise for current in one direction and anti-clockwise for current in the other direction. The strength of the magnetic field decays with increased distance from the wire. This result is highlighted by the effect of placing compasses around a current-carrying wire as shown in Figure 2.42.

Figure 2.42(a) shows the compasses all pointing north in the absence of a magnetic field produced by the wire. In this case, the current is zero. Figure 2.42(b) shows the effect of passing a current through the wire. A circular magnetic field is produced, and this causes the compass needles to

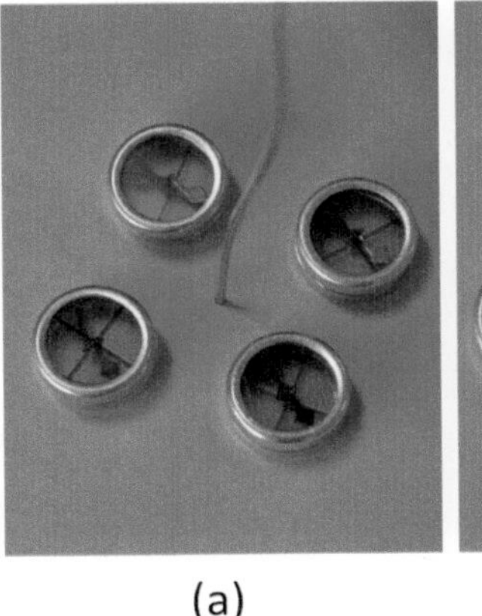

(a)

(b)

Fig 2.42

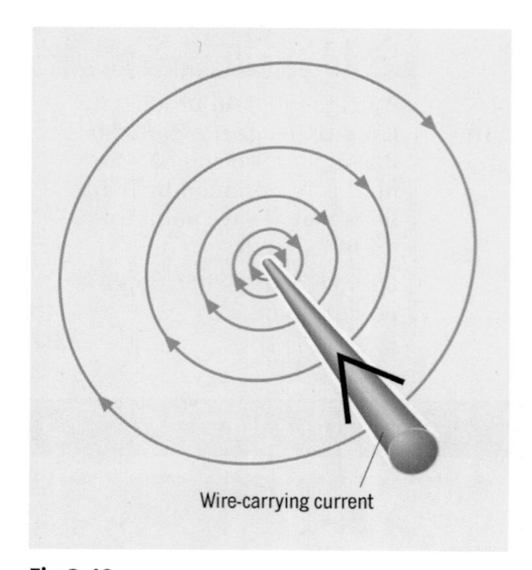

Fig 2.43

point round in a circle so that they are always pointing at 90° to the radius of the circle. This shows that the magnetic field lines around a current-carrying wire are in concentric circles, as shown in Figure 2.43.

To find the magnetic field produced by passing a current along a wire, the right hand grip rule can be used. This is demonstrated in Figure 2.44. It is essential to note that the direction of current assumed here is conventional current – that is, the flow of positive charge from positive to negative within the circuit. This is, however, simply a convention: current is actually the flow of electrons, but when dealing with magnetic fields, the assumption is that the conventional current flows from positive to negative.

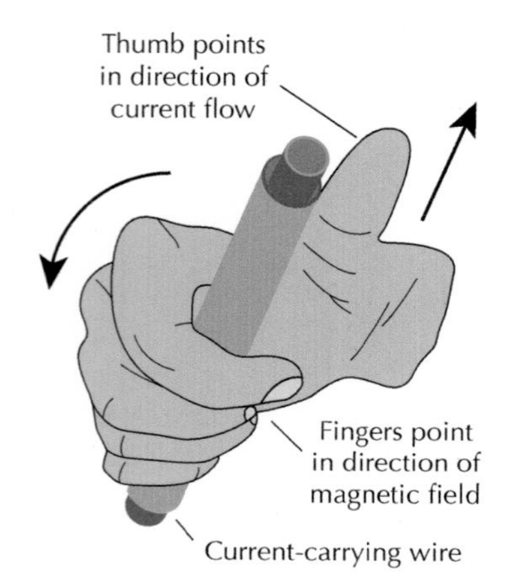

Fig 2.44

Key point

For discussion of magnetic fields produced by current flowing in a wire, always use conventional current, where current flows from positive to negative (as the flow of positive charges).

1.3.3 Charged particles in magnetic fields

A current flowing through a wire produces a magnetic field around the wire, as discussed above. The right-hand grip rule can be used to determine the direction and shape of the magnetic field lines. If this wire is placed into another magnetic field, the magnetic fields will interact. This will result in a force being experienced by the wire.

Key point

A current-carrying wire will produce a magnetic field around it – this magnetic field can be represented by the right-hand grip rule illustrated in Figure 2.44.

Experiment 1.3.3 Current-carrying wire in a magnetic field

This experiment investigates the effects of an external magnetic field on a current-carrying wire.

Apparatus

- Low voltage DC power supply*
- Wires
- Crocodile clips
- Aluminium strip
- Variable resistor
- Horseshoe or bar magnets

* A 1.5 V D-cell battery can also be used in place of the power supply. Take great care when attaching the wires to the end of the battery.

Instructions

1. Connect the aluminium strips to the power supply by using the crocodile clips and wires. Set up a series circuit, as shown in Figure 2.45, incorporating a variable resistor that can be used to control the current flowing in the circuit.
2. Place the aluminium strip between two bar magnets (aligned north and south as shown in Figure 2.46) or between the legs of a horseshoe magnet.
3. Set the power supply to 1 V DC and turn it on. Observe what happens to the aluminium strip.
4. Increase the voltage to 2 V DC. Again, observe what happens to the strip. How does the effect compare to step 3 above?
5. Reverse the connections on the power supply to reverse the current direction. How does this affect the above effects on the aluminium strip?

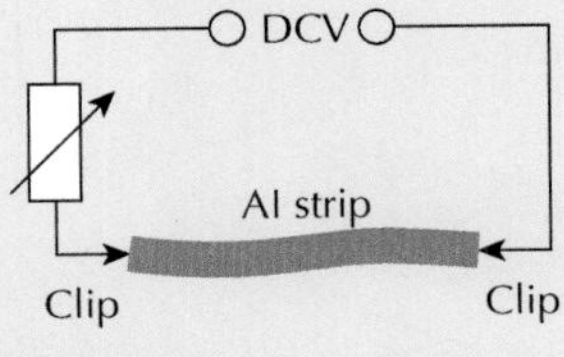

Fig 2.45

Fig 2.46

A current-carrying wire will experience a force in a magnetic field due to the interaction of its own magnetic field with the applied magnetic field. This causes the wire to experience a force and move. The size and direction of this movement depends on the strength of the external magnetic field and the direction and size of the current flowing in the wire.

Key point

A current-carrying wire will experience a force in a magnetic field due to the interaction of the magnetic field it produces with the external magnetic field.

Fig 2.47

Forces on charged particles

Just like charges flowing along a wire, a free charge moving in a magnetic field will experience a force.

Starting with the example above of the wire moving in a magnetic field, a law can be derived that links the force experienced by the wire, and the direction of the current and the magnetic field. This law was developed by Sir John Ambrose Fleming (Figure 2.47) and is called the 'left hand rule'. Importantly, as before, this rule assumes it is positive charges that move; the left hand rule works with conventional current.

Fleming's left hand rule links the current direction (convention current) to the magnetic field and direction the force experiences as shown in Figure 2.48 for a current-carrying wire.

The left hand shows:

- First finger: Field. The index finger shows the direction of the magnetic field, from north to south.
- Second finger: Current. The middle finger shows the direction of the current (conventional current!)
- Thumb: Motion (force). The thumb shows the direction of the force that causes the wire to move.

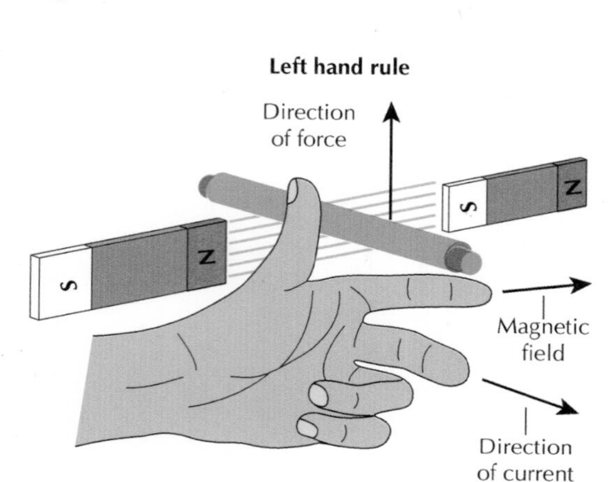

Fig 2.48

The three fingers should be held at right angles to each other as shown in Figure 2.48. Rotating the hand will allow the direction of the force, field and current to be determined for any system. This rule can be extended to moving singular charges. Again, Fleming's left hand rule applies to the flow of positive charges.

When considering the flow of negative charges (electrons), use the right hand rule. The fingers have the same meaning: first finger is the field; second finger is the current; thumb represents the direction of the motion (force) (see Fig. 2.49).

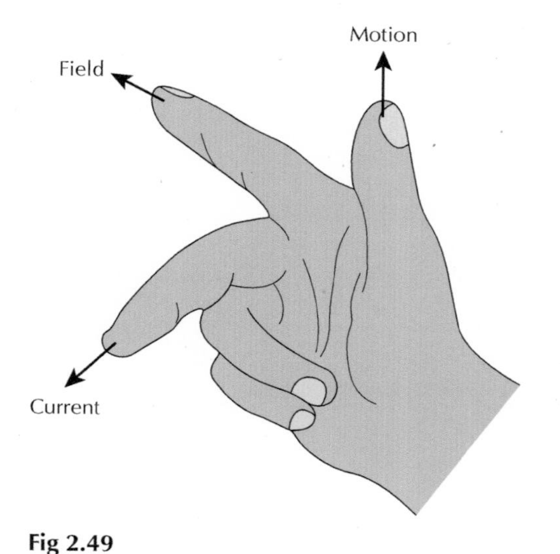

Fig 2.49

> **Key point**
>
> When a charge moves in a magnetic field, it experiences a force. If the charge is positive, the force, field and charge movement are linked by the left hand rule. If the charge is negative, then they are linked by the right hand rule.

Application: the CRT television

The CRT television has already been considered above for the movement of charges in an electric field. However, most cathode ray tube systems actually use magnetic fields to manipulate the direction of the electron. A typical CRT is shown in Figure 2.50.

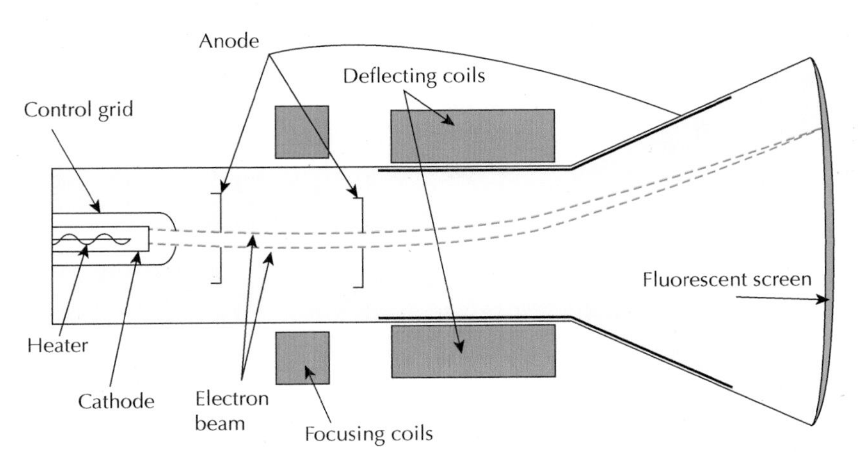

Fig 2.50

The electrons are accelerated as before across a potential difference. This beam of electrons is then controlled using focusing coils and deflecting coils. The right hand rule can be used to work out the direction of the magnetic field required to produce the desired deflection of the electron. The strength of the field can be varied to adjust the magnitude of the deflection. This is in the same way as the electric field causes a deflection of the electrons. A force due to the motion of the particles acts on the electron to change its velocity and hence its deflection.

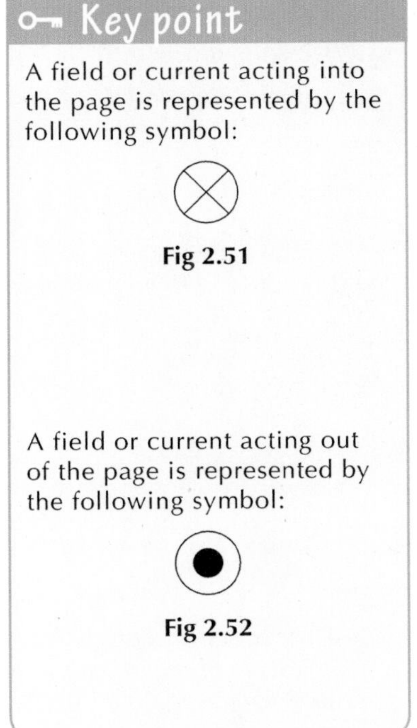

Key point

A field or current acting into the page is represented by the following symbol:

Fig 2.51

A field or current acting out of the page is represented by the following symbol:

Fig 2.52

Worked example

In the cathode ray tube shown below, electrons are accelerated across a potential difference of 1 kV and enter a region of uniform magnetic field. See Experiment 1.2.3.

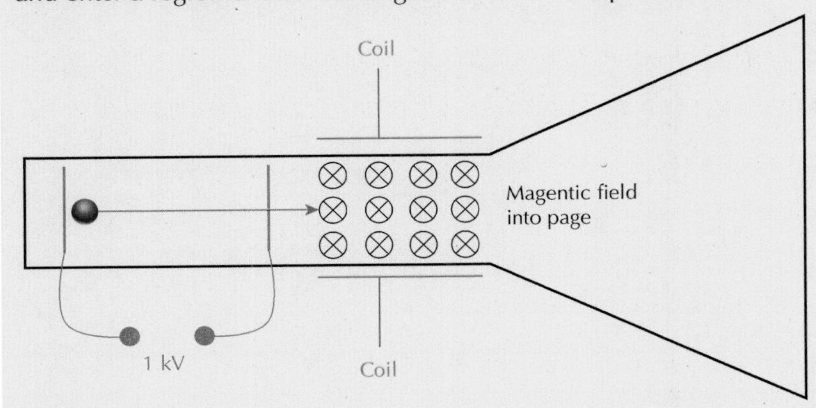

Fig 2.53

a) Calculate the velocity of the electron as it enters the region of the magnetic field.

b) Describe the motion of the electron as it moves through the magnetic field. You must state the direction in which the electron moves.

c) The field is reversed. Describe the effect this has on the motion of the electron.

a) *The velocity of the electron as it enters the region of the magnetic field is found by energy conservation. Find the work done accelerating the electron across the 1 kV potential difference,*

$$E_W = QV$$
$$E_W = (1.6 \times 10^{-19})(1000)$$
$$E_W = 1.6 \times 10^{-16} J$$

Set this equal to the equation for kinetic energy to find the velocity:

$$E_K = \frac{1}{2}mv^2$$
$$1.6 \times 10^{-16} = \frac{1}{2}(9.11 \times 10^{-31})v^2$$
$$v^2 = \frac{2 \times (1.6 \times 10^{-16})}{(9.11 \times 10^{-31})}$$
$$v^2 = 3.51 \times 10^{14}$$
$$v = 1.87 \times 10^7 ms^{-1}$$

b) *As the electron enters the region of the magnetic field, it is travelling to the right. The electron is negatively charged so use the right hand rule. The field is acting into the page. Using the right hand rule (diagram below), you find:*

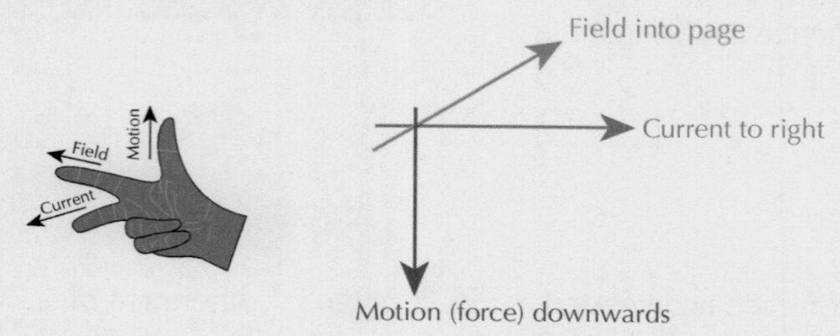

Fig 2.54

The force acts downwards, causing the electrons to be deflected downwards.

c) *Reversing the field so that it is now out of the page gives the following from the right hand rule:*

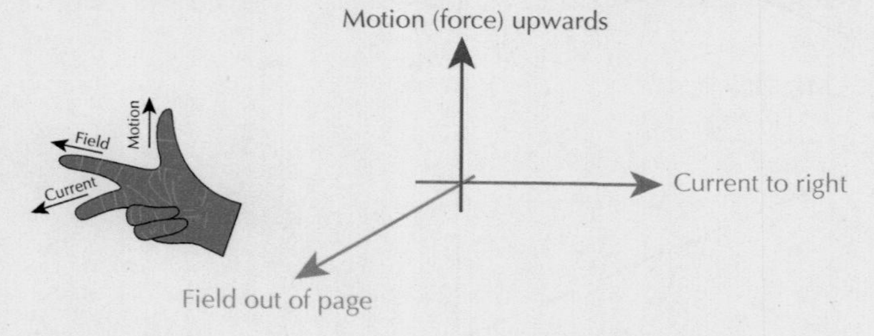

Fig 2.55

The deflection is now upwards as the force now acts upwards.

Exercise 1.3.3 Charged particles in magnetic fields

1 Students are investigating the magnetic fields produced by bar magnets using iron filings and paper.

a) Describe an experiment they could carry out to investigate the magnetic fields around bar magnets.

b) For each of the combinations of bar magnets below, draw the magnetic field lines that the students would observe from the bar magnets.

Fig 2.56

2 When a current passes along a wire, a magnetic field is formed around the wire. Apply the right-hand grip rule to draw a diagram of the wires shown in Fig 2.57 along with the shape and direction of the magnetic field around the wire.

3 A current-carrying wire is found to move when it is in an external magnetic field. Explain this effect in terms of the magnetic fields produced.

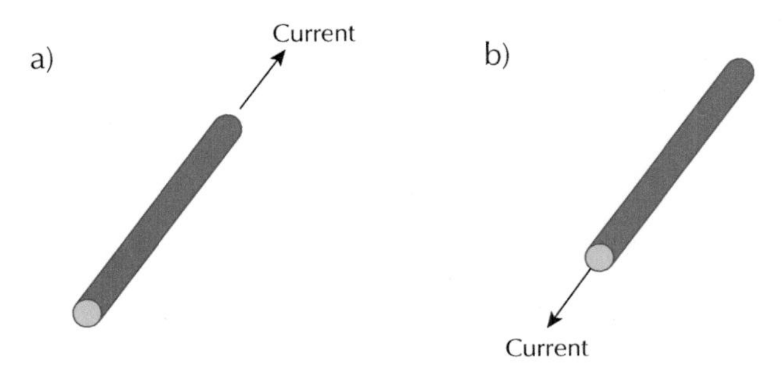

Fig 2.57

4 A student reading about current-carrying wires in a magnetic field sees the following diagram on a website.

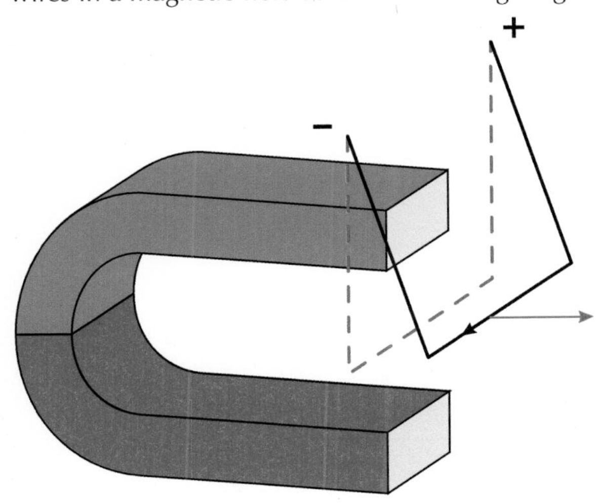

Fig 2.58

a) Explain the effect observed on the current-carrying wire.

b) By using the right hand or left hand rules, state the direction of the magnetic field in the above example (either blue to red, or red to blue).

c) Which of the two coloured bars represents the north pole in the diagram above?

d) Give two ways that the size of the wire's deflection could be increased.

(*continued*)

5 In a CRT oscilloscope, electrons from the cathode are accelerated to a maximum velocity before entering a region of a constant magnetic field produced by the deflecting coils. The electrons follow the path shown in the diagram below.

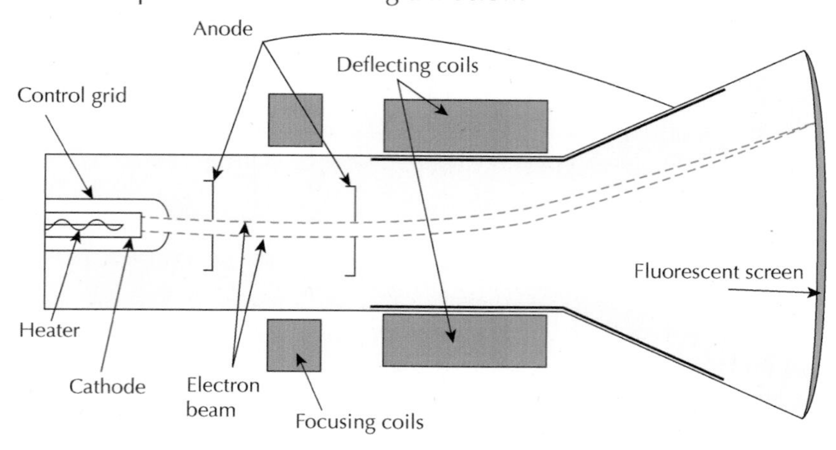

Fig 2.59

a) State the direction of the magnetic field produced by the deflection coils.

b) What path would positively charged protons follow through the deflection coils?

c) State how the deflection of the electrons could be reversed to the downwards direction.

6 For each of the diagrams below, state the direction of the deflection experienced by the charged particle.

a)

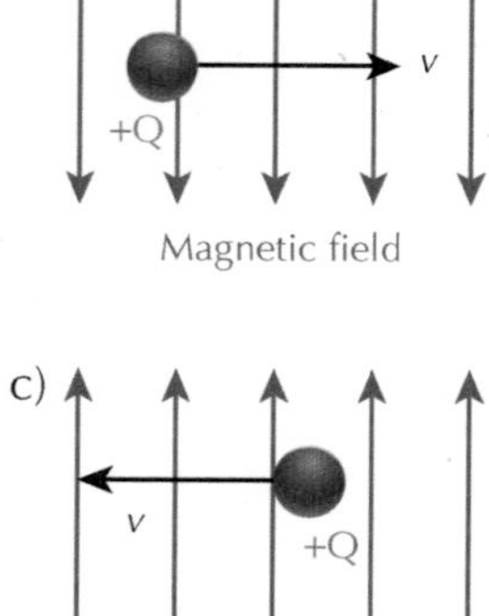

b)

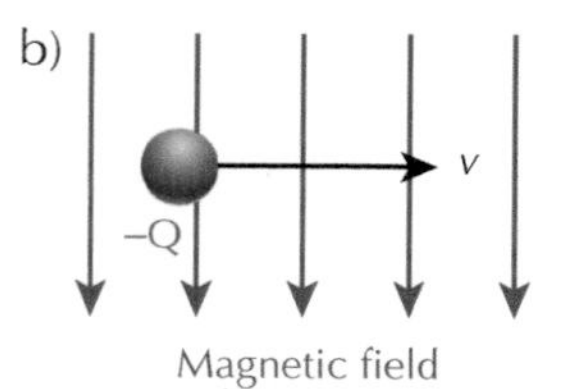

c)

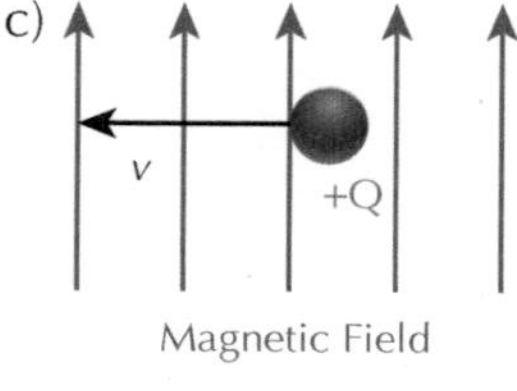

d)

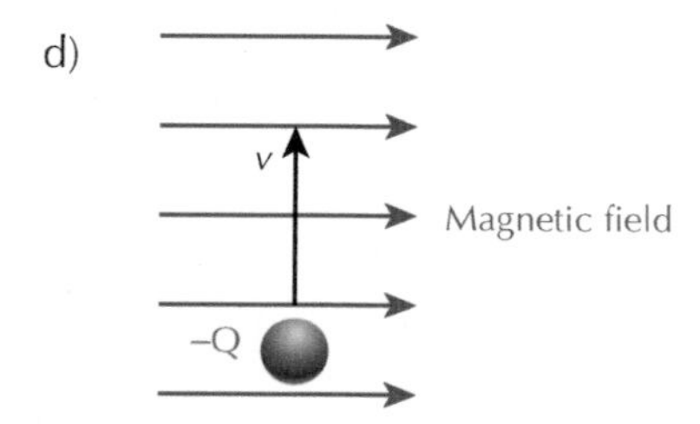

e)

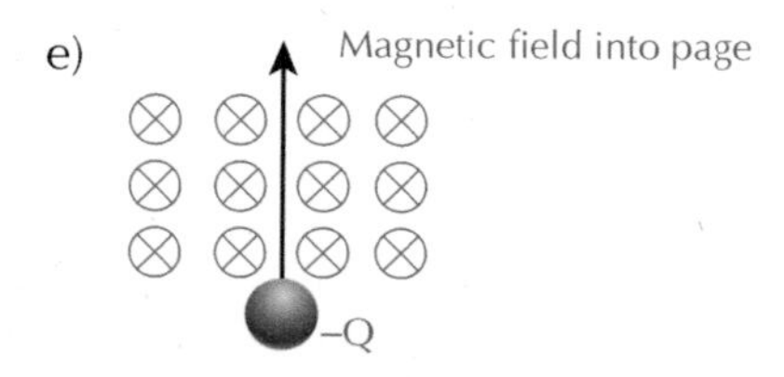

f)

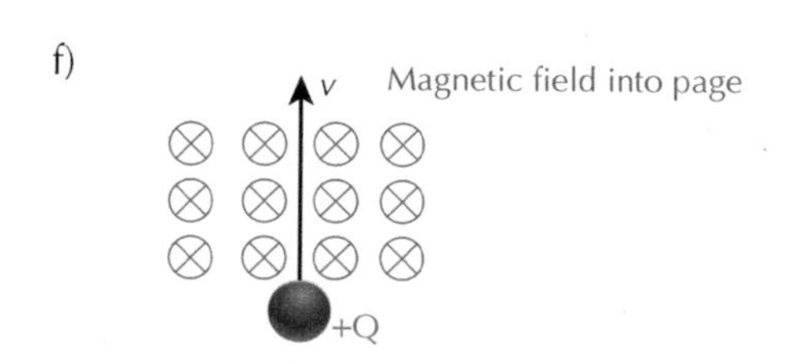

Fig 2.60

1.3.4 The motor effect

A current-carrying wire in an external magnetic field will experience a force. This causes the wire to move. This effect can be used in an electric motor to produce rotational movement. A simple motor can be constructed from a coil of wire, a DC power supply and two magnets to produce a magnetic field. A simple electric motor is shown in the diagram below.

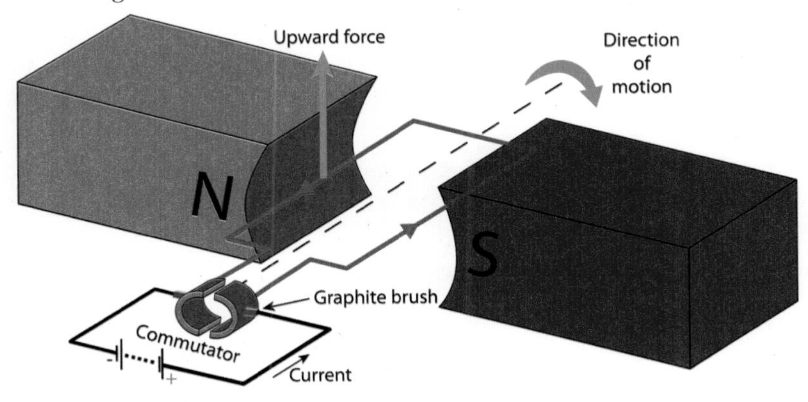

Fig 2.61

The motor here is shown with conventional current so the left hand rule applies to the motion of charges through the wire. The commutator and graphite brush are used to supply current to the coil. A magnetic field is produced by the magnets acting from north to south.

Starting with the left-hand side of the coil, conventional current is coming towards the commutator and the magnetic field is left to right across the page. The left hand rule can be applied (for conventional current) to find the direction of the force. This is shown in the diagram below:

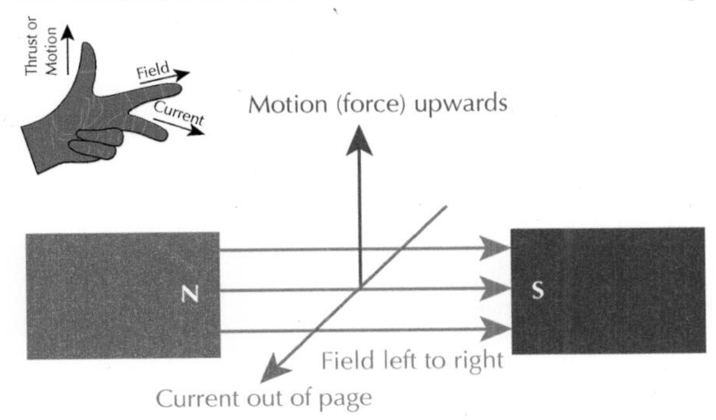

Fig 2.62

Applying the same rule to the other side of the coil shows that the force is downwards. These two forces combine to twist the coil so that it rotates. The commutator is split so that once the coil has rotated half a revolution, the current direction is reversed to keep the coil turning in the same direction.

1.4 Particle accelerators

A particle accelerator is designed to accelerate particles to very high speeds for use in experiments. Famous examples of particle accelerators include the Large Hadron Collider (LHC) at CERN, which has been developed as physicists search for new particles to deepen our understanding of the universe. A basic particle accelerator has already been discussed – the cathode ray tube. This accelerates electrons through a potential difference. The electrons are then directed to a screen by electric and magnetic fields. It is necessary to accelerate the electrons up to a high speed so that pictures on the screen can be produced very quickly.

There are three main families of particle accelerators that will be considered here: linear accelerators, cyclotrons and synchrotrons.

Fig 2.63

1.4.1 Linear accelerators

The first particle accelerators led to the development of the cathode ray tube – they were called Crookes tubes (pictured) after William Crookes, who developed them.

A high voltage is applied across the two electrodes and this causes electrons, which are naturally present inside the tube, to be accelerated towards the positive plate. As they move to the positive plate they collide with gas particles in the tube and give off light.

The Crookes tube, and the cathode ray tube that followed, are examples of linear accelerators. The particles are accelerated through a potential difference from the anode to the cathode in a straight line. In the previous section, the speed of the particles when they reach the anode was derived using energy conservation as:

$$QV = \frac{1}{2}mv^2$$

$$v = \sqrt{\frac{2QV}{m}}$$

The greater the voltage the particle is accelerated across, the greater the velocity of the particle.

An example of a linear accelerator used to probe the properties of matter is the Standard Linear Accelerator Centre (SLAC) in California.

The SLAC linear accelerator is 3 km long. Rather than just using a single cathode and anode like a cathode ray tube, it uses multiple electrodes, which are connected to an alternating voltage. The basic set-up is shown in Figure 2.64.

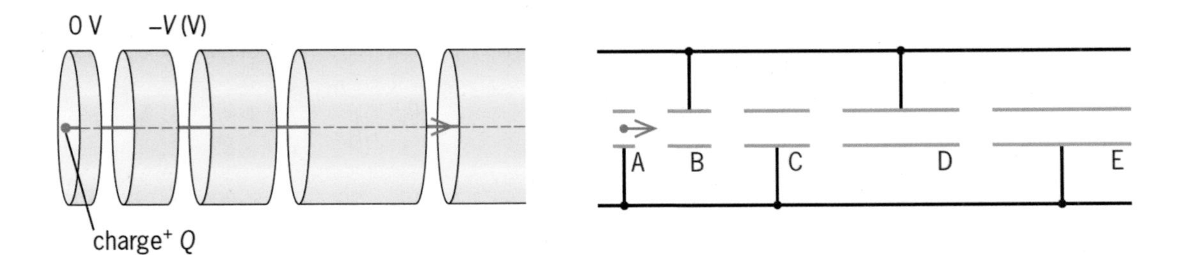

Fig 2.64

Particles such as electrons are accelerated between the gaps of the drift tubes. By connecting the drift tubes to an alternating signal, a potential difference can be set up between the tubes. This potential difference attracts the particles across the gap to the next tube, increasing the velocity. The alternating signal then reverses the potential difference and this attracts the particles to the next drift tube, accelerating them again. This process repeats, increasing the velocity of the particles each time. Inside the drift tubes, the particles move with a constant velocity. The drift tubes get progressively longer as you move along the accelerator to account for the increasing velocity of the particles. As the frequency of the alternating supply is constant, the particles must spend the same time in each drift tube.

A linear accelerator is the cheapest of all the accelerators, and the most efficient at supplying energy to the particles. However, it requires a large space because typical linear accelerators are several kilometres in length.

1.4.2 Cyclotrons

In a cyclotron, the particles being accelerated move in circular paths. Magnetic fields are used to move the particles in a circular path, as discussed in the previous section. By constraining the particles to circular paths, the space the accelerator occupies is much smaller.

A typical cyclotron is shown in Figure 2.65. It consists of two Ds with a gap between them, placed in an external magnetic field.

Similarly to the linear accelerator, the particles are only accelerated in the gaps between the Ds. The Ds are like the drift tubes, but in this case are used to turn the particles around so they return to the gap. The alternating voltage ensures that the particles are always accelerated across the gap between the Ds at the right time. The external magnetic field is used to change the direction of the moving particles as discussed in the section above. The faster the particles go, the greater the distance they will travel in the D; the particles will move around an arc with a greater radius.

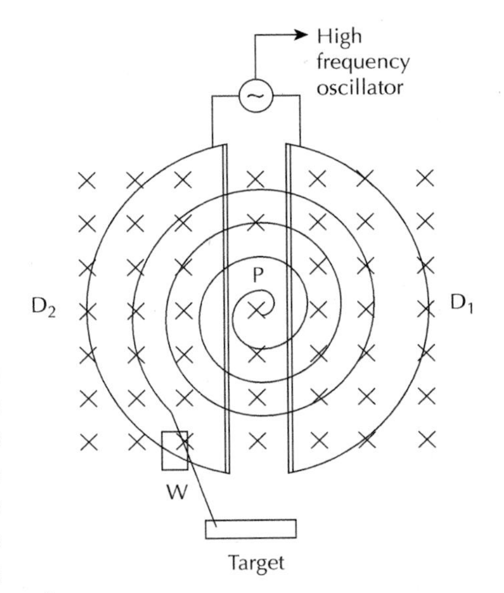

Fig 2.65

1.4.3 Synchrotrons

Particles travelling at high speeds begin to exhibit relativistic effects, which are discussed in Area 1. This means that for higher and higher energies, the speeds no longer increase as would typically be expected.

Instead, the increase in speed is lower than expected, and the particle's mass increases according to the equation (note: this is not a requirement for Higher):

$$m = \frac{m_0}{\sqrt{1 - \frac{v^2}{c^2}}}$$

When the particles reach high enough speeds, their speed stops increasing and it is simply their mass that increases. This means that the design of the cyclotron above, with a constant frequency of alternating voltage, will no longer be suitable as the speed of the particles will not change as expected, and will eventually stop increasing. The synchrotron overcomes this by varying the frequency of the alternative voltage.

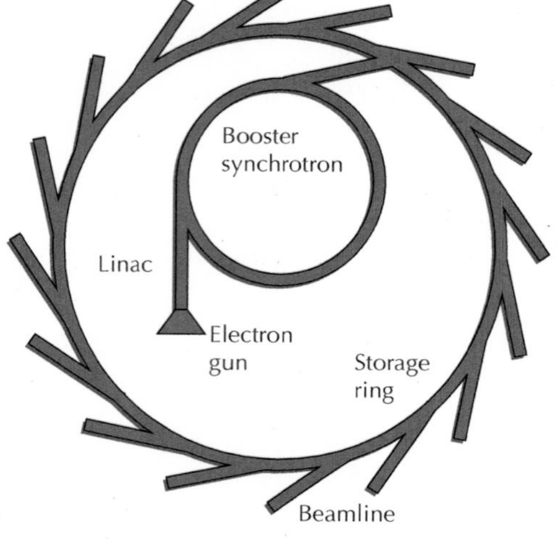

Fig 2.66

Figure 2.66 shows a synchrotron particle accelerator. Particles follow a circular path like a cyclotron; however, the radius of the path is constant.

Particles are initially accelerated by a linear accelerator (linac) before entering a booster section where their speed is increased to nearly the speed of light. As this section is circular, particles can go round many times and continue to gain energy as they do so. Particles are then moved to a storage ring, where they continue to circle until they are sent to one of the experiments, where they will collide with a target to probe the properties of matter on a sub-atomic scale.

1.4.4 The Large Hadron Collider (LHC)

One of the most famous synchrotron particles accelerators is the Large Hadron Collider (LHC) at CERN in Switzerland. It has a circular tunnel that is 27 km in length.

The diagram below shows the basic set up of the LHC.

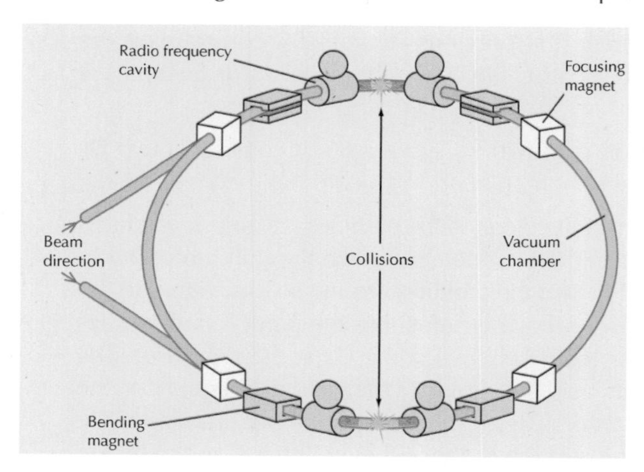

Fig 2.67

The particles are accelerated into the circular loop. In the case of the LHC, they travel in opposite directions. This means that their relative velocity is double their actual velocity, giving much more energetic collisions. For this to work, it requires very precise alignment and positioning of the beam of particles, which is achieved using magnets around the accelerator. The magnetic fields control the path of the particles as discussed previously.

2 The standard model

What you should know (National 5)

- Simple model of the atom consisting of a nucleus (protons and neutrons) and orbiting electrons
- Protons and neutrons have similar masses
- Protons have a positive charge
- Electrons have a negative charge

Learning intentions

- Orders of magnitude:
  - Very small (sub-nuclear)
  - Very large (distances to far celestial bodies)
- Use of scientific notation to represent very large and very small numbers
- The standard model of fundamental particles and interactions
- Evidence for the existence of sub-nuclear particles: the proton, neutron and electron
- Evidence for the existence of anti-matter

2.1 Orders of magnitude

Physics is a subject that studies everything in the known universe – from the very small (sub-atomic) to the very large, including the enormous scale of the universe itself. Such a widely-ranging scale means that physicists have to be able to work with significantly varying orders of magnitude.

2.1.1 Scientific notation

When dealing with very large and very small numbers, physicists use scientific notation. This is where a number is written as two numbers multiplying each other. The second number is a power of 10, which shows the scale of the number you are working with. This makes the numbers both neater to write and more easily read. For example, the speed of light is 300,000,000 $m s^{-1}$. This can be written in scientific notation as $3.00 \times 10^8 ms^{-1}$.

2.1.2 Very small

The approximate diameter of an atom is 0.0000000001 m, and the standard model deals with sizes that are even smaller. Using scientific notation, this is written as 1×10^{-10} m. An example of an even smaller number written in scientific notation is the diameter of an electron, which is 1×10^{-18} m.

When dealing with very high speeds that are travelling short distances, we use the formula

$$v = \frac{d}{t}$$

For example: the time taken for light to travel across a science lab that is 5 m wide is calculated using speed, distance and time:

$$v = \frac{d}{t}$$

$$v = \frac{5}{3 \times 10^8}$$

$$v = 1.67 \times 10^{-8} ms^{-1}$$

2.1.3 Very large

While particle physicists deal with the very small scale, astronomers work on a very large scale. Compared with atomic diameters, which are of order 10^{-10} m, the diameter of the Earth, for example, is of order 10^7 m, while the distances to remote galaxies in the universe are of order 10^{26} m. We can compare the orders of magnitude by subtracting the indexes. For example, the distance to far galaxies is 19 orders of magnitude greater (26 – 7) than the diameter of the Earth. The diameter of the Earth is 17 orders of magnitude greater than the diameter of an atom.

2.2 The standard model

The standard model is a method for physicists to classify particles according to common properties that they may share. It is similar in many ways to the periodic table, which is a method used to classify elements into groups or families that share similar properties, such as metals and non-metals, a group of alkali metals etc. In this section, we begin by considering the main groups and the particles that are members of these groups. More detail on the particles themselves is then given later in the section.

For the standard model, we begin by classifying particles into matter particles and force-mediating particles. The matter particles make up everything around us, while the force-mediating particles are responsible for the fundamental forces of nature (such as gravity or the electromagnetic force).

2.2.1 Matter particles

Matter particles, called fermions, are separated into two main categories: quarks and leptons. There are six of each, which are related in pairs. All matter we see around us is made up of these matter particles, which are shown in Figure 2.68 below.

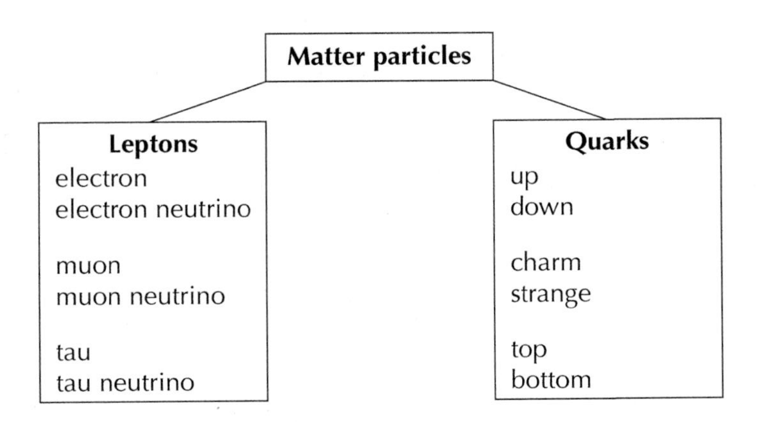

Fig 2.68

All of these particles are fundamental particles – this means that they cannot be broken down into anything smaller.

However, the quarks combine to form larger particles that are not fundamental. Quarks are not observed on their own in nature. Quarks also have a property called colour, and they only combine to form colourless objects (we don't, however, consider this property at Higher level).

The matter particles that are formed by quarks are called hadrons. This group is split into two smaller groups depending on whether the particle is made up of two quarks or three. A meson is a particle made up of two quarks (a quark–anti-quark pair), and a baryon is a particle that is made up of three quarks. This is shown in Figure 2.69 with some typical mesons and baryons given.

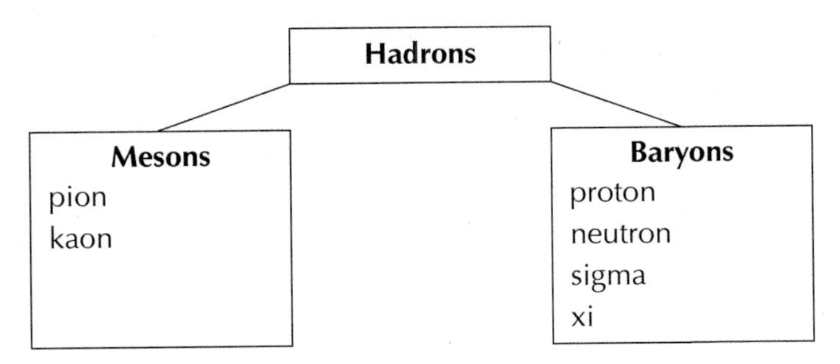

Fig. 2.69

In summary, we can now split our matter particles into fundamental and non-fundamental. Fundamental particles cannot be split into anything smaller and the only ones that are found on their own are leptons such as the electron. Figure 2.70 shows the classification of the fundamental particles.

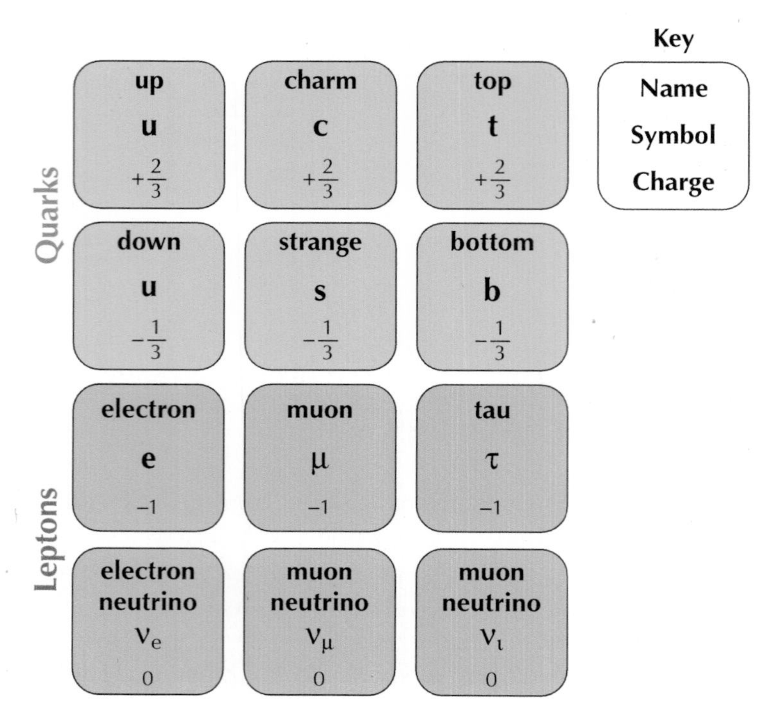

Fig 2.70

Non-fundamental particles are made up of smaller particles called quarks, and are called baryons. These are split into two further families: mesons, which are made up of two quarks; baryons, which are made up of three quarks.

2.2.2 Anti-matter

In 1928, Paul Dirac stated an equation that combines quantum theory and Einstein's theory of relativity for the motion of an electron at relativistic speeds. This equation would allow atoms to be treated according to Einstein's theory of relativity, and Dirac won the Nobel prize for his work in 1930. However, Dirac's equation had a very important outcome.

Consider the equation:

$$x^2 = 4$$

This has two solutions

$$x = \pm 2$$

Dirac's equation has the same consequence and the two solutions were interpreted as the presence of both matter and anti-matter.

Each matter particle has an equivalent anti-matter particle. For example, the electron's anti-particle is the positron and the proton's anti-particle is the anti-proton. Examples are considered in more detail in the next section.

Key point

When discussing particles, it is conventional to discuss the charge in units of the charge on an electron. So, the electron has a charge of −1, the proton a charge of +1.

An anti-particle is identical to its matter particle, except that it has the opposite charge (it also has a different baryon number, but this is outside the scope of the Higher course). For example, an electron has a charge of −1 while the anti-particle, the positron, has a charge of +1.

Anti-nuclei

Once it was discovered that the constituent parts of an atom have anti-particles, it was predicted that any nucleus would have an anti-nucleus. In 1965, the first anti-nucleus was discovered: anti-deuterium. It consists of an anti-proton and an anti-neutron and has the opposite charge properties to that of a deuterium nucleus.

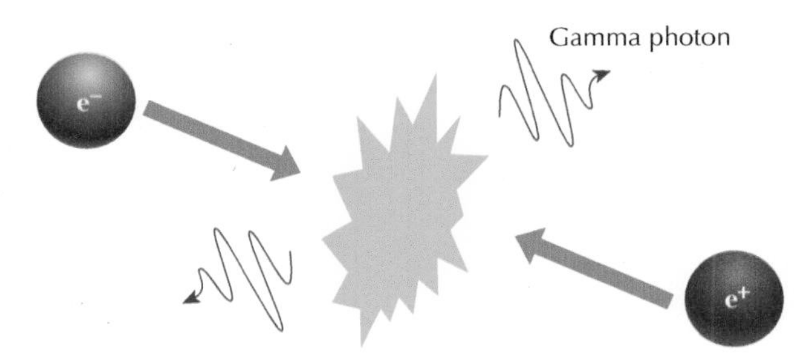

Fig 2.71

Matter–anti-matter collisions

When matter and anti-matter collide, they annihilate each other and produce energy. This destroys the two particles. An example of this is the collision between an electron and a positron, depicted in Figure 2.71.

The energy from the annihilation is emitted as two gamma ray photons. They move off in opposite directions in order to conserve momentum.

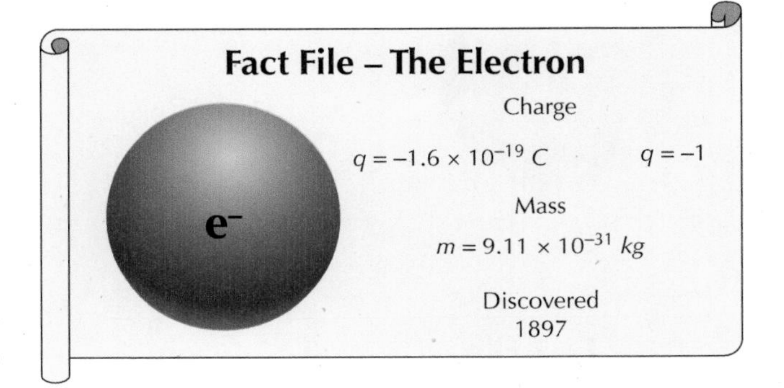

Fig 2.72

2.3 Particle facts

Electron

The electron is a negatively charged particle that is very light in comparison to the proton and neutron. It orbits the nucleus. According to Bohr's model of the atom, the electron orbits the nucleus in discrete orbits. The electron was discovered in 1897 by J.J. Thompson.

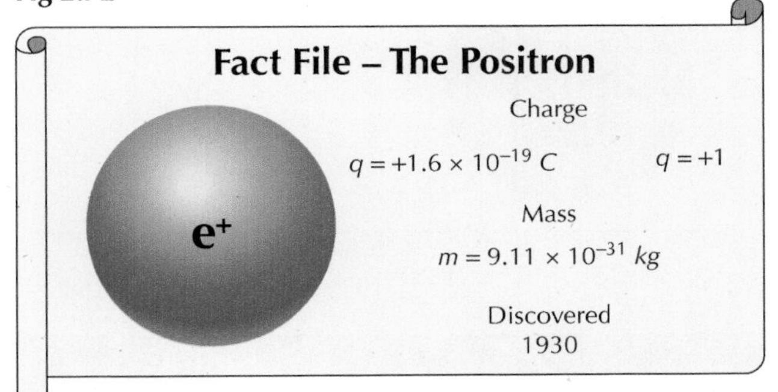

Fig 2.73

The positron

The first anti-matter particle to be discovered was the positron, also known as the anti-electron. This anti-matter particle has the same mass as an electron, but as a charge with opposite magnitude. It was discovered in 1930 by Carl Anderson.

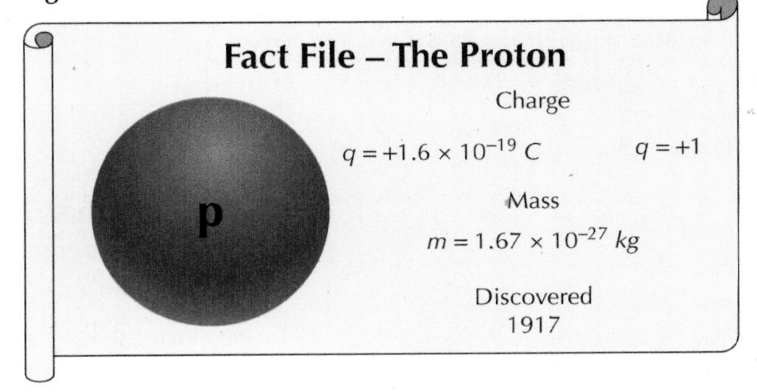

Fig 2.74

Proton

The proton is a positively charged particle with a mass much greater than that of an electron, and similar to a neutron. The proton is found in the nucleus of an atom. The number of protons in a nucleus is the atomic number, which determines which element the atom is. The proton was discovered by Ernest Rutherford in 1917.

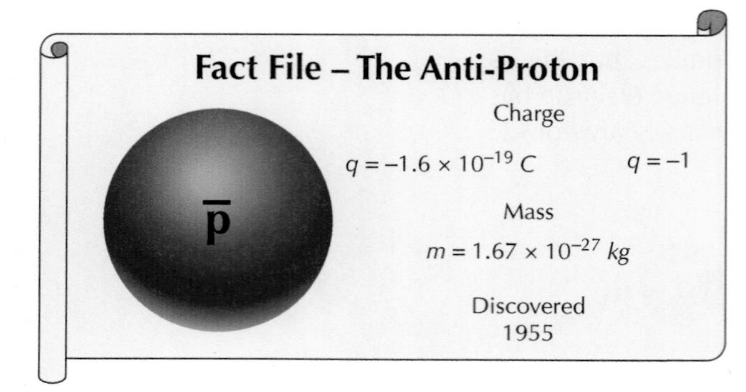

Fig 2.75

The anti-proton

The anti-proton has opposite charge to that of a proton, but the same mass. It was discovered in 1955 by Owen Chamberlain, Emilio Segrè et al.

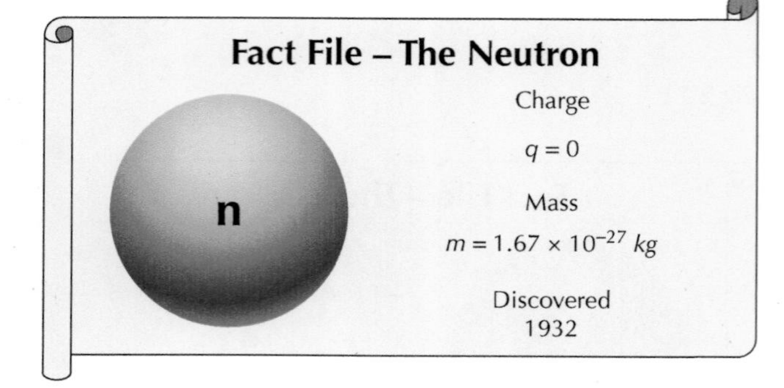

Fig 2.76

Neutron

The neutron has a similar mass to that of a proton, and is a neutral particle, meaning it has no charge. Like a proton, it can be found in the nucleus of an atom. Different numbers of neutrons give rise to different isotopes of the same atom. The neutron was discovered by James Chadwick in 1932.

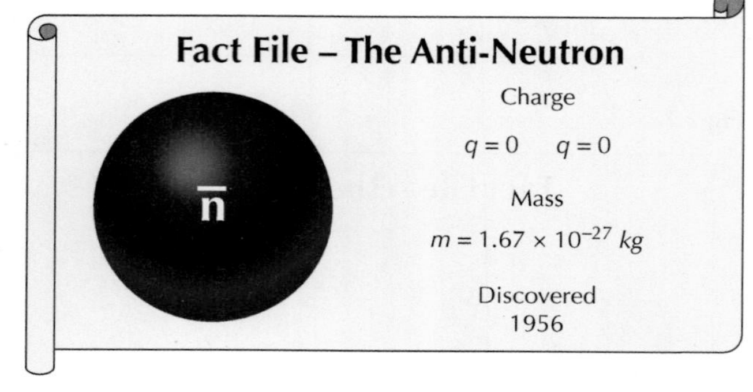

Fig 2.77

The anti-neutron

The anti-neutron was discovered by Bruce Cork in 1956. It is a particle with no charge (like the neutron) and shares the same mass. On the face of it, the anti-neutron would appear to be the same as the neutron. However, other properties, discussed later, were opposite to those of the neutron.

2.3.1 Force-mediating particles

The other main group of particles is the force-mediating particles, which are bosons. There are four fundamental forces of nature. Each force has its own force-mediating particle. These are listed in Table 2.3 below, along with the relative strength of the force and the approximate range over which the force acts.

Force	Particle	Relative strength	Range (m)
Strong force	Gluon	10^{38}	10^{-15}
Weak nuclear force	W and Z bosons	10^{25}	10^{-18}
Electromagnetic	Photon	10^{36}	Infinite
Gravitational	Graviton	1	Infinite

Table 2.3

Electromagnetic force

The electromagnetic force affects any particle with an electric charge and acts over an infinite range.

The particle that is responsible for the electromagnetic force is the photon. This particle has no mass and no charge.

Weak force

The weak force was first revealed by beta decay (described below, and in more detail in Chapter 3, Nuclear reactions). The weak force acts on both quarks and leptons, and it is the only reaction where a quark can change into another quark (or a lepton can change into another lepton). This is referred to as a 'flavour change'. With beta decay, a neutron turns into a proton. For this to happen, a down quark must change to an up quark.

The particles responsible for the weak interaction are the intermediate vector bosons, so-called W and Z particles.

Strong force

The nucleus contains protons, which have a positive charge, and confines them. According to electrostatics, the positively charged protons should repel and the nucleus should be blasted apart! However, this does not happen. Clearly, there is another force responsible for holding the nucleus together – this is the strong force.

The strong force acts on quarks and holds them together to make up particles such as the proton. A residual strong force is then what holds these particles together in the nucleus. The strong force can only act over a very short distance, but unlike the other forces, its strength actually increases with distance.

The particles responsible for the strong force are called gluons (as they 'glue' the nucleus and quarks together). An example of such a particle is the pion.

2.4 The neutrino

Beta decay is studied in detail in Chapter 3 in this area of study. It is the emission of an electron by a nucleus. This is illustrated in the diagram below for Carbon-14.

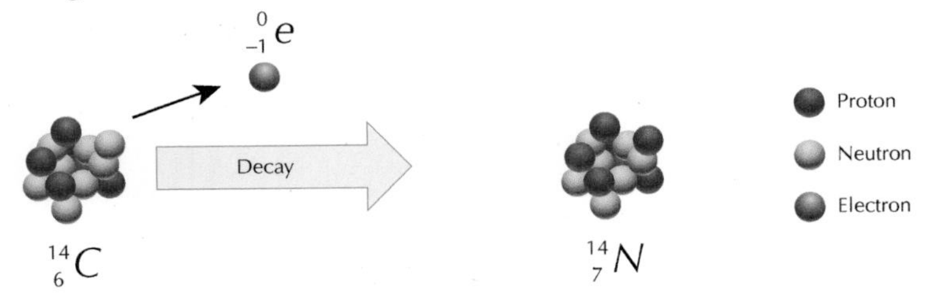

Fig 2.78

In beta decay, the nucleus does not recoil in the usual manner for the emission of an electron. This suggests that momentum is not being conserved. This is highlighted in Figure 2.79.

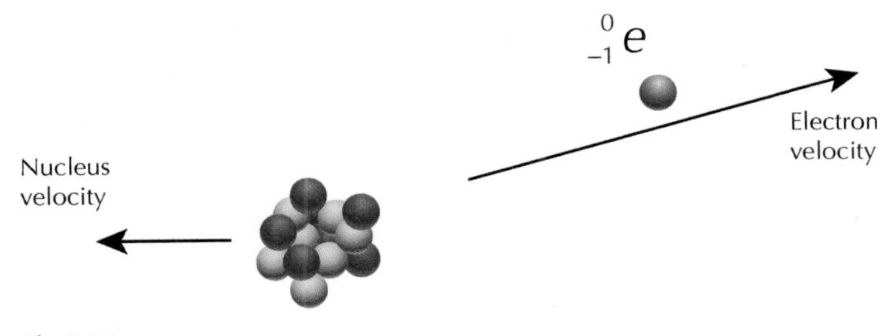

Fig 2.79

Enrico Fermi predicted the existence of another particle involved in the interaction above – the neutrino, which is denoted with the letter ν. The neutrino has no charge, and has a very low mass. Beta decay, with the emission of an electron, also consists of the emission of an anti-neutrino. It is emitted as pictured in Figure 2.80 and shows that the law of conservation and momentum is fulfilled.

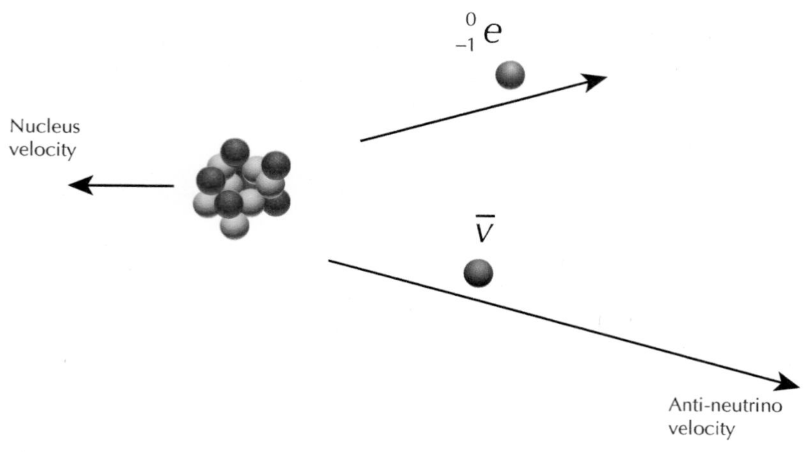

Fig 2.80

2.5 Hadrons

2.5.1 Quarks

In the 1960s, it was discovered that protons and neutrons were made up of even smaller particles, and these smaller particles were named quarks. The discovery of quarks came from high energy collisions such as those that would be found from particle accelerators such as at CERN.

Quarks are thought to be fundamental particles (that is, they are not made up of anything smaller). There are six quarks, each with its own anti-quark. These are listed in Table 2.4 along with their corresponding charge.

Quark	Symbol	Charge
Up	u	$+\frac{2}{3}$
Anti-up	$\bar{u}$	$-\frac{2}{3}$
Down	d	$-\frac{1}{3}$
Anti-down	$\bar{d}$	$+\frac{1}{3}$
Top	t	$+\frac{2}{3}$
Anti-top	$\bar{t}$	$-\frac{2}{3}$

Quark	Symbol	Charge
Bottom	b	$-\frac{1}{3}$
Anti-bottom	$\bar{b}$	$+\frac{1}{3}$
Charm	c	$+\frac{2}{3}$
Anti-charm	$\bar{c}$	$-\frac{2}{3}$
Strange	s	$-\frac{1}{3}$
Anti-strange	$\bar{s}$	$+\frac{1}{3}$

Table 2.4

In experiments, the smallest charge that has been observed is an integer of the electron charge. This means that quarks have never been observed on their own. Quarks only exist in pairs or threes, making up other particles that are observed.

Both baryons and mesons are made up of quarks. Baryons consist of three quarks and mesons contain a quark and an anti-quark. A composite particle that is made up of quarks is called a hadron.

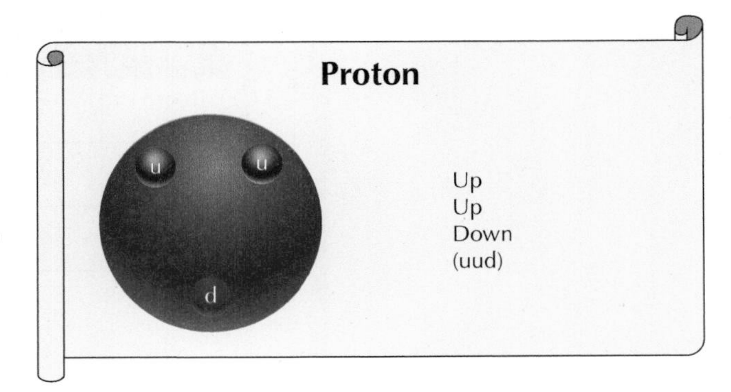

Fig 2.81

Two main examples of baryons are the proton and the neutron. Each is made up of three quarks as detailed in Figures 2.81 and 2.82.

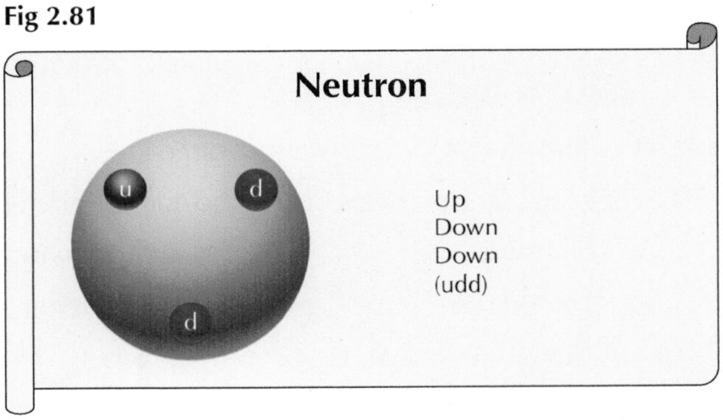

Neutron
u
d
d
Up
Down
Down
(udd)

Fig 2.82

Notice that when the charges of the individual quarks are added together, it gives the overall charge of the baryon. For example, the proton:

$$q = u + u + d$$

$$q = \frac{2}{3} + \frac{2}{3} - \frac{1}{3}$$

$$q = 1$$

Mesons are usually short-lived particles because they consist of a quark and an anti-quark (a matter/anti-matter pair). An example of a meson is the kaon particle, which consists of an up quark and an anti-strange quark (Figure 2.83).

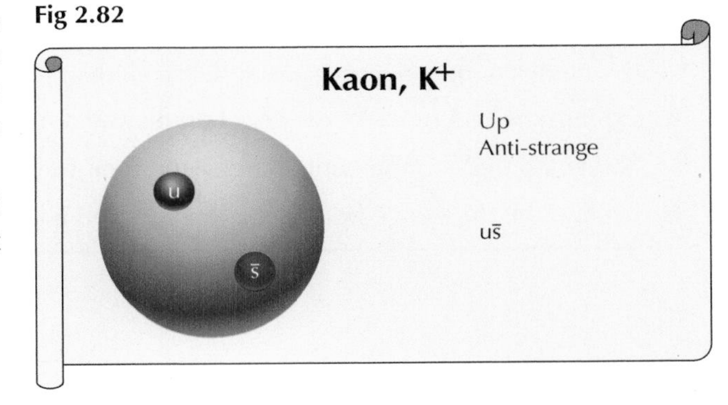

Fig 2.83

As before, the charge of the particle can be found by adding the individual charges of the constituent quarks. For the K^+ particle, the charge is:

$$q = u + \bar{s}$$

$$q = \frac{2}{3} + \frac{1}{3}$$

$$q = 1$$

2.5.2 Leptons

Leptons are fundamental particles; that is, they cannot be broken down into smaller parts. The most common example of a lepton is the electron, which is not made up of smaller particles. There are three main families of lepton: the electron, the muon and the tau. Each lepton has its own neutrino (also a lepton). Both of these particles also have their own anti-particles. There are four particles in each group, and a total of twelve known leptons, as detailed in Table 2.5.

Family	Particle	Neutrino	Anti-particle	Anti-neutrino
Electron	e	v_e	$\bar{e}$	$\bar{v}_e$
Muon	μ	v_μ	$\bar{\mu}$	$\bar{v}_\mu$
Tau	τ	v_τ	$\bar{\tau}$	$\bar{v}_\tau$

Table 2.5

Exercise 2.2 The standard model

1 Describe Bohr's model of the atom, including the three main particles involved, the relative masses and charges.

2 Explain what is meant by anti-matter.

3 For each of the particles below, the quark make-up is given. Determine the charge of each particle.

a) Sigma, Σ: Up Up Strange (uus)

b) Lambda, Λ: Up Down Strange (uds)

c) Charmed Omega, Ω: Strange Strange Charm (ssc)

d) Xi, Ξ: Down Strange Strange (dss)

e) Proton, p: Up Up Down (uud)

f) Neutron, n: Up Down Down (udd)

4 Explain the difference between a baryon and a meson.

5 Name the three fundamental forces and their force-mediating particles.

6 Explain the difference between a fermion and a boson.

3 Nuclear reactions

What you should know (National 5):

- Nature and behaviour of alpha, beta and gamma radiation
- Activity of a radioactive source as the number of radioactive decays in one second
- Absorbed dose as the amount of energy absorbed by 1 kg of material
- Dose equivalent gives a measure of the biological harm of radiation and depends on:
  - Type of radiation
  - Energy absorbed
- Half-life of a radioactive source
- Carbon-dating
- Qualitative description of nuclear fission and nuclear fusion to produce energy in a nuclear reactor

Learning intentions

- Model of the atom, including how to label an atom:
  - Atomic number
  - Mass number
- Binding energy calculated from the mass defect of an atom (difference in mass between component parts of the atom and the actual mass of the atom)
- Nuclear decay equations for alpha, beta and gamma radiation
- Nuclear equations for fission and fusion reactions
- Calculation of energy released during a fission or fusion reaction
- The nuclear fission reactor
- Nuclear fusion in stars

3.1 Model of the atom

The Bohr model of the atom consists of a positively charged nucleus at the centre and negatively charged electrons orbiting it as shown in Figure 2.84.

The orbits of the electrons will be considered in more detail in the spectra section. For nuclear physics, the focus will be on the nucleus of the atom.

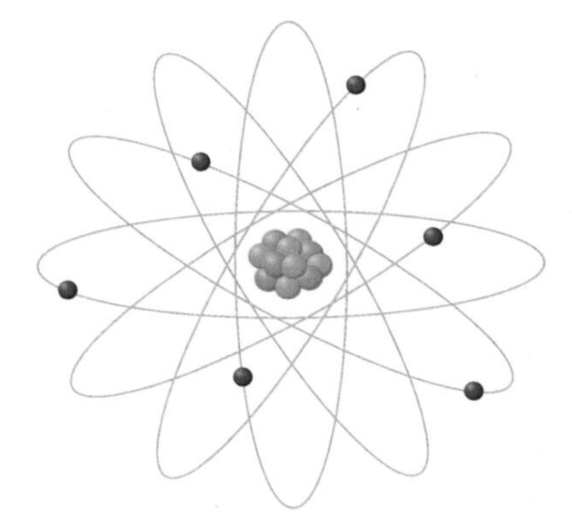

Fig 2.84

3.1.1 Structure of the nucleus

The nucleus of an atom consists of protons and neutrons known as nucleons. Two properties of the nucleus are:

- The total number of protons and neutrons in the nucleus is called the mass number, A.
- The number of protons in the nucleus is called the atomic number, Z.

In a neutral atom, the number of protons equals the number of electrons.

The mass numbers, charges and symbols for protons, neutrons and electrons are given below.

Particle	Mass number	Charge	Symbol
Proton	1	+1	$^{1}_{1}p$
Neutron	1	0	$^{1}_{0}n$
Electron	0 *	–1	$^{0}_{-1}e$

* Electron mass is negligible – 1840 times less than that of a proton.

Table 2.6

An element is specified by its chemical symbol and the atomic and mass numbers:

$$^{A}_{Z}X$$

where X is the chemical symbol. For example, the atom uranium-235 has the symbol:

$$^{235}_{92}U$$

This shows that the atom has 235 nucleons, of which 92 are protons. This means that there are 143 neutrons in the nucleus ($235 - 92 = 143$).

Each element in the periodic table has a different number of protons, hence a different atomic number. Figure 2.85 shows carbon, oxygen and nitrogen atoms. Notice the different number of protons in each nucleus and the different atomic numbers in the symbols below.

The carbon atom at the top has 6 protons and 6 neutrons leading to the symbol $^{12}_{6}C$.

The oxygen atom in the middle has 8 protons and 8 neutrons leading to the symbol $^{16}_{8}O$.

The nitrogen atom at the bottom has 7 protons and 7 neutrons leading to the symbol $^{14}_{7}N$.

Atoms from the same element can have different numbers of neutrons. This leads to different isotopes. An example of this is shown in Figure 2.86 for helium. Each isotope of helium has the same number of protons (2) but different numbers of neutrons.

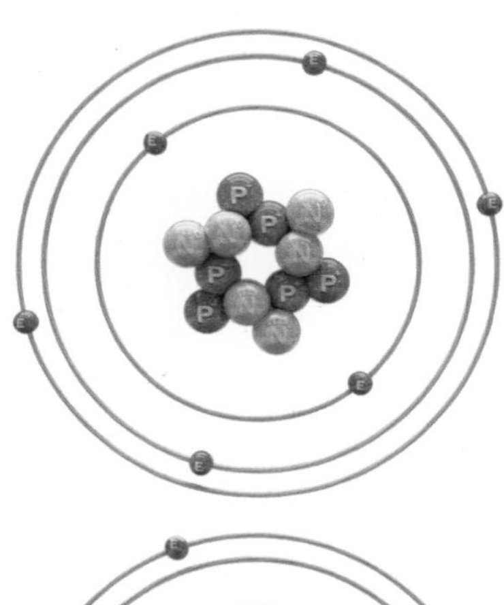

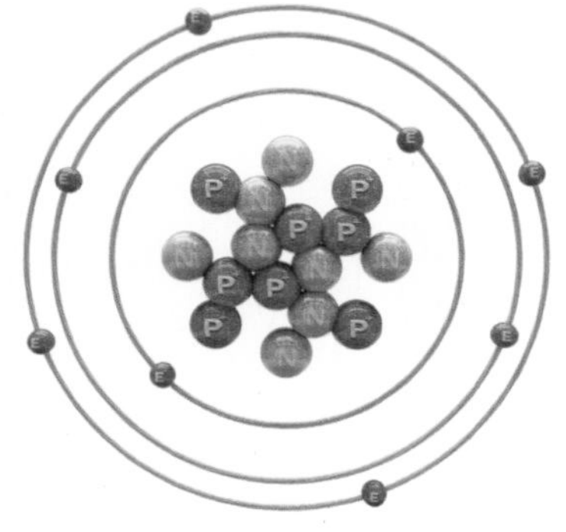

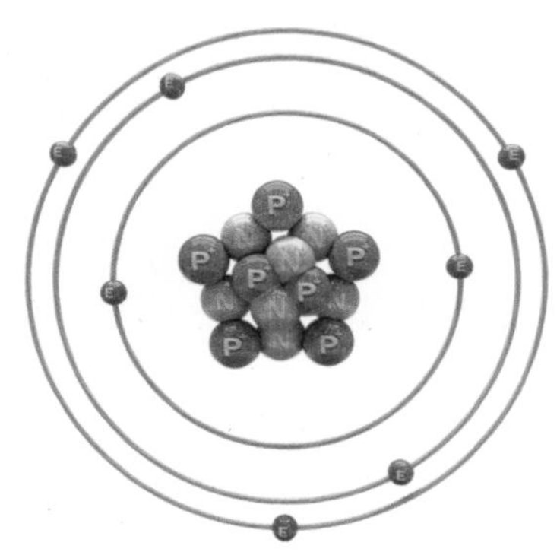

Fig 2.85

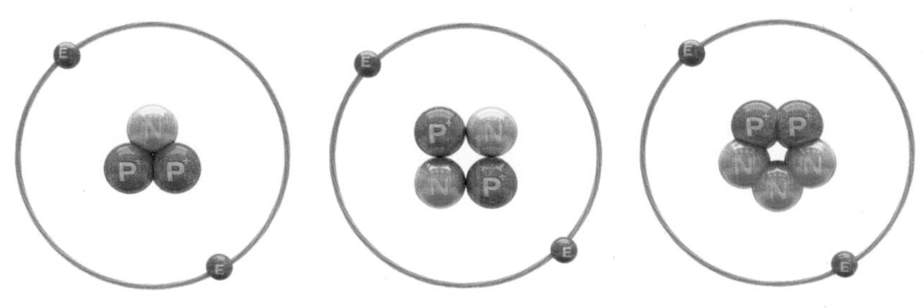

Fig 2.86

This gives the following symbols for the isotopes of helium:

$$^{3}_{2}He \quad ^{4}_{2}He \quad ^{5}_{2}He$$

The different isotopes can be recognised from the same atomic number, but different mass numbers. Some isotopes are said to be stable – that is, they will not decay by emitting radiation; if they do decay, they do so very slowly, with half-lives too long to be measured. Others are said to be unstable – they undergo radioactive decay by emitting particles and/or energy. Of the isotopes of helium shown in Figure 2.86, helium-3 and helium-4 are said to be stable, while helium-5 is unstable.

Key point

Every element has a different number of protons and thus a different atomic number. The same element can have different numbers of neutrons, leading to different isotopes.

Key point

The nucleus of an atom is made up of protons and neutrons. An element is represented by a chemical symbol and its mass number, A, and atomic number, Z.

$$^{A}_{Z}X$$

Worked example

Calculate the number of protons and neutrons in an atom of lead-206, which has the chemical symbol $^{206}_{82}Pb$.

The atomic and mass numbers can be read straight from the chemical symbol:

$$^{206}_{82}Pb$$

Mass number, $A = 206$

Atomic number, $Z = 82$

This means there are 82 protons in the nucleus.

There are $206 - 82 = 124$ neutrons in the nucleus.

Exercise 3.1.1 Structure of the nucleus

1 For each of the elements below, state the number of protons and neutrons in the nucleus.

a) $^{9}_{4}Be$

b) $^{7}_{3}Li$

c) $^{35}_{17}Cl$

d) $^{27}_{13}Al$

e) $^{184}_{74}W$

(continued)

2 For each of the atoms shown in Fig 2.87, state:

a) The number of protons.

b) The number of neutrons.

c) The atomic number.

d) The mass number.

e) The element (a periodic table may be required).

f) The chemical symbol, including mass number and atomic number.

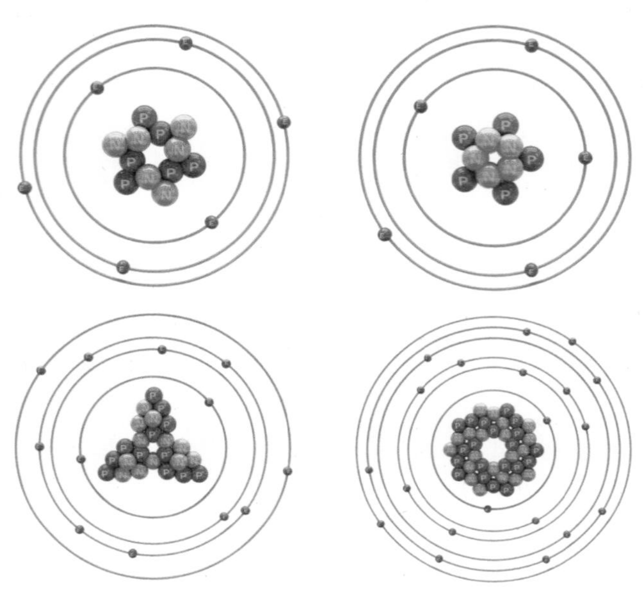

Fig 2.87

3 By considering the atoms in Question 2, what can you say about the number of electrons orbiting the nucleus compared to the number of protons in the nucleus? What does this tell you about the overall charge of the atom?

4 Explain what is meant by the term isotope. In your explanation, make reference to the concept of stability.

5 Shown below are different isotopes of two different elements.

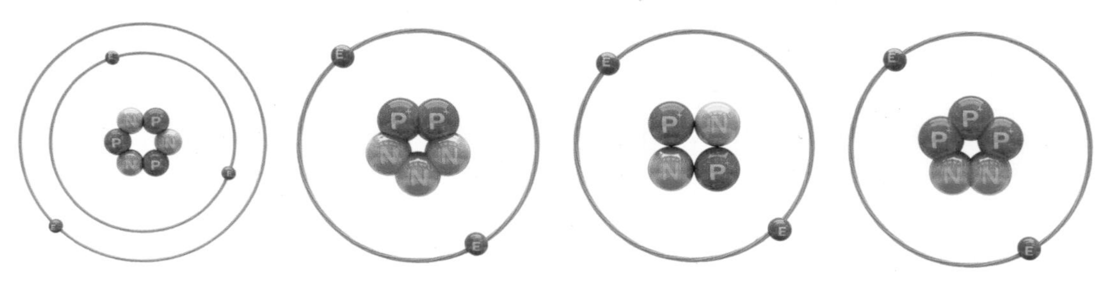

Fig 2.88

a) Group the atoms above into two groups – one for each element.

b) Identify each element.

c) For each atom above, write it in the form ${}^{A}_{Z}X$.

3.1.2 Binding energy

When the nucleus of an atom is considered from an electrostatic point of view, the nucleus should not exist. Like charges repel each other, so the protons in the atom should repel and push the nucleus apart. Clearly there are stronger forces at play within the nucleus, and this leads to the idea of the nuclear strong force discussed in Chapter 2. The amount of energy required to hold the nucleus together (or rather the amount of energy required to break a stable nucleus apart) is known as the 'binding energy'.

Research has allowed physicists to find the mass of a proton and the mass of a neutron:

- Proton: $m_p = 1.6726 \times 10^{-27} kg$
- Neutron: $m_n = 1.6749 \times 10^{-27} kg$

Comparing the mass of the constituent parts of a nucleus to the mass of the nucleus reveals an amazing result – the mass of the nucleus is less than its constituent parts! For example, the mass of an oxygen atom shown in Figure 2.89 is $2.6559 \times 10^{-26} kg$.

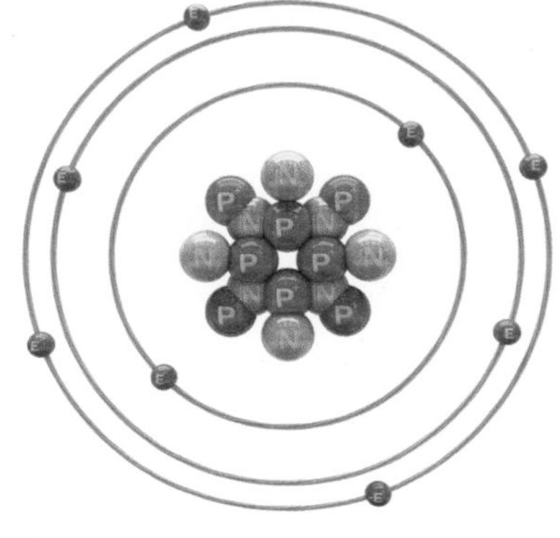

Fig 2.89

The oxygen atom has 8 protons and 8 neutrons, which gives a total mass of:

$$m = 8m_p + 8m_n$$

$$m = (8 \times 1.6726 \times 10^{-27}) + (8 \times 1.6749 \times 10^{-27})$$

$$m = 1.3381 \times 10^{-26} + 1.3399 \times 10^{-26}$$

$$m = 2.6780 \times 10^{-26} kg$$

There is a *missing mass* here. This is known as the mass difference and it is linked to the binding energy by Einstein's famous equation:

$$E = mc^2$$

Here the binding energy is calculated by finding the missing mass:

$$m = 2.6780 \times 10^{-26} - 2.6567 \times 10^{-26}$$

$$m = 2.13 \times 10^{-28} kg$$

which leads to a binding energy of:

$$E = (2.13 \times 10^{-28})(3 \times 10^8)^2$$

$$E = 1.917 \times 10^{-21} J$$

Einstein's equation above has a very important outcome – it demonstrates that mass and energy are equivalent to each other. In other words, mass (the matter that we are made up of) is just condensed energy that was first created at the time of the Big Bang.

The binding energy is different for different atoms. Figure 2.90 shows the binding energy per nucleon, plotted against the number of nucleons present in the nucleus.

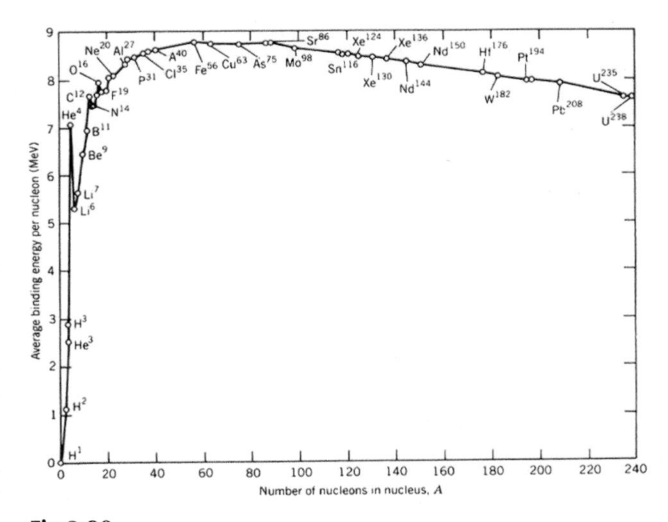

Fig 2.90

> **Key point**
>
> A nucleus is held together by its binding energy, which is a result of the mass difference: the total mass of the nucleus is less than the combined mass of its constituent parts. The binding energy is given by the equation:
>
> $$E = mc^2$$
>
> where m is the mass difference equal to the difference in the nuclear mass, and the total mass of the constituent parts of the nucleus.

It is apparent that the binding energy rises sharply for light atoms as the mass number increases and then levels off to the element iron. Beyond this, the binding energy gets lower for heavier elements. The maximum binding energy for iron is key to the life of large stars, and will be investigated in a later section. The binding energy is also essential to the elements that can be used for producing energy by nuclear fission and nuclear fusion. Both are investigated in the next two sections.

Exercise 3.1.2 Binding energy

1 Explain the terms mass difference and binding energy. How does the binding energy of an atom depend on the number of nucleons for light nuclei and heavy nuclei?

2 For each of the atoms below, calculate:

a) The mass difference

b) The binding energy

(i) Helium-4, $m = 6.6463 \times 10^{-27}$ kg

(ii) Carbon-12, $m = 1.9926 \times 10^{-26}$ kg

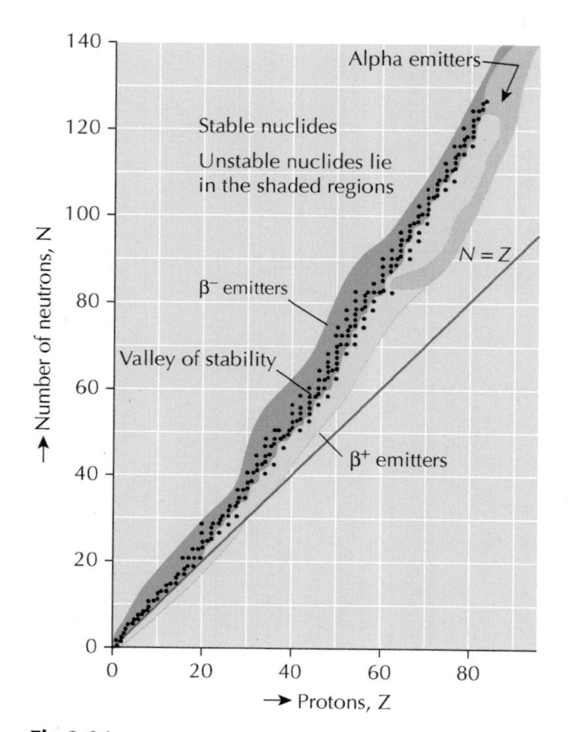

Fig 2.91

3.1.3 Nuclear stability and radioactive decay

Radioactive decay was introduced at National 5 level, where three main types of ionising radiation were considered – alpha, beta and gamma. If an atom is unstable, then it will emit radiation (particles or energy waves) until it becomes stable.

Nuclear stability

An atom is said to be stable if it does not undergo radioactive decay. The atom exists in the same state for a prolonged period of time. An atom that is unstable will continue to undergo radioactive decay until it becomes stable.

Considering the lightweight atoms, the number of protons and neutrons are roughly equal to each other. However, heavier atoms usually have more neutrons than protons in the nucleus. This can be represented on a stability diagram, an example of which is shown in Figure 2.91. The straight line shows the line where the number of neutrons equals the number of protons.

Stable atoms fall into the valley of stability, and atoms that lie outside this are unstable. They will emit radiation (see below) in order to become stable.

Alpha, beta and gamma radiation

Unstable nuclei are called radioisotopes or radionuclides. They will emit radiation to become stable. Three main types of radioactive decay are considered here: alpha, beta and gamma. These are summarised in the table below:

Radiation type	Nature	Symbol
Alpha	Helium nucleus, slow moving	${}^{4}_{2}He$; α
Beta	Fast electron	${}^{0}_{-1}e$; β
Gamma	High frequency EM wave	γ

Table 2.7

The table gives the symbol for each of the decay types along with the atomic and mass numbers. Alpha radiation is a helium-4 nucleus, beta radiation is an electron (mass number 0 means that it is not a part of the nucleus), and gamma radiation is a high-energy electromagnetic wave.

If a nucleus emits an alpha particle, it will change both the mass number and the atomic number of the nucleus. This means that a new element will be produced. An example of this is the decay of uranium-238, which has an atomic number of 92:

$${}^{238}_{92}U$$

When uranium-238 emits an alpha particle, its mass number changes to 234 (it loses 4 nucleons) and its atomic number changes to 90 (it loses 2 protons). This corresponds to thorium-234:

$${}^{234}_{90}Th$$

Figure 2.92 shows this decay:

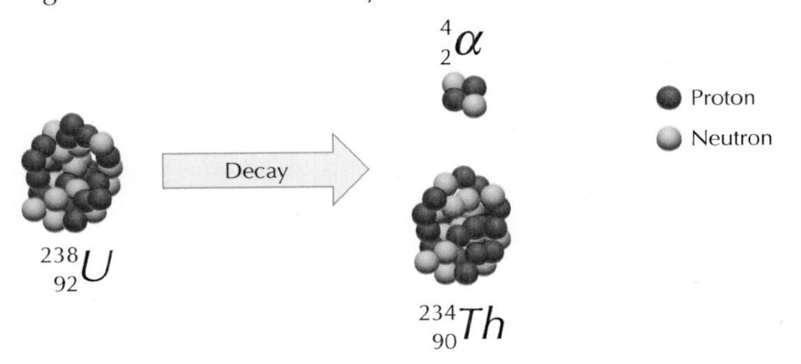

Fig 2.92

The decay is represented with a decay equation:

$${}^{238}_{92}U \rightarrow {}^{234}_{90}Th + {}^{4}_{2}He$$

Notice that the decay equation is balanced – the mass numbers on the right-hand side add up to give the mass number of the left-hand side, and the same applies to the atomic numbers.

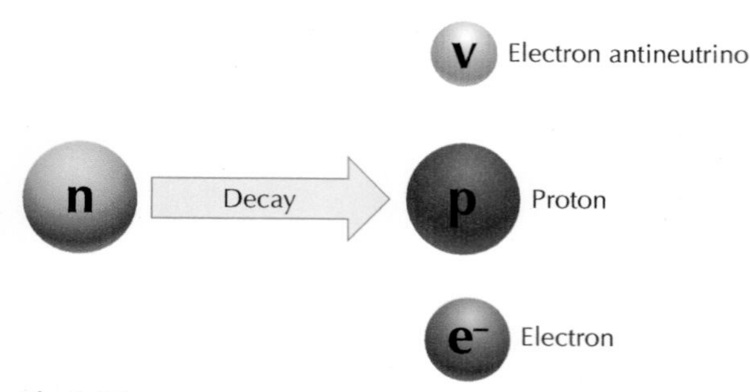

Fig 2.93

Beta decay is different from alpha decay. An electron is emitted by the nucleus. There are no electrons in the nucleus as they orbit the atom. In order to emit an electron, one of the neutrons in the nucleus decays to a proton and an electron – the new proton remains in the nucleus while the electron is emitted as a beta particle. An electron anti-neutrino is also emitted but for this discussion these will be ignored. This concept is shown in Figure 2.93.

A typical beta decay is the decay of carbon-14 into nitrogen-14. One of the neutrons in a carbon-14 nucleus changes to a proton, emitting an electron. This changes the atomic number of the atom and therefore changes the element – in this case to nitrogen. This is highlighted in Figure 2.94a and the decay equation below:

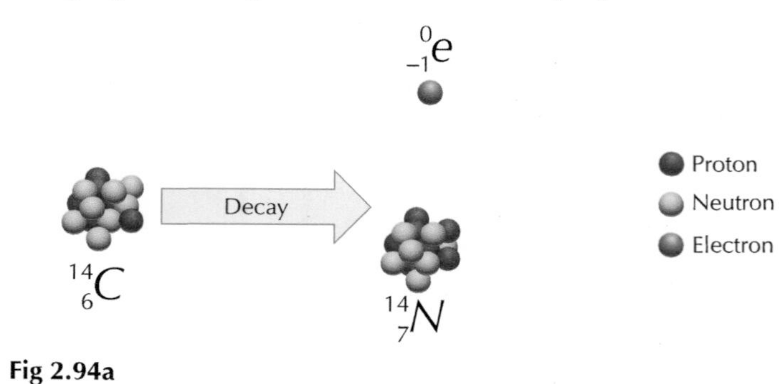

Fig 2.94a

$$^{14}_{6}C \rightarrow {}^{14}_{7}N + {}^{0}_{-1}e$$

As with alpha decay, the equation is balanced.

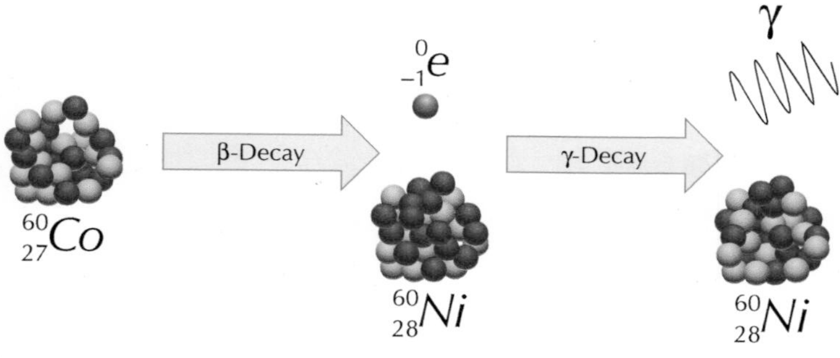

Fig 2.94b

Gamma radiation does not involve the emission of a particle from the nucleus. For this reason, the overall nucleus does not change during the process. Instead, the nucleus simply loses energy as an electromagnetic wave. Gamma radiation is usually emitted following alpha or beta emission to remove excess energy from the nucleus. An example is the decay of cobalt-60 into nickel-60 by beta emission, followed by the emission of a gamma particle to remove excess energy from the atom (see Figure 2.94b).

This can be represented by the following decay equation:

$$^{60}_{27}Co \rightarrow {}^{60}_{28}Ni + {}^{0}_{-1}e \rightarrow {}^{60}_{28}Ni + \gamma$$

As before, the decay equation is balanced with the mass and atomic numbers and they have the same value at each stage.

Key point

Radioactive decay can be represented by a decay equation. The mass and atomic numbers are conserved.

Worked example

Americium-241 decays into neptunium-237 by emitting a radioactive particle as described in the equation below.

$$^{241}_{95}Am \rightarrow {}^{237}_{93}Np + {}^{x}_{y}X$$

Identify the emitted particle and the mass and atomic numbers of the particle, *x* and *y* respectively.

The decay equation must be balanced, so you can find x and y as follows:

$$241 = 237 + x \quad ; \quad 95 = 93 + y$$
$$x = 4 \quad ; \quad y = 2$$

A particle with a mass number of 4 and an atomic number of 2 is an alpha particle, and therefore americium-241 decays by means of alpha emission. Completing the decay equation, you have:

$$^{241}_{95}Am \rightarrow {}^{237}_{93}Np + {}^{4}_{2}\alpha$$

Worked example

Sulphur-35 decays by means of beta emission. Complete the decay equation below, identifying the element that sulphur decays into.

$$^{35}_{16}S \rightarrow {}^{x}_{y}X + {}^{0}_{-1}\beta$$

Using the fact that the decay equation must be balanced allows you to find the mass and atomic numbers of the product of this decay:

$$35 = 0 + x \quad 16 = y - 1$$
$$x = 35 \quad ; y = 17$$

The product has an atomic number of 17, which can be identified to be chlorine using a periodic table. Thus, the decay equation can be written as:

$$^{35}_{16}S \rightarrow {}^{35}_{17}Cl + {}^{0}_{-1}\beta$$

Exercise 3.1.3 Nuclear stability and radioactive decay

1 Explain the difference between a stable and an unstable nucleus.

2 Name the three main types of radiation, describing each one.

3 Study the following radioactive decay equations. Identify the type of radioactive decay, and complete the decay equation.

a) $^{63}_{28}Ni \rightarrow {}^{63}_{29}Cu + {}^{x}_{y}X$

b) $^{235}_{98}Cf \rightarrow {}^{231}_{96}Cm + {}^{x}_{y}X$

c) $^{137}_{55}Cs \rightarrow {}^{137}_{26}Ba + {}^{x}_{y}X \rightarrow {}^{137}_{26}Ba + {}^{x}_{y}X$

d) $^{232}_{90}Th \rightarrow {}^{228}_{88}Ra + {}^{x}_{y}P \rightarrow {}^{228}_{89}Ac + {}^{x}_{y}Q \rightarrow {}^{228}_{90}Th + {}^{x}_{y}R \rightarrow {}^{224}_{88}Ra + {}^{x}_{y}S$

4 Polonium-210 decays by emission of an alpha particle. Complete a decay equation for polonium-210, using a periodic table to identify the product of the decay.

(continued)

5 Plutonium-236 and plutonium-239 are isotopes. Plutonium has an atomic number of 94.

a) Explain what is meant by the term isotope.

b) Plutonium-239 can be used to make nuclear weapons. Plutonium-239 decays by emitting alpha radiation. Use a periodic table to identify what plutonium-239 decays into.

c) Represent the decay of plutonium-239 as a decay equation.

6 Sodium-24 is used to find leaks in industrial pipelines.

a) Use a periodic table to identify the atomic number of sodium.

b) Sodium decays by beta emission. Write down the decay equation for the radioactive decay of sodium-24, using a periodic table to identify the product.

7 Strontium-90 is used as a power source for weather satellites. It has an atomic number of 38.

a) Write the chemical symbol for strontium-90, including the atomic and mass numbers.

b) Write the decay equation for strontium-90, using a periodic table to identify the product of the decay and state the nature of the decay.

Fig 2.95

Fig 2.96

3.2 Nuclear fission

Nuclear fission is responsible for the production of electricity in a nuclear power station by producing the heat required to drive the turbines. Based on Einstein's famous equation, the mass lost during a nuclear fission reaction is converted into energy as discussed below.

3.2.1 The fission reaction

In a nuclear fission reaction, a heavy nucleus splits apart into two lighter nuclei. The result is an emission of energy because the combined mass of the product nuclei is less than the heavy nucleus. As previously discussed, the lost mass is converted into energy.

Nuclear fission can happen spontaneously – that is, a heavy unstable nucleus can split into smaller nuclei and emit energy. An example of this is uranium-236, which splits into krypton-89 and barium-144, which is shown in Figure 2.97.

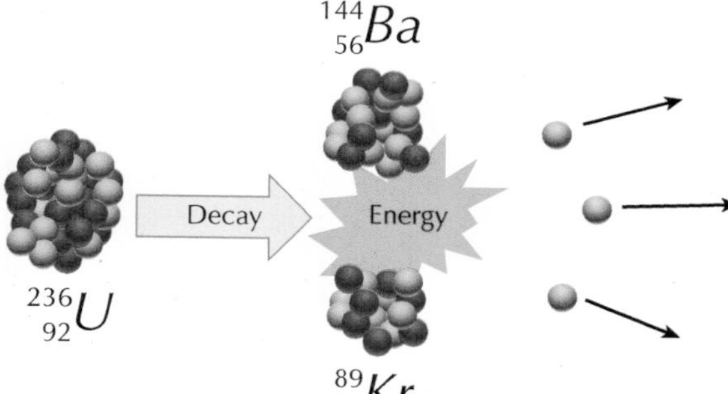

Fig 2.97

As well as splitting into two lighter nuclei, there are also three neutrons released. This is required to ensure that the decay equation is balanced:

$$^{236}_{92}U \rightarrow {}^{144}_{56}Ba + {}^{89}_{36}Kr + 3\,^{1}_{0}n$$

There are other possible combinations that uranium-236 can split into. Another is shown in Figure 2.98 where uranium-236 splits into caesium-143 and rubidium-90.

$$^{236}_{92}U \rightarrow {}^{143}_{55}Cs + {}^{90}_{37}Rb + 3\,^{1}_{0}n$$

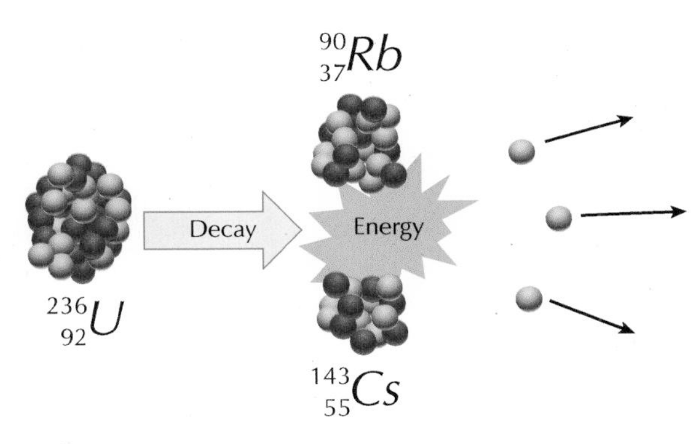

Fig 2.98

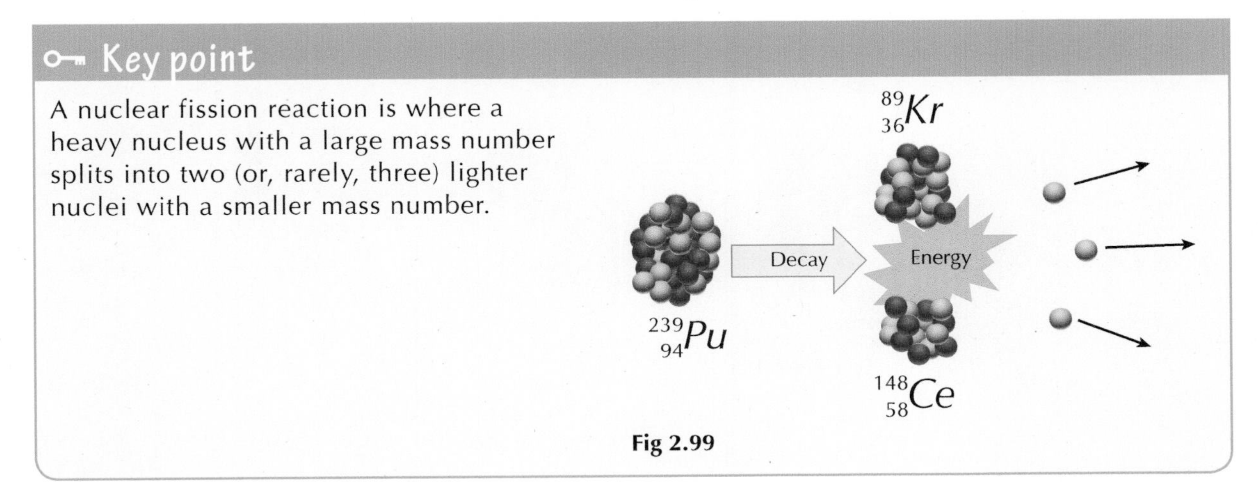

Key point

A nuclear fission reaction is where a heavy nucleus with a large mass number splits into two (or, rarely, three) lighter nuclei with a smaller mass number.

Fig 2.99

Spontaneous vs induced fission

In the above nuclear reaction, uranium-236 split into krypton-89 and caesium-148. It does this spontaneously; there is no external input to cause the uranium to split. This is called **spontaneous** nuclear fission.

Uranium-236 is an unstable nucleus that is very rare on Earth. However, uranium-235 is more readily found. This can be changed to uranium-236 by the addition of a neutron. The uranium-236 can then undergo nuclear fission and split into krypton and barium as described above, releasing energy. The reaction is shown in Figure 2.100:

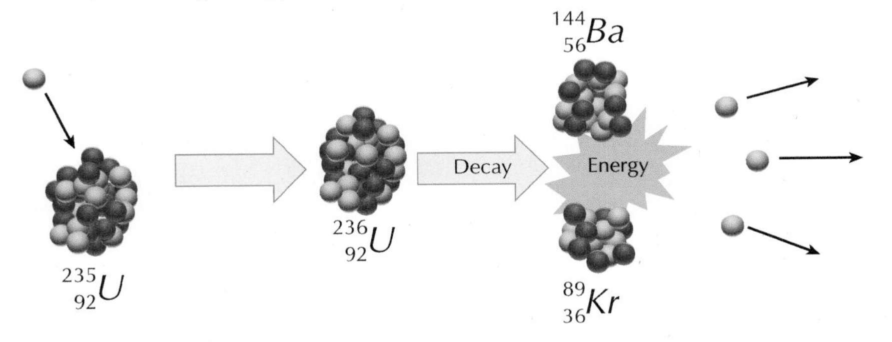

Fig 2.100

$$^{235}_{92}U + ^{1}_{0}n \rightarrow ^{236}_{92}U \rightarrow ^{144}_{56}Ba + ^{89}_{36}Kr + 3^{1}_{0}n$$

This is an example of **induced** nuclear fission. A neutron has been added to uranium-235 to turn it into uranium-236, which then undergoes fission. Usually, this decay is represented in the following way, omitting the stage with uranium-236:

$$^{235}_{92}U + ^{1}_{0}n \rightarrow ^{144}_{56}Ba + ^{89}_{36}Kr + 3^{1}_{0}n$$

Another example of a material that will undergo nuclear fission when hit with a neutron is plutonium-238. A sample reaction of plutonium-238 is shown in Figure 2.101:

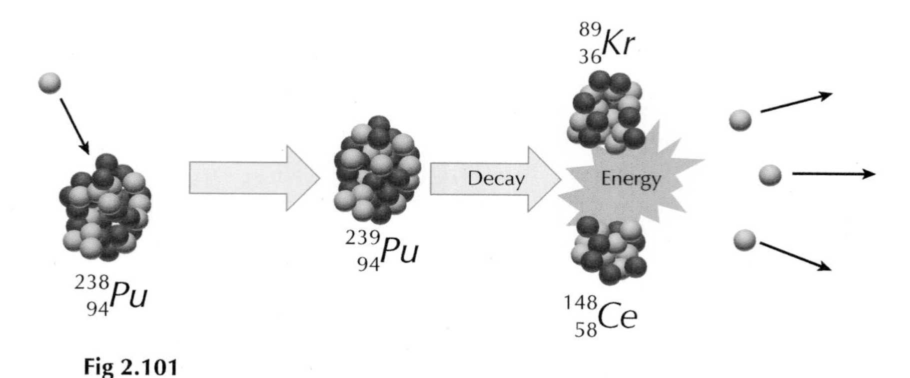

Fig 2.101

$$^{238}_{94}Pu + ^{1}_{0}n \rightarrow ^{239}_{94}Pu \rightarrow ^{148}_{58}Ce + ^{89}_{36}Kr + 3^{1}_{0}n$$

Simplified, this decay equation can be written as:

$$^{238}_{94}Pu + ^{1}_{0}n \rightarrow ^{148}_{58}Ce + ^{89}_{36}Kr + 3^{1}_{0}n$$

In all of the above examples, there are three additional neutrons emitted. These neutrons can go forward and collide with further nuclei to stimulate more fission reactions. This process is known as a 'chain reaction'.

Key point

Nuclear fission can be either spontaneous or induced. Spontaneous nuclear fission happens with no external influence, and can be recognised by there being only one term on the left-hand side of the decay equation:

$$^{236}_{92}U \rightarrow ^{143}_{55}Cs + ^{90}_{37}Rb + 3^{1}_{0}n$$

Induced nuclear fission takes place with the bombardment of a neutron, and can be recognised by the presence of a neutron on the left-hand side of the decay equation:

$$^{238}_{94}Pu + ^{1}_{0}n \rightarrow ^{148}_{58}Ce + ^{89}_{36}Kr + 3^{1}_{0}n$$

3.2.2 Energy from nuclear fission

In the above examples of nuclear fission reaction, the combined mass of the products of the fission reaction are *less* than the nuclei at the beginning. This means that mass has been lost during the fission reaction. This 'lost mass' results in the production of energy according to Einstein's famous equation:

$$E = mc^2$$

This equation has already been applied to the analysis of binding energy, where the mass of the nucleus is less than its constituent parts. In the example of nuclear fusion, the lost mass must be worked out by finding the total mass both before and after the fission reaction. This lost mass is then converted into energy using the equation above. This is highlighted in the following worked example.

Worked example

In a nuclear reactor, uranium-235 is bombarded with a single neutron and undergoes nuclear fission, splitting into rubidium-90 and caesium-143 as shown by the decay:

$$^{1}_{0}n + ^{235}_{92}U \rightarrow ^{143}_{55}Cs + ^{90}_{37}Rb + 3^{1}_{0}n$$

The masses of the elements and particles are shown below:

Element	Mass (kg)
Uranium-235	3.95256×10^{-25}
Caesium-143	2.37336×10^{-25}
Rubidium-90	1.49307×10^{-25}
Neutron	1.67492×10^{-27}

Table 2.8

Calculate the energy released by the above fission reaction.

The energy released by the fission reaction will depend on the mass lost during the reaction. First, work out the mass before the reaction:

$$m = 3.95256 \times 10^{-25} + 1.67492 \times 10^{-27}$$
$$m = 3.9693092 \times 10^{-25} kg$$

Now work out the mass after the reaction:

$$m = 2.37336 \times 10^{-25} + 1.49307 \times 10^{-25} + 3(1.67492 \times 10^{-27})$$
$$m = 3.9166776 \times 10^{-25} kg$$

The lost mass is found by calculating the difference between the mass before and the mass after:

$$\Delta m = 3.9693092 \times 10^{-25} - 3.9166776 \times 10^{-25}$$
$$\Delta m = 5.26316 \times 10^{-27} kg$$

Einstein's equation can now be used to work out the energy from the reaction:

$$E = mc^2$$
$$E = 5.26316 \times 10^{-27} \times (3 \times 10^8)^2$$
$$E = 5.26316 \times 10^{-27} \times 9 \times 10^{16}$$
$$E = 4.74 \times 10^{-10} J$$

Notice that the masses were not rounded during the calculation – due to the small mass differences, it is necessary to carry all of the decimal places.

Key point

The mass before nuclear fission is greater than the mass after. The lost mass is converted to energy according to the following equation:

$$E = mc^2$$

where m is the lost mass (difference in masses before and after) and c is the speed of light.

Exercise 3.2.2 Energy from nuclear fission

The table below can be used for answering the following questions:

Element	Mass (kg)
Plutonium-239	3.96955×10^{-25}
Plutonium-238	3.95291×10^{-25}
Uranium-236	3.91963×10^{-25}
Uranium-235	3.90300×10^{-25}
Barium-144	2.38990×10^{-25}
Caesium-143	2.37336×10^{-25}
Xenon-143	2.37349×10^{-25}
Lanthanum-143	2.37318×10^{-25}
Caesium-140	2.32338×10^{-25}
Tellurium-137	2.27370×10^{-25}
Iodine-131	2.17376×10^{-25}
Molybdenum-100	1.65900×10^{-25}
Molybdenum-97	1.62576×10^{-25}
Bromine-90	1.49333×10^{-25}
Rubidium-90	1.49307×10^{-25}
Strontium-90	1.49295×10^{-25}
Krypton-89	1.47651×10^{-25}
Yttrium-89	1.47632×10^{-25}
Neutron	1.67492×10^{-27}

Table 2.9

Fig 2.102

1 The following nuclear reaction takes place in the reactor of a nuclear power station:

$$^{235}_{92}U + ^{1}_{0}n \rightarrow ^{x}_{y}Cs + ^{93}_{37}Rb + 2^{1}_{0}n$$

a) Determine the atomic and mass numbers of caesium in the reaction above.

b) Determine the energy released by the reaction.

2 A nuclear power station used plutonium-239 as a fuel. It undergoes nuclear fission as follows:

$$^{239}_{94}Pu + ^{1}_{0}n \rightarrow ^{x}_{y}Te + ^{100}_{42}Mo + 3^{1}_{0}n$$

a) Calculate the atomic and mass numbers of tellurium in the reaction above.

b) Determine the energy released by the reaction.

c) If the power station is to produce a power of 40 GW, calculate the number of reactions required in one second, assuming the power station is 100% efficient.

3.3 Nuclear fusion

A nuclear fusion reaction releases energy in the same manner as a nuclear fission reaction – mass is lost during the reaction, which is converted into energy. Nuclear fusion takes place at the core of every star, including our sun. We are therefore bathed in energy produced by nuclear fusion every day!

3.3.1 The fusion reaction

In a nuclear fusion reaction, two lighter nuclei join together to form a heavier nucleus. The combined mass of the nuclei before is greater than the mass of the resulting nucleus. An example of a nuclear fusion reaction is deuterium and tritium, shown in the equation below and in Figure 2.103.

$$^2_1H + ^3_1H \rightarrow ^4_2He + ^1_0n$$

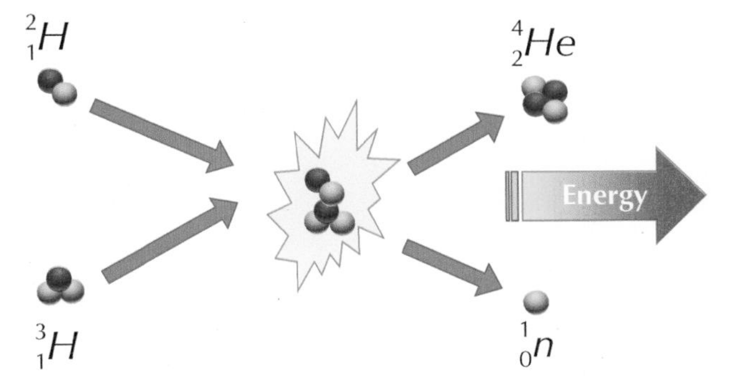

Fig 2.103

As with nuclear fission, the mass and atomic numbers are conserved in the above reaction.

3.3.2 Energy from nuclear fusion

There is a loss of mass that occurs during a nuclear fusion reaction. This lost mass is converted to energy according to Einstein's equation:

$$E = mc^2$$

The following worked example shows the typical energy released from a nuclear fusion reaction.

Worked example

In a star, deuterium (^2_1H) and tritium (^3_1H) collide at very high speeds and undergo nuclear fusion to form helium (^4_2He) and a released neutron.

Given the masses of the particles shown in the table below, find the energy released by this fusion reaction.

Element	Mass (kg)
Deuterium (hydrogen-2)	3.34449×10^{-27}
Tritium (hydrogen-3)	5.00827×10^{-27}
Helium-4	6.64648×10^{-27}
Neutron	1.67492×10^{-27}

Table 2.10

Fig 2.104

(continued)

First, the combined mass before the reaction must be calculated:

$$m = 3.34449 \times 10^{-27} + 5.00827 \times 10^{-27}$$
$$m = 8.35276 \times 10^{-27}\,kg$$

Now, calculate the mass after the reaction:

$$m = 6.64648 \times 10^{-27} + 1.67492 \times 10^{-27}$$
$$m = 8.3214 \times 10^{-27}\,kg$$

To find the energy, the mass difference must then be calculated:

$$\Delta m = 8.35276 \times 10^{-27} - 8.3214 \times 10^{-27}$$
$$\Delta m = 3.136 \times 10^{-29}\,kg$$

The energy released by the reaction is then given by Einstein's equation:

$$E = mc^2$$
$$E = 3.136 \times 10^{-29} \times (3 \times 10^8)^2$$
$$E = 2.82 \times 10^{-12}\,J$$

Key point

Energy is released by a nuclear fusion reaction because the mass of the atoms before the reaction is greater than the mass after. The energy released is given by the equation:

$$E = mc^2$$

Exercise 3.3.2 Energy from nuclear fusion

1 Describe the main difference between nuclear fission and nuclear fusion.

2 Explain why nuclear fission is more easily achieved on Earth than nuclear fusion.

The following table can be used to answer the remaining questions:

Element	Mass (kg)
Deuterium (hydrogen-2)	3.34449×10^{-27}
Tritium (hydrogen-3)	5.00827×10^{-27}
Helium-3	5.00823×10^{-27}
Helium-4	6.64648×10^{-27}
Beryllium-8	1.32931×10^{-26}
Carbon-12	1.99265×10^{-26}
Oxygen-16	2.65602×10^{-26}

Element	Mass (kg)
Neon-20	3.31982×10^{-26}
Magnesium-24	3.98281×10^{-26}
Silicon-28	4.64568×10^{-26}
Sulphur-31	5.14428×10^{-26}
Chromium-52	8.62492×10^{-26}
Iron-56	9.28821×10^{-26}
Neutron	1.67492×10^{-27}

Table 2.11

3 At the core of a star, two tritium atoms can fuse to produce helium-4 and two additional neutrons. The reaction is shown in Figure 2.105.

a) Write down a decay equation to represent the nuclear fusion reaction.

b) Calculate the mass difference for the nuclear fusion reaction.

c) Calculate the energy released by the nuclear fusion reaction.

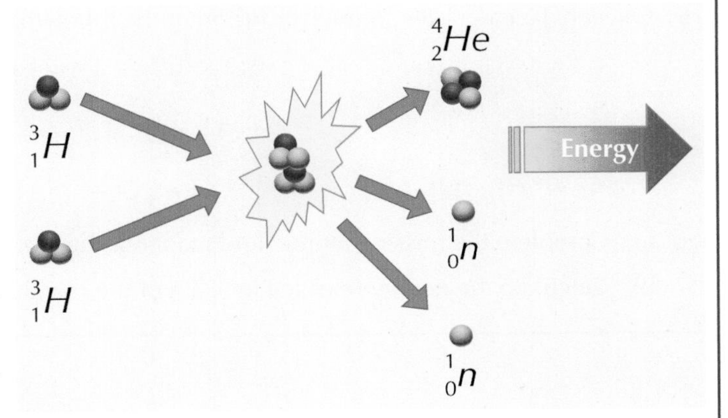

Fig 2.105

4 Helium can be fused into heavier elements. An example of this is the fusion to beryllium, shown in Fig 2.106.

a) Represent this reaction with a decay equation.

b) Calculate the energy released.

c) Beryllium can fuse with another helium nucleus as shown in Figure 2.107.

(i) Identify the element produced by this reaction, including its atomic and mass numbers. You may assume that no additional neutrons are released in the reaction.

(ii) Calculate the energy released.

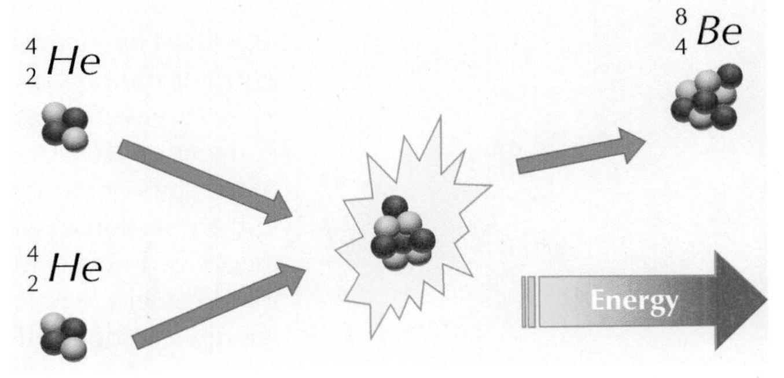

Fig 2.106

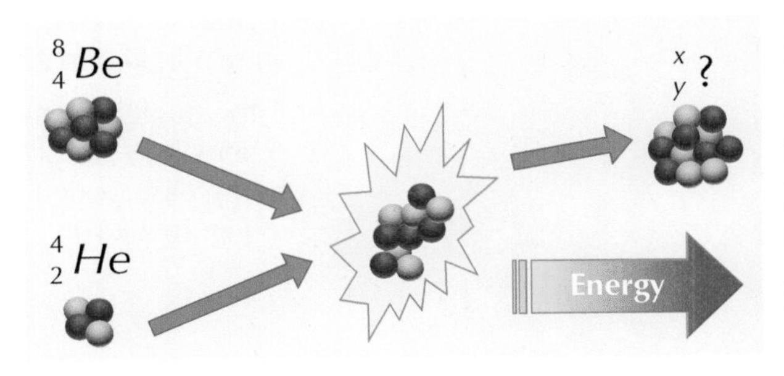

Fig 2.107

(continued)

5 Carbon-12 can undergo nuclear fusion in a reaction shown in Figure 2.108.

Calculate the energy released by this reaction.

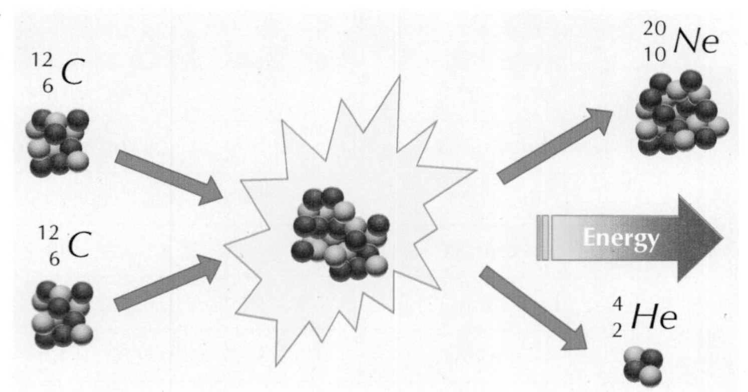

Fig 2.108

6 Oxygen-16 can undergo nuclear fusion in the following reactions.

$$^{16}_{8}O + ^{16}_{8}O \rightarrow ^{24}_{p}Mg + 2^{4}_{2}He$$

$$^{16}_{8}O + ^{16}_{8}O \rightarrow ^{q}_{14}Mg + ^{4}_{2}He$$

$$^{16}_{8}O + ^{16}_{8}O \rightarrow ^{r}_{s}S + ^{1}_{0}n$$

a) Complete the missing atomic and mass numbers p, q, r and s.

b) Calculate the energy released by each of the reactions.

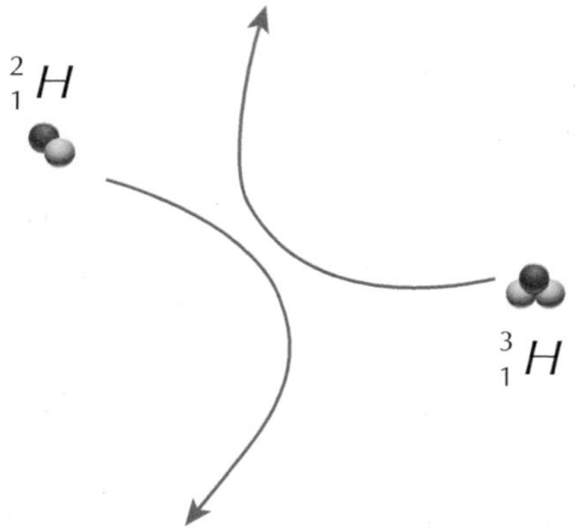

Fig 2.109

3.3.3 Nuclear fusion in stars

Stars bathe the universe in light energy. Without stars, life on Earth as well as life in the universe would not exist. Once born, a star can burn for millions, or even billions of years. When in-depth studies of stars were carried out, physicists quickly realised that conventional fuel sources could not explain either the energy given off by stars or the length of time that a star could burn. The answer to the question of what fuels stars was provided by Einstein and the theory of nuclear fusion. At its core, a star fuses lighter elements into heavier elements in nuclear fusion, releasing huge amounts of energy as it does this.

In order for lighter atoms to fuse, it requires them to be moving at very high velocities. Usually, the nuclei of atoms repel each other, as they are both positively charged – a typical path of two nuclei is shown in Figure 2.109.

If the nuclei are moving at very high speeds, they cannot avoid each other – they smash into each other, and fuse to form a heavier nucleus. In the case of deuterium and tritium, helium-4 is produced as shown in Figure 2.110.

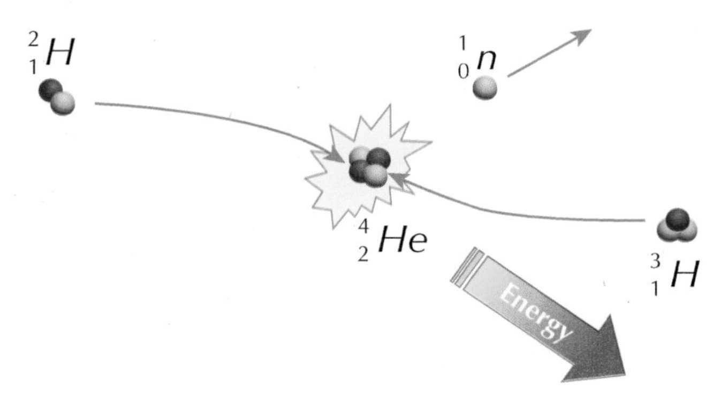

Fig 2.110

To maintain the high speeds required for nuclear fusion, very high temperatures are required. The high velocities (temperatures) are produced in a star using the gravitational field. The sheer size of a star means that the force of gravity is very large, and this causes the nuclei to move at very high speeds – high enough for nuclear fusion. The helium-4 nuclei in the above example can then fuse to give heavier elements again by nuclear fusion. Larger stars are capable of fusing heavier elements, up to iron-26, due to the greater force of gravity.

While the sheer size of stars can produce gravitational fields that generate pressures and temperatures required for nuclear fusion, it is much more difficult to achieve nuclear fusion on Earth. The very high temperatures required are hard to produce and maintain. An example of where they are produced is in a device called a tokamak – this is a giant magnetic bottle that contains a plasma with magnetic fields where nuclear fusion can take place for a short time.

The release of energy from nuclear fusion is the light, heat (and other energies) that reach us on Earth. A star will continue to release energy by nuclear fusion until one of two things happens:

- The star runs out of hydrogen – nuclear fusion shuts down, gravity takes over and the star dies.
- The star fuses heavier atoms into iron. Iron cannot be fused into heavier elements to release energy so nuclear fusion stops, gravity takes over and the star dies in a supernova explosion.

4 Inverse square law

What you should know (National 5):

- The irradiance of light depends on the power of the light and the area over which the light is spread

Learning intentions

- Concept of irradiance as the power per unit area
- Inverse square law for light from a point source

4.1 Irradiance

The further away from a light source you stand, the dimmer the light source appears to be. This is because the irradiance of the light decreases as the distance increases. The brightness of the light source depends on the amount of energy that is entering your eye, and as shown in the section on photoelectric effect, this will be linked to the number of photons entering your eye. For a point source of light, such as a light bulb or a star, the observed irradiance obeys an inverse square law. This can be used in astronomy to work out the distance to a star.

Irradiance, I, of light is a measure of the power that is incident per unit area:

$$I = \frac{P}{A}$$

where P is the power and A is the area in square meters. The unit of irradiance is Wm^{-2}.

The concept of irradiance is very similar to that of pressure. Pressure is defined as the force per unit area and can be thought of as the amount of damage done to the area. The smaller the contact area, the greater the pressure – think about the damage a stiletto heel can do to a floor or a drawing pin can do to a solid wall.

Fig 2.111

The irradiance from a light source depends on both power and the area illuminated. The smaller the area illuminated, the greater the irradiance. A typical light bulb will spread its light over a large area, so the irradiance will be low. A laser concentrates its power over a very small area, making the irradiance very large. For some lasers, the irradiance is large enough to cut through metal and even diamond! Figure 2.112 shows a laser being used in industry to cut sheet metal:

Fig 2.112

Converting to square metres

Irradiance is defined as the power per unit area, where the area is measured in square metres. For many sources of light, such as a laser, the beam area will be very small and often quoted in square millimetres. This must be converted to square metres before being used to find the irradiance.

To convert from square millimetres into square metres:

$$1\,mm^2 = 1 \times 10^{-6}\,m^2$$

To work this out, 1 square millimetre is equal to:

$$1\,mm^2 = 1\,mm \times 1\,mm$$
$$1\,mm^2 = 1 \times 10^{-3}\,m \times 1 \times 10^{-3}\,m$$
$$1\,mm^2 = 1 \times 10^{-6}\,m^2$$

Key point

The irradiance of a radiation depends on the power, P, and the surface area, A, according to:

$$I = \frac{P}{A}$$

This is very similar to the equation for pressure. For this reason, irradiance is sometimes referred to as radiation pressure.

Worked example

A laser beam has a beam area of 4×10^{-6} m^2. The power of the laser beam is 0.1 W. Calculate the irradiance of the laser beam.

$$I = \frac{P}{A}$$
$$I = \frac{0.1}{4 \times 10^{-6}}$$
$$I = 25{,}000\ Wm^{-2}$$

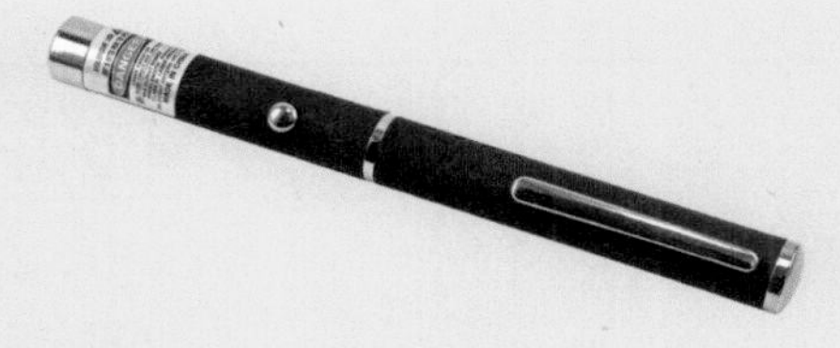
Fig 2.113

Exercise 4.1 Irradiance

1 The light from a 60 W light bulb is spread across the surface area of a small room with an area of 15 m^2. Calculate the irradiance of the bulb.

2 Car headlights have a combined power of 50 W. They spread their light over an area of 5 m^2. What is the irradiance of the headlights?

3 The accepted value for irradiance of the Sun's radiation by the Earth's upper atmosphere is 1368 Wm^{-2}. Calculate:

a) The power of the Sun's energy absorbed by an area of 5 square metres.

b) The energy absorbed in a time of one minute.

4 460 Wm^{-2} of visible light is incident on the Earth's surface from the Sun. A solar cell can convert 45% of this light to electrical energy (45% efficient).

a) Calculate the power incident on a solar cell of surface area 14 square meters.

b) How much electrical power does this solar cell produce?

c) How much electrical energy is produced by the solar cell in a 12-h time period?

5 By considering the typical area of a laser beam (1 mm^2) to the area over which the light from a light bulb is spread, explain why a laser beam with a very low power can do more damage than a high-powered light bulb.

6 A laboratory laser produces an irradiance of 1200 Wm^{-2} over an area of 2 mm^2. It is used to supply energy to a special crystal that will emit different colours of light when illuminated by laser light.

a) Calculate the power of the laser incident on the crystal.

b) If the crystal needs to be supplied with 10 J of energy, find the length of time it needs to be illuminated by the laser.

Fig 2.114

4.2 Inverse square law

It has already been shown that a light bulb spreading its light over a large surface area has a smaller irradiance than a laser beam that spreads its light over a very small area. A light bulb is an example of a point source of light. A laser is not a point source of light because it only spreads its light over a very small area in a directed beam.

GO! Experiment 4.2 Irradiance from a point source of light

This experiment investigates the irradiance from a point source of light as its distance is varied from the sensor.

Apparatus

- Dark room
- Metre stick
- 12 V lamp
- Light sensor and meter

Instructions

1. Connect the lamp to a suitable 12 V power supply and ensure that it lights up. The voltage to the lamp must be kept constant throughout the experiment.
2. Position the light sensor at 0.1 m away from the lamp and measure the irradiance.
3. Repeat the irradiance measurements for different distances away from the lamp, using a metre stick to measure the distance.
4. Record your results in Table 2.12.

Distance, d (m)	Light irradiance, I (units)	d^2 (m^2)	$\frac{1}{d^2 (m^{-2})}$
0.1			

Table 2.12

5. Plot a graph of irradiance vs distance from the bulb.
6. Plot a graph of irradiance vs inverse distance from the bulb.
7. Use your graphs to obtain a relationship between irradiance and the distance from the bulb.

Light from a point source travels in all directions, covering a spherical area as shown in Figure 2.115.

The surface area of a sphere is given by:

$$A = 4\pi r^2$$

Therefore, as the distance, d, away from the source increases, the surface area the light covers increases as $4\pi d^2$. Here d is the radius of the sphere. The irradiance must therefore decrease as $4\pi d^2$ for a given power:

$$I = \frac{P}{A}$$

$$I = \frac{P}{4\pi d^2}$$

Fig 2.115

For a given source, the power is constant. The value of 4π is also constant, so:

$$I = \frac{k}{d^2}$$

$$I \propto \frac{1}{d^2}$$

This is an example of an inverse square law. The irradiance from a point source of light decreases as the square of the distance away from the light. For example, the irradiance of a light source decreases by a factor of four when the distance doubles.

The above inverse square law can be written as follows:

$$I = \frac{k}{d^2}$$

$$Id^2 = k$$

For the same light source, the irradiance multiplied by the distance from the source squared will always be the same value. Thus:

$$I_1d_1^2 = I_2d_2^2$$

Key point

The irradiance of light from a point source decreases with distance away from the source according to the inverse square law:

$$I \propto \frac{1}{d^2}$$

Thus:

$$I_1d_1^2 = I_2d_2^2$$

Worked example

The irradiance a distance of 3 m away from a light source is measured to be 45 Wm^{-2}. Calculate the irradiance a distance of 5 m away from the light source.

$$I_1d_1^2 = I_2d_2^2$$

$$45 \times 3^2 = I_2 \times 5^2$$

$$405 = 25I$$

$$I = 16.2\ Wm^{-2}$$

Exercise 4.2 Inverse square law

1 The irradiance from a light bulb is being measured. At a distance of 2 m away from the bulb, the irradiance is found to be 12 Wm^{-2}.

a) State the relationship between irradiance and distance from a point source of light.

b) Calculate the irradiance at a distance of 1 m from the bulb.

c) Calculate the irradiance at a distance of 2.5 m from the bulb.

2 The irradiance from a point source of light is found to be 15 Wm^{-2} at a distance of 4 m from the source. Find the distance away from the source where the irradiance is 30 Wm^{-2}.

3 Students set up an experiment to investigate the relationship between irradiance and distance from a point source as shown in Table 2.13:

Distance from source, d (m)	1.0	1.5	2.3	3.5	5.0
Measured irradiance, I (Wm^{-2})	200	89	38	12	8

Table 2.13

a) Use all of the data to establish a relationship between the irradiance and the distance from the light source.

b) Calculate the irradiance at a distance of 4 m from the source.

c) At what distance away from the bulb would the irradiance of the light be 70 Wm^{-2}?

4 An engineer measures the energy produced by a solar cell with an area of 4 m^2. The engineer finds an energy of 3400 J is produced by the cell in 30 s. The solar cell is assumed to be 100% efficient.

a) Calculate the power incident on the solar cell.

b) Find the irradiance of light incident on the solar cell.

c) The solar cells are 20 m away from a test source of light, which is a point source.

(i) Calculate the irradiance at a distance of 10 m away from the source of light.

Fig 2.116

(ii) Find the new power of light incident on the cell.

(iii) Find the energy produced by the cell at a distance of 10 m away in a time of 30 s.

5 A student is comparing a 40 W light bulb to a 5 mW laser pen. She measures the irradiance from each of the sources at a distance of 2 m away from the source.

Fig 2.117

a) Which of the above sources of light is a point source? Explain your choice.

b) The beam area of the laser pen at a distance of 2 m away is found to be 4 mm^2. Find the irradiance from the laser pen at this distance.

c) Assuming the light bulb is 100% efficient, calculate the irradiance of the bulb at a distance of 2 m away. (Hint – think about the area the light is spread over at a distance of 2 m!)

d) Compare your answers to parts b) and c) and comment on why a laser is considered to be more dangerous to eyesight than a bulb, despite its much lower power.

6 A 100 W light bulb is 15% efficient. It can be considered as a point source.

a) Find the irradiance of the bulb at a distance of 4 m.

b) Calculate the total power incident on a surface area of 0.5 m^2 at a distance of 4 m from the bulb.

c) Find the amount of light energy incident on this area in a time of 1 hour.

5 Wave–particle duality

What you should know (National 5):

- The colour of light depends on the frequency of the light wave
- The energy of an electromagnetic wave depends on the frequency of the wave; the higher the frequency, the greater the energy

Learning intentions

- Relationship between the frequency of an electromagnetic wave and the energy carried by a photon
- Concept of light travelling as **photons**
- Photoelectric effect as evidence for the particle nature of light:
  - Only photons with enough energy can eject electrons from the surface of a metal
  - There is a one to one interaction: an electron can absorb **only one photon**
  - The **work function** of a metal is the minimum energy required to free an electron
  - The maximum kinetic energy of a released electron is equal to the difference between the energy of the photon freeing the electron and the work function of the metal

5.1 Quantum theory of light

In our studies of physics so far, we have considered light to be a continuous wave. The frequency of the light can change, and these changes result in a change of colour of the light. This classical picture of light was used to explain experimental results until the late 19th and early 20th centuries. However, scientists such as Albert Einstein, Philipp Lenard and Arthur Compton made important discoveries that shook this classical understanding. A new theory of light was required.

5.1.1 Classical vs quantum theory

Classical theory of light describes light travelling as a continuous wave through space. This theory was supported by the fact that light produces an interference pattern when shone through Young's slits or a diffraction grating. Interference is a property of waves. A continuous wave is shown in Figure 2.118, along with its key properties.

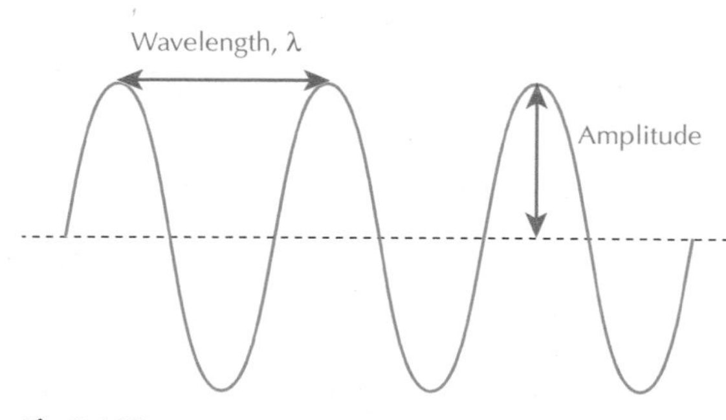

Fig 2.118

The frequency, f, wavelength, λ, and velocity, v, of the wave are linked by the equation:

$$v = f\lambda$$

The speed of light in air is 3×10^8 m s^{-1}, and is a physical constant that depends on the permittivity (the ability of a substance to store electrical energy in an electric field) and permeability of free space (for more information, see the Special relativity section in Area 1, Chapter 5).

The amplitude of the wave is related to the energy carried by the wave – the greater the amplitude, the more energy the wave carries. In classical wave theory, the greater the amplitude of the wave, the brighter the light source.

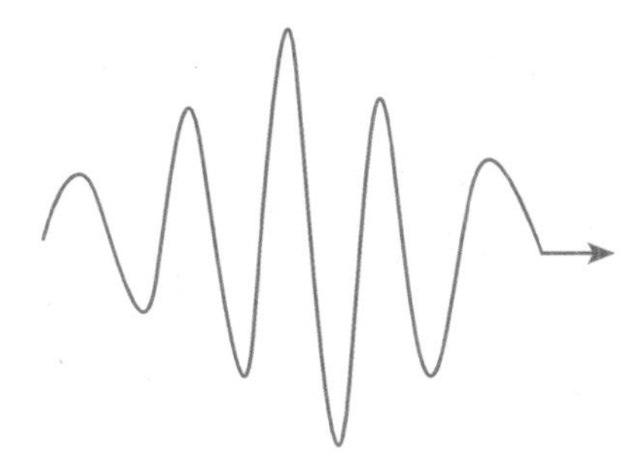

Fig 2.119

In the early 20th century, Albert Einstein proposed a different theory for light to explain experiments carried out by Heinrich Hertz and Philipp Lenard. He proposed that light was a stream of tiny wave packets (particles) called quanta or photons.

Just like classical wave theory, a photon has a velocity, wavelength and frequency that are linked by the following equation:

$$v = f\lambda$$

However, a key difference to classical wave theory is that the energy does not depend on the amplitude. Instead, the energy of the photon is linked to the frequency.

Key point

A photon is a wave packet – a small 'chunk' of light that combines both particle- and wave-like behaviour.

5.1.2 Energy of a photon

The energy of the photon depends on its frequency, where the greater the frequency, the greater the energy:

$$E \propto f$$

The constant of proportionality here was named after Max Planck (Figure 2.120), and is known as Planck's constant, h:

$$h = 6.63 \times 10^{-34}\ Js$$

This gives the following equation for the energy carried by a photon of light:

$$E = hf$$

Fig 2.120

As well as showing that the energy of a photon is related to the frequency (and therefore the colour), the equation above has another important result. It shows that the total energy is quantised – that is, it is not continuous, but instead a discrete quanta of energy with multiples of Planck's constant. It is this quantisation that lends itself to the name of 'quantum theory'.

In practice, brighter light carries more energy – for example, brighter light will produce more energy from a solar cell. This is because a bright light source will have many photons leaving it in a given time, whereas a dimmer light source will have fewer photons leaving it in a given time. The greater the number of photons per second, the greater the energy per second and hence the brighter the light.

Key point

The energy of a photon depends on the frequency of the photon:

$$E = hf$$

where h is Planck's constant:

$$h = 6.63 \times 10^{-34}\ Js$$

Worked example 1

Calculate the energy of a photon of red light that has a wavelength of 633 nm.

In order to work out the energy, you first need to know the frequency of the light. This can be found using the wave equation:

$$v = f\lambda$$

Substitute what is known (remember, the speed of light is 3×10^8 m s^{-1}) and solve for the frequency:

$$3 \times 10^8 = f \times 633 \times 10^{-9}$$
$$f = 4.74 \times 10^{14} Hz$$

Now you can work out the energy of the photon using the following equation:

$$E = hf$$

Substitute what you know and solve for energy:

$$E = 6.63 \times 10^{-34} \times 4.74 \times 10^{14}$$
$$E = 3.14 \times 10^{-19} J$$

Exercise 5.1.2 Energy of a photon

1 Describe the key differences between the classical wave theory of light and the quantum theory of light.

2 Explain what a photon is.

3 Calculate the energy of the following photons of light:

a) Blue light with a wavelength of 480 nm

b) Red light with a wavelength of 700 nm

c) Green light with a wavelength of 540 nm

d) Violet light with a wavelength of 400 nm

e) Ultraviolet light with a wavelength of 240 nm

4 A photon has an energy of 2.56×10^{-20} J.

a) Calculate the frequency of the photon.

b) Calculate the wavelength of the photon if it is travelling through air.

5 A red laser emits light with a wavelength of 633.6 nm. Find:

a) The frequency of the laser light.

b) The energy of each photon of the laser light.

Fig 2.121

6 Consider the energies of different photons from the electromagnetic spectrum.

a) Which member of the electromagnetic spectrum emits photons with the longest wavelength? Explain your answer.

b) Which member of the electromagnetic spectrum emits photons with the shortest wavelength? Explain your answer.

c) By considering the diagram below, which shows members of the electromagnetic spectrum, state why UV radiation from the sun is more dangerous to skin than visible light. Why do we need to take precautions against UV radiation?

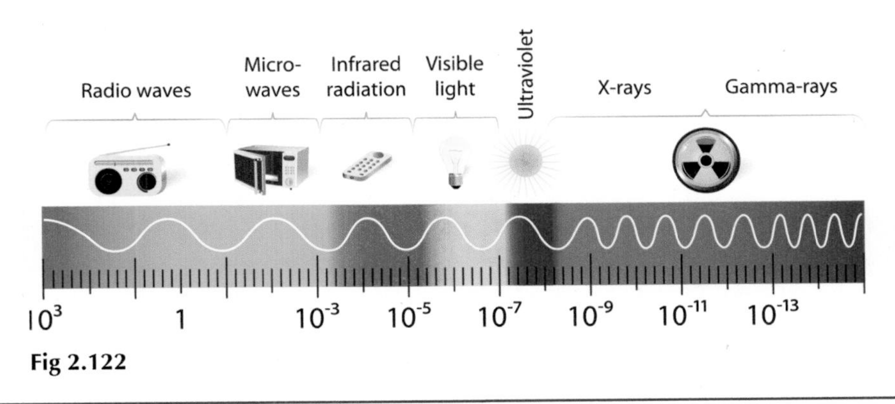

Fig 2.122

5.2 The photoelectric effect

One of the main pieces of evidence for the particle theory of light is the photoelectric effect.

5.2.1 The photoelectric effect: Hertz's observations

The photoelectric effect was first discovered by Heinrich Hertz in the late 19th century when he was experimenting with his spark-gap apparatus, shown in Figure 2.123.

Hertz was conducting experiments to investigate the electromagnetic radiation that would carry a spark from the transmitter to the receiver. An induction coil was used to produce a high voltage spark on the spark gap of the transmitter. This produced electromagnetic waves, which travelled to the receiver and produced a spark on the receiver (metal ring). This was used by Hertz to demonstrate the presence of electromagnetic waves.

The spark-gap apparatus was quite difficult to work with, with the air gap having to be very small for the spark to be reproduced on the receiver. This meant that the experiment would only work over distances of a few metres – no more than 15 metres in Hertz's work. In an attempt to better see the spark on the receiver, Hertz carried out the experiment in a dark room. This was found to reduce the distance over which the experiment worked. Illuminating the experiment with visible light, or ultraviolet light had the effect of increasing the sensitivity of the experiment, allowing the spark to be observed at greater distances from the transmitter. Without knowing it, Hertz had discovered an important effect that would later shake the world of classical physics and introduce the quantum world. In 1897, these observations were shown by J.J. Thompson to be due to the light pushing electrons off the metal

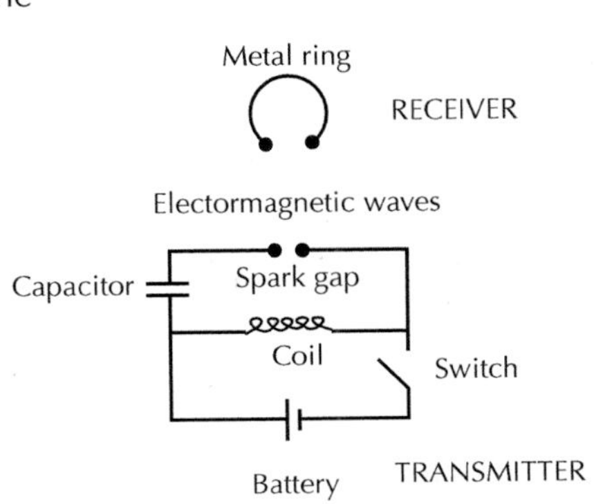

Fig 2.123

surface. The free electrons would allow for a spark to be observed with weaker electromagnetic radiation, i.e. at greater distances from the transmitter.

5.2.2 The photoelectric effect: detailed experiments

The first detailed experiments studying the photoelectric effect were carried out by Philipp Lenard in 1902 and then by Robert Millikan.

GO! Experiment 5.2.2 Demonstrating the photoelectric effect

This experiment investigates the basic properties of the photoelectric effect.

Apparatus

- A gold leaf electroscope
- A clean zinc plate
- Charged rods and a duster
- Various light sources, including visible and ultraviolet

Fig 2.124

Instructions

1. Attach the zinc plate to the top of the gold leaf electroscope.
2. Charge the zinc plate with a negative charge (excess electrons) using the charged rods. This will charge the electroscope and the golf leaf will rise.
3. Illuminate the zinc with each of your light sources – including visible and ultraviolet – and note the effect this has on the gold leaf.
4. Which light source is capable of discharging the electroscope?

Hertz's original observations with the spark-gap apparatus revealed that visible or ultraviolet light improved the sensitivity of the detector. This was predicted to be due to the electromagnetic radiation releasing electrons from the surface of the metal. More detailed experiments have found that not every colour of light (and therefore not every frequency of light) is capable of removing the electrons from metals. Furthermore, when electrons are released from the metal they have a certain kinetic energy – the value of this energy does not depend on the brightness of the light, but instead on the frequency of the light.

Philipp Lenard's photoelectric experiments

In 1902, Philipp Lenard carried out detailed experiments into the photoelectric effect using the apparatus shown in Figure 2.125.

When light of the correct frequencies illuminates the metal plate, a small current is detected in the ammeter. This current is produced by electrons being ejected from the metal surface, as shown in Figure 2.126. These electrons are attracted to the positive plate, which produces a current.

This current increases as the irradiance of the light increases, which is to be expected. However, the fact that only certain colours of light are capable of releasing electrons is confusing.

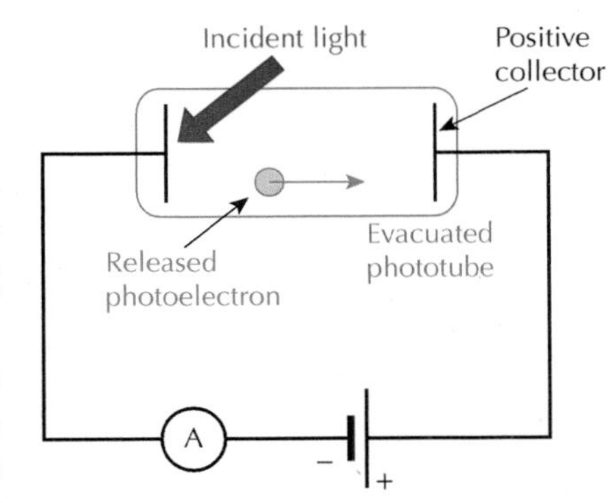

Fig 2.125

This result is compounded by a subtle alteration to the above apparatus (see Figure 2.127). The collector for the electrons is now connected to the negative terminal of the power supply, resulting in a negatively charged collector.

Rather than having the released electrons attracted to a positive plate, they are repelled by a negative plate. This means the electrons need to have enough kinetic energy when released to make it across the gap. The energy required depends on the potential difference across the gap (V) – the greater the potential difference, the greater the kinetic energy required. This comes from the equation for work done in an electric field:

$$E_W = QV$$

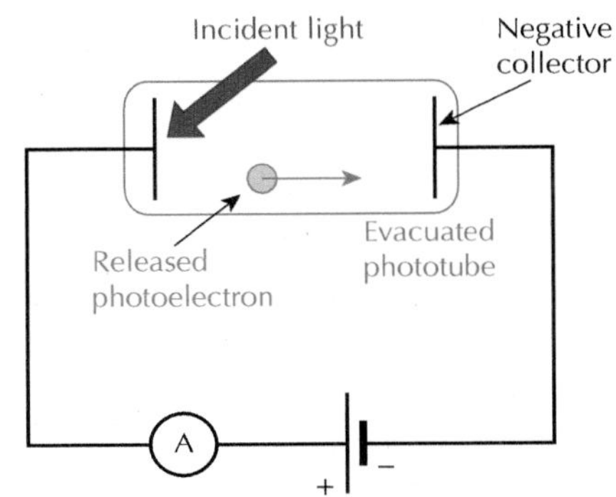

Fig 2.126

where Q here is the charge on the electron. Increasing the potential difference across the gap will slowly reduce the observed current to zero. At this point, the maximum kinetic energy of the released electrons,

$$E_K = \frac{1}{2}mv^2$$

is equal to the work done and the electrons cannot make it across the gap. This effect is similar to attempting to throw a ball fast enough (i.e. with enough kinetic energy) to hit the ceiling. It can be observed that for a given colour of light, the voltage required to reduce the current to 0 (known as the stopping voltage) is independent of the irradiance of the light. This is contrary to what would be expected classically, where the irradiance of the light is linked to its energy.

Robert Millikan's photoelectric experiments

Robert Millikan extended Lenard's experimental observations to find that there was a colour dependence (and hence a frequency dependence) on the emission of photoelectrons. He found that light with a frequency below a certain value would not remove electrons from the metal plate, regardless of how bright the light was. In other words, the irradiance of the light (classically related to the energy) had no effect on whether electrons were emitted or not.

The minimum frequency required to remove electrons from a surface is known as the 'threshold frequency'. This implies that for a given material, there is a minimum energy required to remove an electron. This minimum energy is different for different materials and is referred to as the 'work function', ϕ, of the material. These results are summarised in Figure 2.128.

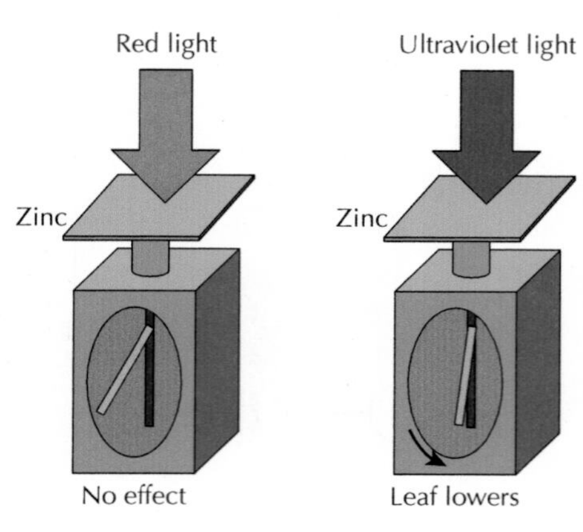

Fig 2.127

Summary of observations

From Hertz's initial experiments to Lenard's and Millikan's, there were several important observations that highlighted major problems with the classical wave theory of light. These observations are:

- Light (electromagnetic radiation) is capable of removing electrons from a metal surface.
- The frequency of the light has to be high enough (above the threshold frequency, f_0) to eject electrons.
- The photocurrent (number of electrons released per second) depends on the irradiance of the light – the greater the irradiance, the greater the current.
- The maximum kinetic energy of the ejected electrons depends on the frequency of the incident light and not the irradiance.

5.2.3 Einstein's photoelectric equation

In order to explain the effects observed in the photoelectric effect, Einstein proposed a new theory for light in 1905.

Einstein explained that the classical theory of light as a continuous wave remains applicable to situations such as reflection, refraction and interference. However, it does not account for the observations seen in many other experiments, including the photoelectric effect. Instead, the results can be accounted for by considering light travelling as a particle. The concept of the photon emerged after Einstein's paper was published, and this is now the accepted theory for the nature of light.

Fig 2.128

As discussed in the previous section, each photon of light has an energy that depends on its frequency:

$$E = hf$$

This energy can be used to eject an electron from a metal. However, only one photon can interact with one electron. So, if the energy of the photon is not great enough to free the electron, it will remain trapped in the metal. The electron cannot absorb energy from more than one photon – there is a one-to-one correspondence between the photon and the electron. This explains the threshold frequency for the emission of electrons from a metal. This also explains the observation of increased photocurrent for increased irradiance above a certain frequency – the greater the number of incident photons in a given time, the greater the number of emitted electrons.

Excess energy that the electron does not use to escape from the metal is kinetic energy. This kinetic energy depends only on the frequency of the incident radiation, and not the irradiance. This shows that the classic theory of energy being linked to amplitude (irradiance) of a continuous wave is not accurate. The energy of the light depends on the frequency.

The photoelectric effect equation

Einstein described the photoelectric effect with the following relationship:

$$E_K = hf - \phi$$

where E_K is the maximum kinetic energy of a released electron, hf is the energy of the incident photon and ϕ is the work function of the material. The work function of the material is also expressed as:

$$\phi = hf_0$$

where f_0 is the threshold frequency. This gives the following relationship:

$$E_K = hf - hf_0$$

This relationship can be explained using the simple energy diagram shown in Figure 2.130.

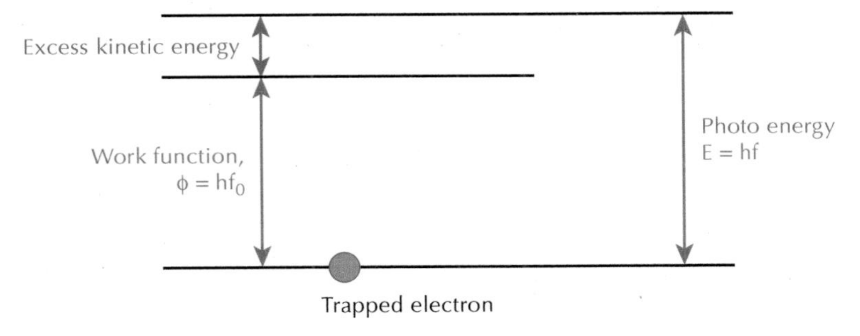

Fig 2.129

If a photon with energy greater than the work function is incident on the metal, electrons will be released. The excess energy will be kinetic energy of the electron.

It is important to remember that the kinetic energy in the above equation and diagram is the maximum kinetic energy. Some electrons are bound by greater energies and thus require photons with a greater frequency to be released. This means that as the frequency of the light is increased, the number of electrons released will increase. Electrons that are bound by greater energies can be released from the metal. Think about a hot cup of coffee. If you blow gently over the top of the coffee, you can 'skim off' the top layer of particles. However, if you blow harder then you can blow away particles that are deeper. This effect is similar to electrons being ejected by photons. This results in a relationship between the photon energy and the photocurrent being observed as shown in Figure 2.131.

The photoelectric equation demonstrates that as the photon energy increases, the maximum kinetic energy also increases. It is the equation of a straight line:

$$E_K = hf - hf_0$$
$$y = mx + c$$

Key point

Only one photon can interact with one electron at a time – there is a one-to-one correspondence.

An electron can therefore only be released from a metal if the energy of the incident photon is great enough – in other words, if the frequency of the incident photon is greater than or equal to the threshold frequency.

Key point

If electrons are being emitted from the metal by photons, then the irradiance of the light will affect the number of electrons emitted in a given time – the greater the number of photons, the greater the number of electrons emitted.

The kinetic energy of the emitted electrons does not depend on the irradiance of the incident light.

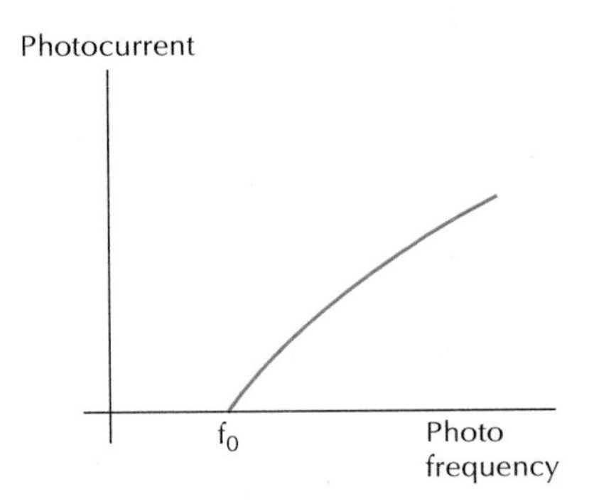

Fig 2.130

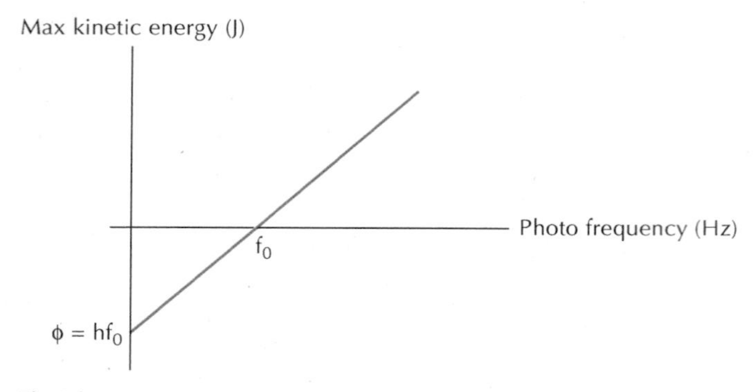

Fig 2.131

When the maximum kinetic energy is plotted against the photon frequency, the results can be shown in a graph (Fig 2.132).

The gradient of the line gives Planck's constant, h and the y-intercept is the work function.

If the photon energy and work function for a metal are known, it is possible to work out whether photoelectrons will be released, and what the maximum kinetic energy of these photoelectrons will be.

Key point

The maximum kinetic energy of an electron released from a metal by a photon with frequency f is given by:

$$E_K = hf - hf_0$$

where f_0 is the threshold frequency and h is Planck's constant:

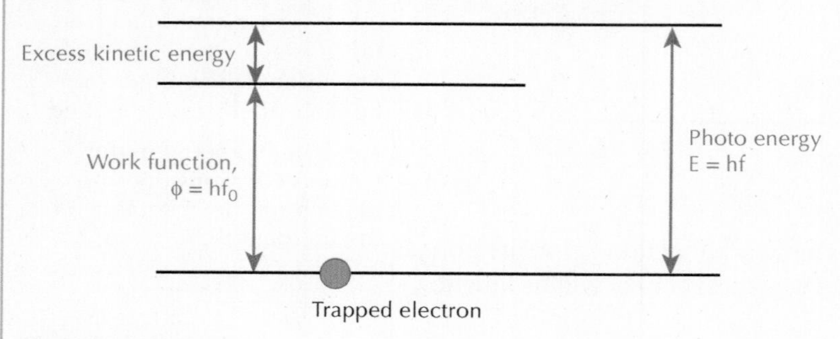

Fig 2.132

If the photon energy is below the threshold frequency, no electrons will be released from the metal.

Worked example 1

The work function of a metal is 1.5×10^{-19} J. If light of wavelength 450 nm in air is incident on the metal, determine:

a) Whether or not a photoelectron is released from the metal.

b) If an electron is released, what its maximum kinetic energy will be.

In order to work out whether a photoelectron will be released or not, you need to know the energy of the photon, which is given by:

$$E = hf$$

You know the wavelength of the photon, so must use the wave equation,

$$v = f\lambda$$

to find the frequency. Substitute what you know and solve for the frequency:

$$3 \times 10^8 = f(450 \times 10^{-9})$$

$$f = \frac{3 \times 10^8}{450 \times 10^{-9}}$$

$$f = 6.67 \times 10^{14} Hz$$

The energy of the photon can now be calculated:

$$E = hf$$

$$E = 6.63 \times 10^{-34} \times 6.67 \times 10^{14}$$

$$E = 4.42 \times 10^{-19} J$$

This energy is greater than the work function of the metal, so photoelectrons will be released.

The maximum kinetic energy of the photoelectrons is the difference between the photon energy and the work function of the metal, according to Einstein's photoelectric effect equation:

$$E_K = hf - hf_0$$

$$E_K = 4.42 \times 10^{-19} - 1.5 \times 10^{-19}$$

$$E_K = 2.92 \times 10^{-19} J$$

Exercise 5.2.3 Einstein's photoelectric effect equation

1 Explain the discovery that led Heinrich Hertz to find that electrons could be 'pushed off' metals using electromagnetic radiation.

2 Philipp Lenard and Robert Millikan made important discoveries that led Albert Einstein to conclude that light was not a continuous wave. Lenard demonstrated that the maximum kinetic energy of electrons emitted from a metal does not depend on the brightness (irradiance) of the light, and Millikan showed that photoelectrons were only emitted when light above a certain frequency was used.

a) Explain the model of light that Einstein proposed in light of Lenard's and Millikan's discoveries.

b) Expain why Lenard's experimental results cannot be accounted for by a continuous wave model of light.

c) Explain why Millikan's experimental results cannot be accounted for by a continuous wave model of light.

d) Explain what is meant by the terms 'threshold frequency' and 'work function of a metal'.

3 Einstein's photoelectric effect equation is used to describe the photoelectric effect.

a) Write down Einstein's photoelectric effect equation.

b) Explain the meaning of each of the terms in Einstein's photoelectric equation.

(continued)

4 The following experiment is used in the laboratory to demonstrate the photoelectric effect. A clean zinc plate is placed on top of a gold leaf electroscope and is given an excess negative charge. This causes the leaf in the electroscope to rise.

a) Describe why charging the zinc plate with electrons causes the gold leaf to rise.

b) The zinc plate is illuminated with visible light and the gold leaf does not change its position. When illuminated with ultraviolet light, the gold leaf is observed to fall.

(i) Explain, in terms of the electrons on the plate, why the gold leaf falls when illuminated with ultraviolet light.

(ii) Why does visible light not cause the gold leaf to fall?

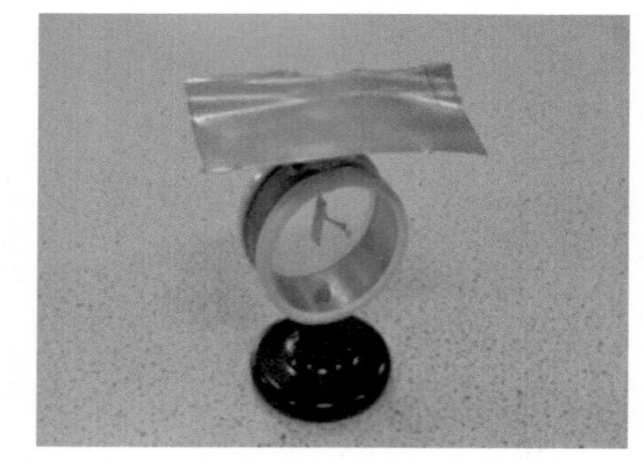

Fig 2.133

5 In an experiment to demonstrate the photoelectric effect, an uncharged plate of zinc is placed onto a gold leaf electroscope. When illuminated with UV light, the gold leaf was found to rise steadily. By considering the movement of charge from the zinc plate, explain this observation.

6 The work function of a particular metal is 3.4×10^{-19} J. Light with wavelengths below are incident on the metal. Determine whether or not each wavelength is capable of ejecting photoelectrons:

a) 900 nm

b) 700 nm

c) 400 nm

d) 200 nm

7 The circuit in Figure 2.135 is set up to demonstrate the photoelectric effect.

When light of a high enough frequency is incident on the metal surface, electrons are released and a current is detected on the ammeter.

The work function of the metal is 2.7×10^{-19} J.

a) Explain why, when electrons are ejected from the metal surface, a current is detected on the ammeter.

b) Calculate the minimum frequency and corresponding maximum wavelength that will produce a current on the ammeter.

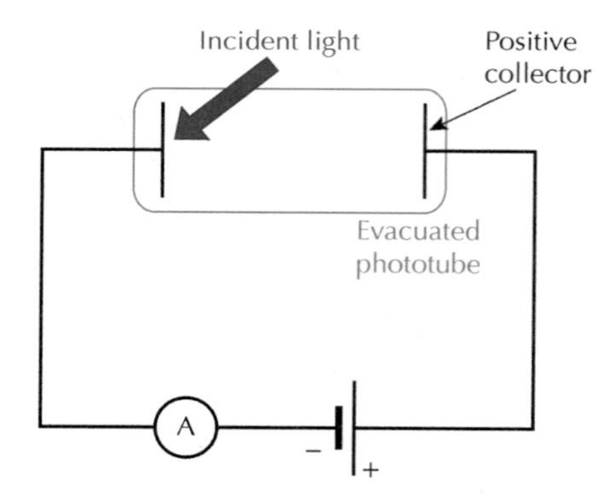

Fig 2.134

c) Light with a wavelength of 260 nm is incident on the metal plate.

(i) Find the maximum kinetic energy of the ejected photoelectrons.

(ii) The irradiance of the incident light is increased. Describe and explain the effect this has on the current observed.

d) The polarity of the power supply is now reversed. Explain why electrons can still make it across the gap to produce a current on the ammeter so long as the voltage of the power supply is small enough.

8 Light of wavelength 450 nm is incident on a metal. Electrons with a maximum kinetic energy of 1.4×10^{-20} J are found to be ejected from the metal.

a) Find the work function of the metal.

b) Calculate the threshold frequency for this metal.

9 The apparatus shown in Figure 2.136 is used to investigate the photoelectric emission from different metals.

Light is incident on different metals inside the evacuated tube. Test both zinc and caesium, the work functions for which are:

- Zinc: work function = 6.88×10^{-19} J
- Caesium: work function = 3.36×10^{-19} J

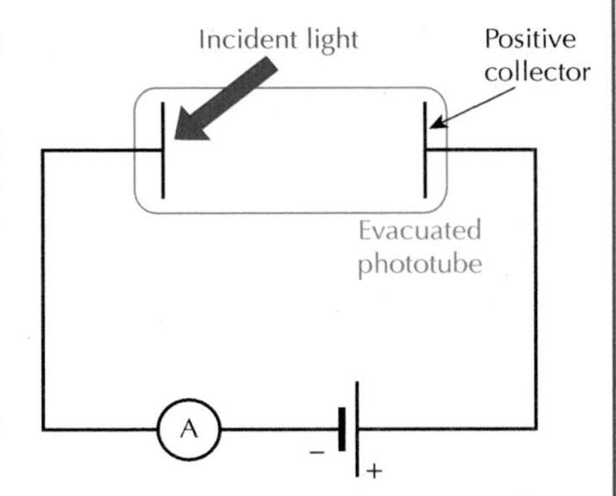

a) Explain what is meant by the term 'photoelectric emission'.

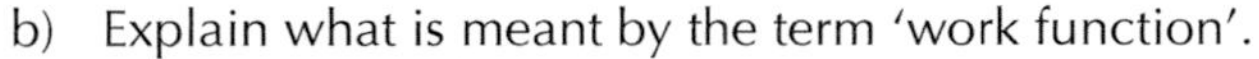

b) Explain what is meant by the term 'work function'.

Fig 2.135

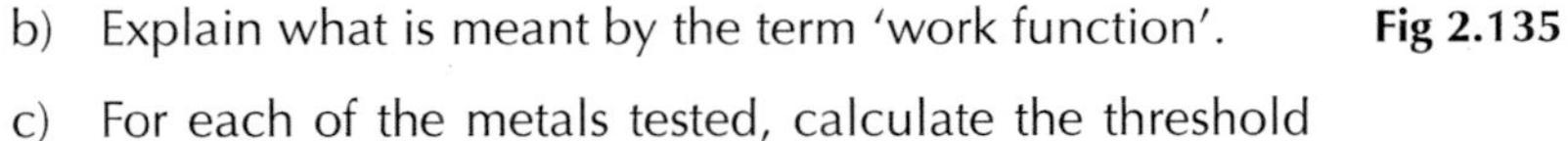

c) For each of the metals tested, calculate the threshold frequency for photoelectric emission.

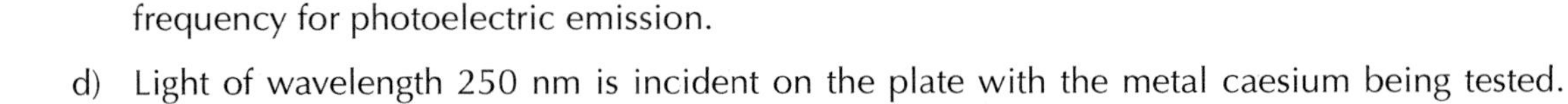

d) Light of wavelength 250 nm is incident on the plate with the metal caesium being tested. A current is recorded on the ammeter.

(i) Calculate the maximum kinetic energy of a photoelectron released from caesium.

(ii) State and explain the effect on the ammeter reading of increasing the irradiance of the incident light.

10 The apparatus in Question 9 is used to investigate the photoelectric emission of electrons from a metal plate. Light with a wavelength of 400 nm is used to eject the electrons. The electrons leave the plate with a maximum kinetic energy of 2.1×10^{-19} J. Find the work function of the metal.

11 The apparatus shown below is used to investigate the work function of a metal.

Incident light releases electrons from the metal. They are repelled by the negative collector.

a) Explain why the released electrons can still produce a current despite being repelled by the negative collector. Under what conditions would the current be produced?

b) Students find that a voltage of 200 V is sufficient to reduce the current to zero.

(i) Calculate the work done against the electron as it moves across a potential difference of 200 V.

(ii) State the kinetic energy of the electron as it is released from the plate, assuming 200 V reduces the current to zero.

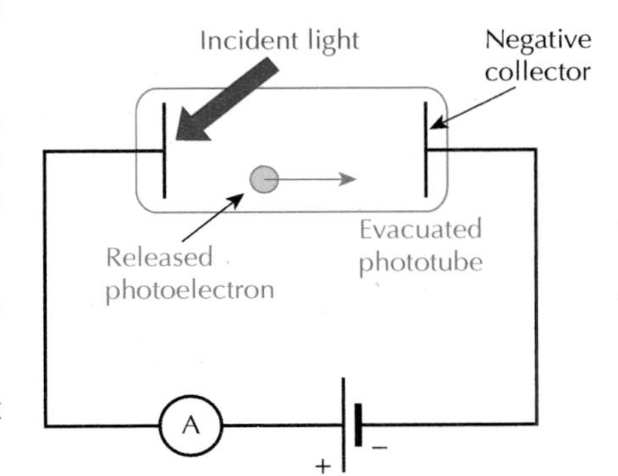

Fig 2.136

(iii) If the incident light has a wavelength of 400 nm, find the work function of the metal.

6 Interference

What you should know (National 5):

- Wave properties:
  - Wavelength
  - Frequency
  - Amplitude
  - Speed
- Diffraction is the curvature of a wavefront when it passes an obstacle

Learning intentions

- Conditions for constructive and destructive interference
- Concept of coherence: waves that have a constant phase relationship, the same wavelength and velocity
- Explanation of constructive and destructive interference in terms of the phase difference between two waves
- Interference of two waves from coherent sources
- Link between path difference and constructive and destructive interference
- Relationship for a diffraction grating linking slit separation, spacing between grating and screen and wavelength of the light.

6.1 Review of waves

6.1.1 Properties of a wave

Figure 2.137 shows a typical wave. The wave oscillates about a centre point (shown as a dashed line). A peak or crest is a maximum above the line and a trough is a minimum below the line. The wave has two main physical properties:

- Wavelength, λ, which is the distance between two identical adjacent points on the wave; for example, between two adjacent crests. It is length, so is measured in metres.
- Amplitude, A, which is the height of the wave from the middle to the top of a crest or the bottom of a trough. Amplitude is linked to the energy carried by the wave.

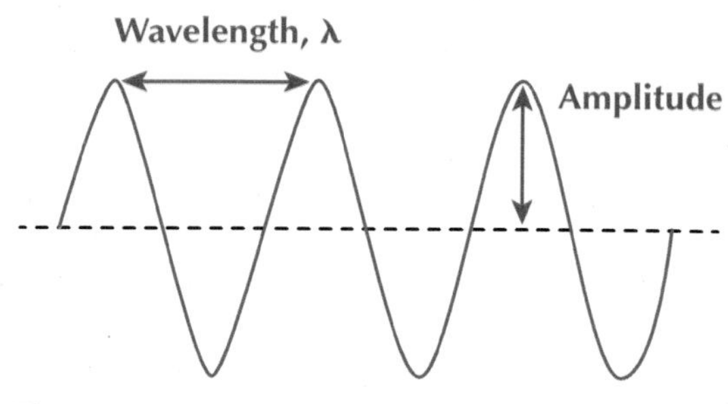

Fig 2.137

As well as the physical properties of a wave, the following properties are also present:

- The frequency, f, of a wave is the number of waves produced in one second. Frequency is measured in hertz (Hz), where 1 Hz is equal to 1 wave per second. Frequency is defined as the number of waves, N, produced in a given time, t:

$$f = \frac{N}{t}$$

- The period, T, of a wave is the length of time taken to produce one complete wave. It is measured in seconds and directly linked to the frequency of the wave as follows:

$$T = \frac{1}{f}$$

- The velocity, v, of a wave is the distance the wave covers in a given time. It is the speed at which the energy moves and the standard distance, speed, time equation can be applied, providing the speed of the wave is constant:

$$v = \frac{d}{t}$$

- The velocity, v, frequency, f, and wavelength, λ of a wave are linked by the following equation:

$$v = f\lambda$$

6.2 Interference of waves

When two waves meet, they combine to produce a new wave pattern. This process is known as interference, and it is a key characteristic of a wave. If an interference pattern is produced, then this is evidence that the energy is behaving like a wave.

When discussing interference, it is useful to introduce the concept of phase difference of a wave. A phase difference refers to waves being 'out of sync' with each other. If the phase difference between two waves is zero, then the waves are doing the same thing at the same time, as shown in Figure 2.138.

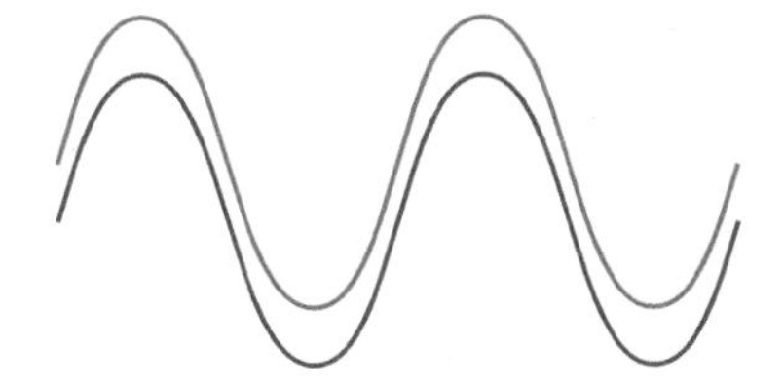

Fig 2.138

A phase difference of half a wavelength is shown in Figure 2.139. Measuring this phase difference as an angle, the waves are 180° out of phase.

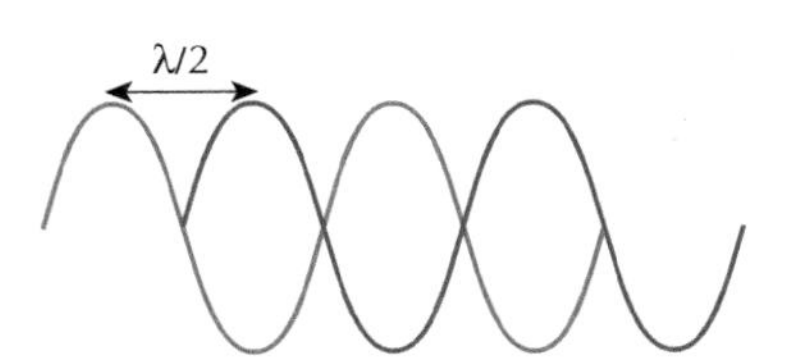

Fig 2.139

Any multiple of a wavelength can occur as a phase difference (any angle), as illustrated in Figure 2.140.

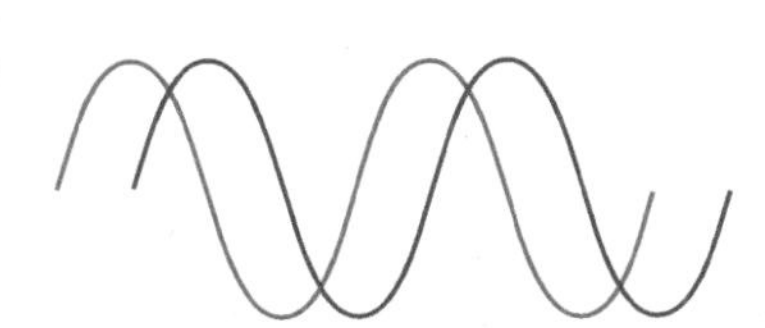

Fig 2.140

Waves that have a constant phase difference (the phase difference is the same along all points of the wave at all times) are said to be coherent.

6.2.1 Constructive and destructive interference

When two waves combine, they interfere to cause an interference pattern.

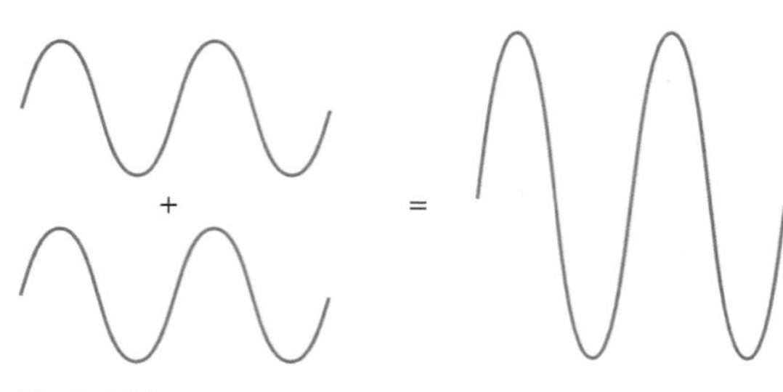

Fig 2.141

Constructive interference

For constructive interference, two coherent waves meet such that a peak meets a peak and a trough meets a trough, as shown in Figure 2.141.

These waves add constructively to give a wave with a bigger amplitude. These waves are in phase. The phase difference between the waves is either equal to zero or an integer number of wavelengths.

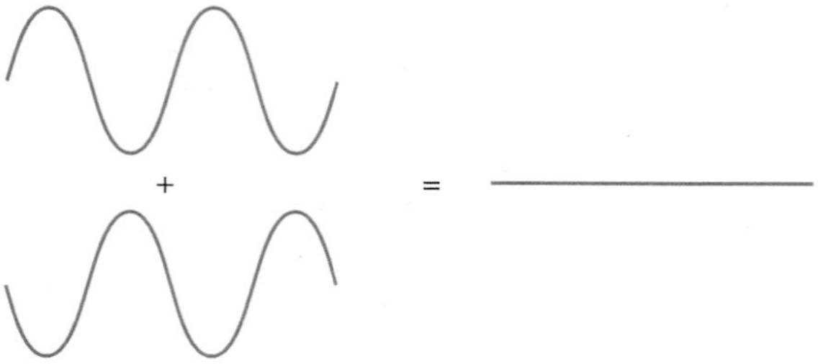

Fig 2.142

Destructive interference

For destructive interference, two coherent waves meet such that a wave peak meets a wave trough as shown in the Figure 2.142.

If the individual amplitudes are identical then the two waves cancel each other out, and the result is a wave with zero amplitude (no energy). Here the waves are half a wavelength (180°) out of phase with each other. They still have the same wavelength.

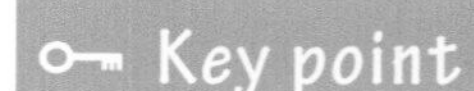

Key point

Two coherent waves can interfere either constructively or destructively.
Constructive interference: both waves are in phase.

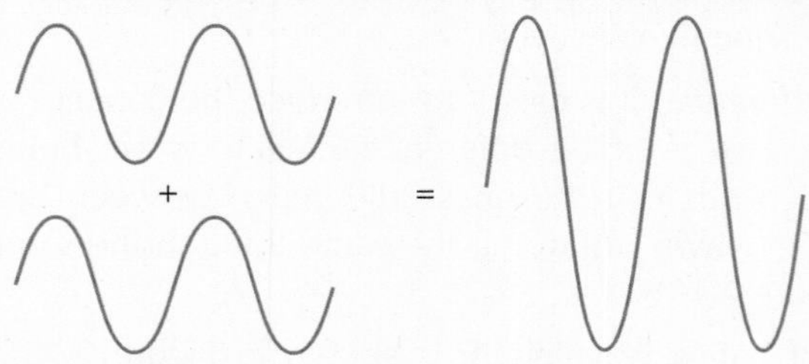

Fig 2.143

Destructive interference: both waves are out of phase by half a wavelength (180°):

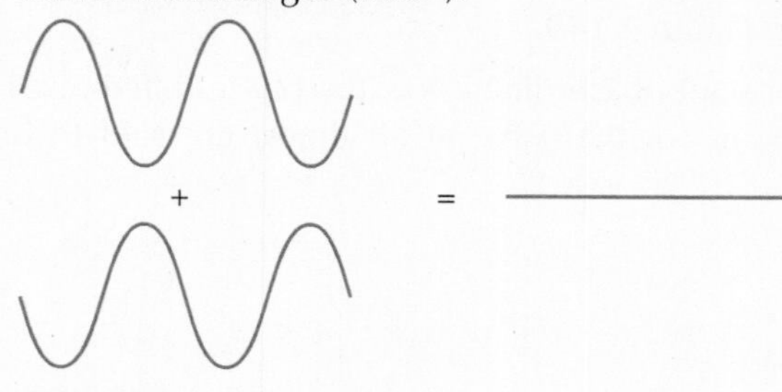

Fig 2.144

Exercise 6.2.1 Constructive and destructive interference

1 Explain the term 'phase difference' in the context of waves.

2 For each of the pairs of waves in a) – d) below, state whether or not the waves are in phase or out of phase.

a) b)

c) d)

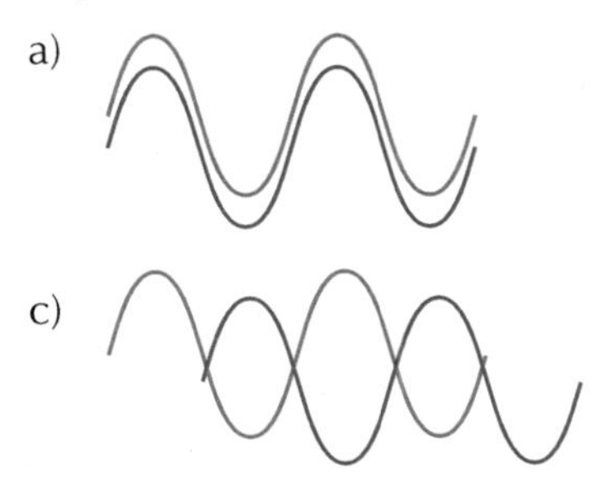

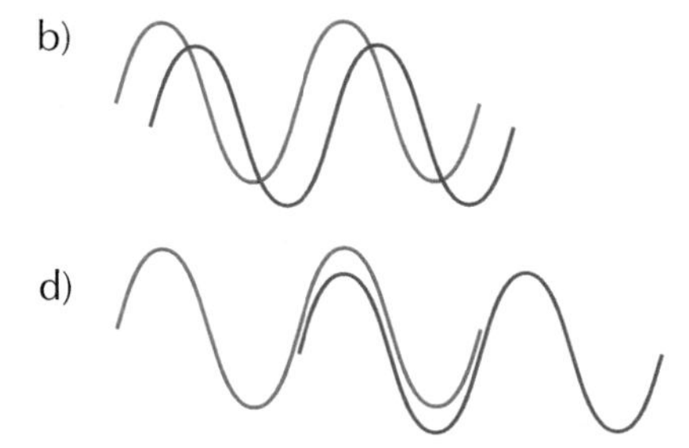

Fig 2.145

3 When two waves combine, the resultant wave produced has double the amplitude.

a) What type of interference is this?

b) What is the phase difference between the two waves?

4 When two waves combine, the result is complete cancellation of the waves.

a) What type of interference is this?

b) What is the phase difference between the two waves?

6.2.2 Double-source interference

The above examines the effect of two waves combining to give either constructive or destructive interference. This property of waves can be observed with sound, water and light. Here, interference effects from two sources are considered. These sources must be coherent – they must emit waves with the same wavelength that are in phase with each other. The easiest way to achieve this would be to use a single source and a double slit.

GO! Experiment 6.2.2a Double-source interference: water waves

This experiment investigates the production of an interference pattern from two coherent sources using water waves.

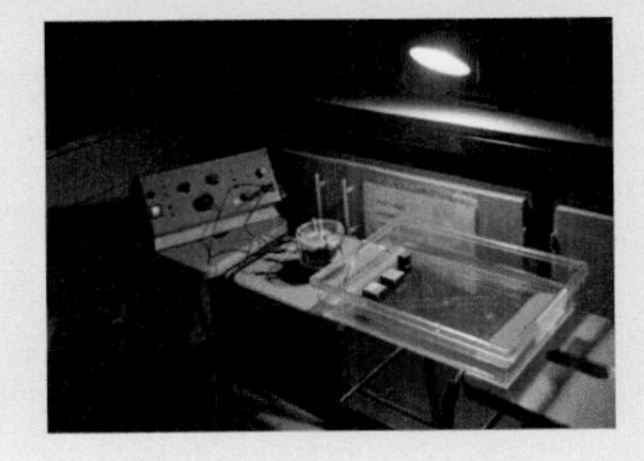

Fig 2.146

Apparatus

- A wave generator (connected to a signal generator)
- Ripple tank
- Water
- Obstacles
- Overhead light

Instructions

1 Set up the experiment as shown in the photograph above. Ensure that the wave generator is submerged in the water and produces waves when the signal generator is switched on. Create two sources by constructing an obstacle that has two slits as shown in Fig 2.147.

2 Switch on the signal generator and observe the pattern produced by the two slits. These slits produce two coherent sources because they come from the same source of waves. Do you see regions of calm water and rough water?

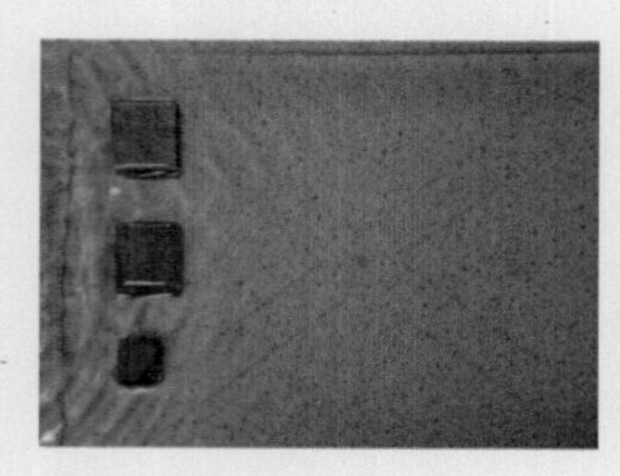

Fig 2.147

3 Comment on the results observed above, clearly stating any regions where you see constructive or destructive interference.

4 Use the signal generator to adjust the frequency of the waves. This will change the wavelength of the waves. Comment on the separation of the regions of calm water and the regions of rough water as you change the wavelength of the waves.

Experiment 6.2.2b Double-source interference: sound waves

This experiment investigates the production of an interference and the interference effects produced by two coherent sources of sound waves.

Apparatus

- Two loudspeakers
- A signal generator

Instructions

1 Connect two loudspeakers to the same signal generator and place the loudspeakers approximately 1 m apart. The loudspeakers will be two coherent sources.

2 Switch on the signal generator and set it to produce a sound of 1000 Hz.

3 Walk around the laboratory listening to the volume of the sound produced. Stick your finger in the ear that is not directed towards the source in order to reduce the adverse effects of reflection. Do you hear changes in volume? Take note of the separation between the quiet regions in the room.

4 Now change the frequency to 2000 Hz and repeat the experiment. Comment on how the separation of the quiet regions changes with the increase in frequency.

5 Repeat the above experiment with the frequency set to 500 Hz and again comment on the separation of the quiet regions in the room.

6 Write a short report that describes the above experiment and the type of interference that is observed. Include the effect of changing the frequency of sound on the separations of the quiet regions in the lab in your report. Why do you think that there were no regions where the sound level was completely quiet?

Double-source interference can be explained in terms of the wave fronts emerging from the sources. Consider two slits with wave fronts passing through them. The waves must be coherent – that is, they are emitted from the slits in phase. The slits cause the waves to be diffracted. When a crest of a wave meets another crest, constructive interference results. Where a crest meets a trough, destructive interference results. This leads to regions in space with an intense signal and others with no signal, as shown in Figure 2.148.

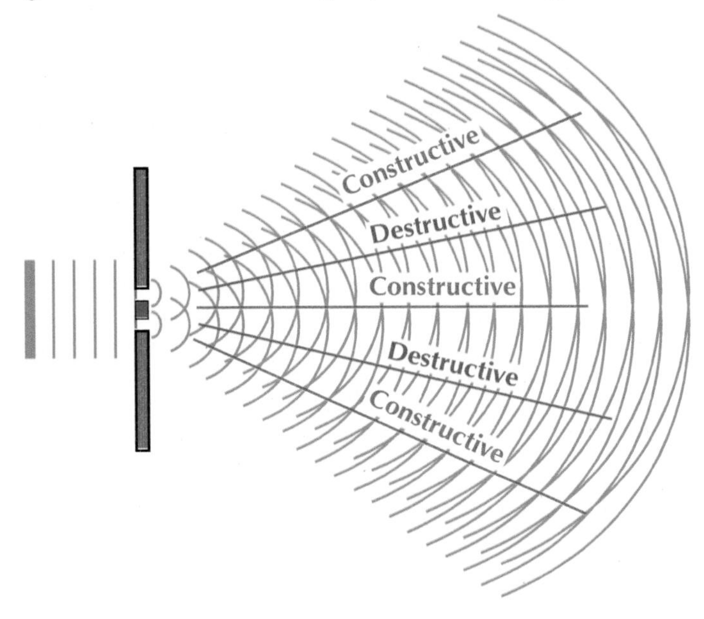

Fig 2.148

Changes in the wavelength of the waves will result in a change in the separations of the regions of constructive and destructive interference. Measuring the separations will therefore give information about the wavelength of the source used. The pattern of loud and quiet that is produced is known as an interference pattern. It is produced by constructive and destructive interference.

If two coherent sources of light are used (light shone at two close-together slits known as Young's slits), you can see an interference produced on a distant screen.

With this experiment, regions of light and dark are observed on the screen as shown in Figure 2.149:

Fig 2.149

Incident
light
S1
S2
Double
slit
Screen

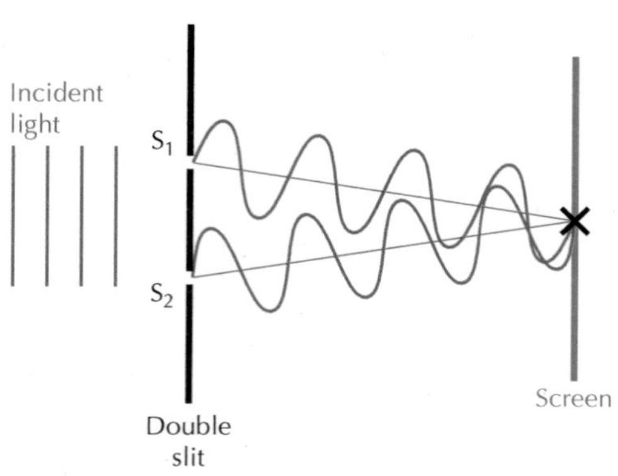

Fig 2.150

Just like sound and water in the experiments above, the light forms regions of light and dark that correspond to regions of constructive and destructive interference, respectively. The positions of the light and dark regions will depend on the wavelength of the light used. Measuring the separation of the light or dark fringes can therefore be used to calculate the wavelength of the light. This can also be done for sound and water waves.

Consider a point, X, on the screen that is the same distance from both of the slits, S_1 and S_2.

The light follows the same path length regardless of the slit. Therefore, the path difference between the two paths taken by the light is zero:

$$\textit{Path difference} = S_2X - S_1X = 0$$

Incident
light
S1
S2
Double
slit
Screen

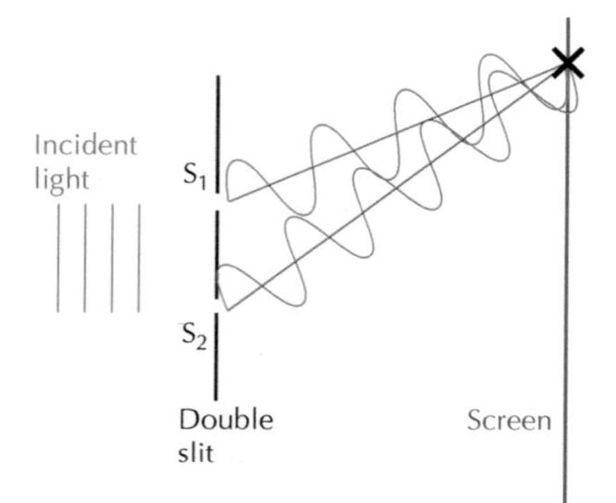

Fig 2.151

As the length of path is the same, the same number of wavelengths of light will be fitted into each path. In Figure 2.150, four complete wavelengths are fitted into each path. This means that the waves arrive in phase at point X, producing constructive interference. This leads to a bright fringe being observed.

There are other positions on the screen where the waves arrive in phase. A second example is shown in Figure 2.151, where the path from S_2 to X is five wavelengths long and the path from S_1 to X is four wavelengths long.

As the waves arrive in phase, constructive interference occurs and therefore a bright fringe is observed at the new position X. The path difference for this is one wavelength:

$$\textit{Path difference} = S_2X - S_1X = \lambda$$

More generally, if the path difference is equal to an integer number of wavelengths, then constructive interference will be observed because the waves arrive at the screen in phase. For constructive interference:

$$\textit{Path difference} = m\lambda$$

where m is an integer.

Incident
light
S1
S2
Double
slit
Screen

Fig 2.152

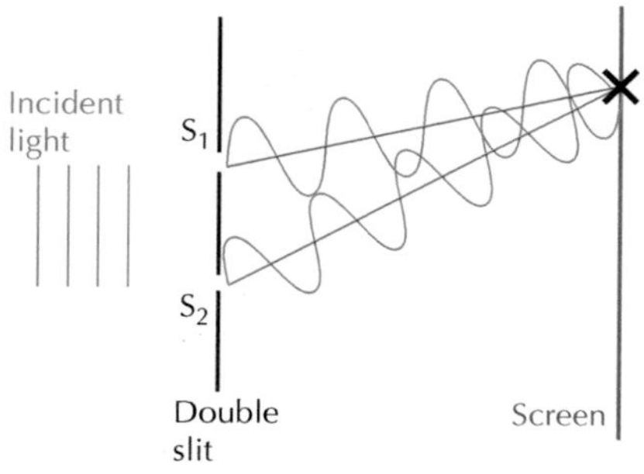

Destructive interference can also be produced, which will result in dark areas on the screen. This will occur when the waves arrives half a wavelength out of phase as discussed above. The first minimum will occur when the path difference is equal to one half of a wavelength. This is shown in Figure 2.152, where the path from S_2 to X is four and a half wavelengths long, and the path from S_1 to X is four wavelengths long.

This highlights that destructive interference will be obtained if:

$$\text{Path difference} = \frac{1}{2}\lambda$$

More generally, destructive interference will result for one half wavelength, one and a half wavelengths, two and a half wavelengths and so on, where m is an integer:

$$\text{Path difference} = \left(m + \frac{1}{2}\right)\lambda$$

Key point

Coherent waves emitted from two sources can interfere either constructively or destructively on a distant screen. Constructive interference is given when:

$$\text{Path difference} = m\lambda$$

Destructive interference is given when:

$$\text{Path difference} = \left(m + \frac{1}{2}\right)\lambda$$

Worked example

Students investigating the interference of sound waves set up two loudspeakers connected to the same signal generator. The set up for the experiment is shown in Figure 2.153. At a position in the laboratory where paths to the loudspeaker are shown as in the diagram, the student observes the first maximum away from the central position.

Calculate the wavelength and frequency of the sound emitted by the loudspeakers. You may assume the speed of sound in air to be 340 m s^{-1}.

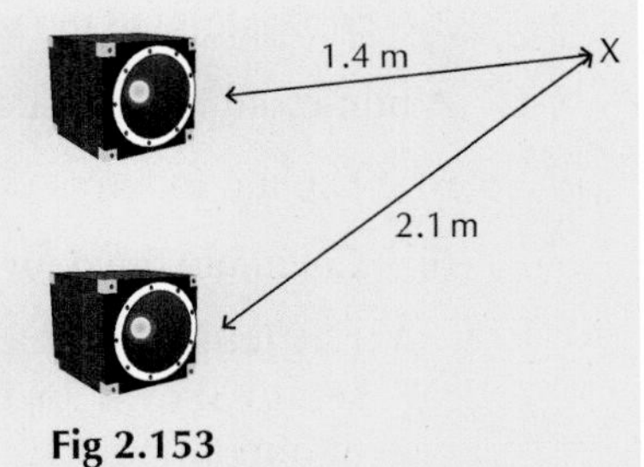

Fig 2.153

At the point X, you have constructive interference so the following equation applies:

$$\text{Path difference} = m\lambda$$

As this is the first maximum away from the centre position, you can say:

$$m = 1$$

The path difference is found by subtracting the short path from the long path:

$$\text{Path difference} = 2.1 - 1.4$$
$$\text{Path difference} = 0.7\ m$$

Hence, you can now find the wavelength of the sound waves:

$$\text{Path difference} = m\lambda$$
$$0.7 = (1)\lambda$$
$$\lambda = 0.7m$$

The frequency of the wave can be found using the wave equation:

$$v = f\lambda$$
$$340 = f \times 0.7$$
$$f = \frac{340}{0.7}$$
$$f = 486\ Hz$$

Exercise 6.2.2 Double-source interference

1 Students investigating the properties of light use Young's slits to produce an interference pattern from the red light produced by a HeNe laser.

a) Explain in terms of waves how an interference pattern is produced by the Young's slits on a distance screen.

b) Explain why the two sources of light must be coherent.

2 The experiment shown in Figure 2.154 is used to investigate the interference pattern produced by water waves.

Fig 2.154

a) A maximum amplitude of the water waves is found at a position that is equal distances away from both sources. Explain in terms of waves why a maximum is produced at this position.

b) The next maximum produced by the waves is found to be at a position where the path difference between the two sources is 3.6 cm. Calculate the wavelength of the water waves.

c) A minimum is found between the two maxima described in parts a) and b) above.

(i) Explain in terms of waves how a minimum is produced.

(ii) Calculate the path difference at this minimum.

(iii) It is found that there are still waves at the minimum. By considering the experimental set-up shown in the photograph above, explain why waves are still detected at the minimum points.

3 An experiment to find the wavelength of sound is conducted. The results obtained by the pupils are shown in Figure 2.155, where X represents the third minimum away from the centre.

Calculate the wavelength and the frequency of the sound waves. You may assume that the speed of sound in air is 340 m s^{-1}.

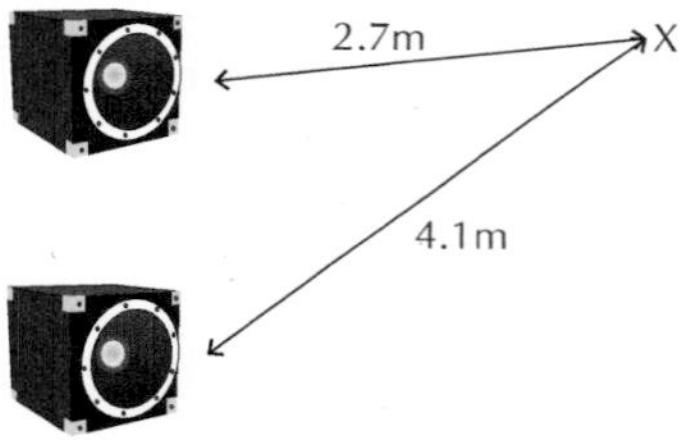

2.7m
X
4.1m

Fig 2.155

4 A Young's slit experiment is set up as shown in Figure 2.156. Maxima and minima are observed on a distant screen as shown.

Use the information given in the diagram to calculate the wavelength of the HeNe laser light.

5 Two loudspeakers are connected to the same signal generator. The signal generator produces a signal with a frequency of 700 Hz. Calculate the path difference to the third maximum away from the central maximum for the sound produced by the loudspeakers.

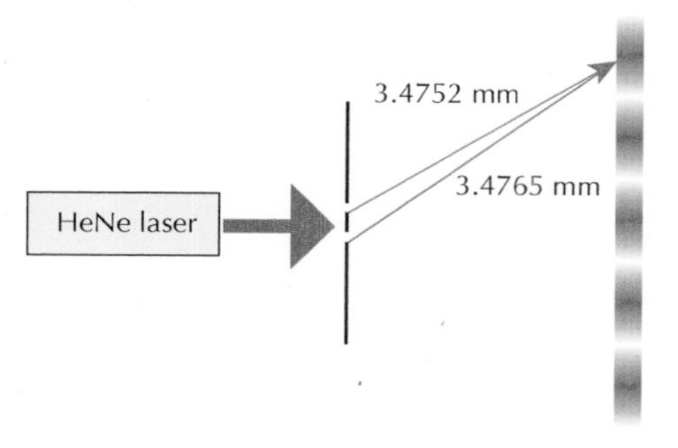

Fig 2.156

6 Students are investigating the interference of water waves being passed through a pair of slits. They set up the experiment shown in Figure 2.157.

a) The wave generator produces waves with a wavelength of 4.5 cm that travel at a speed of 1.4 $m s^{-1}$.

(i) Calculate the frequency of the waves produced by the wave generator.

(ii) Calculate the path difference between the two sources to a position where the second maximum is observed.

b) The frequency of the water waves is now reduced. The speed remains constant. State and explain the effect this has on the path difference that would be measured to the second maximum.

Fig 2.157

7 A microwave generator is set up in front of a metal screen. The screen has two slits in it, producing two coherent sources of microwaves.

Figure 2.158 shows the position of the central maximum (P) and the first order maximum (Q) for different frequencies of microwaves. For each diagram, find the wavelength and frequency of the microwaves (microwaves travel at the speed of light).

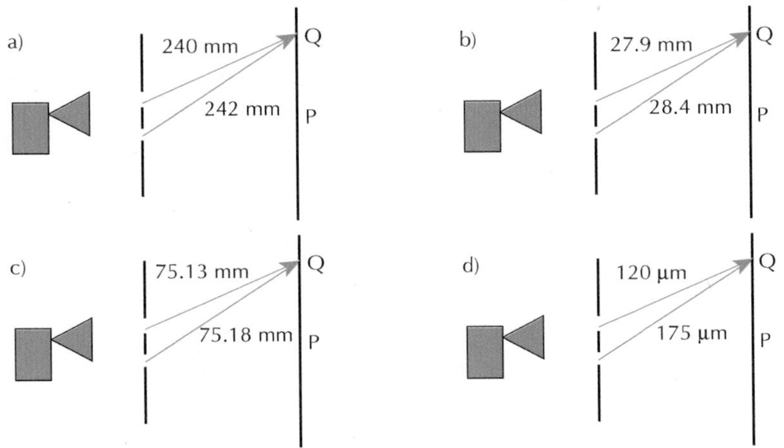

Fig 2.158

8 Students investigating the wavelength of a microwave source set up an experiment where the microwaves are directed to a pair of slits to produce two coherent sources. A central maximum is located at P by using a microwave detector. As the detector is moved upwards, a series of minima and maxima are found. At the fourth minimum, the readings in Figure 2.159 were obtained:

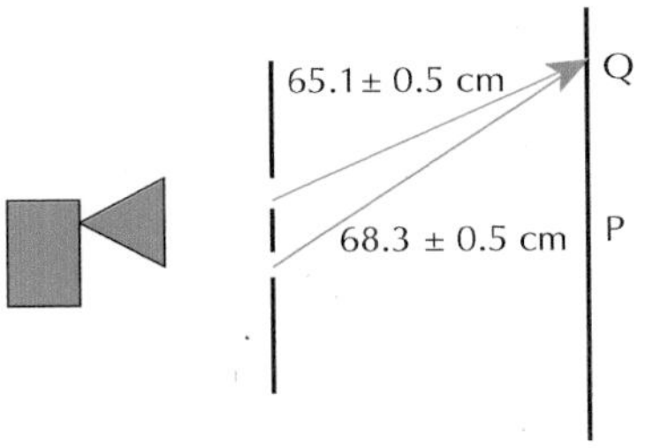

a) Calculate the percentage uncertainty in the reading for each path difference.

b) Calculate the total percentage uncertainty in the measurement of wavelength.

c) Calculate the wavelength of the microwaves, giving your answer in the form:

Wavelength ± Random Error

65.1 ± 0.5 cm
Q
68.3 ± 0.5 cm
P

Fig 2.159

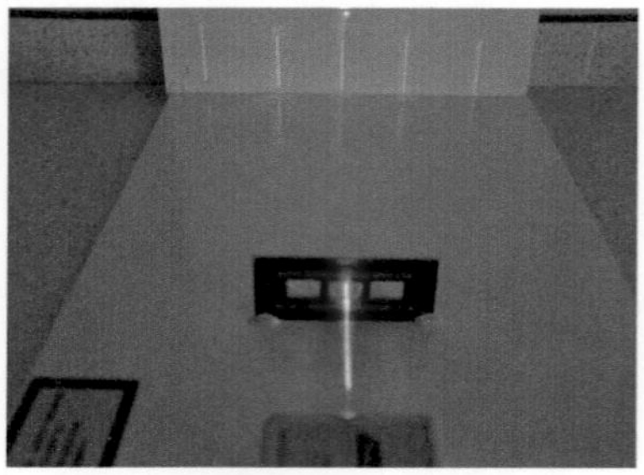

Fig 2.160

6.2.3 Diffraction grating

Two coherent sources of waves can produce an interference pattern. This concept can be extended to having many coherent sources. This can be produced by using a diffraction grating. Rather than having two closely spaced sources as with Young's slits, a diffraction grating has tens or hundreds of closely spaced sources over a small length. This set-up also produces an interference pattern. A typical example is shown in Figure 2.160 for red light.

Unlike the interference pattern from Young's slits, the diffraction grating gives well-defined maxima and very little variation of brightness between the light regions and the dark regions. This is due to there being multiple sources involved in the production of the interference pattern.

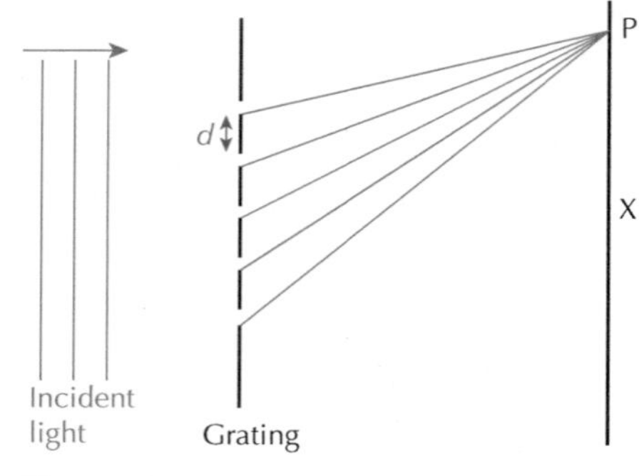

Fig 2.161

The diffraction grating equation

An equation for the position of the maxima from a diffraction grating can be derived by considering the light passing through the individual slits of the grating. Consider Figure 2.161, which shows adjacent slits of the grating. Adjacent slits are separated by a distance d. The point X represents the central maximum and the point P represents the first maximum. The first maximum is called the first order maximum.

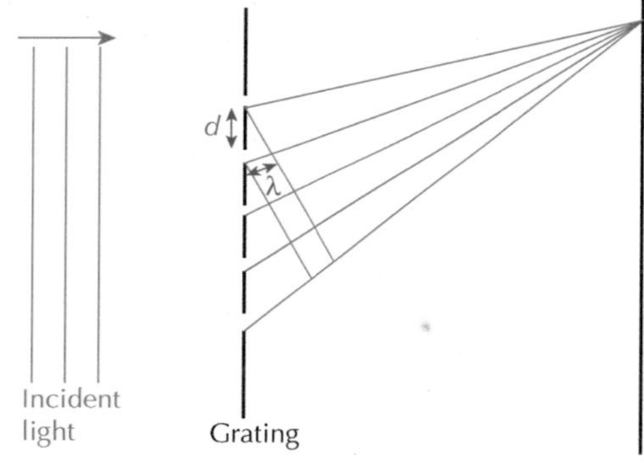

Fig 2.162

The path difference between two adjacent slits and the point P where the first maximum occurs is equal to one complete wavelength:

$$\textit{Path difference} = \lambda$$

This is shown in Figure 2.162.

More generally, the path difference between adjacent slits to the m'th maximum is given by:

$$\textit{Path difference} = m\lambda$$

Zooming in to two adjacent slits, and considering the angle between the path to the central maximum and the path to the m'th maximum as θ, the grating equation can be derived. Consider this in Figure 2.163.

A right-angled triangle is formed that links $m\lambda$ to d. Using SOH-CAH-TOA:

$$\sin\theta = \frac{m\lambda}{d}$$

$$m\lambda = d\sin\theta$$

the above equation shows the angle to different maxima for a specific wavelength of light and slit separation.

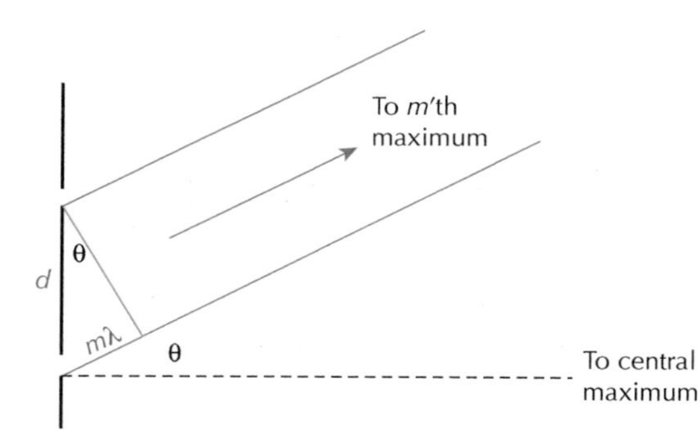

Fig 2.163

Key point

The difference between the m'th order maximum and the central order maximum from a diffraction grating with slit separation d is given by:

$$m\lambda = d \sin \theta$$

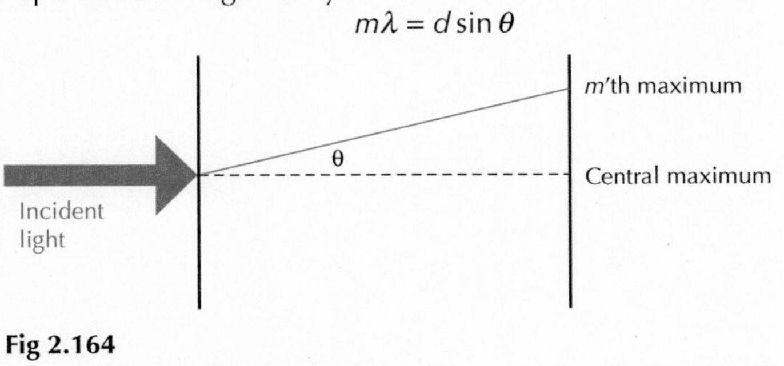

Fig 2.164

Experiment 6.2.3a Effects of slit separation for a diffraction grating

This experiment investigates the effects of changing the slit separation on the interference pattern produced by a diffraction grating.

Apparatus

- A monochromatic light source (laser)
- Diffraction gratings with different slit separations
- Screen

Instructions

Fig 2.165

1. Set up the experiment as shown in Figure 2.165, using sticky putty to affix the diffraction grating to the same spot and the screen to the same spot. This ensures the distance between the screen and the grating is kept constant.
2. Calculate the slit separation using the equation shown below where d is the separation, N is the number of lines per metre:
$$d = \frac{1}{N}$$
3. Direct the laser light at the first diffraction grating and note the position the maxima form. If the screen is a white board, a white board marker can be used for this.
4. Repeat steps 2 and 3 with different diffraction gratings, noting the positions of the maxima.
5. Write a short report that describes the effect of changing the slit separation on the separation of the maxima produced. You should include a qualitative relationship between slit separation and maxima separation.

Experiment 6.2.3b Using a diffraction grating to find the wavelength of light

This experiment uses a diffraction grating to measure the refractive index of light from a monochromatic source such as a laser.

Apparatus

- A monochromatic light source (laser)
- Diffraction grating
- Screen
- Ruler
- Protractor

Instructions

1 Set up the experiment as shown in Figure 2.166 and adjacent diagram, ensuring the diffraction grating and the screen are securely fixed to the surfaces they are resting on.

Fig 2.166

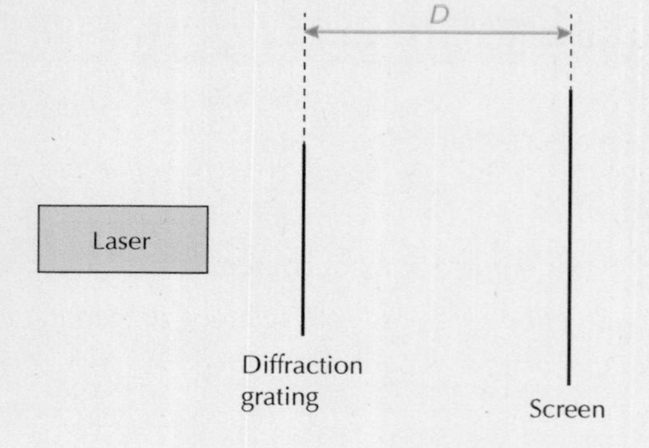

2 Measure the distance between the diffraction grating and the screen (D in the diagram above).

3 Calculate the separation of the slits using the equation:

$$d = \frac{1}{N}$$

where d is the separation, N is the number of lines per metre.

4 Illuminate the grating with the laser light and note the positions of the central maximum and the maxima on either side of this. *Hint – use a diffraction grating with the smallest slit separation you can find.*

5 Measure the distance between the central maximum and the first order maximum as shown in Figure 2.167.

6 Use trigonometry to calculate the size of angle θ in Figure 2.167:

$$\tan\theta = \frac{L}{D}$$

For small angles, $\tan\theta$ is approximately equal to sine θ.

7 Use the diffraction grating equation to calculate the wavelength of the light:

$$m\lambda = d\sin\theta$$

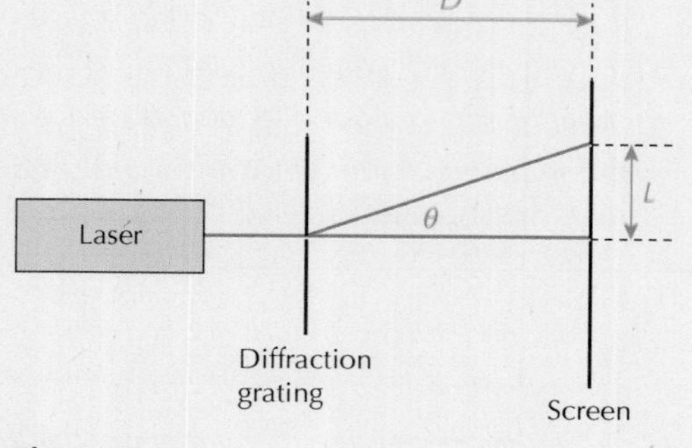

Fig 2.167

Consequences of the diffraction grating equation

A diffraction grating can be used to find the wavelength of light that is incident on it. The angle to the m'th maximum depends on the wavelength of the light and if the slit separation is known, then the equation:

$$m\lambda = d \sin \theta$$

can be used to find the wavelength. The above equation has important consequences for the positions of the maxima formed by the grating.

The first of these is the effect of the slit separation on the separation of the maxima observed. Rearranging the above equation gives:

$$\sin \theta = \frac{m\lambda}{d}$$

The smaller the value of the slit separation, d, the larger the value of sin θ, and thus the larger the value of θ. This is illustrated in practice by Figure 2.168. The slit separation has been increased (by having fewer lines per metre) for each picture, moving to the right:

Fig 2.168

A further observation is the effect of changing the wavelength (colour) of the light. The greater the wavelength of the light (the closer to red), the greater the separation. This is shown in Figure 2.169.

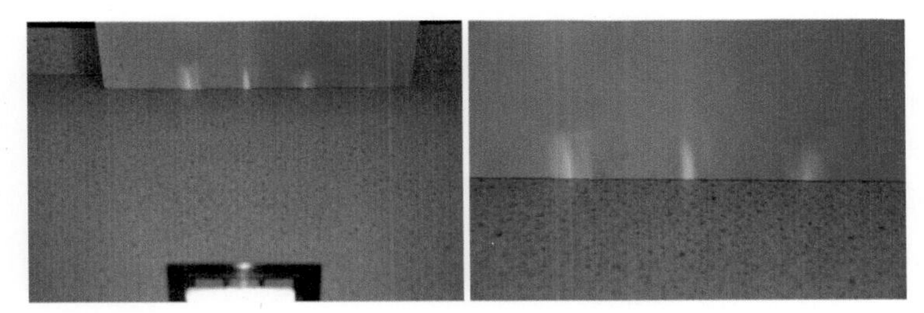

Fig 2.169

Key point

The slit separation, d, for a diffraction grating can be found using the following equation:

$$d = \frac{1}{N}$$

N in the above equation is the number of lines per metre. Many gratings quote the number of lines per millimetre. In this case, the number of lines per metre can be found by multiplying this value by 1000.

Worked example

A diffraction grating has 500 lines per millimetre. Light of wavelength 630 nm is incident on the grating. Calculate the angle between the central maximum and the third order maximum.

Firstly, calculate the slit separation using the equation:

$$d = \frac{1}{N}$$

The number of lines per metre, N, is given by:

$$N = 500 \times 1000$$
$$N = 500{,}000 \text{ lines}$$

Hence, the slit separation is given by:

$$d = \frac{1}{500{,}000}$$
$$d = 2 \times 10^{-6} m$$

Now, use the diffraction grating equation to calculate the angle θ between the central and third order maxima:

$$m\lambda = d \sin \theta$$

Here, m = 3 because you are considering the third order maximum so:

$$(3)(630 \times 10^{-9}) = (2 \times 10^{-6}) \sin\theta$$
$$\sin\theta = \frac{(3)(630 \times 10^{-9}}{(2 \times 10^{-6})}$$
$$\sin\theta = 0.945$$
$$\theta = \sin^{-1} 0.945$$
$$\theta = 70.9°$$

Worked example

A diffraction grating with 500 lines per millimetre is used to view the light emitted by a monochromatic source. The diffraction grating is placed 1.50 m away from the viewing screen as shown in Figure 2.170. The distance between the central and first order maximum is measured to be 0.50 m as shown. Use this information to calculate the wavelength of the light.

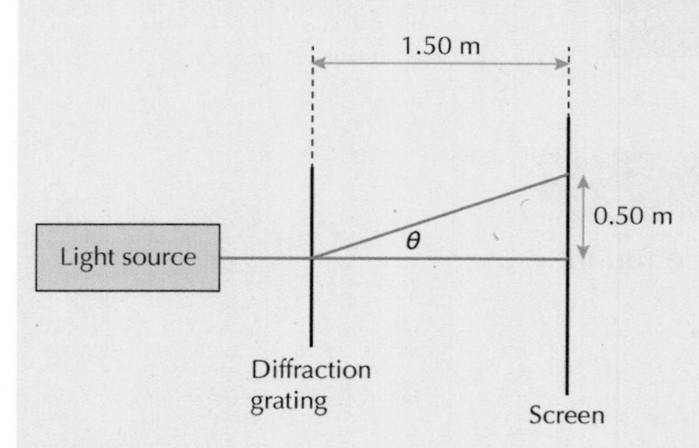

Fig 2.170

The wavelength of the light will be found by using the diffraction grating equation:

$$m\lambda = d\sin\theta$$

The slit separation is first found by using the equation:

$$d = \frac{1}{N}$$

$$d = \frac{1}{500{,}000}$$

$$d = 2\times10^{-6}m$$

The angle between the central and first order maxima is found by using trigonometry for a right-angled triangle (SOH-CAH-TOA):

$$\tan\theta = \frac{o}{a}$$

$$\tan\theta = \frac{0.5}{1.5}$$

$$\theta = \tan^{-1}\left(\frac{0.5}{1.5}\right)$$

$$\theta = 18.4°$$

The wavelength of the light can now be calculated:

$$m\lambda = d\sin\theta$$

$$(1)\lambda = (2\times10^{-6})\sin 18.4°$$

$$\lambda = 6.32\times10^{-7}m$$

$$\lambda = 632\ nm$$

Exercise 6.2.3 Diffraction gratings

1 Calculate the slit separation for the following diffraction gratings:

a) 100 lines per mm

b) 500 lines per mm

c) 500 lines per cm

d) 900 lines per mm

e) 10 lines per μm

2 Calculate the number of lines per millimetre for diffraction gratings with the following slit separations:

a) 5×10^{-6} m

b) 200 μm

c) 5 μm

d) 250 nm

e) 90 nm

(continued)

3 A diffraction grating with 400 lines per mm is used to produce an interference pattern from a source of monochromatic light. The angle between the central and second order maximum is found to be 31.5°.

a) Calculate the wavelength of the source.

b) If a diffraction grating with 600 lines per mm is used, calculate the new angle between the central and second order maxima.

c) State the effect of increasing the number of lines per mm on the diffraction grating to the angle between the maxima of the interference pattern.

4 Red light with a wavelength of 640 nm is passed through a diffraction grating with 200,000 lines per metre.

a) Calculate the separation between the slits of the diffraction grating.

b) Calculate the angle between the zero and first order maxima.

5 The experiment shown in Figure 2.171 is used to ascertain the wavelength of a monochromatic source of light. A diffraction grating with 100 lines per millimetre is used in the experiment. The angle to the third maximum is measured to be 35°.

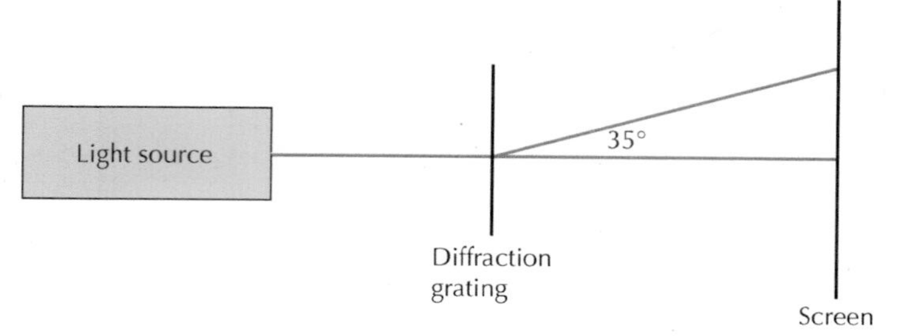

Fig 2.171

a) Calculate the separation of the slits for the diffraction grating.

b) Find the wavelength of the light.

c) Explain why a more accurate measurement of the wavelength can be made using the third order maximum rather than the first order maximum.

6 Two students are using diffraction gratings to measure the wavelengths of light from different LEDs shown in Figure 2.172.

One student suggests that they should use the diffraction grating with the greatest number of lines per millimetre, in order to obtain the most accurate measurement of the wavelength. Use your knowledge of physics to comment on the validity of this statement.

Fig 2.172

7 White light consists of wavelengths ranging from 440 nm to 730 nm. White light from a filament lamp is collimated and passed through a diffraction grating with 400 lines per millimetre.

a) Describe and explain the pattern that would be observed on a screen. (*Hint – remember that the central maximum will be white.*)

b) Calculate the angle between the central maximum and the second order maximum for light with a wavelength of 730 nm.

c) Calculate the angle between the central maximum and the second order maximum for light with a wavelength of 440 nm.

d) The diffraction grating is replaced with one that has 100 lines per millimetre. What effect does this have on the separation of the colours of the white light spectrum?

8 Special glasses that have diffraction gratings for lenses can be purchased. A white light source when viewed with these glasses produces the pattern shown in Figure 2.173.

Explain why this pattern is produced by the glasses.

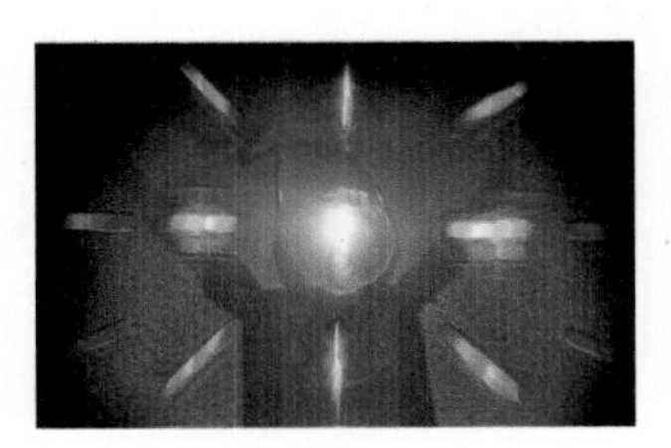

Fig 2.173

9 A diffraction grating is used to view collimated white light from a 12 V filament lamp in the lab. The experiment is set up as shown in Figure 2.174.

The diffraction grating has 500 lines per millimetre. A viewing screen is placed 2.0 m away from the diffraction grating. A white fringe is viewed at the central maximum and spectra are viewed along the length of the screen at the first and second order maxima as shown in the diagram.

a) Calculate the slit separation of the diffraction grating.

b) Calculate the angle, θ, to the first order maximum for deep red light (730 nm) and violet light (440 nm).

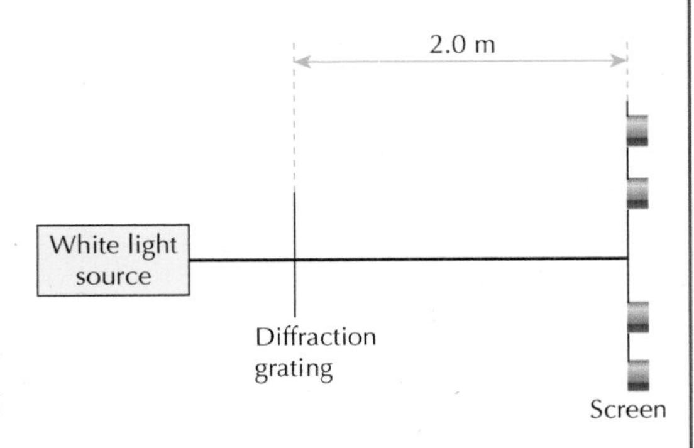

Fig 2.174

c) Use your results from part b) to find the width of the spectrum at the first order maximum. (*Hint – you will need to calculate the vertical displacement for both red and violet light.*)

d) Repeat steps b) and c) for the second order maximum. How does the width of the spectrum compare to the first order maximum?

7 Refraction of light

What you should know (National 5):

- Identification of the difference between continuous spectra and line spectra
- Use of spectral data to identify the elements present in a star

Learning intentions

- Difference between emission and absorption spectra
- How spectra are produced, including the link between energy level transitions and wavelength of light emitted/absorbed
- Bohr model of the atom – link to emission and absorption spectra from movement of electrons between energy levels
- Photon emission from the atom
- Link between Planck's constant, energy levels and wavelength of the photon emitted
- Use of absorption spectrum from the Sun to identify elements present in the Sun's atmosphere

Fig 2.175

The history of spectroscopy can be traced back to the 17th century, to experiments carried out by Isaac Newton using prisms. The word 'spectrum' was introduced to describe the 'rainbow' of colours produced when white light was passed through the prism. Newton demonstrated that the colours produced were from splitting the white light into the different colours that make it up, rather than the prism producing the colours. The experiments were replicated in nature with the rainbow, where water droplets split the white light from the Sun into different colours.

Fig 2.176

In the early 19th century, Joseph von Fraunhofer (Figure 2.176) replaced the prism with a diffraction grating, which was discussed in Chapter 5. This resulted in a greater separation of the colours of white light and increased the amount of detail that could be found in spectra. The field of detailed spectroscopy was born.

7.1 Emission spectra

When an electrical current is passed through the filament of a light bulb, it glows white. The filament emits light. If this light is passed through a prism or diffraction grating, it will be split into a spectrum of colours ranging from red at one side through to violet at the other. This spectrum is emitted by the light bulb, so it is referred to as an emission spectrum. Any light source will give out a range of colours, which is called an emission spectrum. Different light sources will give out different spectra.

There are two main types of emission spectra that will be considered here: a continuous spectrum and a line spectrum.

Experiment 7.1a Emission spectra

This experiment investigates the emission spectra given out by different light sources, highlighting the difference between the two main types of spectra – line and continuous.

Apparatus

- Spectrometer*
- White light source such as a filament bulb
- Gas discharge lamps for various elements
- Power supplies

* Although a prism can be used to split light into its individual colours, it does not separate the colours enough to be clearly analysed. A diffraction grating or a spectrometer will provide a clearer spectrum.

Instructions

1. Begin with the filament bulb. Connect the bulb to a power supply and ensure that the lamp illuminates.
2. Use your spectrometer to view the spectrum produced by the bulb. Describe clearly what you see.
3. Now connect a gas discharge lamp to a power supply and switch on. Allow the lamp suitable time to warm up and produce its correct colour.
4. View the light from the discharge lamp using the spectrometer. Describe clearly what you see.

Extra instructions

1. In Chapter 6, Area 1 (blackbody radiation) it was explained that the light emitted by a blackbody emitter depends on its temperature. The temperature of a bulb can be changed by varying the voltage of the power supply connected to it. Use your spectrometer to view the emission spectrum from a filament bulb for different supply voltages. Describe carefully what you see.

Continuous spectrum

When the light from a filament bulb (white light) is viewed through a spectrometer, a spectrum similar to the one shown in Figure 2.177 is observed.

Fig 2.177

This spectrum is called a continuous spectrum. All of the colours are visible with no large gaps between them. There is a continuous spectrum of colour.

Line spectrum

When light emitted by a sodium gas discharge light is viewed through a spectrometer, it does not reveal a continuous spectrum of all of the colours. Instead, discrete colours are seen, similar to the spectrum shown in Figure 2.178:

Fig 2.178

The same is true for other elements used in gas discharge lamps, such as helium, neon and mercury. This type of spectrum is called a line emission spectrum because it is made up of individual lines. As discussed in the model of the atom section, these lines are specific to the atoms giving off the light. The spectrum is a signature that shows the presence of an element.

GO! Experiment 7.1b Bohr and Kirchhoff emission spectra

This experiment demonstrates the spectra emitted by heating different elements in a Bunsen flame, using a diffraction grating to view the spectral lines. This experiment was first carried out by Robert Bunsen and Gustav Kirchhoff in the mid-19th century and showed a link to the absorption spectra from the Sun discovered by Wollaston and Fraunhofer.

Apparatus

- Bunsen burner
- 1 mol/litre hydrochloric acid
- Wooden splints
- Different metal salts
- Diffraction grating or spectrometer

Instructions

1. Light the Bunsen burner and set it to a blue flame (air hole fully or partly open). Remember to ensure that the Bunsen burner is not left unattended!
2. Dip a wooden splint into the hydrochloric acid and then into the metal salt – this will form the metal chloride; for example, lithium chloride.
3. Place the splint into the Bunsen flame and observe the colour emitted (shown in Figure 2.179).
4. Use the diffraction grating to view the spectrum produced by burning the salt in the Bunsen flame. Note down the following observations:
   - Is the spectrum a line spectrum or continuous?
   - What is the main colour observed?
   - What lines are observed with the diffraction grating?
   - How do the spectra differ from each other?

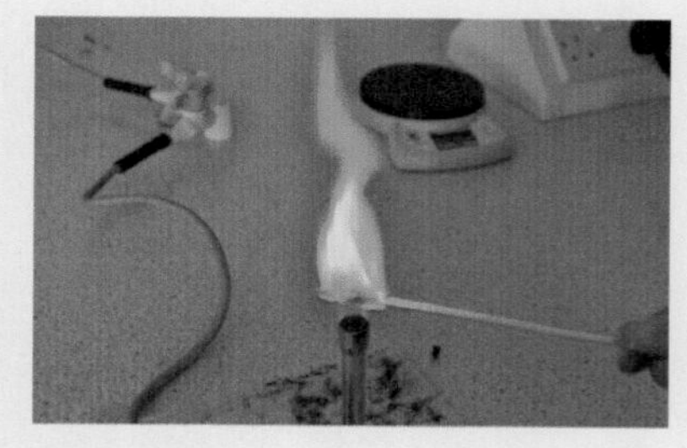

Fig 2.179

7.2 Absorption spectra

As well as getting a gas to emit light through exciting the gas, the same gas can also absorb light. If you pass white light, which contains all of the colours (a continuous spectrum), through a gas, that gas will absorb specific colours. This will result in a continuous spectrum that is missing certain lines. An example of this is shown below:

Fig 2.180

The black regions in the spectrum correspond to colours that have been absorbed by the gas. The line spectrum and the absorption spectrum from a gas will correspond with each other. If a gas emits certain colours, it will absorb the same colours. This is highlighted by Figure 2.181. Notice that the missing colours in the absorption spectrum correspond to the emitted colours in the line spectrum:

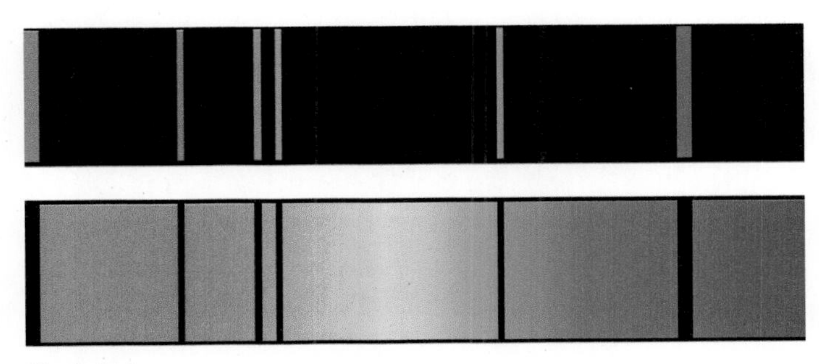

Fig 2.181

7.2.1 Fraunhofer lines

The spectrum produced by the Sun appears to be a continuous spectrum. (Be careful – never stare directly at sunlight!) This is visible in nature in the form of a rainbow, as well as when a prism is used to split the light into its individual colours. However, close inspection of the spectrum from the Sun shows that there are dark regions in the spectrum – just like the absorption spectra above. These dark lines were first discovered in the early 19th century by William Hyde Wollaston and then later by Joseph von Fraunhofer.

Fraunhofer carried out a detailed study of the lines missing from the spectrum, and observed some of these lines to be missing from the spectra of stars, as well as our Sun. Consequently, the lines became known as Fraunhofer lines. The main features that were observed by Fraunhofer were lines A through to K as shown in Figure 2.182:

Key point

When white light is passed through a gas, some of the colours are absorbed. This produces an absorption spectrum. The colours that are absorbed depend on the elements present in the gas. The absorbed colours represent a signature of the elements present and can be used to identify them.

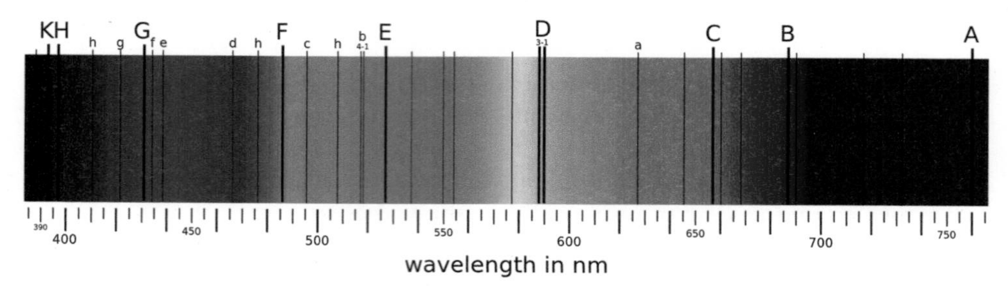

Fig 2.182

The missing lines were attributed to gases in the atmosphere of the Sun that were absorbing certain colours. Work by Robert Bunsen and Gustav Kirchhoff in the mid-19th century demonstrated different emission spectra from elements when they were heated. The emission lines from elements were found to correspond with the dark regions in the Sun's spectrum. This could then be used to identify elements present in the Sun's atmosphere.

Worked example

The emission spectra of different elements are shown in Figure 2.183:

Fig 2.183

Use the emission spectra in Figure 2.183 to identify the elements present in the gas that produces the following absorption spectrum from white light:

Fig 2.184

Using the emission spectra, the black lines in the absorption spectra are compared to show that the elements present are Q and S:

Fig 2.185

Exercise 7.2 Spectra

1 Two spectra are shown in the diagram below.

a)

b)

Fig 2.186

State which of the above spectra is a continuous spectrum and which is a line spectrum. Give an example of what may produce each type of spectrum.

2 Students view the white light produced by a filament lamp through a diffraction grating. Use your knowledge of spectra to describe what the students will see, and explain what this shows about white light.

3 In an experiment to investigate emission spectra, different metal salts are placed into a Bunsen burner flame. Different colours are seen.

a) Explain the observation of different colours for different metal salts.

b) Describe what would be seen if the flames were viewed using a diffraction grating. Would you expect a continuous or a line spectrum?

4 When the spectrum from the Sun (solar spectrum) was viewed through a diffraction grating, black lines were observed in the continuous spectrum. These lines were called Fraunhofer lines, after the scientist who investigated them in detail.

a) Name the type of spectrum observed.

b) Explain the presence of dark lines in the solar spectrum.

c) Describe how the dark lines in the Sun's spectrum can be used in conjunction with the emission spectra of elements to identify elements present in the solar atmosphere.

5 The emission spectra from four different elements (P, Q, R and S) are shown in Figure 2.187.

Identify the elements present in the gases that produce the following absorption spectra from white light a) – c) in Figure 2.188.

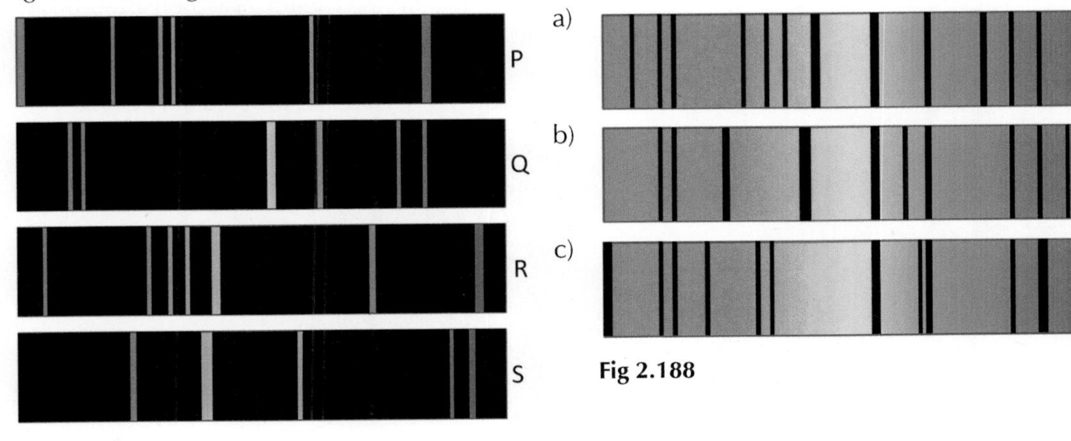

Fig 2.188

Fig 2.187

7.3 The model of the atom

In the previous section, it was shown that different elements can emit difference colours of light. This means that the model of the atom for the element must take into account the ability to emit light, and must allow for different elements to give off different colours of light.

The concept of matter being made up of many small units has been around for thousands of years. The name 'atom' dates back to 400 BC, to the Greek philosopher Democritus, who postulated that matter was made up of multiple small particles. These particles were described as 'uncuttable' because they could not be reduced into smaller chunks. This gave rise to the word atom – from the Greek word *atomos* for uncuttable. Democritus' model for the atom was simple – an atom was a round solid ball, as in Figure 2.189.

Fig 2.189

Matter was made up of many of these atoms packed together, as shown in Figure 2.190.

Fig 2.190

John Dalton built on this model hundreds of years later when he proposed that atoms of different elements could combine as complete units to form compounds. (See Figure 2.191).

Fig 2.191

These could then join to produce matter as shown in Figure 2.192.

Fig 2.192

At this point, the presence of sub-atomic particles, such as electrons and protons, was not known. The models did not account for them. In the late 19th century, J.J. Thompson discovered the electron and proposed a new model for the atom – the plum pudding model. This consisted of the atom being made up of mostly positive material, with electrons free to move around it as shown in Figure 2.193. This highlights where the name *plum pudding model* came from!

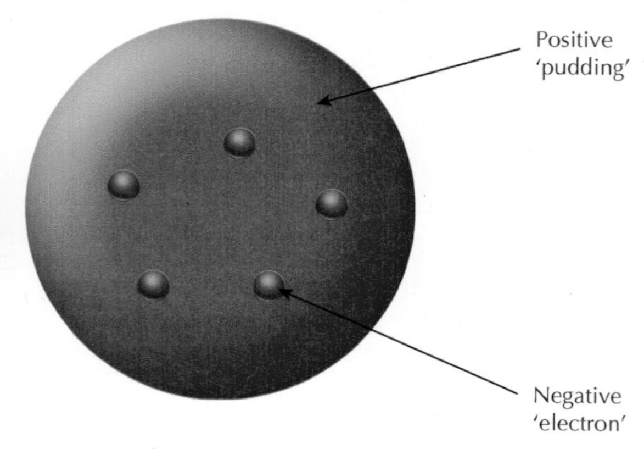

Fig 2.193

The atom is neutral overall. The negative charge of the electrons is balanced out by the positive charge of the 'pudding'. J.J. Thompson predicted that the electron was 1000 times smaller than the whole atom, a fact later proved by Robert Millikan with the Millikan oil drop experiment. However, it was not long before this model of the atom was disproved and a different model put forward – the Bohr model.

7.4 Bohr model of the atom

The Bohr model of the atom was a refinement of Ernest Rutherford's model of the atom. Niels Bohr was his student, and Rutherford's model inspired Bohr to develop a model for the atom that is still used to this day. It explains the quantum mechanical effects such as spectra that have been discussed so far.

7.4.1 Rutherford scattering experiment (Geiger–Marsden experiment)

J.J. Thompson's model of the atom was ultimately disproved by experiments conducted in Ernest Rutherford's lab, which investigated the scattering of alpha particles (positively charged) by gold foil. The experiments were initially carried out by Hans Geiger and Ernest Marsden, who found that when alpha particles were fired at gold foil, a small number were found to be deflected by a large amount.

Fig 2.194

The experiment consisted of a source of alpha particles directed at a thin metal (gold) film. A fluorescent screen (zinc-sulphide) was used to detect the alpha particles. Many were found to pass straight through the foil with little or no deflection. However, some underwent deflections of greater than 140°! This result astonished Rutherford, pictured here in Figure 2.194, who at the time stated:

'It was quite the most incredible event that ever happened to me in my life. It was almost as incredible as if you had fired a 15-inch shell at a piece of tissue paper and it came back and hit you.'

The results observed in these experiments could not be predicted by Thompson's plum pudding model, where the positive charge was evenly distributed across the nucleus. The large angle scattering observed would only be possible if the alpha particles had hit something with a large mass, and that also had a positive charge. This led to Rutherford's model of the atom.

7.4.2 Rutherford model

Based on the Geiger–Marsden experiments with metal films, Rutherford postulated a model of the atom where the positive charge was all concentrated in one place (the nucleus) and electrons surrounded this nucleus. Most of the nucleus was empty space. The model is shown in the diagram in Figure 2.195.

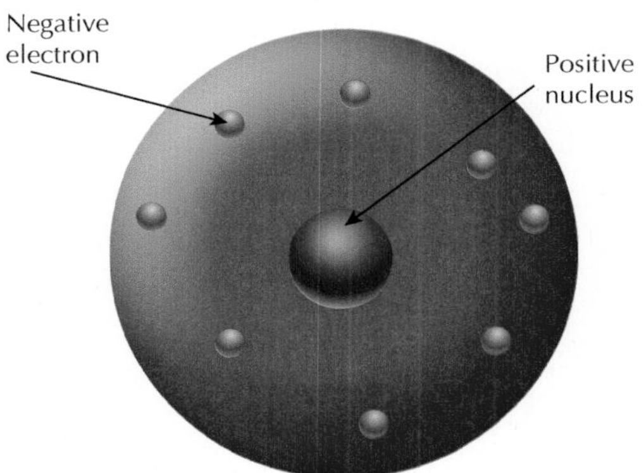

Fig 2.195

The fact that only a small number of alpha particles underwent large angle scattering highlighted that the nucleus of the atom only occupies a small space. The large angle scattering showed that the nucleus is both positively charged and has a large mass. The studies of momentum and collisions in Area 1 show that for the alpha particle to undergo a deflection of greater than 90°, it must have collided with an object of greater mass.

7.4.3 Bohr model

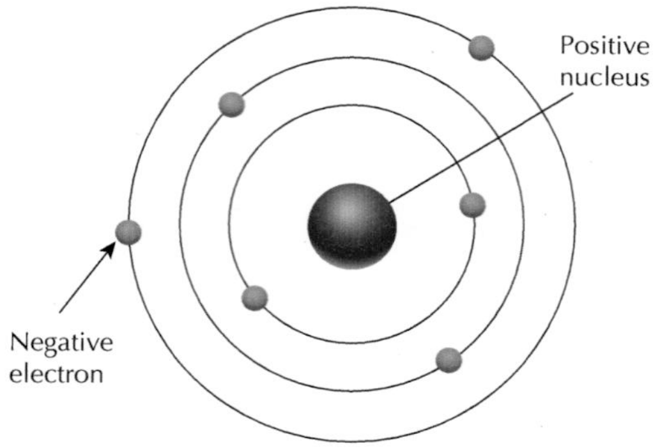

Positive nucleus

Negative electron

Fig 2.196

Niels Bohr refined Rutherford's model of the atom to account for some of the early quantum mechanical discoveries of the time, such as emission and absorption spectra. Bohr stated that the electrons orbiting the nucleus would only occupy certain discrete orbits, as shown in Figure 2.196.

The discrete orbits correspond to discrete energy levels. That is, electrons occupying a certain orbit have a certain energy. The position of these orbits will depend on the charge of the nucleus, and will therefore be different for different elements. This was used to explain absorption and emission spectra considered in the previous section.

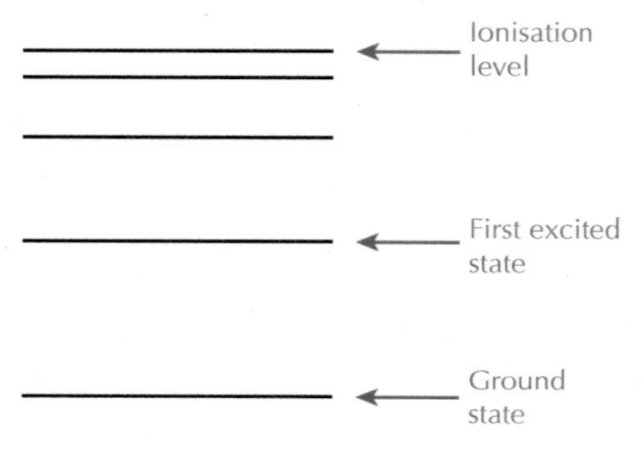

Fig 2.197

If an electron is supplied with enough energy, it can be removed from the atom – this process is known as ionisation. The minimum energy required to remove an electron is known as the ionisation energy. This can be shown on an energy level diagram. Typically, the circular orbits of the electrons are represented by horizontal lines (Figure 2.197).

An important effect of the ionisation level being considered as zero energy is that the ground and excited states of the atom must be negative. This is because energy is supplied to the electron to get it to zero energy ionisation level. Typical energies are shown in Figure 2.198. The ground state is the lowest energy state and is usually given the symbol E_0. The first excited state is given the symbol E_1, the second excited state is E_2 and so on.

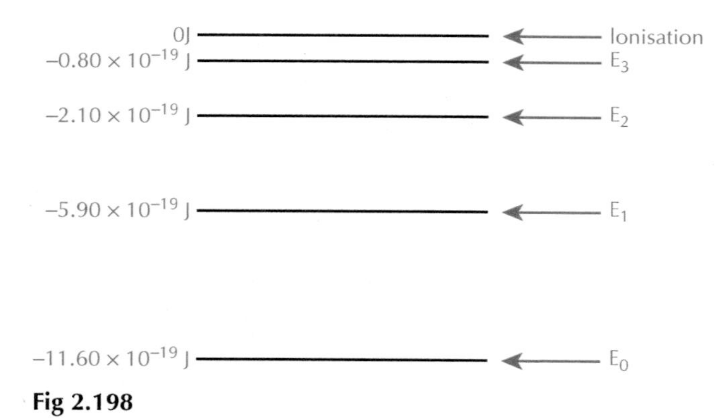

Fig 2.198

The key features of the Bohr model of the atom are:

- Atom consists of positively charged nucleus at the centre with negatively charged electrons orbiting in discrete orbits.
- Positively charged nucleus contains the vast majority of the mass of the atom.
- Most of the atom is empty space.
- The electrons orbit the nucleus in discrete orbits that correspond to discrete energy levels.
- The ionisation level is where the electron has zero potential energy. All energy levels are therefore negative.

Key point

The Bohr model of the atom consists of a positively charged nucleus at the centre with the electrons orbiting in discrete circular orbits as shown in Figure 2.200.

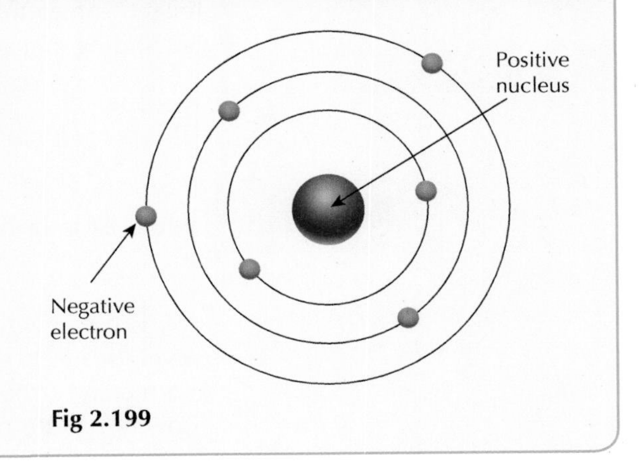

Fig 2.199

7.5 Photon emission and absorption

The Bohr model of the atom shows that the electrons can occupy discrete orbits that correspond to discrete energy levels. The position of the orbits, and thus the energy of the energy levels depends on the charge of the nucleus. This means that the atoms of different elements will have different energy levels.

Electrons are able to make transitions between the energy levels of an atom. If enough energy is supplied to an electron, it can be removed from the atom altogether in a process called ionisation. However, the electron can also be moved between energy levels, and this process will result in the production of emission or absorption spectra depending on the direction in which the electron moves.

7.5.1 Photon emission

An electron can be stimulated to jump from a lower energy state to an upper energy state by absorbing energy from a single photon. However, the electron will not stay in the upper energy state for long. It will move back towards the lower energy state, and when it does so it will emit a photon of light. This is illustrated by Figure 2.200.

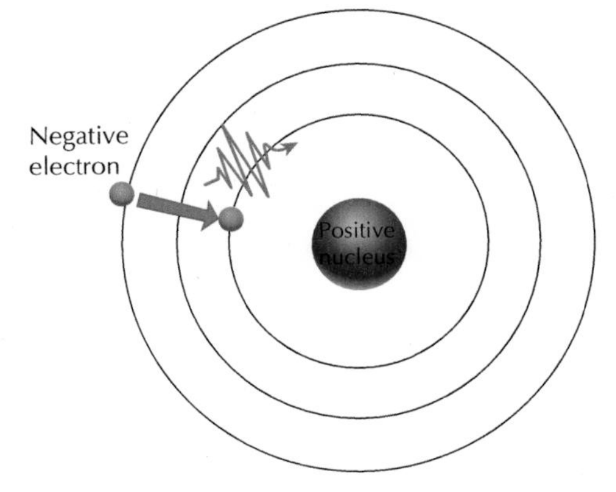

Fig 2.200

The energy of the emitted photon depends on the difference between the energy levels that the electron moves between. This is made clearer by Figure 2.201:

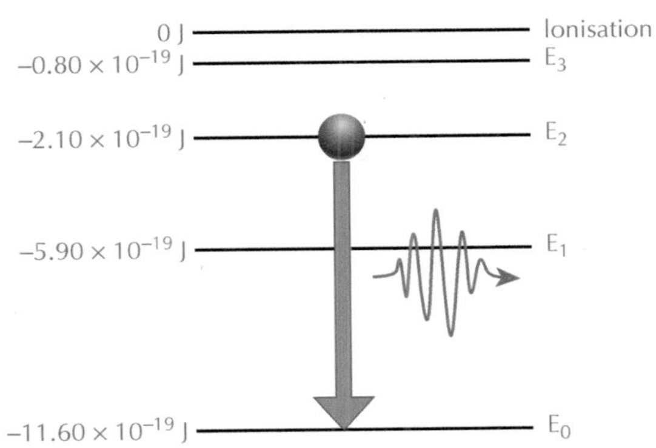

Fig 2.201

The energy of the emitted photon above will be equal to the difference in energy between levels E_2 and E_0. As the studies of the photoelectric effect show, this energy will correspond to a certain frequency according to the equation:

$$E = hf$$

This process can be summarised with the following equation:

$$\Delta E = hf$$

where ΔE is the difference in energy of levels the electron falls between:

$$\Delta E = E_2 - E_1$$

W_2 is the energy level the electron drops into and W_1 is the energy level the electron falls from. It is essential that you substitute these the correct way round into the equation. For emission, the change in energy will be negative, which shows the atom loses energy (in the form of a photon).

Key point

When an electron moves from a high-energy state to a lower-energy state, it emits a photon with an energy equal to the difference in energy between the states:

$$\Delta E = E_2 - E_1 = hf$$

W_2 is the final energy and W_1 is the initial energy.

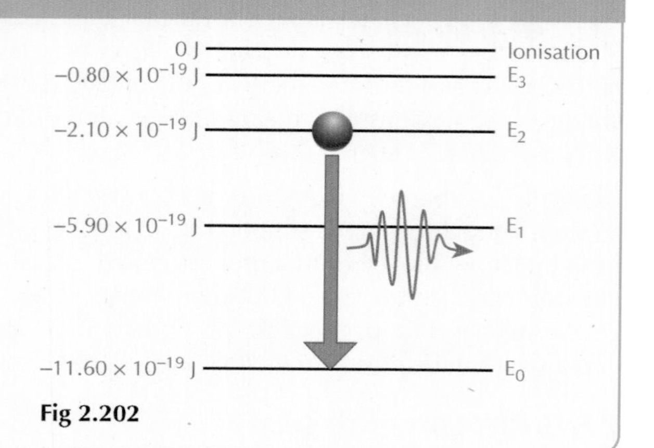

Fig 2.202

Worked example

Calculate the frequency of the photon emitted by making the transition from energy level E_3 to E_0 shown in Figure 2.203.

Firstly, calculate the energy of the emitted photon by finding the difference in energy between the upper and lower levels:

$$\Delta E = E_2 - E_1$$

$$\Delta E = \left(-14.60 \times 10^{-19}\right) - \left(-0.90 \times 10^{-19}\right)$$

$$\Delta E = -13.7 \times 10^{-19} J$$

Now, find the frequency of the photon:

$$\Delta E = hf$$

$$13.7 \times 10^{-19} = 6.63 \times 10^{-34} f$$

$$f = \frac{13.7 \times 10^{-19}}{6.63 \times 10^{-34}}$$

$$f = 2.1 \times 10^{15} Hz$$

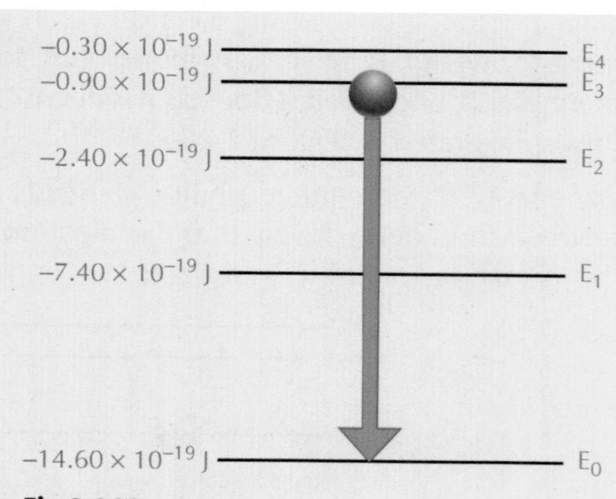

Fig 2.203

7.5.2 Emission spectra explained

In the previous discussion, it has been shown that different elements will emit different colours in their emission spectra. Some elements will emit many colours, while others will emit fewer. The colour of the line is linked to its frequency. We have seen that the frequency of the photon depends on the difference between the higher and lower energy levels that the electron moves between:

$$\Delta E = W_2 - W_1 = hf$$

This means that atoms with different energy levels will emit photons with different energies and thus, different colours. This explains the results seen in the spectra experiments where different elements emit different colours. The spectrum is a fingerprint of the element, dictated by the energy levels in the atoms of that element.

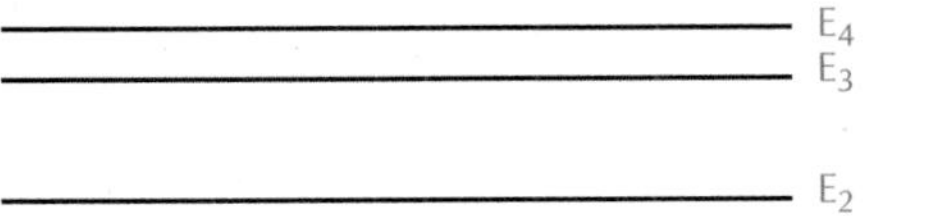

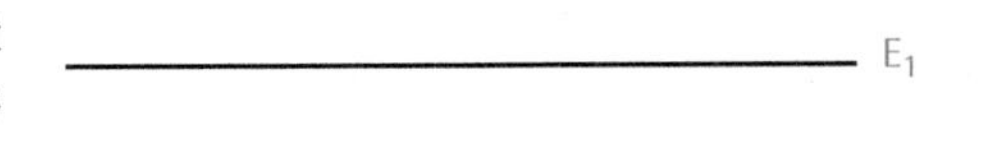

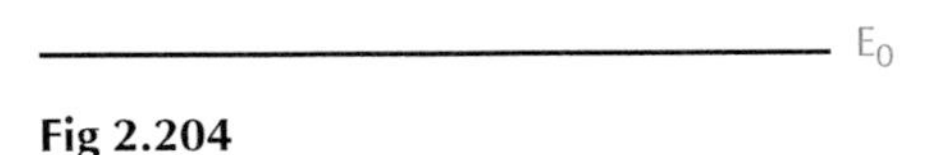

Fig 2.204

Consider the energy levels shown in Figure 2.204. There are a variety of different transitions that an electron can make, each of which will result in a different line in the emission spectrum. Each possible transition is shown in Figure 2.205.

There are 10 different transitions possible, leading to 10 different frequencies of photon, and therefore 10 different lines in the emission spectrum. However, some of these photons may not be in the visible part of the spectrum.

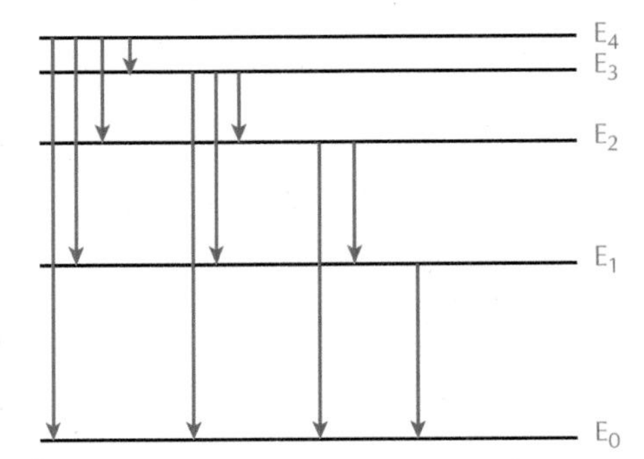

Fig 2.205

7.5.3 Photon absorption

As well as emitting a photon when the electron loses energy, the opposite process of gaining energy by absorbing a photon can also occur. The energy of the photon must be exactly the same as the difference in energy between the two levels:

$$W_2 - W_1 = hf$$

This equation has the same form as the one used for emission spectra because it is simply dealing with the reverse process. Here, the difference in energy will be positive, showing that the atom gains energy equal to the energy of the photon that is absorbed. Remember that Einstein stated that photons can only be emitted or absorbed in complete units, so only photons with the correct energy will be able to excite an electron to jump to a higher energy level. The concept is illustrated in Figures 2.206 and 2.207.

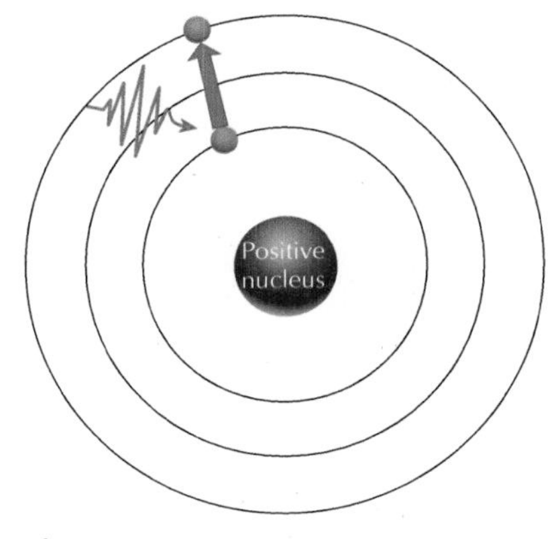

Fig 2.206

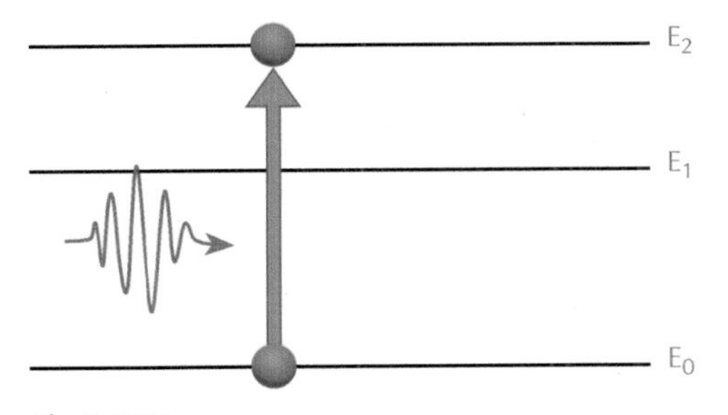

Fig 2.207

Key point

If a photon of the correct energy is incident on an atom, it can stimulate an electron to jump to a higher energy level, provided its energy exactly matches the difference in energy between the two states. The stimulating photon would be absorbed by this process and its energy used to make the electron jump to a higher energy level. This is governed by the equation:

$$W_2 - W_1 = hf$$

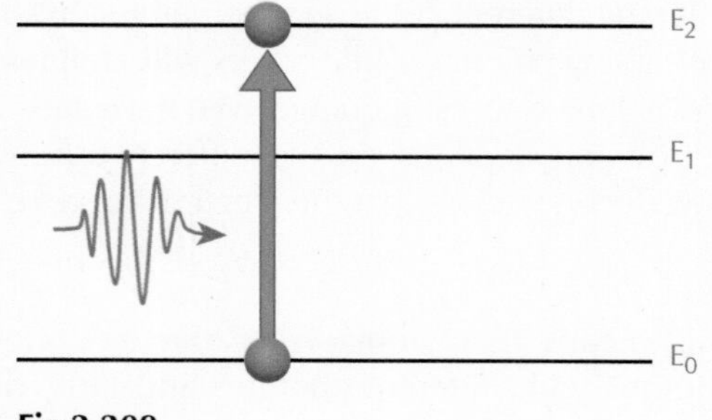

Fig 2.208

where W_2 is the energy of the final level, W_1 is the energy of the initial level and f is the frequency of the absorbed photon.

Worked example

A photon incident on an atom causes electrons orbiting the nucleus to occupy higher energy states. One transition that is found is shown in Figure 2.209.

Calculate the wavelength in a vacuum of the photon that produced the above transition.

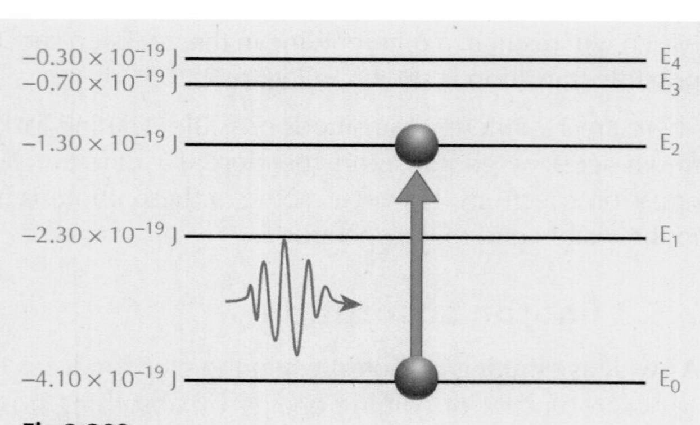

Fig 2.209

The approach to this problem is very similar to the emission worked example above. First, work out the energy difference between the two levels:

$$\Delta E = W_2 - W_0$$
$$\Delta E = (-1.30 \times 10^{-19}) - (-4.10 \times 10^{-19})$$
$$\Delta E = 2.80 \times 10^{-19} J$$

Then find the frequency of the photon:

$$\Delta E = hf$$
$$2.80 \times 10^{-19} = 6.63 \times 10^{-34} f$$
$$f = \frac{2.80 \times 10^{-19}}{6.63 \times 10^{-34}}$$
$$f = 4.22 \times 10^{14} Hz$$

This frequency is then used to find the wavelength of the photon:

$$v = f\lambda$$
$$3 \times 10^{8} = 4.22 \times 10^{14} \lambda$$
$$\lambda = \frac{3 \times 10^{8}}{4.22 \times 10^{14}}$$
$$\lambda = 710\ nm$$

7.5.4 Absorption spectra explained

It has now been demonstrated that the spectrum observed from the sun is not continuous, but rather, has dark lines in it. This is due to the elements present in the solar atmosphere. Photons of the correct frequency (with the correct energy) are absorbed by the atoms of these elements. This removes this particular colour from the spectrum, which gives rise to the dark lines of the absorption spectrum.

As the absorption of a photon is the reverse process of the emission of a photon, the dark lines in the absorption spectrum will correspond to the colours seen in the emission spectrum in Figure 2.210:

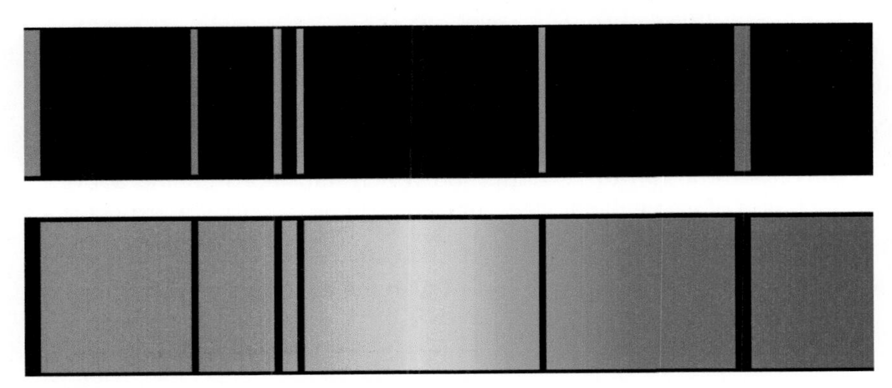

Fig 2.210

Exercise 7.5 Photon emission and absorption

1 Describe the Bohr model of the atom. Include a sketch of the atom in your diagram.

2 In the Bohr model of the atom, electrons are said to occupy discrete orbits. Using a sketch of the model of the atom, show:

a) How the movement of an electron between levels can result in the emission of a photon of light.

b) How a photon of light can be absorbed to excite an electron between levels.

c) Describe the spectrum produced by the cases in part a) and part b) above.

3 The energy levels for a specific atom are shown in Figure 2.211.

a) State the number of possible transitions that would produce a photon of light.

b) Which transition would produce a photon with the highest energy?

c) Which transition would produce a photon with the lowest energy?

d) Calculate the frequency of the photon produced by the electron moving from E_3 to E_1.

e) Calculate the frequency of the photon produced by the electron moving from E_2 to E_0.

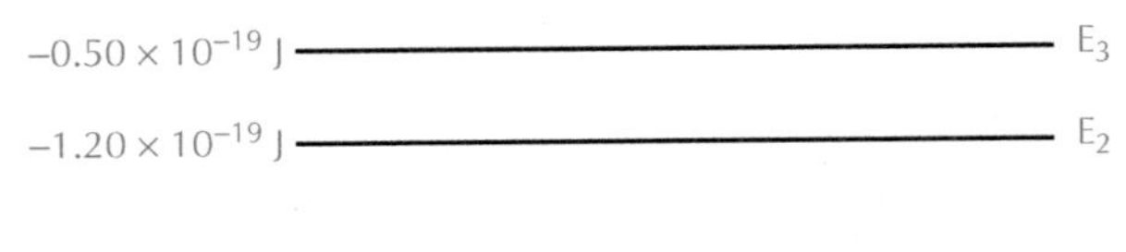

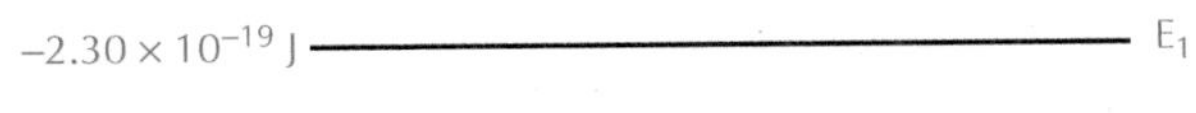

Fig 2.211

4 An absorption spectrum is observed when white light is passed through a gas containing one type of atom. The energy levels for this atom are shown in Figure 2.212:

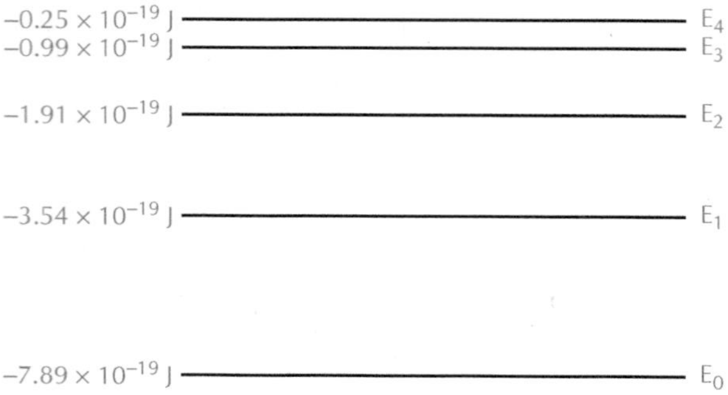

-0.25×10^{-19} J — E_4
-0.99×10^{-19} J — E_3
-1.91×10^{-19} J — E_2
-3.54×10^{-19} J — E_1
-7.89×10^{-19} J — E_0

Fig 2.212

a) How many lines would you expect in the absorption spectrum (including those not in the visible spectrum)?

b) Explain how an absorption spectrum is formed by passing white light through a gas containing atoms.

c) Describe how the absorption spectrum can be used to identify the element present in the gas. Your answer should make reference to the model of the atom and the unique energy levels possessed by the atoms of different elements.

d) Find the wavelength of the photon absorbed that causes an electron to move from E_0 to E_2.

e) If the visible spectrum runs from 400 nm to 700 nm, find which transitions for the above spectrum correspond to a photon of visible light. How many black lines would you expect to see in the visible absorption spectrum?

5 In the visible spectrum for an element, three lines are observed with the following wavelengths:

- 450 nm
- 630 nm
- 800 nm

a) If three lines were observed, how many energy levels are there in the atom?

b) Calculate the energy difference between the levels responsible for producing the photons with the above wavelengths.

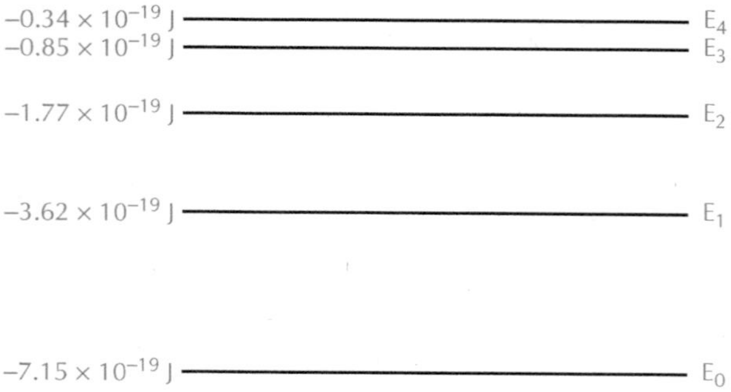

6 The energy levels for an atom are shown in Figure 2.213.

-0.34×10^{-19} J — E_4
-0.85×10^{-19} J — E_3
-1.77×10^{-19} J — E_2
-3.62×10^{-19} J — E_1
-7.15×10^{-19} J — E_0

Fig 2.213

a) Which transition corresponds to a photon with the shortest wavelength? Calculate this wavelength.

b) Which transition corresponds to a photon with the longest wavelength? Calculate this wavelength.

c) How many transitions give rise to a photon in the visible region of the spectrum? State the transitions and the corresponding wavelengths of the transitions.

7 When a gas consisting of one type of element is heated, the spectrum shown in Figure 2.214 is produced:

Fig 2.214

(continued)

a) If one of the lines has a wavelength of 550 nm, find:

(i) The corresponding frequency of the emitted photons.

(ii) The energy of the emitted photons.

b) Explain, using your knowledge of the Bohr model of the atom, how a spectrum like the one above is produced.

8 The Geiger–Marsden alpha-scattering experiments showed that some alpha particles underwent large angle deflections when fired at a thin gold film. However, most of the alpha particles passed through undetected.

a) Use your knowledge of physics to explain how the above observation led Rutherford to his model of the atom consisting of a positive nucleus with electrons surrounding it.

b) What refinements did Bohr make to Rutherford's model of the atom?

c) How do the refinements made by Bohr to the model of the atom account for photon emission and absorption?

8 Refraction

What you should know (National 5):

- Refraction is the process of light changing speed and wavelength (and, as a consequence, direction) when going from one medium to another
- When light travels from air into glass, it slows down and bends towards the normal
- Total internal reflection is where light does not exit a medium (e.g. glass) but instead reflects back into the glass
- Total internal reflection occurs when the angle that light hits the glass is greater than or equal to the critical angle

Learning intentions

- Definition of refractive index as the ratio of sine, the angle of light in air divided by sine, and the angle of light in the material
- Snell's law – linking the refractive index to the change in wavelength and change in velocity of light travelling from one medium to another
- Application of Snell's law to determine the path of a light ray when entering a medium with a simple shape until it exits the medium
- Variation of refractive index with the frequency of the material, linked to the dispersion of white light
- Definition of critical angle and total internal reflection
- Derivation of the link between critical angle and the refractive index of a material

Fig 2.215

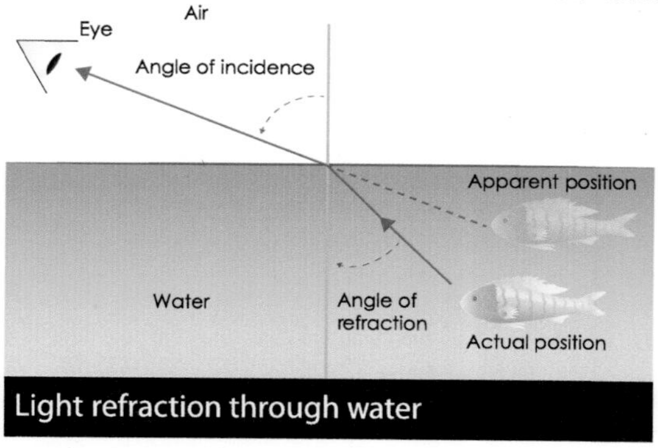

8.1 Snell's law

When light travels from one material into another, it can appear to change direction. When you place a straw into a beaker of water, you will notice that the straw appears to be bisected at the surface of the water, as shown in Figure 2.215.

The reason for this observation comes down to the way a ray of light travels between different materials. The effect has many important mechanisms – an example of which is the change in apparent position of objects in water. The eye assumes all light is travelling in a straight line. As shown in Figure 2.216, this leads to the eye seeing objects as higher in water than they actually are.

Eye
Air
Angle of incidence
Apparent position
Water
Angle of refraction
Actual position
Light refraction through water

Fig 2.216

8.1.1 Refractive index

Previous studies of physics have looked at the effect materials have on the path light takes through them, i.e. refraction. At National 5 level, refraction is defined as the process of light changing speed as it goes from one medium to another. An effect of this speed change is a change in direction of the light, if it is not being shone along the normal line (at 90° to the interface). The magnitude of the speed (and direction) change depends on the material. Specifically, it depends on a property of the material called 'refractive index'.

Experiment 8.1.1 Determining the refractive index of Perspex

This experiment determines the refractive index of Perspex by comparing the angle of incidence to the angle of refraction for light. A relationship between these angles will also be derived.

Apparatus

- Ray box
- Perspex semicircular block
- Protractor
- Ruler

Instructions

1. Place the semi-circular block on a sheet of paper and draw around the block.
2. Use a protractor to draw on a normal line (at 90°) to the flat surface of the semi-circle on the centre of the straight edge (so that the ray passes straight through undeviated).
3. Using a protractor, measure the angle of incidence (θ_1) of 20° and mark the position of the incident ray with a pencil.
4. Using the ray box, shine a ray of monochromatic light along the incident ray into the Perspex block. Mark the position of the refracted ray.
5. Use a ruler to draw on the refracted ray.
6. Use a protractor to measure the angle of refraction, θ_2.
7. Record your results in the following table:

Angle of incidence (θ_{air})	Angle of refraction (θ_{glass})	$\sin(\theta_{\text{air}})$	$\sin(\theta_{\text{glass}})$	$\frac{\sin(\theta_{\text{air}})}{\sin(\theta_{\text{glass}})}$
20				
30				
40				
50				
60				
70				
80				

Table 2.14

8. What do you notice above the value of $\frac{\sin \theta_{air}}{\sin \theta_{glass}}$ in the above equation?
9. Plot a graph of $\sin \theta_{air}$ vs $\sin \theta_{glass}$ and use this to derive a relationship between them.
10. The gradient of the line for the above graph will give the refractive index of the Perspex (this should also equal the value of $\frac{\sin \theta_{air}}{\sin \theta_{glass}}$ for your results).

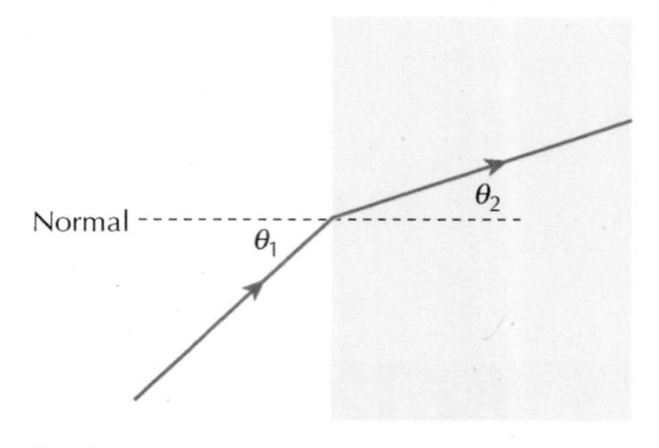

Fig 2.217

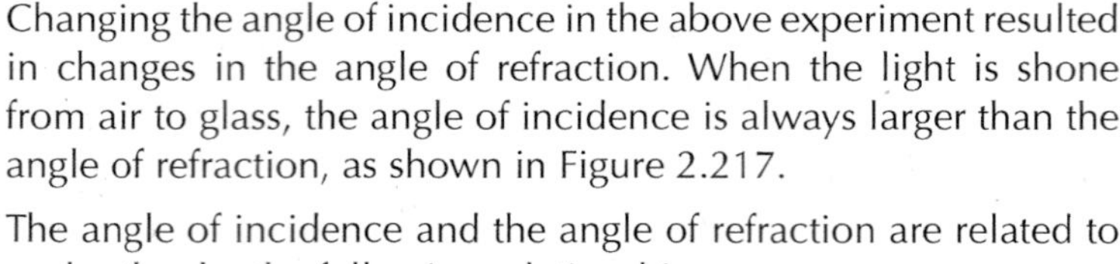

Changing the angle of incidence in the above experiment resulted in changes in the angle of refraction. When the light is shone from air to glass, the angle of incidence is always larger than the angle of refraction, as shown in Figure 2.217.

The angle of incidence and the angle of refraction are related to each other by the following relationship:

$$\sin\theta_1 \propto \sin\theta_2$$

$$\frac{\sin\theta_1}{\sin\theta_2} = k$$

where k is a constant. The constant here is the refractive index of the material, n. This gives the following equation for refractive index:

$$n = \frac{\sin\theta_1}{\sin\theta_2}$$

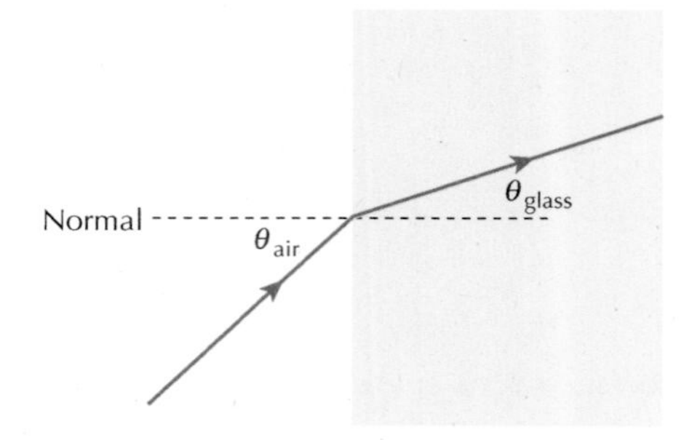

Fig 2.218

The refractive index of a material is always greater than 1. For simplicity, the angles in the above equation are renamed to θ_{air} and θ_{glass} as shown in Figure 2.218.

This gives the following equation for the refractive index:

$$n = \frac{\sin\theta_{air}}{\sin\theta_{glass}}$$

Key point

The refractive index of any material is a measure of the ratio of the velocity of light in the material to the velocity of light in a vacuum (air):

$$n = \frac{\sin\theta_{air}}{\sin\theta_{glass}}$$

Care must always be taken to measure the size of the angle between the normal and each ray.

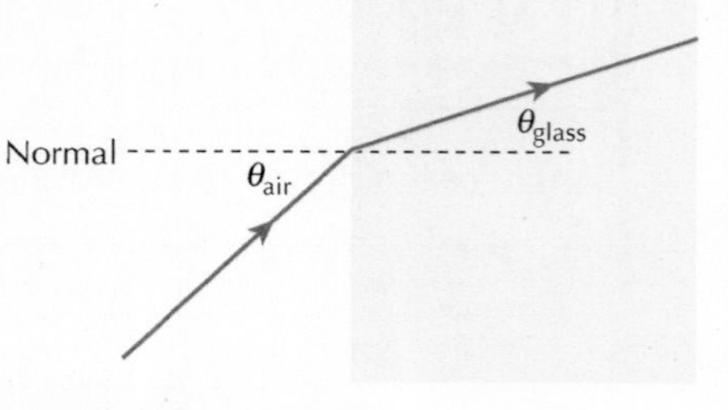

Fig 2.219

Worked example

Red light is shone into a rectangular block made of clear plastic, as shown in Figure 2.220.

Use information from the diagram to find:

a) The refractive index of the plastic.

b) The size of angle P.

c) The size of angle Q.

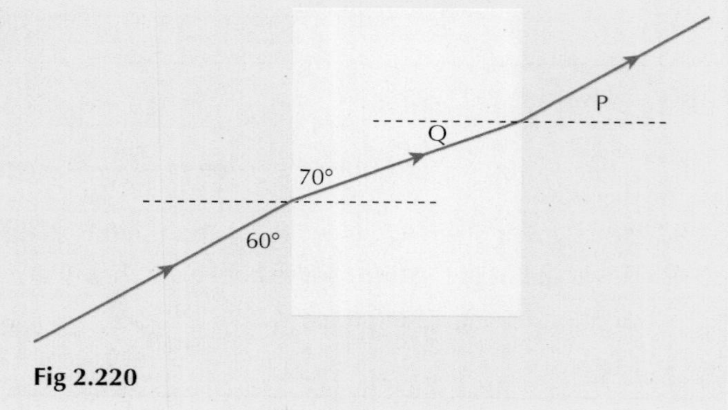

Fig 2.220

a) *The refractive index of the block will be given by,*

$$n = \frac{\sin \theta_{air}}{\sin \theta_{glass}}$$

Here, you must first find θ_{air} and θ_{glass} as the angles given above are not measured between the ray and the normal. Hence, use the complement of the angles given (angles that add up to 90°) to find θ_{air} and θ_{glass}:

$$\theta_{air} = 90 - 60 = 30°$$
$$\theta_{glass} = 90 - 70 = 20°$$

This can now be used to work out the refractive index of the plastic:

$$n = \frac{\sin 30°}{\sin 20°}$$
$$n = \frac{0.5}{0.34}$$
$$n = 1.47$$

b) *The missing angle is the same as the angle of incidence going from the air to the block; hence, $P = 30°$*

c) *The missing angle Q is the same as the angle of refraction inside the block; hence, $Q = 20°$*

Exercise 8.1.1 Refractive index

1 For each of the blocks below, calculate the refractive index of the material. Take care to ensure you use the correct angles in your calculation.

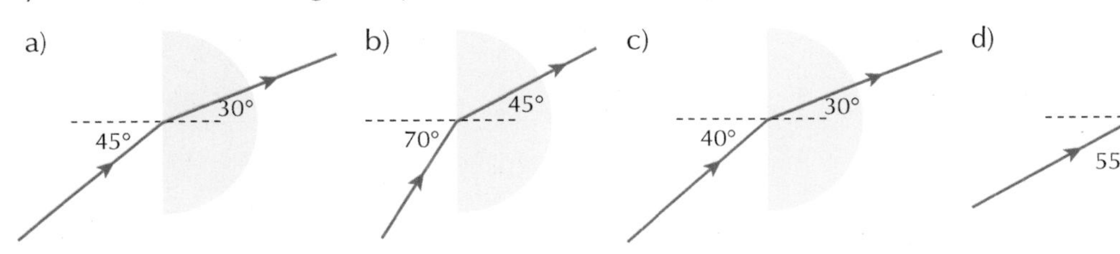

Fig 2.221

2 A material has a refractive index of 1.65. It has been used to make the rectangular shape shown in Figure 2.222. Light is incident on the block at an angle of 40°, as shown. Find the values of the missing angles P, Q and R.

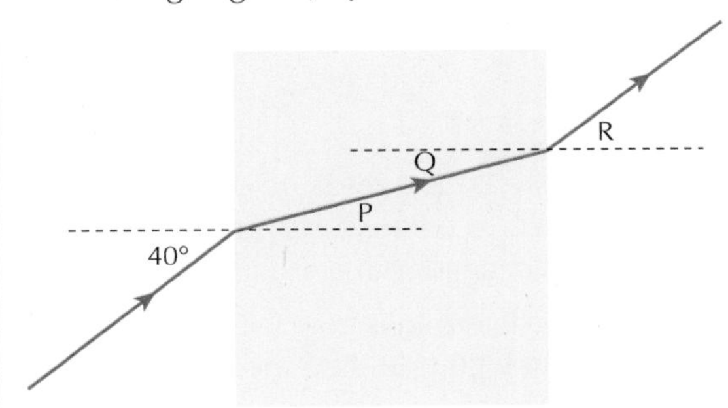

Fig 2.222

3 The diagrams in Figure 2.223 show light travelling through two different materials.

Which of the blocks has the greatest refractive index? Explain your answer.

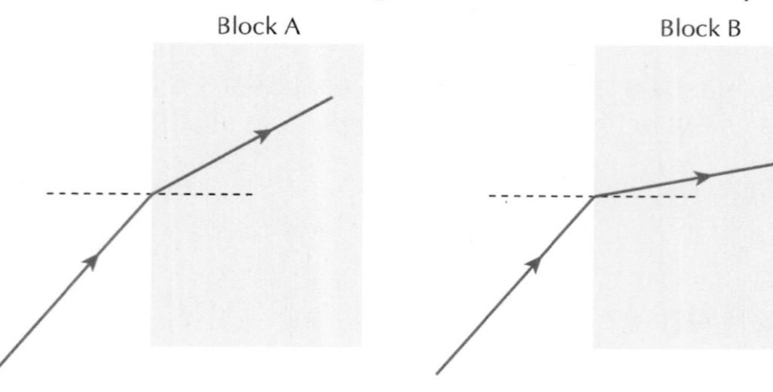

Fig 2.223

4 For each of the blocks shown in Figure 2.224, find the values of the missing angles.

a)

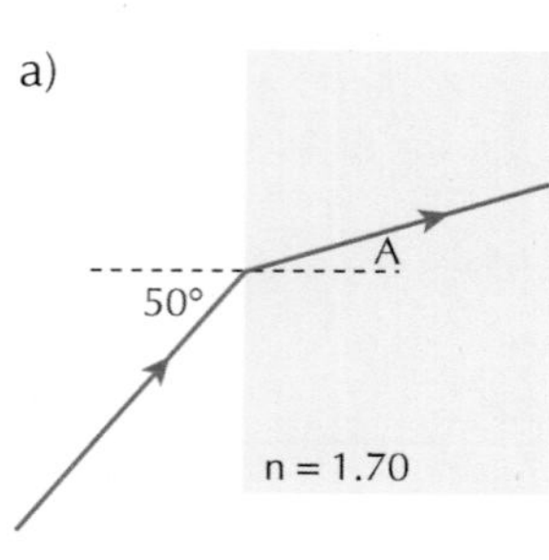

b)

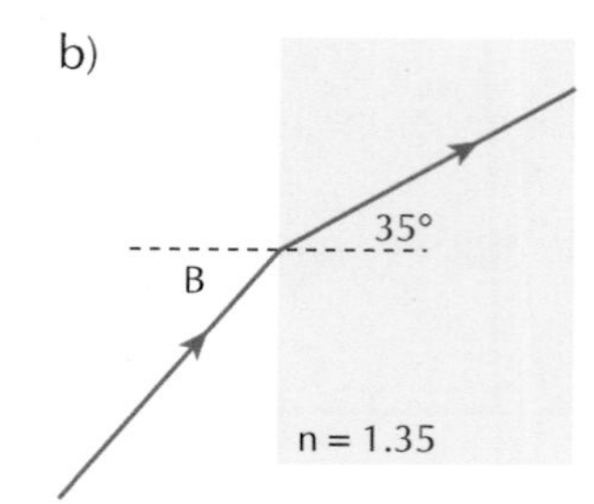

c)

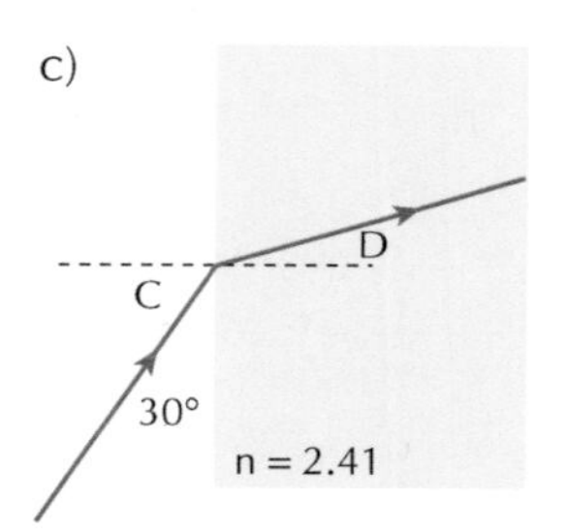

d)

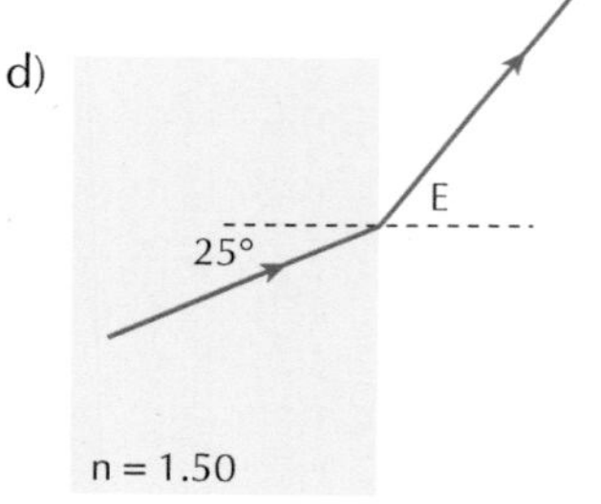

Fig 2.224

5 Use your knowledge of refraction to explain the effect observed when a straw is placed into a beaker of water as shown in Figure 2.225.

Fig 2.225

8.1.2 Snell's law

The refractive index, *n*, of the material is a measure of how much the material causes the light to refract. In other words, it is a measure of how much light slows down and changes direction.

The full version of Snell's law also determines how the wavelength and velocity of the wave change as they go from one medium to another. As light moves from air to glass:

- the velocity reduces
- the wavelength reduces

The amount by which these reduce depends on the refractive index. Consider the wave fronts moving into a glass block shown in Figure 2.226.

In the diagram, θ_{air} is represented as θ_a, θ_{glass} as θ_g, λ_{air} as λ_a and λ_{glass} as λ_g.

Use trigonometry for right-angled triangles to express the wavelength in air in terms of the angle in air:

$$\sin\theta_{air} = \frac{\lambda_{air}}{AB}$$

Likewise, trigonometry can be used to express the wavelength in glass in terms of the angle in glass and the same line AB:

$$\sin\theta_{glass} = \frac{\lambda_{glass}}{AB}$$

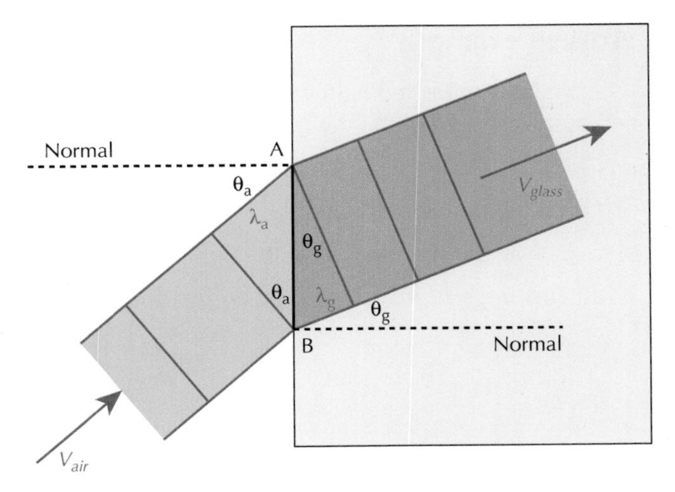

Fig 2.226

The refractive index was previously defined as:

$$n = \frac{\sin\theta_{air}}{\sin\theta_{glass}}$$

Hence, you can substitute your equations from above to get:

$$n = \frac{\left(\frac{\lambda_{air}}{AB}\right)}{\left(\frac{\lambda_{glass}}{AB}\right)}$$

$$n = \frac{\lambda_{air}}{AB} \times \frac{AB}{\lambda_{glass}}$$

$$n = \frac{\lambda_{air}}{\lambda_{glass}}$$

This gives an extension to Snell's law from the previous section, so you can now state:

$$n = \frac{\sin\theta_{air}}{\sin\theta_{glass}} = \frac{\lambda_{air}}{\lambda_{glass}}$$

As already discussed, the frequency of the light remains constant – this defines the colour, and the colour of monochromatic light does not change when it enters a different material. You can therefore use the link between wavelength and frequency to find the final part of Snell's law. You have:

$$v = f\lambda$$

where f is a constant. Hence:

$$v \propto \lambda$$

Therefore, the relationship for velocity mirrors the relationship for wavelengths and you obtain:

$$n = \frac{\sin\theta_{air}}{\sin\theta_{glass}} = \frac{\lambda_{air}}{\lambda_{glass}} = \frac{v_{air}}{v_{glass}}$$

Notice that both the wavelength and the velocity are always greater in air than in another medium such as water or glass.

Key point

Snell's law states that the velocity, wavelength and angle of light going from one medium to another change according to:

$$n = \frac{\sin\theta_1}{\sin\theta_2} = \frac{\lambda_1}{\lambda_2} = \frac{v_1}{v_2}$$

Where the light travels between air and glass, the equation can be modified as follows:

$$n = \frac{\sin\theta_{air}}{\sin\theta_{glass}} = \frac{\lambda_{air}}{\lambda_{glass}} = \frac{v_{air}}{v_{glass}}$$

Worked example

A HeNe laser emits red light with a wavelength of 633 nm. Light from this laser is directed into a beaker of water as shown in the diagram below. The refractive index of water is 1.33.

a) Calculate the angle of refraction, Q.

b) Calculate the velocity of the light in water.

c) Calculate the wavelength of the light in water.

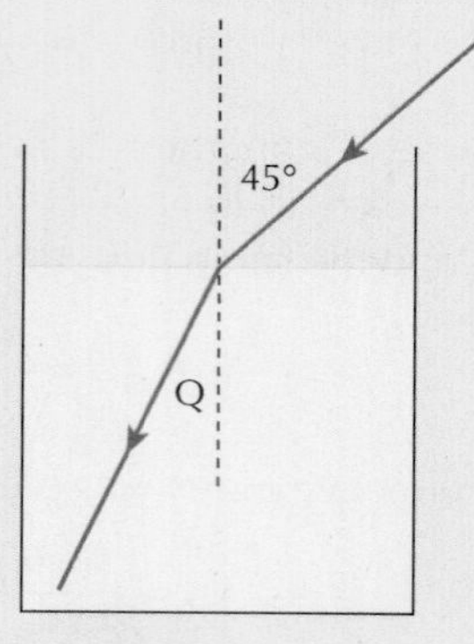

Fig 2.227

a) *The angle of refraction is given by using Snell's law for the angles:*

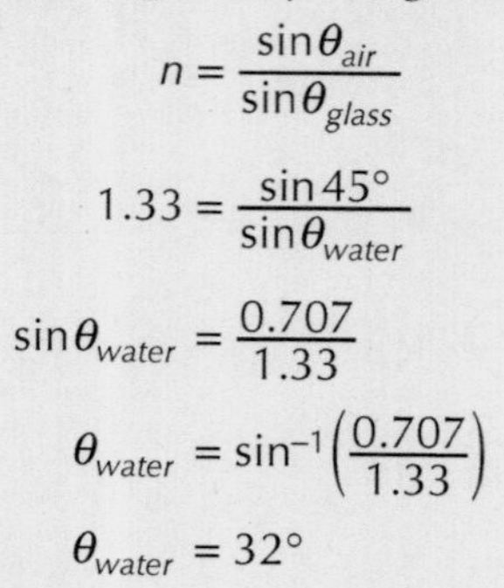

$$n = \frac{\sin\theta_{air}}{\sin\theta_{glass}}$$

$$1.33 = \frac{\sin 45°}{\sin\theta_{water}}$$

$$\sin\theta_{water} = \frac{0.707}{1.33}$$

$$\theta_{water} = \sin^{-1}\left(\frac{0.707}{1.33}\right)$$

$$\theta_{water} = 32°$$

b) *The velocity of light in the water is given by using Snell's law for velocities:*

$$n = \frac{v_{air}}{v_{glass}}$$

$$1.33 = \frac{3\times10^{8}}{v_{glass}}$$

$$v_{glass} = \frac{3\times10^{8}}{1.33}$$

$$v_{glass} = 2.26\times10^{8} ms^{-1}$$

c) *The wavelength of the light in the water is given by using Snell's law for wavelengths:*

$$n = \frac{\lambda_{air}}{\lambda_{glass}}$$

$$1.33 = \frac{633\times10^{-9}}{\lambda_{glass}}$$

$$\lambda_{glass} = \frac{633\times10^{-9}}{1.33}$$

$$\lambda_{glass} = 476\times10^{-9} m$$

$$\lambda_{glass} = 476\ nm$$

Exercise 8.1.2 Snell's law

1 Light of wavelength 560 nm is incident on the blocks of material as shown in Figure 2.228.

a)

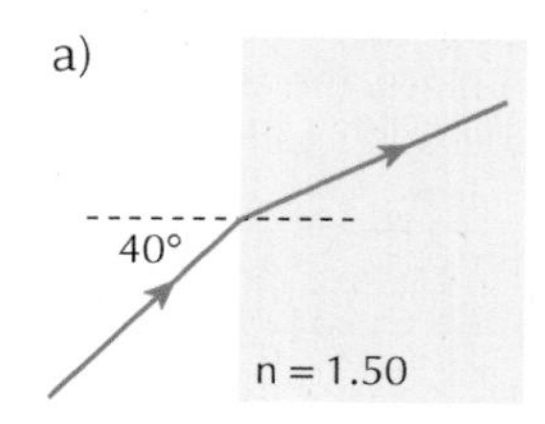

b)

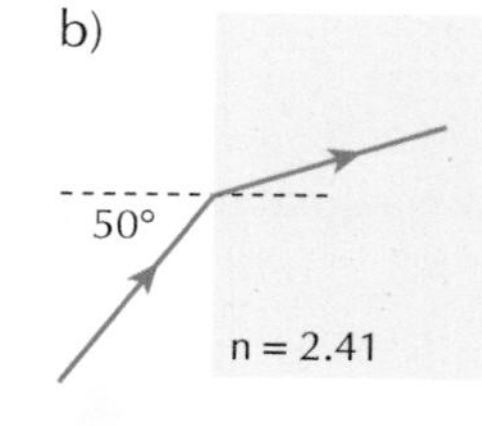

c)

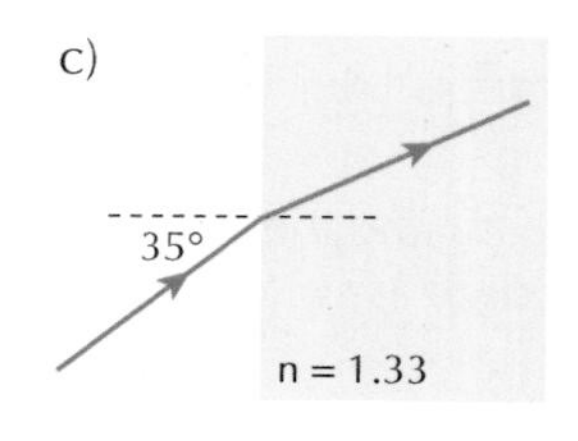

d)

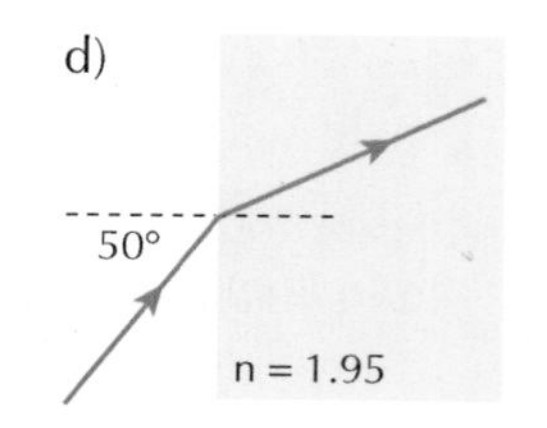

Fig 2.228

2 Red light from a HeNe laser has a wavelength of 640 nm in air. It is shone into a glass prism as shown in Figure 2.229.

a) Calculate the refractive index of the glass.

b) Calculate the velocity of the light inside the glass.

c) Calculate the wavelength of the light inside the glass.

d) Calculate the frequency of the light inside the glass. How does this value compare to the frequency of the light in air? (You may wish to confirm your answer by calculation.)

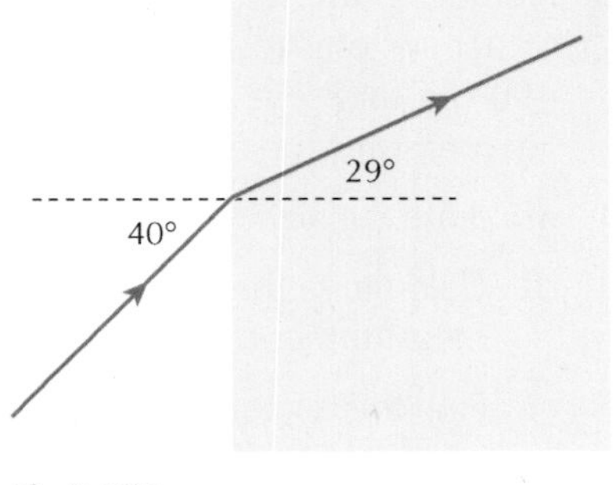

Fig 2.229

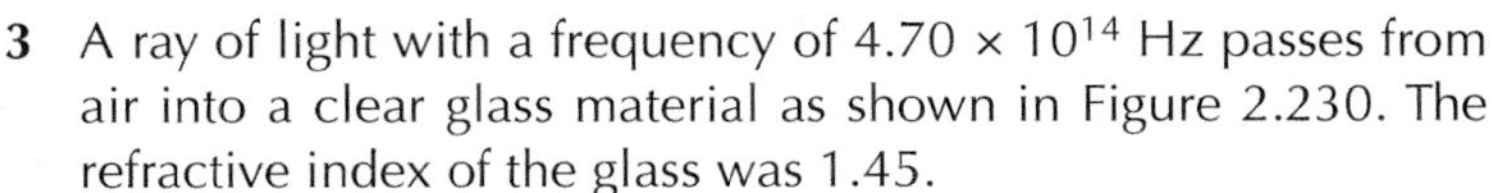

3 A ray of light with a frequency of 4.70×10^{14} Hz passes from air into a clear glass material as shown in Figure 2.230. The refractive index of the glass was 1.45.

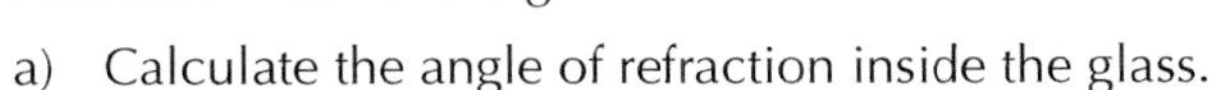

a) Calculate the angle of refraction inside the glass.

b) Calculate the wavelength of the light in the air.

c) Calculate the wavelength of the light in the glass.

d) Calculate the velocity of the light in the glass.

e) State the frequency of the light in the glass.

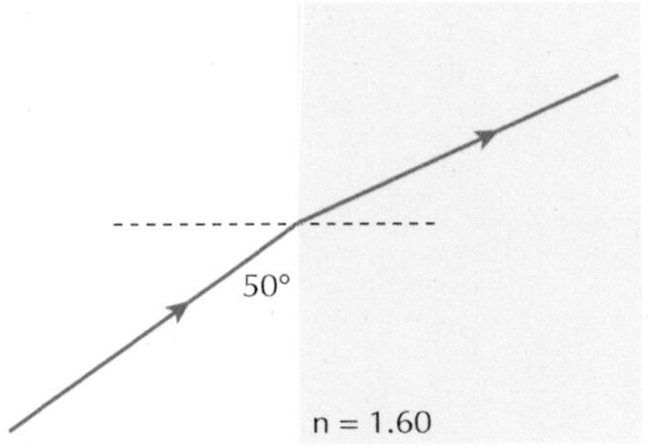

Fig 2.230

4 A student is investigating the refractive index of water. She fills a semi-circular plastic container with water and shines a ray of monochromatic light into the water as shown in Figure 2.231. The angle of incidence in air and refraction in water could be measured.

The student records the results shown in the table below.

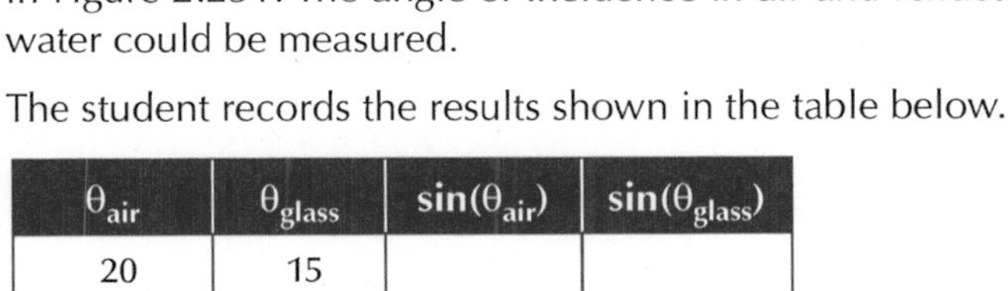

θ_{air}	θ_{glass}	$\sin(\theta_{air})$	$\sin(\theta_{glass})$
20	15		
30	21		
40	30		
50	35		
60	40		

Table 2.14

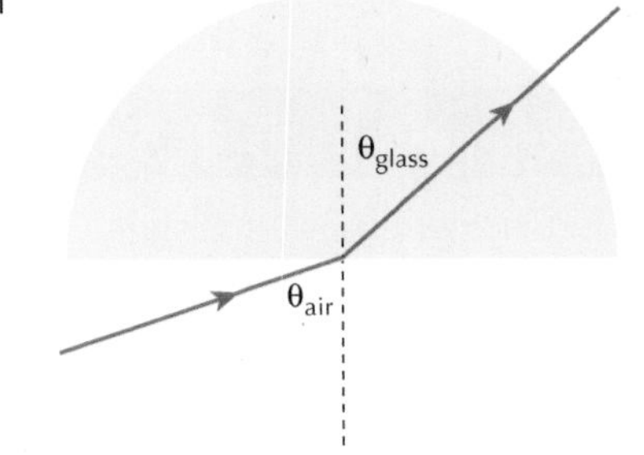

Fig 2.231

a) Copy the above table, completing the two missing columns.

b) Plot a graph of $\sin(\theta_{air})$ versus $\sin(\theta_{glass})$ and use it to determine the refractive index of water from the student's results.

c) If the wavelength of light in air was 610 nm, calculate the wavelength of the light in water.

d) Calculate the speed of light in water.

5 Optical fibres have been implemented in broadband communication systems to replace copper cables because they can carry more signals at once. However, the signals travel more slowly along an optical fibre than they do along a copper cable. Use your knowledge of Snell's law to explain why.

Fig 2.232

6 A student carrying out an investigation into the refractive index of a new plastic material shines light with a wavelength of 490 nm into a semi-circular block as shown in Figure 2.233. He measures the angles shown.

a) State the size of the reading error due to using a protractor.

b) Calculate the percentage uncertainty in each angle measurement.

c) Find the refractive index of the plastic material. Express your answer in the form *refractive index ± percentage error.*

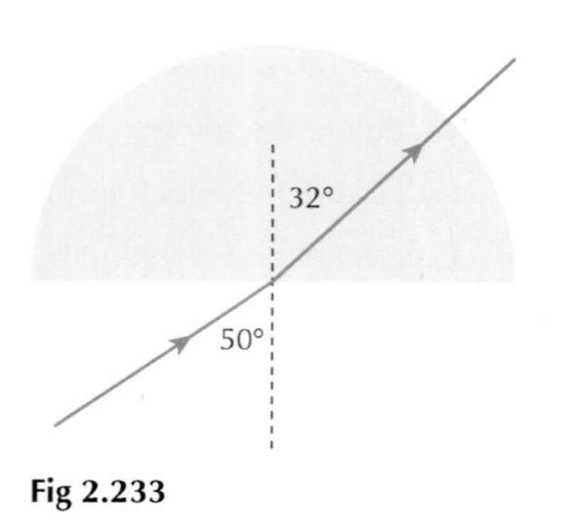

Fig 2.233

8.1.3 Dispersion

For monochromatic light, a material has a single value for the refractive index. However, for different colours of light, the value of the refractive index changes – this means that different colours of light are refracted by different amounts. If a source of light consisting of many colours is incident on a glass block, the light may be separated into its component colours due to this dependence of refractive index on wavelength. This process is called dispersion.

GO! Experiment 8.1.3 Dispersion of white light

This experiment uses a triangular prism to split white light into its component colours to demonstrate the principle of dispersion.

Apparatus

- Glass or Perspex prism
- Ray box (white light)
- Sheet of paper
- DC power supply

Instructions

1 Set the experiment up as shown in Figure 2.234, so that the ray box directs a ray of white light at the glass prism.

2 Hold a sheet of paper vertically approximately 0.5 m away from the prism to observe the ray of light leaving the prism. What do you notice on the sheet of paper? Record which colour is observed at the top (refracted the least) and which colour is observed at the bottom (refracted the most).

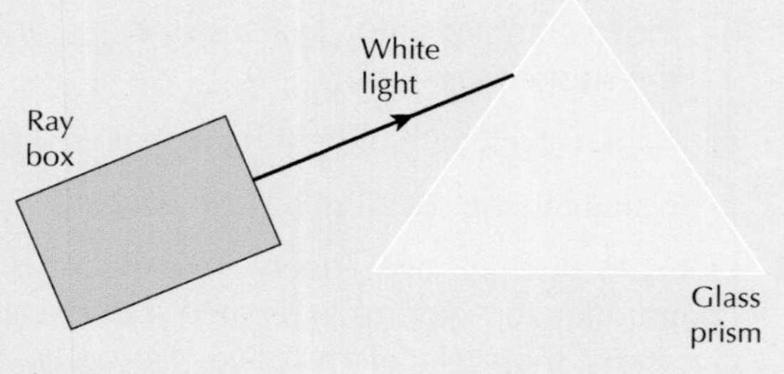

Fig 2.234

White light comprises all of the colours of the visible spectrum, from the long wavelengths of red through to the shorter wavelengths of blue and violet. When white light is passed through a triangular prism, it is observed to split into its component colours as shown in Figures 2.235 and 2.236.

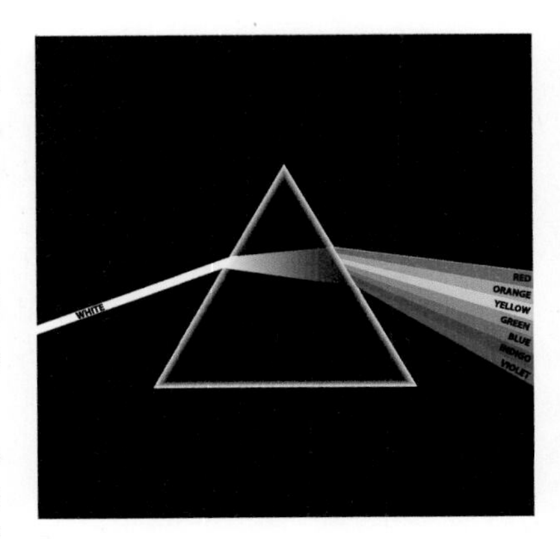

Fig 2.235

It is important to note that white light does not only consist of the seven colours of the rainbow (see Figure 2.235). It consists of countless colours as shown in Figure 2.236. The seven main ones are named as red, orange, yellow, green, blue, indigo and violet.

Fig 2.236

Notice that red light is at the top and violet light is at the bottom. The red light has been refracted least by the prism, and violet light the most. This means that the refractive index of the glass must be different for the red light than it is for the violet light; refractive index is frequency-dependent. This frequency dependence results in white light being split into its component colours. This effect is called dispersion. The greater the frequency, the greater the refractive index.

Typical refractive index differences for a material are given in Figure 2.237.

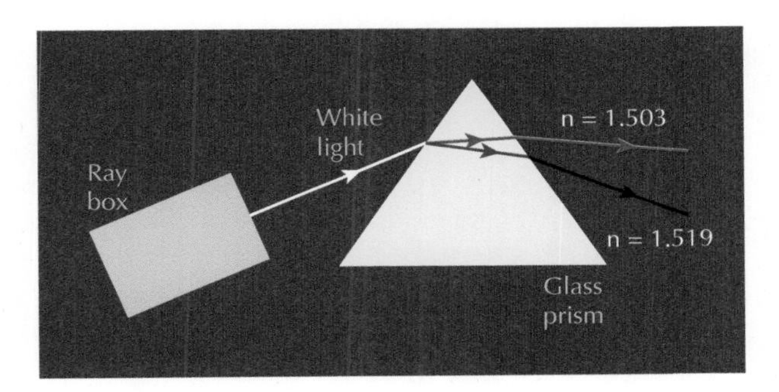

Fig 2.237

The differences in refractive index depend on the material but are typically very small. Thus, although white light can be split into its component colours, the effect is not significant enough to generate a spectrum that could be analysed in the same way as diffraction grating.

Key point

The refractive index of a material depends on the frequency of light illuminating it. The greater the frequency, the greater the refractive index. This means that white light can be split into its component colours in an effect called dispersion.

Worked example

White light is incident on a semi-circular glass block as shown in Figure 2.238. The light is incident on the prism at an angle of 40°.

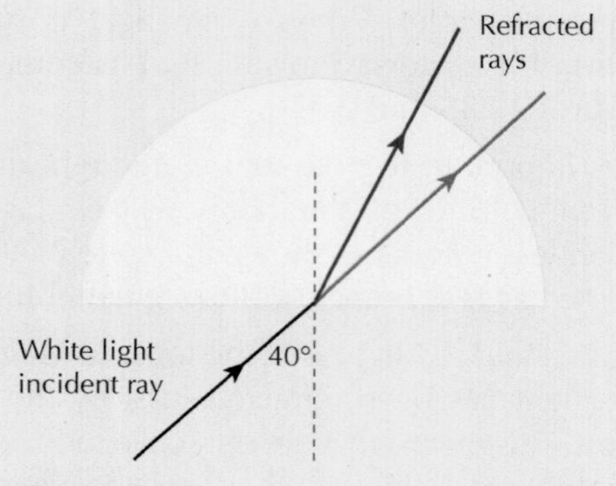

Fig 2.238

If the refractive index for red light is 1.51 and the refractive index for violet light is 1.55, find the angle between the red and violet rays emerging from the block.

To solve this problem, you need to find the angle of refraction for both colours of light using the refractive indexes given and then subtract to find the difference in angles.

For red light, this is:

$$n = \frac{\sin\theta_{air}}{\sin\theta_{glass}}$$

$$1.51 = \frac{\sin 40°}{\sin\theta_{glass}}$$

$$\sin\theta_{glass} = \frac{\sin 40°}{1.51}$$

$$\sin\theta_{glass} = \frac{0.64}{1.51}$$

$$\theta_{glass} = \sin^{-1}\left(\frac{0.64}{1.51}\right)$$

$$\theta_{glass} = 25.1°$$

For violet light, it is:

$$n = \frac{\sin\theta_{air}}{\sin\theta_{glass}}$$

$$1.55 = \frac{\sin 40°}{\sin\theta_{glass}}$$

$$\sin\theta_{glass} = \frac{\sin 40°}{1.55}$$

$$\sin\theta_{glass} = \frac{0.64}{1.55}$$

$$\theta_{glass} = \sin^{-1}\left(\frac{0.64}{1.55}\right)$$

$$\theta_{glass} = 24.4°$$

Hence, the angle between the two colours of light is:

$$\theta = 25.1 - 24.4$$

$$\theta = 0.7°$$

Exercise 8.1.3 Dispersion

1 When white light is incident in a glass prism as shown in Figure 2.239, it is split into its component colours.

a) Explain why the white light is split into its component colours by the glass prism.

b) Which colour is refracted least by the glass prism?

c) Which colour is refracted most by the glass prism?

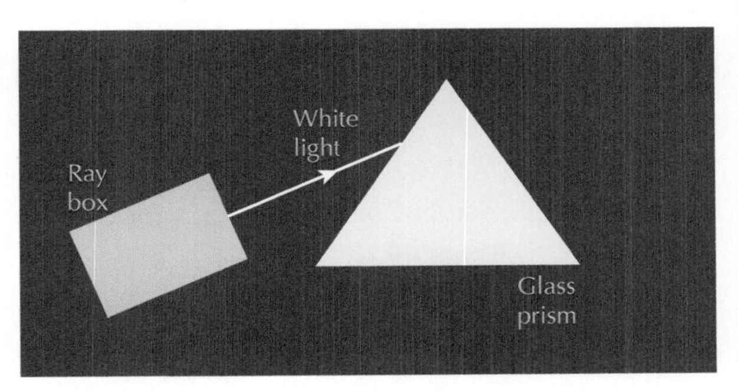

Fig 2.239

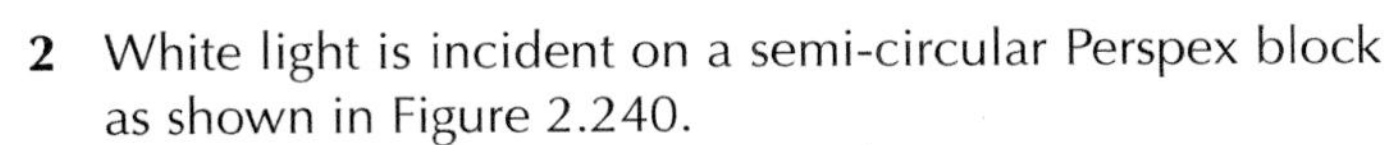

2 White light is incident on a semi-circular Perspex block as shown in Figure 2.240.

The refractive index of the Perspex for red light is 1.45 and for violet light it is 1.49.

a) Calculate the angle of refraction inside the Perspex block for red light.

b) Calculate the angle of refraction inside the Perspex block for violet light.

c) Find the size of the angle between the red and violet rays.

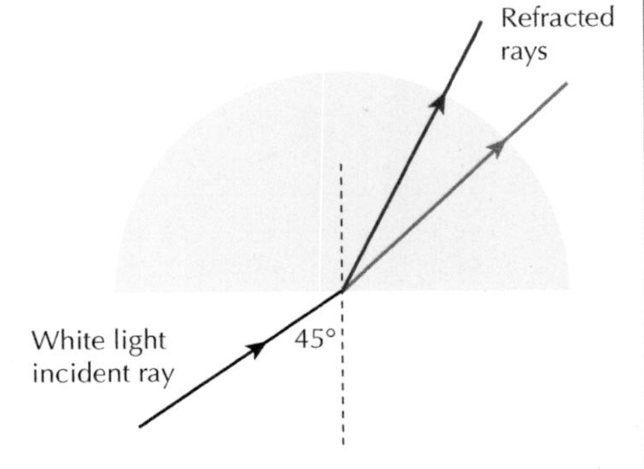

Fig 2.240

3 Magenta light (made up of red and blue) is shone through a semi-circular glass block as shown in Figure 2.241. It is incident on the glass–air boundary at an angle of 30°. Two rays are found to emerge from the block.

a) State the colour of ray P.

b) State the colour of ray Q.

c) Find the size of the angle X, the angle between rays P and Q.

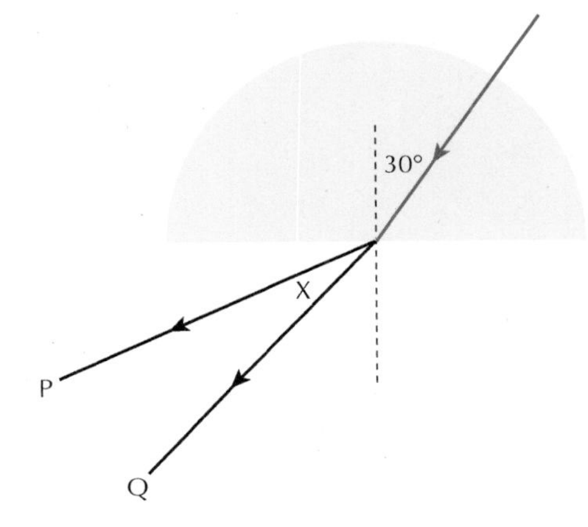

Fig 2.241

4 A ray of white light is incident on a triangular glass prism as shown in Figure 2.242. The refractive index for violet light for the material is found to be 1.64. The angle between the red and violet rays inside the prism is 1.1° as shown.

Use the information on the diagram to calculate the refractive index of the glass for red light.

5 A student wishes to investigate the spectrum produced by a filament lamp. He suggests that he could pass the light through a glass prism and analyse the

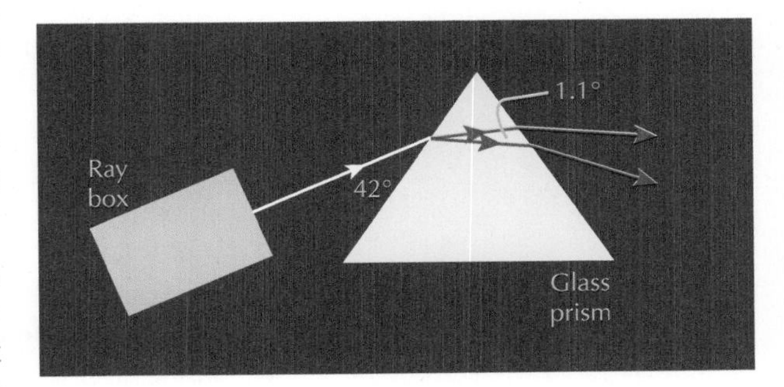

Fig 2.242

spectrum produced. Use your knowledge of dispersion to comment on the effectiveness of this procedure for looking at fine detail within a spectrum. Can you suggest an alternative experiment for investigating the finer detail of a spectrum?

6 White light incident on a diamond produces colours as shown in Figure 2.243.

Use your knowledge of physics to comment on this observation and explain the production of colour by the diamond when illuminated by white light.

Fig 2.243

8.1.4 Applications of Snell's law

Refraction of light has many important applications and allows the explanation of many phenomena, such as rainbows and mirages.

Applying refraction: the lens

There are two main types of lens – convex and concave. The effect of each of these lenses on the path of light through them is shown in Figure 2.244.

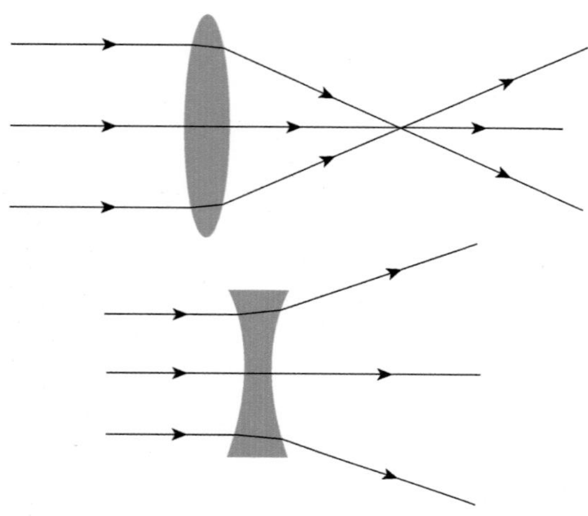

Fig 2.244

The direction of the light ray is changed due to refraction, and this is governed by Snell's law. The light changes speed, and as a result, changes direction when passing from air into glass. It changes again when passing from glass back into air. It is important to note here that refraction occurs at both boundaries of the lens.

Applying Snell's law: apparent depth

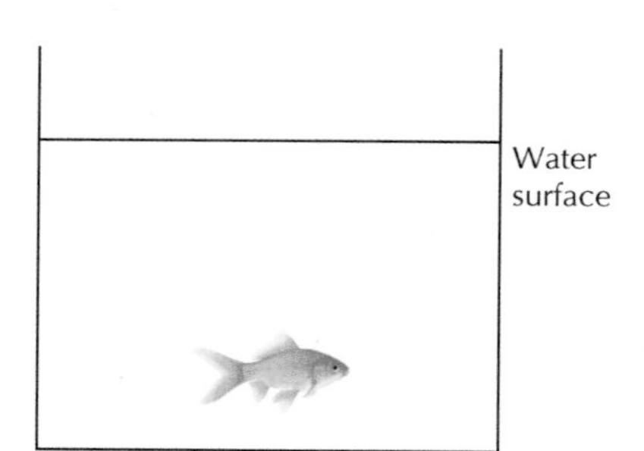

Fig 2.245

The eye always 'assumes' that light rays are travelling in straight lines and do not change their direction. Where refraction has taken place, for example at the boundary between air and water, the eye is not capable of detecting the change in direction of light. This leads to an optical illusion when viewing the apparent depth of an object in water.

Consider a fish swimming in a bowl of water, as shown in Figure 2.245. When viewed from above, the light rays leaving the top of the fin of the fish refract at the water surface. The eye assumes the light rays are travelling in a straight line – following the dotted line in Figure 2.246. Where these dotted lines meet is the point where the eye thinks the fin is in the water; notice that this is actually higher up in the water than the fish really is.

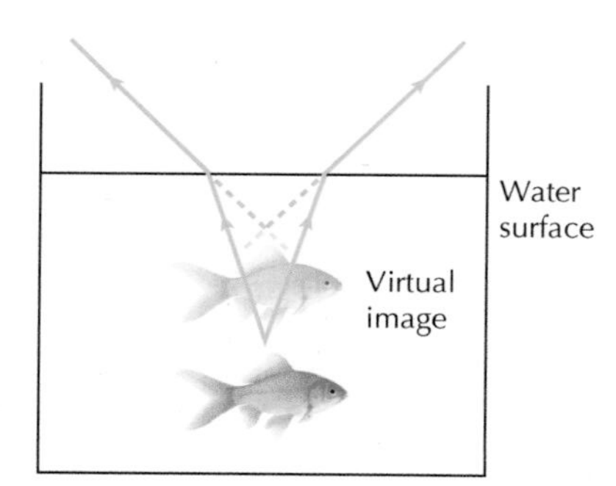

Fig 2.246

This effect can be very dangerous. A deep pool of water can look shallower than it actually is.

Applying Snell's law: the mirage

A mirage is an optical illusion where there appears to be water on the ground, whilst in reality there is no water. The appearance of water is due to an apparent reflection of the surroundings associated with water on the ground. An example of this is shown in Figure 2.247.

The production of a mirage relies on the refraction of light rays passing through the atmosphere. Close to the Earth's surface the density of the air is lower due to higher temperatures. This results in a change of refractive index. As a light ray passes through the changing refractive index, it changes its direction according to Snell's law. This is illustrated in Figure 2.248.

Fig 2.247

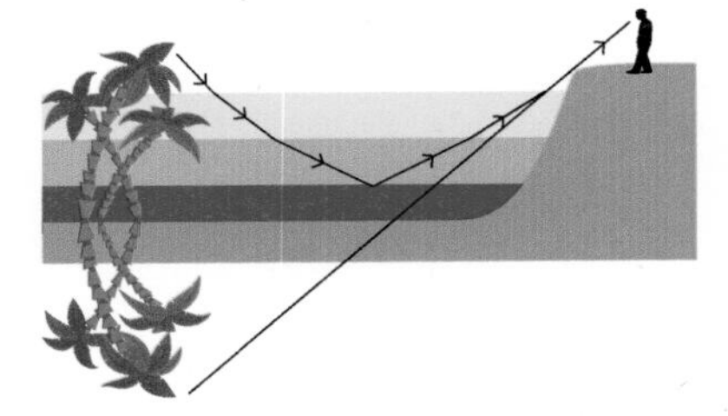

Fig 2.248

Applying Snell's law: the rainbow

A rainbow is formed in the sky when a shower of rain is illuminated by sunlight. The visual effect can be very dramatic, with a beautiful array of colour, as shown in Figure 2.249.

A rainbow relies on the process of refraction and dispersion. White light, made up of many different colours, refracts through the raindrops. As different wavelengths refract by different amounts, the white light is split by the raindrop into its constituent colours as shown in Figure 2.250.

Although each water droplet splits the Sun's white light into its component colours, only one colour is seen from each water droplet. This colour depends on the droplet's position in the sky. Droplets that are higher in the sky refract red light into our eyes, while droplets lower in the sky refract violet light into our eyes. This means red light is seen at the top and violet at the bottom of the rainbow. This is highlighted in Figure 2.251.

Fig 2.249

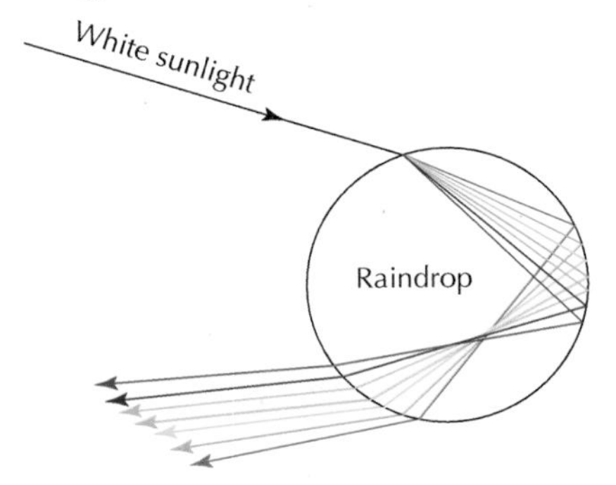

Fig 2.250

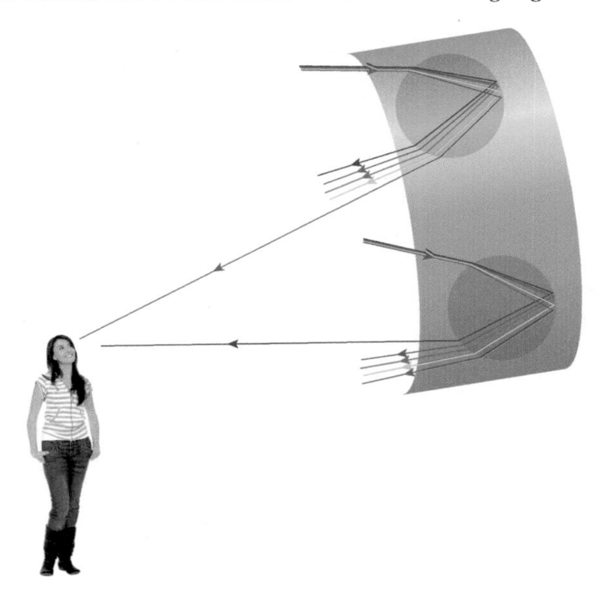

Fig 2.251

8.2 Total internal reflection

As discussed above, when light travels from one medium to another, the velocity and wavelength of the light are changed. This can result in a change in the direction of light. For the most part, the effect of light travelling from air into another medium such as glass or water has been considered so far. Here, the light ray bends towards the normal as shown in Figure 2.252.

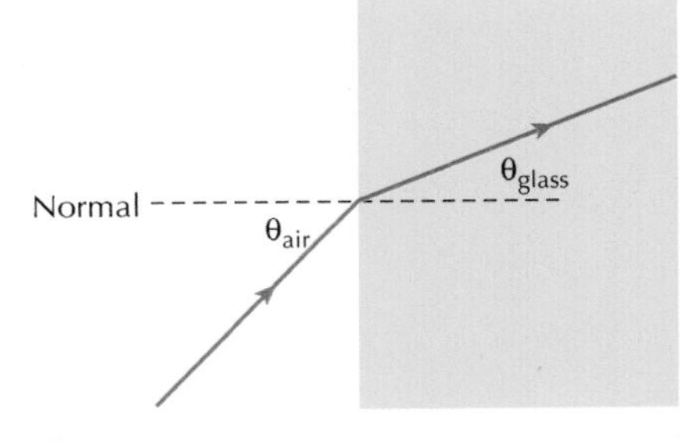

Fig 2.252

However, light can also travel from glass into air, where it will act to refract away from the normal. In these cases, an important effect can take place at the boundary between the glass and the air, which has become the key to high-speed global communication.

8.2.1 Critical angle and total internal reflection

In accordance with Snell's law, when light travels from a dense medium such as glass to a less dense medium such as air, it will speed up and refract away from the normal. This is due to the difference in refractive index of the materials. Under certain circumstances, the light will remain trapped inside the glass.

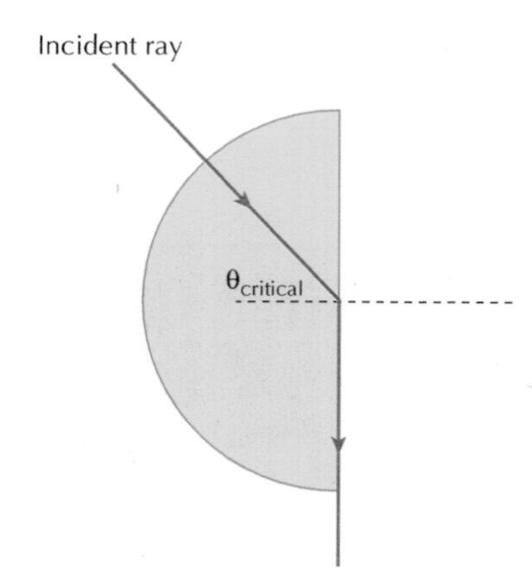

Fig 2.253

Relationship between critical angle and refractive index

The critical angle, θ_c, is the angle in the material for which light emerges into air with an angle of 90°, as shown in Figure 2.253.

Snell's law can be used to derive a relationship between the critical angle and refractive index. Starting with Snell's law:

$$n = \frac{\sin(\theta_{air})}{\sin(\theta_{glass})}$$

For the critical angle, θ_c, which is the angle in glass, the angle of the light in air is 90°. This gives,

$$n = \frac{\sin 90°}{\sin(\theta_c)}$$

$$n = \frac{1}{\sin \theta_c}$$

which can be rearranged for the critical angle,

$$\theta_c = \sin^{-1}\left(\frac{1}{n}\right)$$

At angles smaller than the critical angle, some of the light will be refracted and escape from the glass block. Some of the light will be internally reflected. At angles above the critical angle, all of the light is trapped inside the block – it is all reflected. This means that the light is totally internally reflected. This is shown in Figure 2.254.

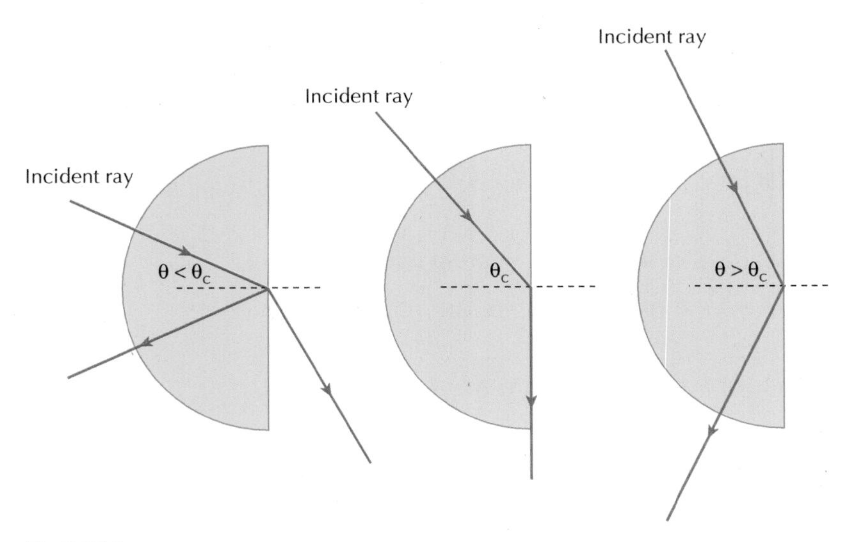

Fig 2.254

Worked example

Calculate the critical angle for water, which has a refractive index of 1.33.

Here, the definition of critical angle (derived from Snell's law) is used:

$$n = \frac{1}{\sin\theta_c}$$

$$\theta_c = \sin^{-1}\frac{1}{1.33}$$

$$\theta_c = 49°$$

Worked example

Light is incident on the inside of a Perspex block at an angle of 45°, as shown in the diagram below. If the refractive index of Perspex is 1.40, determine whether the ray will emerge from the block or undergo total internal reflection.

The path of the ray of light will depend on whether it is incident on the boundary at an angle greater than or smaller than the critical angle. First determine the critical angle, using:

$$n = \frac{1}{\sin\theta_c}$$

You have:

$$\theta_c = \sin^{-1}\frac{1}{1.40}$$

$$\theta_c = 45.6°$$

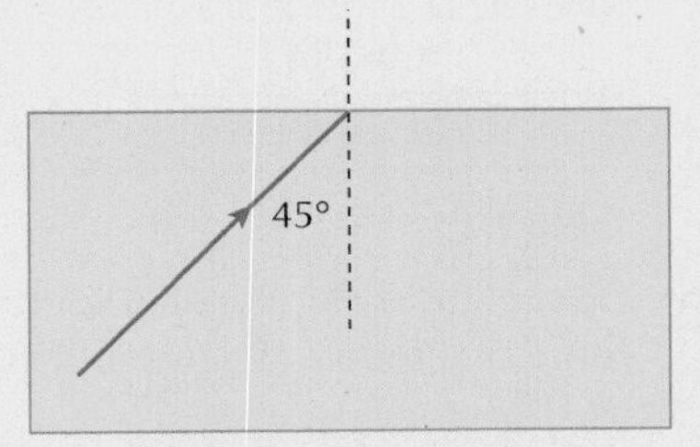

Fig 2.255

As the light ray is incident at an angle smaller than the critical angle:

$$\theta = 45° < \theta_c$$

you can say that the light will be partly refracted at the boundary and will escape. Some of the light will also be reflected.

Exercise 8.2.1 Critical angle and total internal reflection

1 When a light ray is incident on a material–air boundary at a large enough angle, this means that the light is totally internally reflected.

a) Explain what is meant by total internal reflection. Support your answer with an appropriate diagram.

b) State the name of the smallest angle for which total internal reflection occurs.

2 By considering Figure 2.256, derive an appropriate expression for the critical angle of a material.

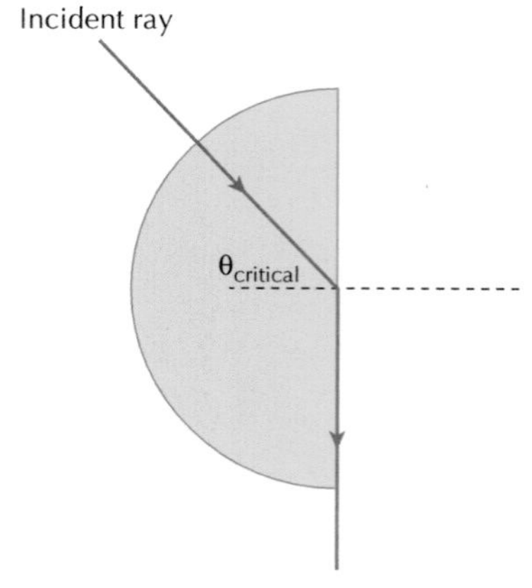

Fig 2.256

3 Calculate the critical angle for each of the materials in the table below:

Material	Refractive index
Amber	1.55
Ice	1.31
Titanium dioxide	2.50
Diamond	2.41
Crown glass	1.50
Flint glass	1.60
Silicone oil	1.45

Table 2.15

4 For each of the materials shown in Table 2.16, the critical angle is given. Calculate the refractive index of each material.

Material	Critical angle
Polycarbonate	39.3°
Acrylic	42.2°
Crystal quartz	40.2°
Silver chloride	29.7°
Magnesium fluoride	46.4°
Quinoline	37.0°
Olive oil	43.2°

Table 2.16

5 Figure 2.257 shows an experiment being conducted to investigate the critical angle of water. The refractive index of water is 1.33. The student directs light into the block at two angles, 25° and 60° as shown.

a) Calculate the critical angle for water.

b) Copy and complete the diagrams in Figure 2.257, showing clearly the path taken by the ray of light in each case. Numerical values for angles are required.

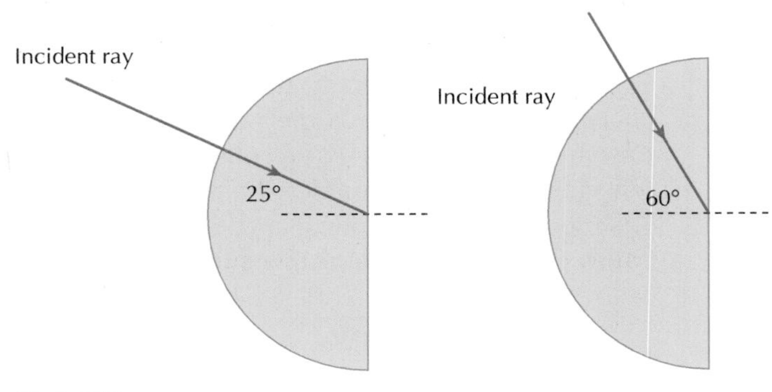

Fig 2.257

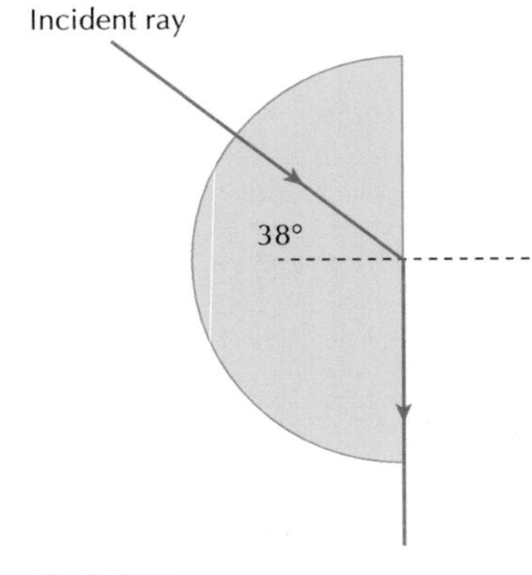

Fig 2.258

6 Use the information in Figure 2.258 to calculate the refractive index of the material.

7 A ray of red light is incident on a crown glass prism as shown in Figure 2.259. The prism is equilateral.

a) Explain why the light does not change direction at point X, where it enters the prism.

b) Does the light undergo total internal reflection at Y? You must justify your answer by calculation.

c) Copy Figure 2.259 and complete the direction of the ray of light until it exits the prism. You must include all relevant angles on your diagram.

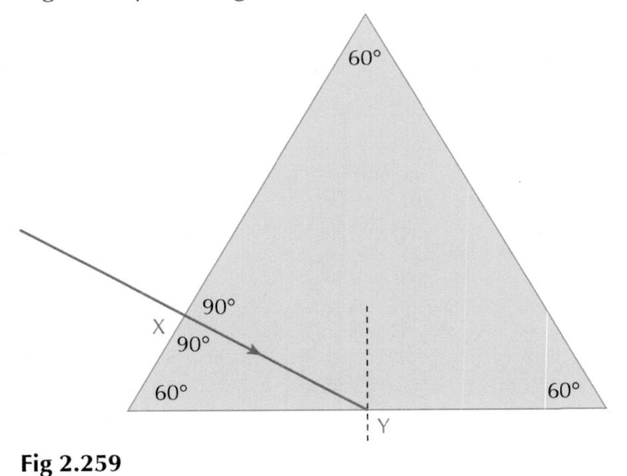

Fig 2.259

8.2.2 Applications of total internal reflection

As seen above, if light is incident on a material–air boundary at a large enough angle, it will be totally internally reflected. This can be applied alongside Snell's law to find the path of light through a medium, and explain how optical fibres work and why a diamond sparkles.

Determining the path of a light ray

The path taken by light travelling through different materials can be determined by applying Snell's law and the concept of total internal reflection for angles above the critical angle.

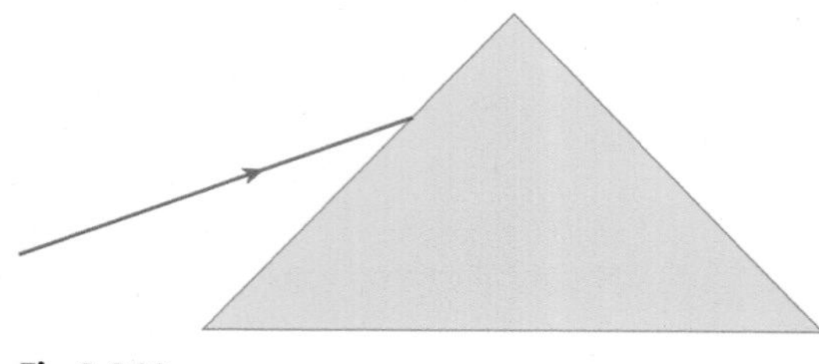

Fig 2.260

Consider illuminating a glass triangular prism with monochromatic red light as shown in Figure 2.260.

The light ray will refract at the air–glass boundary and change direction towards the normal. When it reaches the other side of the prism it will either be totally internally reflected or it will be refracted and exit the prism. This will depend on the critical angle for the glass prims, which depends on its refractive index according to:

$$n = \frac{1}{\sin\theta_c}$$

If the light is totally internally reflected, then the angle of reflection is equal to the angle of incidence according to the law of reflection:

$$\theta_i = \theta_r$$

This is illustrated in Figure 2.261:

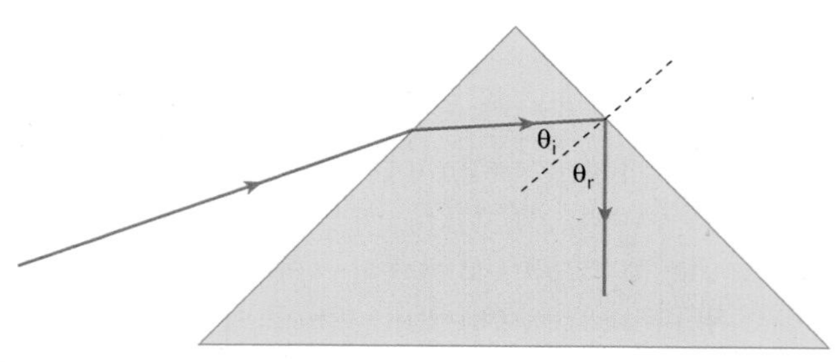

Fig 2.261

If the light is incident at an angle less than the critical angle then it will escape from the prism and the angle will be given by Snell's law:

$$n = \frac{\sin\theta_{air}}{\sin\theta_{glass}}$$

This is illustrated in Figure 2.262:

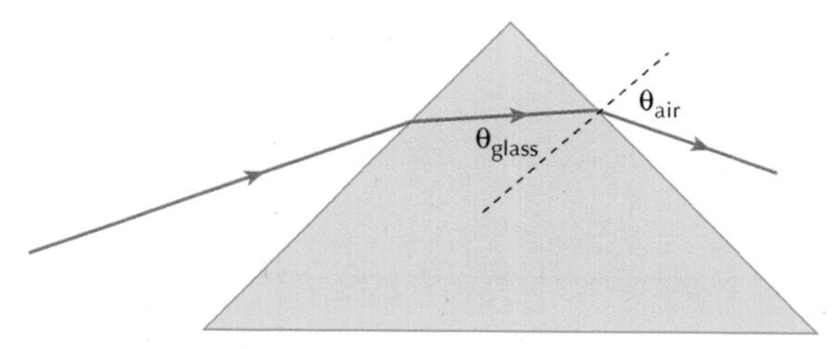

Fig 2.262

Worked example

Red light is incident on an equilateral prism made from flint glass, which has a refractive index of 1.60. It is incident on the left face of the glass at an angle of 30° as shown in Figure 2.263.

Copy Figure 2.263 and complete the path of the ray of light until it exits from the prism. All angles are required on the diagram.

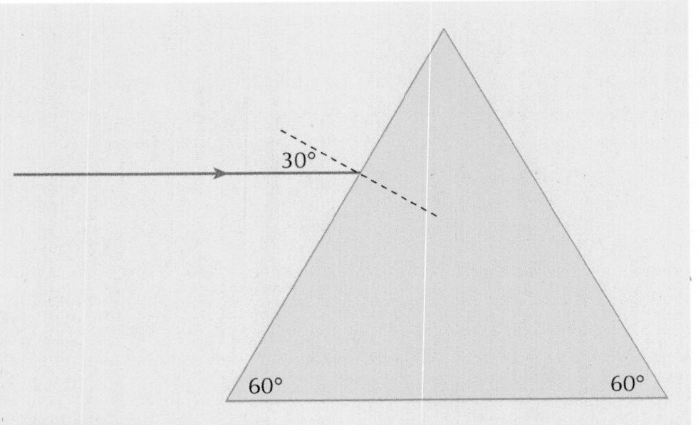

Fig 2.263

The path of the ray of light will depend on Snell's law and the critical angle for the flint glass. First of all, calculate the critical angle for the flint glass:

$$n = \frac{1}{\sin\theta_c}$$

$$1.60 = \frac{1}{\sin\theta_c}$$

$$\sin\theta_c = \frac{1}{1.60}$$

$$\theta_c = \sin^{-1}\left(\frac{1}{1.60}\right)$$

$$\theta_c = 38.7°$$

If the light is incident at angles greater than this inside the flint glass prism, it will be totally internally reflected.

To find the angle of refraction inside the glass block, apply Snell's law:

$$n = \frac{\sin\theta_{air}}{\sin\theta_{glass}}$$

$$1.60 = \frac{\sin 30°}{\sin\theta_{glass}}$$

$$\sin\theta_{glass} = \frac{\sin 30°}{1.60}$$

$$\sin\theta_{glass} = \frac{0.50}{1.60}$$

$$\theta_{glass} = \sin^{-1}\left(\frac{0.50}{1.60}\right)$$

$$\theta_{glass} = 18.2°$$

This allows you to complete the first stage of the path of the light (Figure 2.264). When completing a question in the examination of this sort, include arrows on each ray of light.

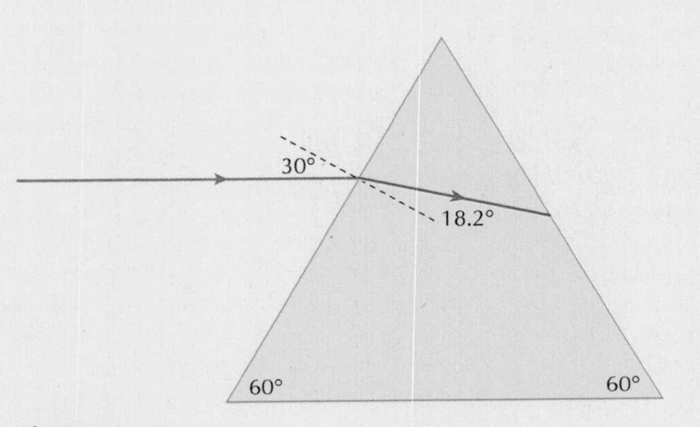

Fig 2.264

Using the fact that angles inside a triangle add up to 180° and angles on a right angle add up to 90°, you can complete the angles shown in Figure 2.265 to find the angle of incidence inside the glass.

(continued)

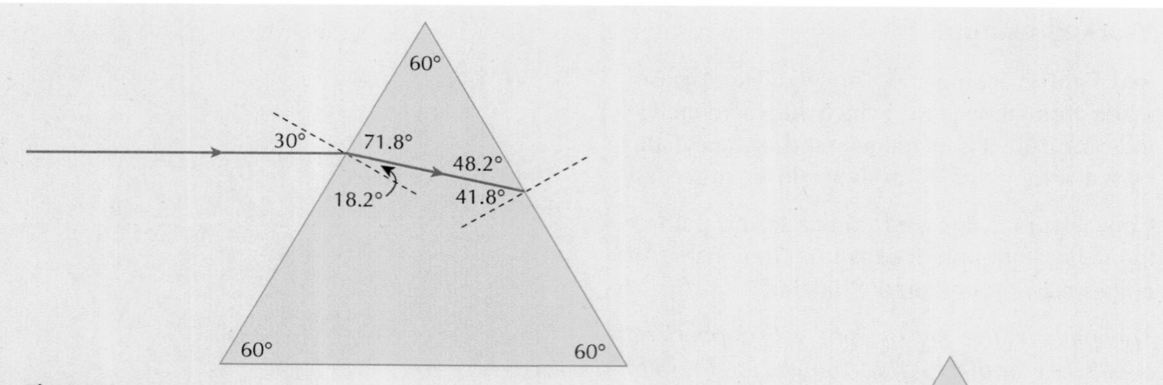

Fig 2.265

As the angle of incidence on the glass–air boundary is greater than the critical angle, the light will be totally internally reflected. This results in the following path of the ray of light; the additional angles have been completed as shown in Figure 2.266.

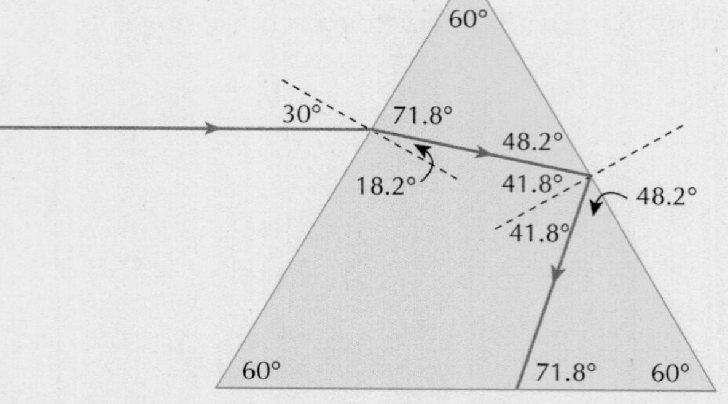

Fig 2.266

Considering the final surface in detail makes it possible to work out the angle of incidence in glass at the bottom surface to be:

$$\theta = 96.4 - 90$$
$$\theta = 6.4°$$

This angle is smaller than the critical angle for flint glass, so you can use Snell's law to work out the angle of the light in air as the ray escapes from the prism:

$$n = \frac{\sin\theta_{air}}{\sin\theta_{glass}}$$
$$1.60 = \frac{\sin\theta_{air}}{\sin 6.4°}$$
$$\sin\theta_{air} = 1.60 \sin 6.4°$$
$$\sin\theta_{air} = 0.18$$
$$\theta_{air} = \sin^{-1}(0.18)$$
$$\theta_{air} = 10.3°$$

Hence, you can now plot the final path of the ray of light as in Figure 2.267.

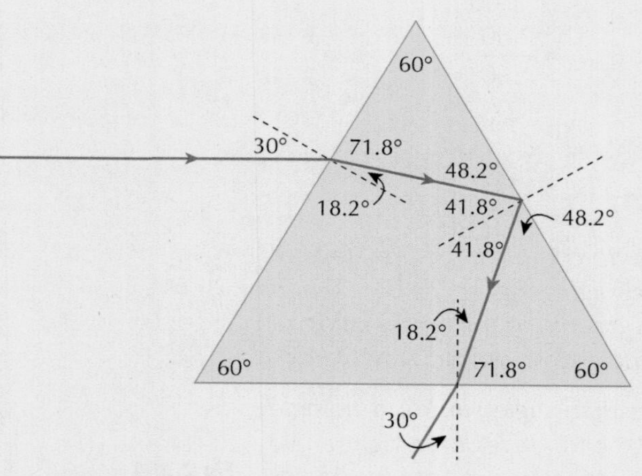

Fig 2.267

Optical fibres

Broadband internet has been made possible due to the introduction of optical fibres. Unlike electrical cables, which can carry only a few signals at a time, optical fibres can carry vast amounts of data (hence 'broadband') over long distances.

Fig 2.268

An optical fibre relies on total internal reflection to work – the light signal is trapped inside the fibre as shown in the diagram below.

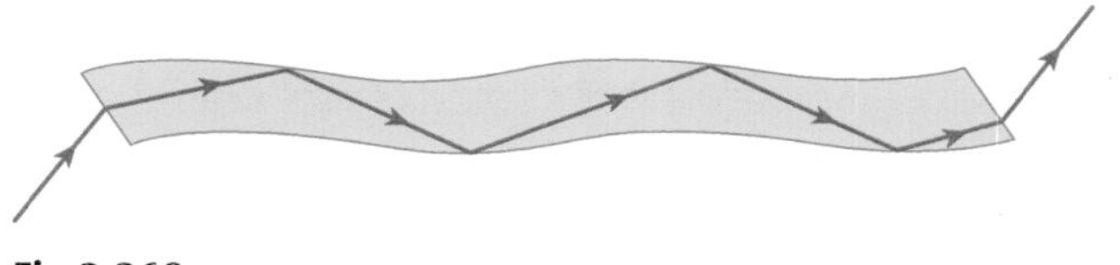

Fig 2.269

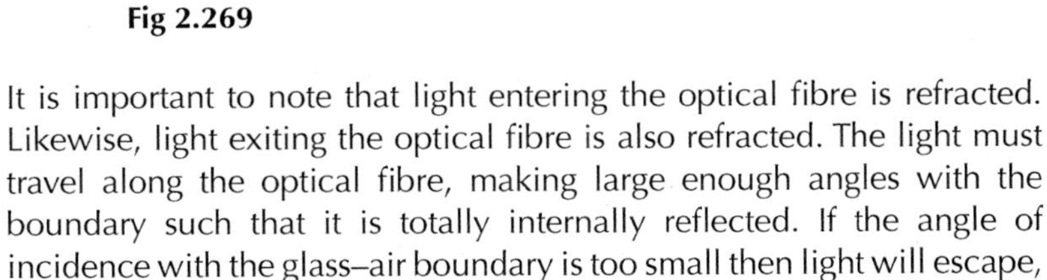

It is important to note that light entering the optical fibre is refracted. Likewise, light exiting the optical fibre is also refracted. The light must travel along the optical fibre, making large enough angles with the boundary such that it is totally internally reflected. If the angle of incidence with the glass–air boundary is too small then light will escape, and refract according to Snell's law. This is shown in Figure 2.271.

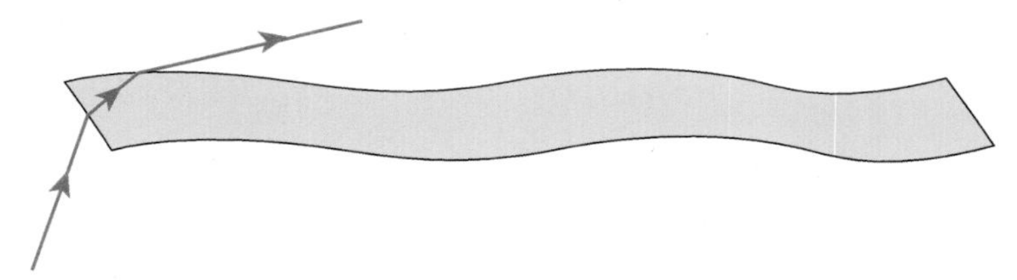

Fig 2.270

Diamonds

A diamond's beauty is in the way it refracts light – a diamond will sparkle. The reason for this sparkling effect is down to total internal reflection within the diamond. As diamond has a high refractive index of 2.41, the critical angle is small. This means the light is trapped within the diamond and only exits at certain points, giving rise to the sparkling effect shown in Figure 2.271.

Fig 2.271

Exercise 8.2.2 Applications of Snell's law

1 Red light is incident on the shapes shown in Figure 2.272 below. Copy each diagram, completing the path of light until it exits from the block. You must show all relevant angles. The refractive index of each block is given.

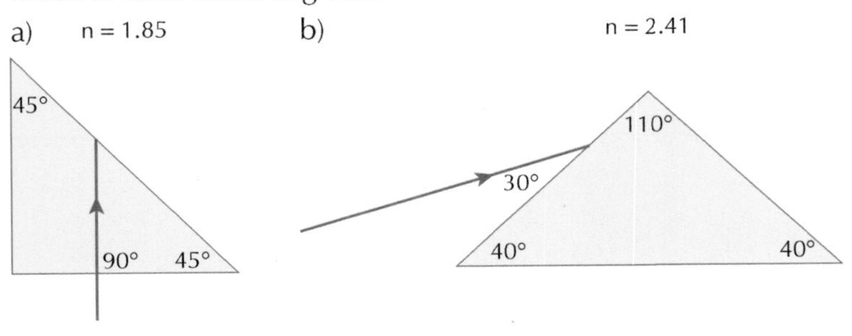

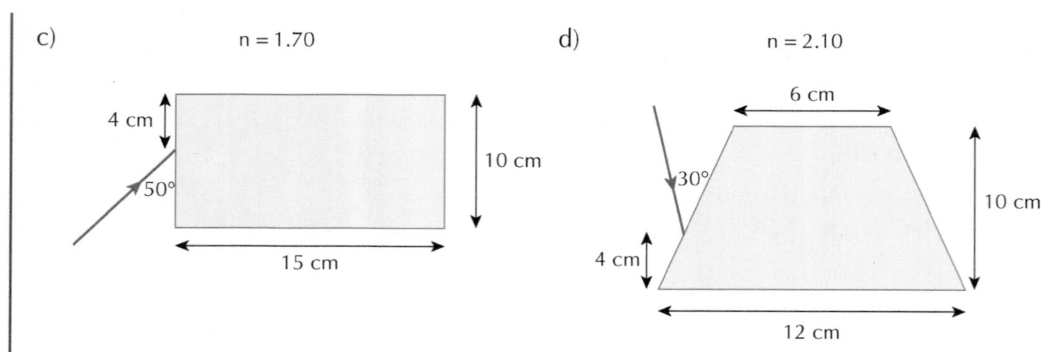

Fig 2.272

2 Diamond has a refractive index of 2.41. Use your knowledge of physics to explain why a diamond sparkles.

3 An optical fibre is used to carry light signals from one place to another. A cross-section of an optical fibre is shown below:

Fig 2.273

Use your knowledge of physics to describe how the light is carried along an optical fibre. What considerations are there for the way light enters the fibre and the materials chosen for its construction?

4 The core of an optical fibre is being tested in air with no cladding, as shown in Figure 2.274. The refractive index of the core is found to be 1.60.

Fig 2.274

a) Calculate the critical angle for the core material.

b) If the light is incident on the glass at an angle of $\theta = 30°$:

(i) Find the angle of refraction inside the core.

(ii) Find the angle of incidence on the core–glass boundary.

(iii) Determine whether the light is guided by the fibre or escapes.

1. **Monitoring and measuring AC**
    - Alternating current
    - Measuring and monitoring AC
2. **Current, potential difference, power and resistance**
    - Circuits revision
    - The potential divider
3. **Electrical sources and internal resistance**
    - Sources of electricity
    - Internal resistance
4. **Capacitors**
    - Defining capacitance
    - Energy storage
    - Capacitors in AC circuits
    - Practical applications
5. **Semiconductors and p–n junctions**
    - Conductors, insulators and semiconductors
    - The p–n junction
    - The LED

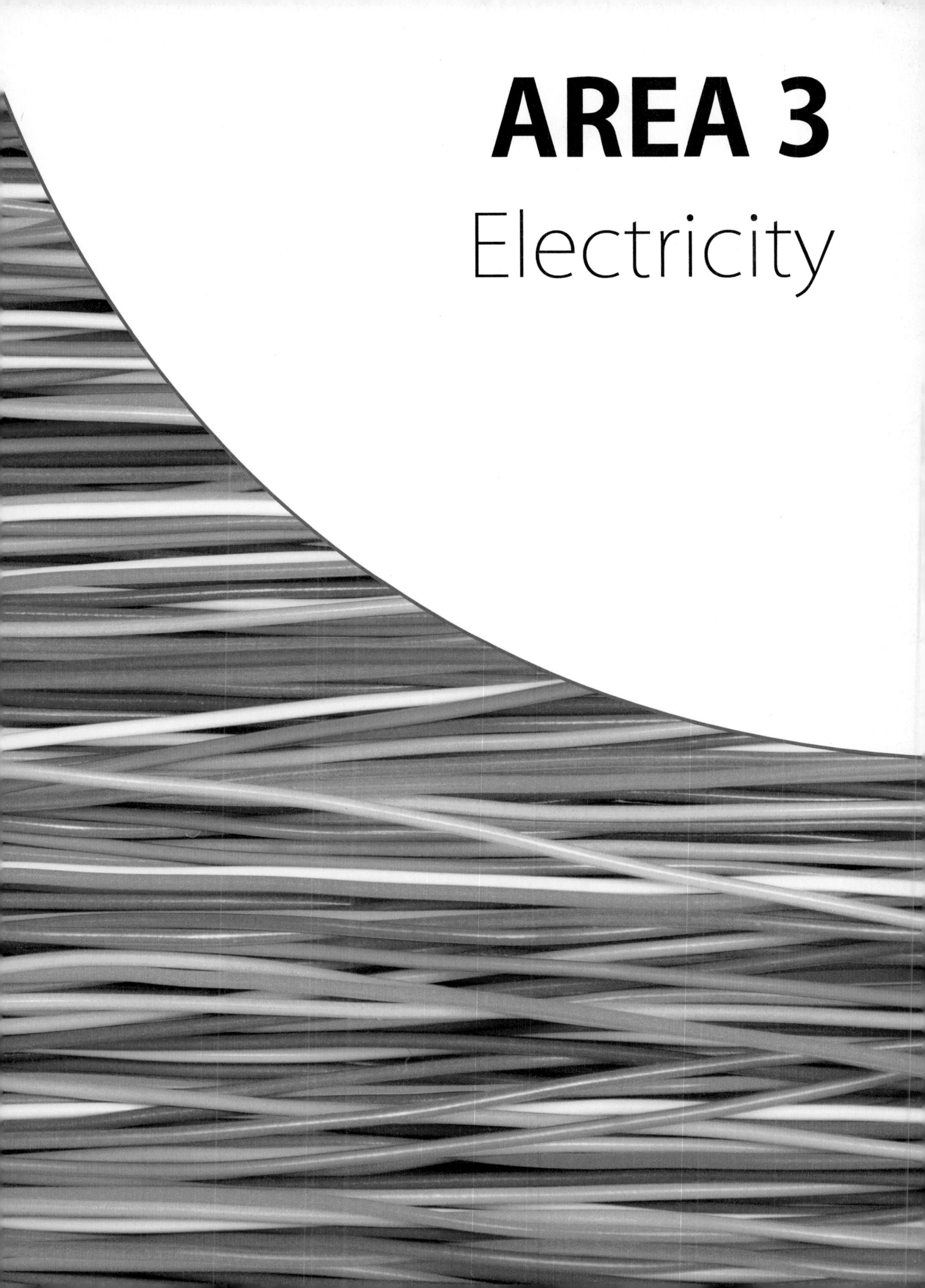

AREA 3

Electricity

1 Monitoring and measuring AC

What you should know (National 5)

- Direct current comes from batteries – it is current that flows in only one direction
- Alternating current comes from the mains supply – the current flows in both directions

Learning intentions

- Definition of alternating current as current that changes direction and instantaneous value with time
- Carry out calculations involving peak and RMS values
- Determine the frequency, peak voltage and RMS values from graphical data

1.1 Alternating current

Many experiments in the physics lab using electricity involve the use of either a battery or a direct current (DC) power supply. However, nearly all of the electrical appliances typically found within the home are designed to work with alternating current (AC).

1.1.1 Producing alternating current

Alternating current is supplied from wall sockets in homes and laboratories. It is produced in power stations by large electrical generators, which function like electric motors working in reverse. Experiment 1.1.1a demonstrates how an electric motor can be used to produce an alternating current by working in reverse.

Fig 3.1

Experiment 1.1.1a Producing alternating current

This experiment uses a simple electrical generator (or electric motor) to generate an alternating current. This current can be viewed using an oscilloscope.

Fig 3.2

Apparatus

- AC generator or AC motor*
- Wires
- Oscilloscope

* A desk fan can be used for this experiment, as described in the following instructions. If an AC motor is used, connections can be made directly to the motor.

Instructions

1. Connect the live and neutral pins of the desk fan plug to the input of an oscilloscope or AC voltmeter as shown in Figure 3.3 using crocodile clips and wires.
2. Use a wind source (for example, a second desk fan) to rotate the fan blades, and observe what is produced on the oscilloscope.
3. Vary the speed of the fan blades and note the effect on the signal produced on the oscilloscope.
4. Write a short report to describe your observations in the above experiment.

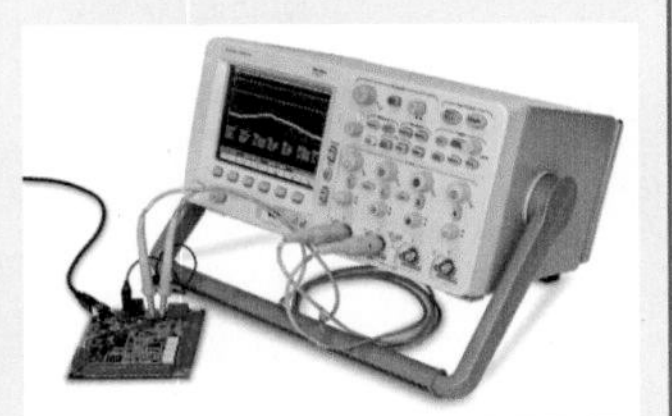
Fig 3.3

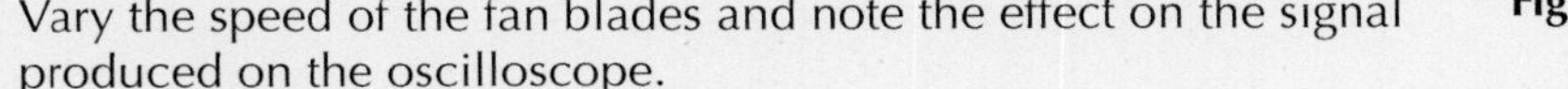

Experiment 1.1.1b Producing alternating current

This experiment uses the motion of a magnet inside a coil to produce an alternating current. The magnet is allowed to move up and down through the coil to produce a current.

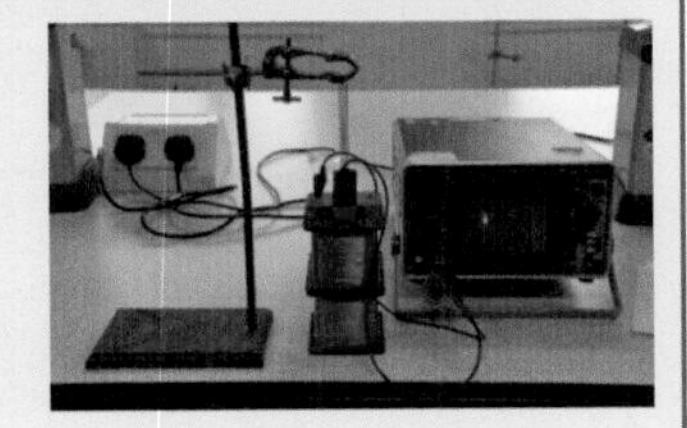
Fig 3.4

Apparatus

- Bar magnet
- Clampstand
- Spring
- Coil of wire
- Oscilloscope or AC voltmeter

Instructions

1. Suspend the bar magnet from a spring attached to a clampstand, as shown in Figure 3.4. Ensure that the bar magnet can easily pass in and out of the coil of wire.
2. Connect the coil of wire to a voltmeter or an oscilloscope, as shown in Figure 3.5.
3. Pull the magnet into the coil and let it go so that it oscillates inside the coil. Describe the effect that you see on the oscilloscope screen. What happens when the magnet moves different distances or different speeds?

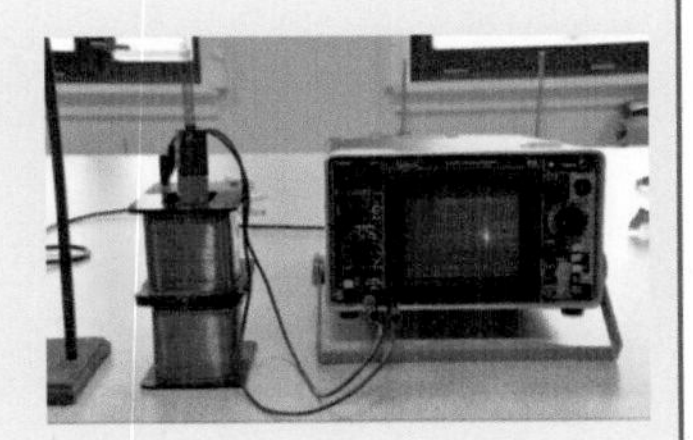
Fig 3.5

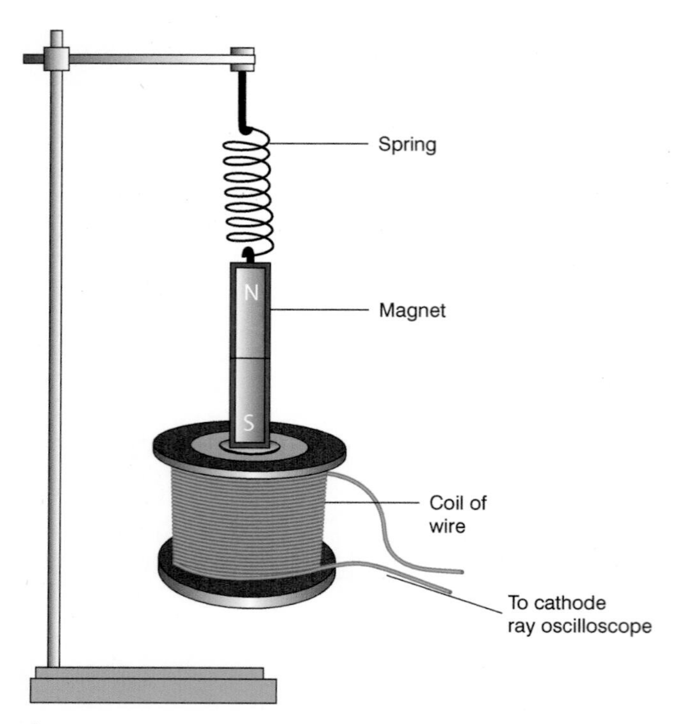

Fig 3.6

In an electric motor, the movement of charge through a magnetic field produces a force according to the left hand rule (for conventional current), or the right hand rule for the flow of electrons. Here, two of the three (current and field) produce the third (force for movement). A generator uses movement through a magnetic field to produce a current.

If a magnet is moved in and out of a coil, it will generate a current in that coil. A typical set up for this is shown in Figure 3.6.

When the magnet moves in one direction, a current in one direction is produced. When the magnet moves in the opposite direction, the direction of the current is reversed. This results in current initially one way, and then the other – in other words, it results in an alternating current. Plotted on a graph, an alternating current must cross the zero line as shown on the diagram and photograph (Figures 3.7 and 3.8).

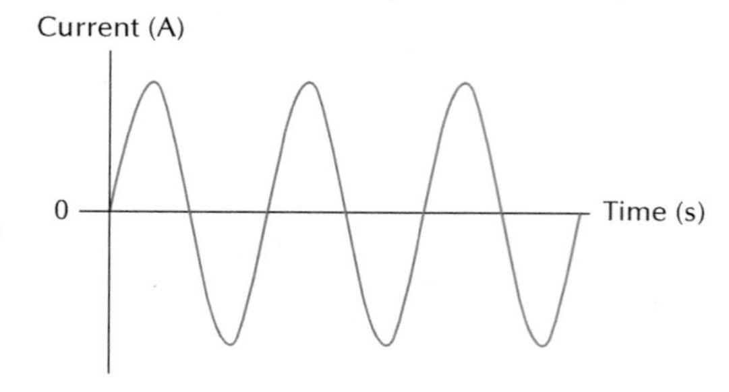

Fig 3.7

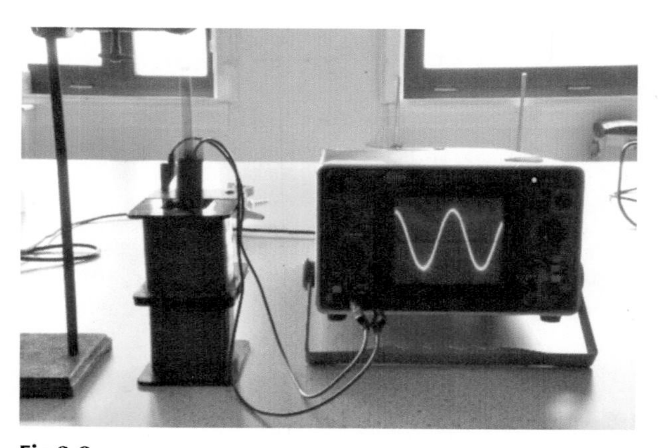

Fig 3.8

The amplitude (amount of energy) of the alternating current can be increased by:

- having a coil with more turns;
- having a stronger magnetic field;
- moving the magnet in and out of the coil at a greater speed.

Key point

Alternating current (AC) is current that reverses the direction of flow of charge at regular intervals.

1.1.2 Peak vs RMS

An alternating current (and voltage) will produce a waveform. The properties of waves have been investigated in detail in Area 2 and are reviewed here for electrical signals.

Consider the AC signal shown in Figure 3.9. The amplitude of the wave is measured from the middle (zero) point to either the top of a crest, or bottom of a trough. The amplitude of the wave corresponds to the peak voltage or peak current. In Figure 3.9, the peak voltage is 10 V.

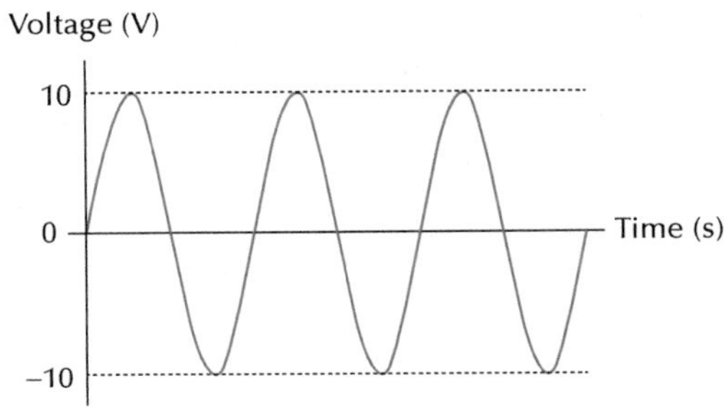

Fig 3.9

An electrical appliance connected to an AC supply will only receive the peak voltage and current for a very small amount of the time. For this reason, the root mean square (RMS) voltage and current are also defined, to give a better indication of the real effect of the supply. The peak and root mean square values are linked as follows:

$$V_{peak} = \sqrt{2}V_{rms}$$

$$I_{peak} = \sqrt{2}I_{rms}$$

On a graph, the RMS value of the voltage or current is a constant positive value with a magnitude of 0.707 ($\frac{1}{\sqrt{2}}$) that of the peak value:

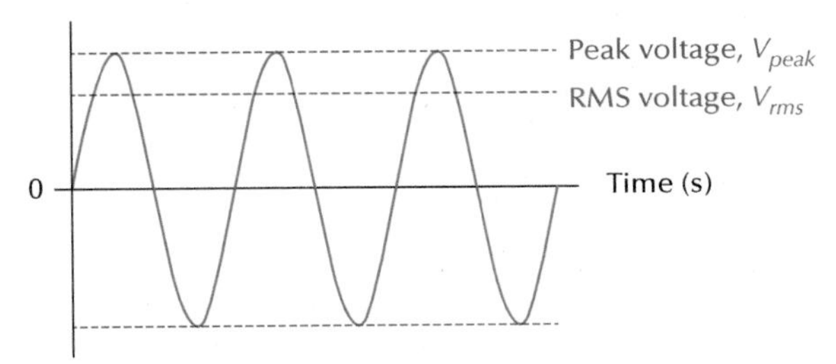

Fig 3.10

An AC supply RMS voltage of 8 V would be the equivalent in terms of power dissipated in the circuit as using a DC voltage of 8 V. The peak voltage of the AC supply would be greater than 8 V.

Key point

The peak voltage or current of an AC signal is measured from the zero point to the top of a peak or bottom of a trough of the signal.

The root mean square (RMS) value of the voltage or current is a measure of the real effect of the signal and is given by:

$$V_{peak} = \sqrt{2}V_{rms}$$

$$I_{peak} = \sqrt{2}I_{rms}$$

Worked example

A student measures the AC signal from a small generator using an oscilloscope. The signal measured is shown in Figure 3.11.

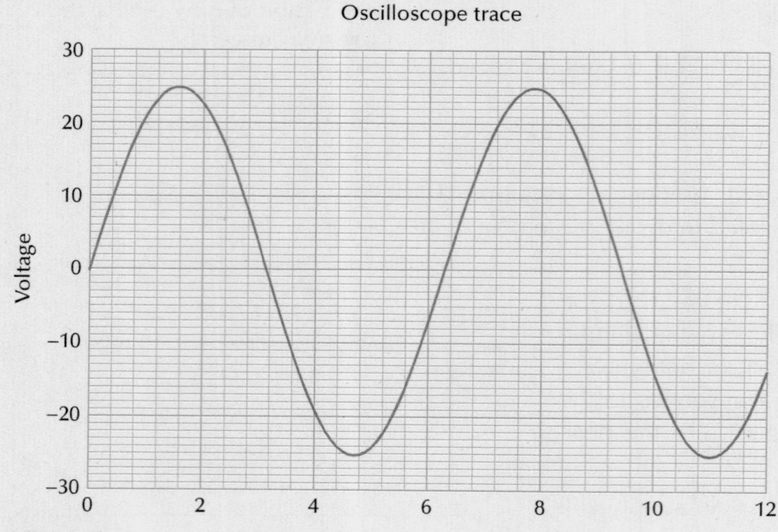

Fig 3.11

Calculate the peak and RMS voltage of the signal.

The peak voltage can be read directly from the graph and is the value from the middle to the top of a peak:

$$V_{peak} = 25V$$

The RMS voltage can then be calculated using the following equation:

$$V_{rms} = \frac{V_{peak}}{\sqrt{2}}$$

Substituting the value for peak voltage and solving gives,

$$V_{rms} = \frac{25}{\sqrt{2}}$$

$$V_{rms} = 17.7\ V$$

Ohm's law for AC

Ohm's law was covered in detail at National 5 level, relating the current in a circuit to the resistance of the circuit and the voltage applied:

$$V = IR$$

For AC, this equation can be written as follows:

$$V_{rms} = I_{rms}R$$

As stated above, the RMS voltage is the equivalent to having a DC supply of the same voltage, allowing the substitution into the above equation. The peak voltage and peak current are also linked by Ohm's law as follows:

$$V_{peak} = I_{peak}R$$

It is important to note these above equations do not take into account any frequency dependence of the components within the circuit.

Key point

The frequency of an AC signal is given by:

$$f = \frac{1}{T}$$

where T is the length of time taken for one complete cycle of the signal.

1.1.3 Frequency of an AC signal

At National 5 level, the frequency of a wave was defined as the number of waves produced in a second. Frequency was measured in hertz (Hz). When considering alternating current, frequency has the same meaning – it is the number of complete cycles of the signal in one second. A cycle of the signal is considered to be one wavelength, as shown in Figure 3.12.

The frequency of an AC signal can therefore be calculated by working out the period of the signal (time for one complete signal) and then using the following relationship:

$$f = \frac{1}{T}$$

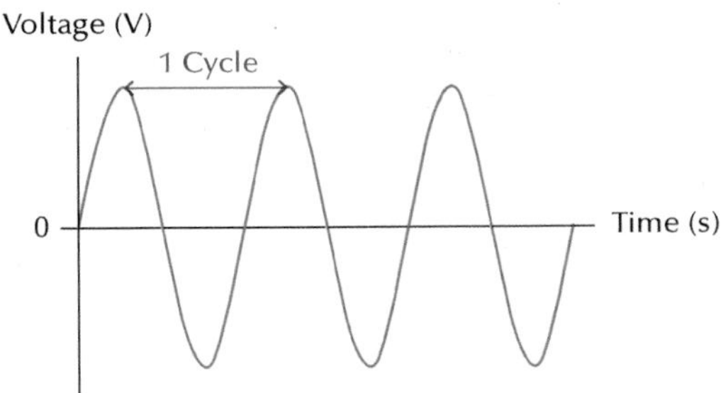

Fig 3.12

Worked example

A student investigating the frequency of a source of electricity connects the source to an oscilloscope. The trace in Figure 3.13 is observed.

Calculate the frequency of the supply.

To find the frequency, the period of the signal must be found. There are 3 waves in 60 ms, so the period of the signal is:

$$T = \frac{60\,ms}{3}$$

$$T = 20\,ms$$

The frequency of the signal can now be found to be:

$$f = \frac{1}{T}$$

$$f = \frac{1}{20 \times 10^{-3}}$$

$$f = 50\,Hz$$

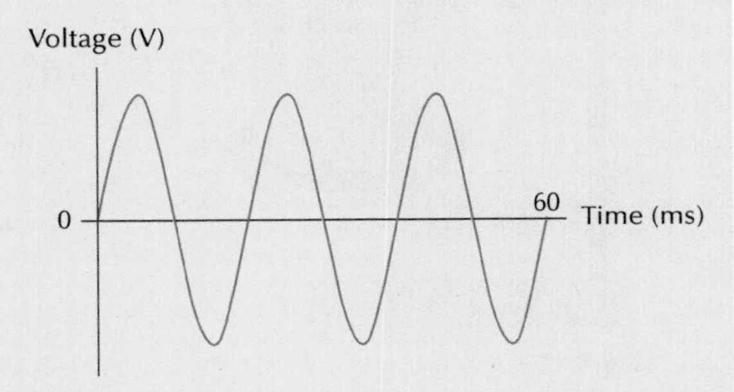

Fig 3.13

Exercise 1.1 Alternating current

1 Explain what is meant by alternating current. Your answer should make reference to the flow of charges within the circuit.

2 A student connects an oscilloscope to an electricity supply and observes the trace shown in Figure 3.14.

He concludes that the signal is alternating current. Use your knowledge of physics to comment on the validity of this statement.

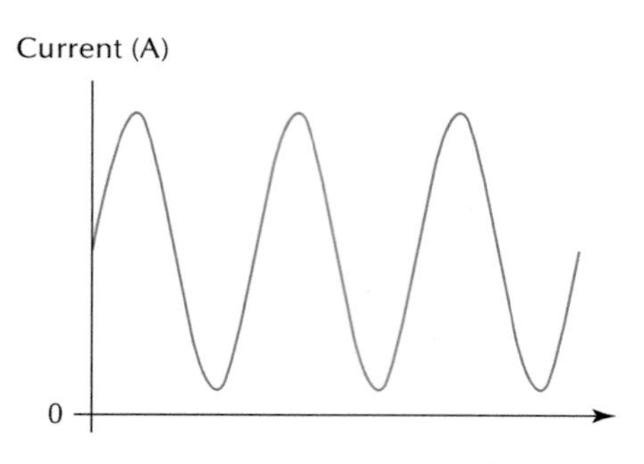

Fig 3.14

3 A student produces alternating current using the experimental set-up shown in Figure 3.15. She pulls the magnet down and allows it to oscillate inside the coil of wire.

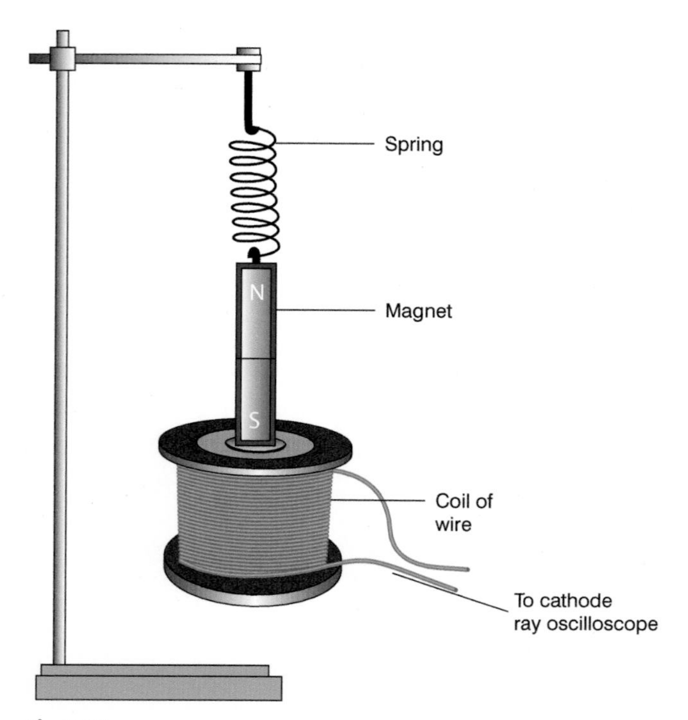

Fig 3.15

a) Use your knowledge of physics to explain why the above set-up produces an alternating current in the coil.

b) Give two ways that the magnitude of the alternating current can be increased.

4 Explain the difference between peak voltage and RMS voltage. Which of these is the equivalent to a DC voltage in the circuit?

5 The UK mains supply is 230 V RMS with a frequency of 50 Hz.

a) Calculate the peak voltage from the mains.

b) Calculate the period of the signal from the mains.

c) If a circuit with a resistance of 30 Ω is connected to the mains, find:

(i) The RMS current in the circuit.

(ii) The peak current in the circuit.

Fig 3.16

6 Students are investigating different AC supplies using an oscilloscope. They connect the oscilloscope to the supplies and observe the following signals:

i)

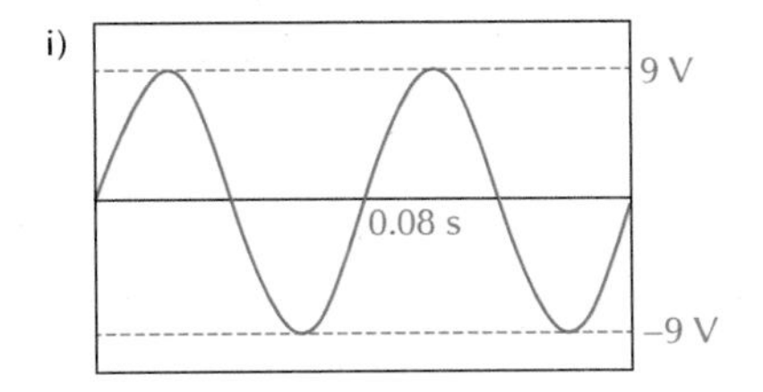

ii)

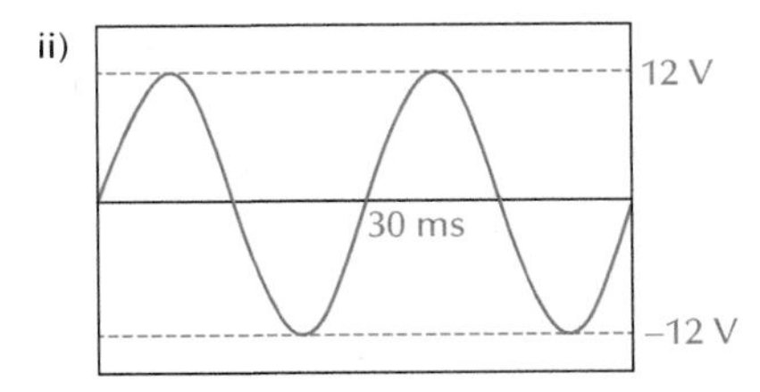

iii)

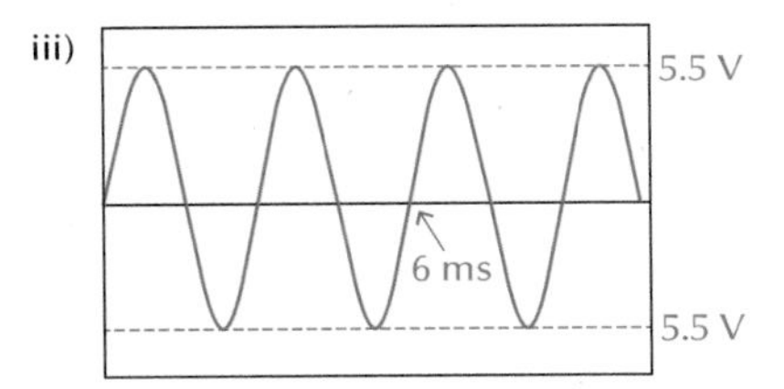

iv)

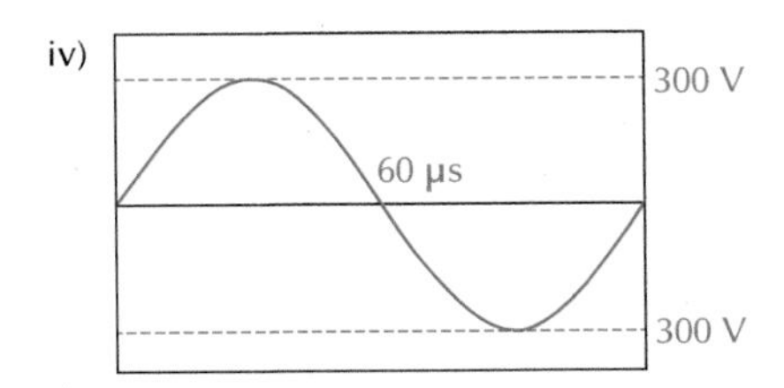

Fig 3.17

a) State the peak voltage of each signal.

b) State the period of each signal.

c) Calculate the RMS voltage of each signal.

d) Calculate the frequency of each signal.

7 A kettle is designed to be used connected to the UK mains supply. It is rated to have a current of 9 A (RMS) flowing through it.

a) Calculate the peak current in the kettle.

b) Find the resistance of the kettle.

8 The electric motor of a vacuum cleaner has a resistance of 70 Ω. Calculate the peak and RMS current in the motor if it is connected to the UK mains supply.

Fig 3.18

(continued)

9 Figure 3.19 shows an alternator that generates alternating current.

Use your knowledge of physics to describe how rotational movement of the spinning coil generates alternating current.

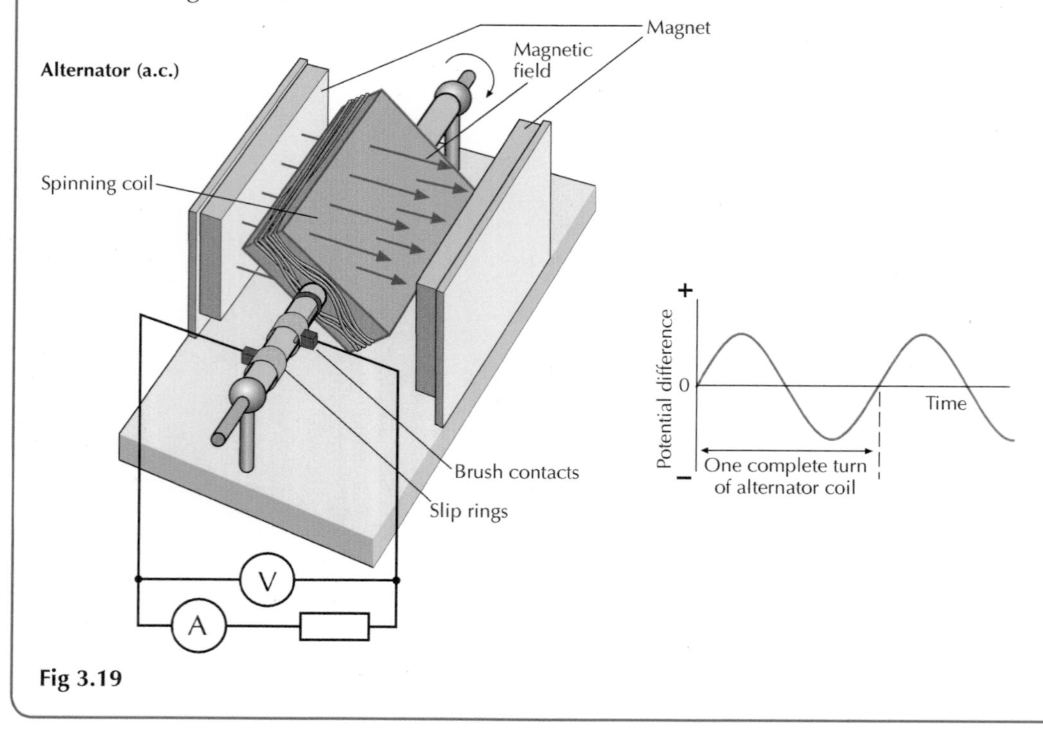

Fig 3.19

1.2 Measuring and monitoring AC

Alternating current has already been measured in the experiments previously. In this section, the techniques used to measure AC are formalised. The techniques include the use of a multimeter to read the RMS current or voltage, and the oscilloscope to read both the peak and RMS values.

Fig 3.20

1.2.1 Using a multimeter

Modern multimeters are favoured when investigating electric circuits because a single device can serve multiple purposes. This means that you do not need to have separate ammeters, voltmeters and ohmeters. Multimeters are capable of reading both AC or DC but you must ensure that the meter is set up correctly for what you are measuring!

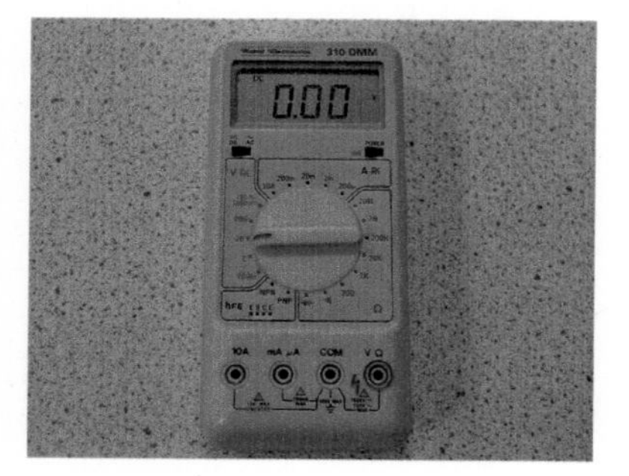

Fig 3.21

The multimeter

You may have used a multimeter to measure DC current or voltage in previous studies of physics or engineering. A typical multimeter is shown in Figure 3.21.

The meter is divided into three main parts: a digital display for allowing readings to be made, a rotary switch to select the desired mode of operation, and terminals where the meter can be connected to the circuit.

Measuring DC voltage

To measure a DC voltage in a circuit, the meter must first be set to the correct settings. This is achieved using the rotary switch to set the meter to DCV (some meters also have a toggle switch to switch between AC and DC).

Next, the correct range of values must be chosen so that the meter is not overloaded, but will also give the reading to the desired accuracy. If the values of the voltages being measured are not known, it is best to start with as large a range as possible and then reduce as required to get the desired accuracy.

The meter is then connected to the circuit using the terminals: COM (common) and V (for voltage). As discussed at National 5 level, the voltmeter is connected in parallel across the component being measured. Figure 3.22 shows the voltage being measured across a lamp.

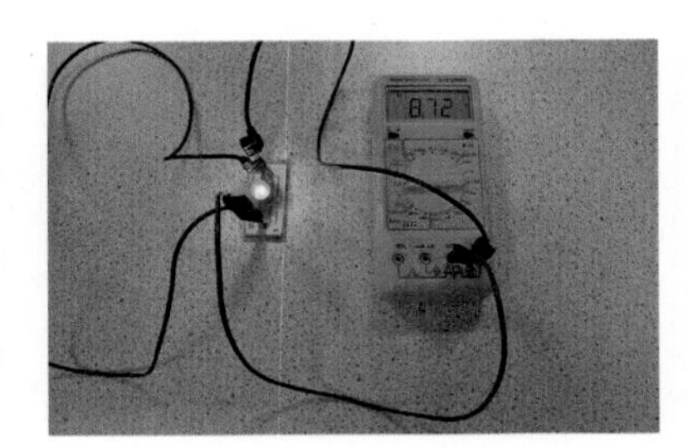

Fig 3.22

Measuring DC current

To measure the DC current in a circuit, the rotary switch must be set to DCA. As with DC voltage, choose a large range if the values of current being measured are not known, and reduce as necessary to measure the current to the desired accuracy. This is important to ensure that you do not break a fuse in the meter.

The meter must be connected into the circuit using the terminals: COM (common) and either mA (for small currents) or 10 A (for larger currents). It is vital to choose the correct terminals and settings, as passing too large a current through a meter set to read small currents will result in a fuse being blown. As discussed at National 5 level, the ammeter must be connected in series in the circuit. Figure 3.23 shows the measurement of current in a series circuit containing a power supply and a lamp.

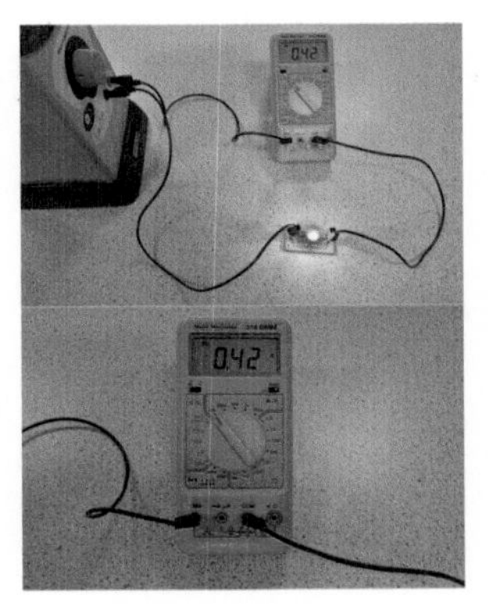

Fig 3.23

Measuring AC voltage

To measure the AC voltage across a component, the multimeter must be connected into the circuit as described above for DC voltage. The main difference is that the rotary switch must be set to ACV. The multimeter will give a single value for the AC voltage, which corresponds to the RMS value.

Measuring AC current

To measure AC current in a circuit, connect the multimeter as for measuring DC, but with the rotary switch set to ACA. As with AC voltage, the value given by the multimeter is a single value that corresponds to the RMS value of the current.

> **Key point**
>
> When measuring AC, a multimeter gives a single value that corresponds to the RMS current or RMS voltage. The peak values can be found using:
>
> $$I_{peak} = \sqrt{2}I_{rms}$$
>
> $$V_{peak} = \sqrt{2}V_{rms}$$

1.2.2 Using an oscilloscope

A digital multimeter can easily be set up in a circuit to measure the voltage across a component – either DC or AC. However, the multimeter

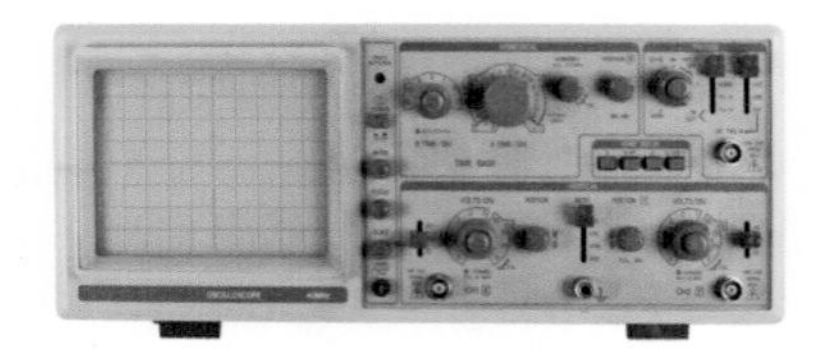

Fig 3.24

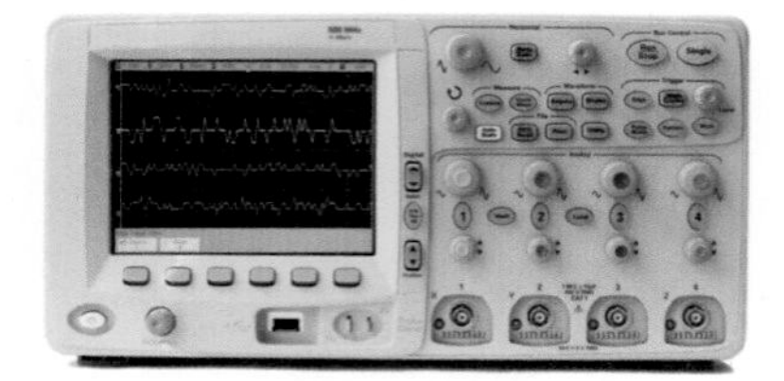

Fig 3.25

only gives a single value for the AC voltage and gives no further indication as to the frequency of the signal. To fully investigate a signal in a circuit, an oscilloscope can be used.

Oscilloscopes come in many shapes and sizes, and at widely varying costs. Some oscilloscopes designed for reading very high frequencies across multiple channels can cost tens or hundreds of thousands of pounds. General-use oscilloscopes are less expensive and can be used for the majority of applications. A typical analogue oscilloscope is shown in Figure 3.24.

A typical digital oscilloscope is shown in Figure 3.25.

Regardless of the type of oscilloscope used, the basic functions of the device remain the same. The oscilloscope screen can be regarded as a sheet of graph paper. The horizontal axis represents the time and the vertical axis represents the voltage. The scales can be changed to suit the signal being investigated, using the time base and voltage gain controls.

Voltage gain

The scale of the vertical axis is adjusted using the V-gain (sometimes called the Y-gain) control. This controls the voltage that each box on the screen represents, and is given the unit V/div. Consider the signal in Figure 3.26.

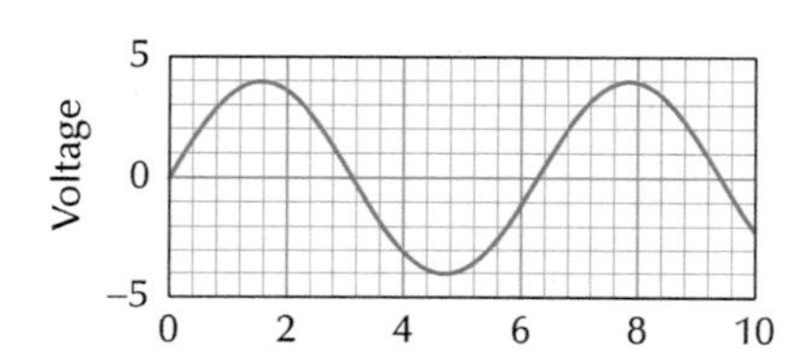

Fig 3.26

The signal is four divisions high from the mid-point. If the oscilloscope setting is, for example, 2 V/div, then the amplitude of this signal is 8 V. The peak voltage is therefore 8 V.

Frequency

The frequency of a signal can be measured using an oscilloscope by measuring the length of time taken for one complete cycle of the signal. The scale of the time axis is adjusted using the time base control. Consider the signal shown in Figure 3.27:

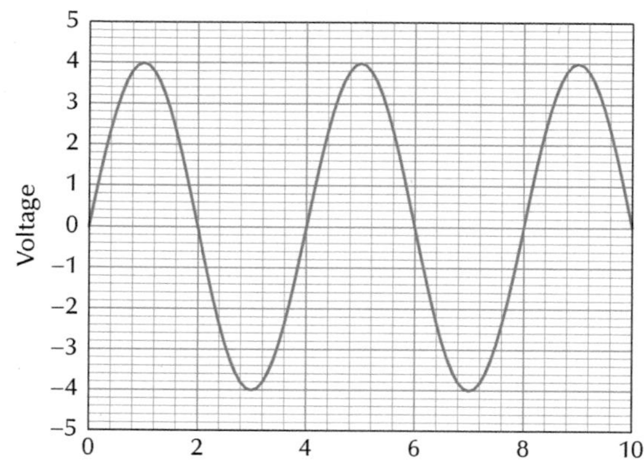

Fig 3.27

One cycle of the signal is four divisions long. If the time base of the signal was set to 5 ms/div for example, then the time for one cycle of the signal is 20 ms. The frequency of the signal can then be calculated using:

$$f = \frac{1}{T}$$

Worked example

An engineer is investigating the AC signal generated by a test wind turbine in a lab. She connects the output of the turbine's generator to an oscilloscope and observes the trace shown in Figure 3.28.

The time base is set to 50 ms/div and the voltage control is set to 10 V/div.

a) Calculate the peak voltage of the signal.

b) Calculate the RMS voltage of the signal.

c) Calculate the frequency of the signal.

a) *The peak voltage is found by first finding the amplitude of the signal – 3.5 boxes. The peak voltage is then given by:*

$$V_{peak} = 3.5 \times 10$$

$$V_{peak} = 35V$$

b) *The RMS voltage is found using:*

$$V_{rms} = \frac{V_{peak}}{\sqrt{2}}$$

Substitute the value for peak voltage found in part a) and solve:

$$V_{rms} = \frac{35}{\sqrt{2}}$$

$$V_{rms} = 24.7V$$

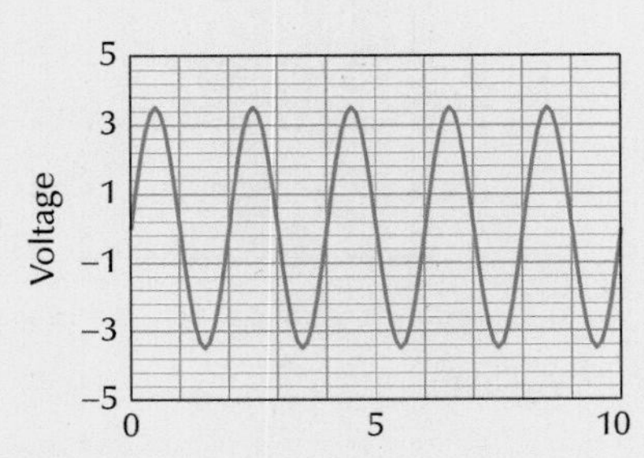

Fig 3.28

c) *To find the frequency of the signal, the period must first be found. One cycle is 2 boxes long, so the period of the signal is:*

$$T = 2 \times 50$$

$$T = 100\,ms$$

The frequency is then found using:

$$f = \frac{1}{T}$$

Substitute and solve:

$$f = \frac{1}{100 \times 10^{-3}}$$

$$f = 10Hz$$

Exercise 1.2 Measuring and monitoring AC

1 An electrician is asked to check the function of an electric heater. The heater should have a total resistance of 150 Ω with a tolerance of 10%. The engineer uses a multimeter to assess the current drawn from the mains supply by the heater.

a) Explain what setting the engineer should use for the multimeter.

b) The current drawn by the heater is 1.7 A RMS. Is the heater operating within its tolerance for resistance? You must justify your answer by calculation.

c) Calculate the peak current drawn by the heater.

2 A student using a multimeter in an electrical circuit measures an AC current of 4.5 A RMS. Calculate the peak current measured.

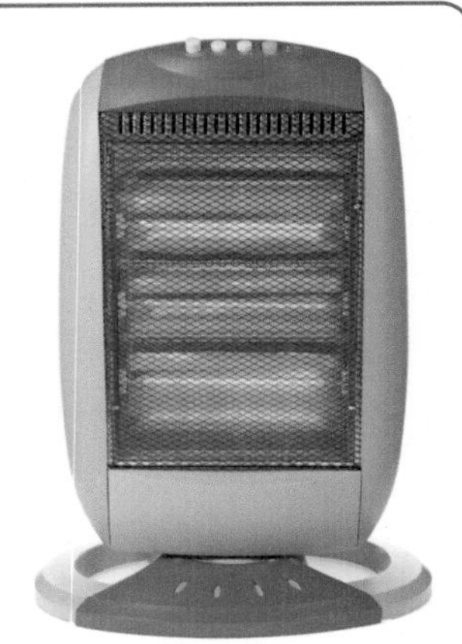

Fig 3.29

(continued)

3 For each of the oscilloscope traces shown in Figure 3.30, find:

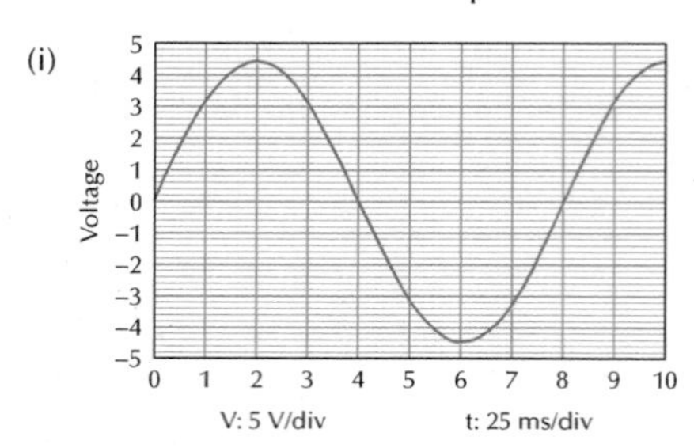

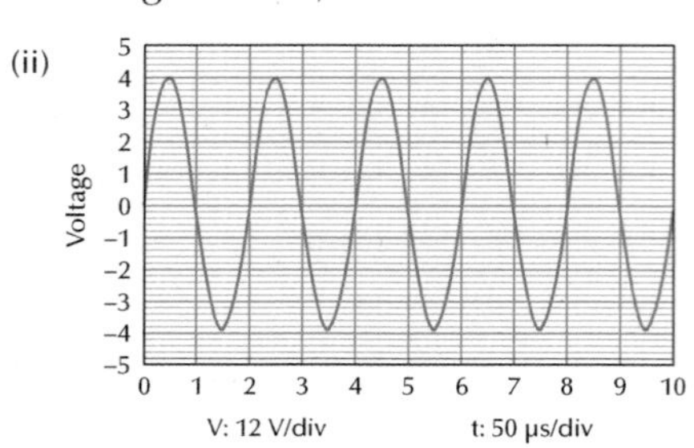

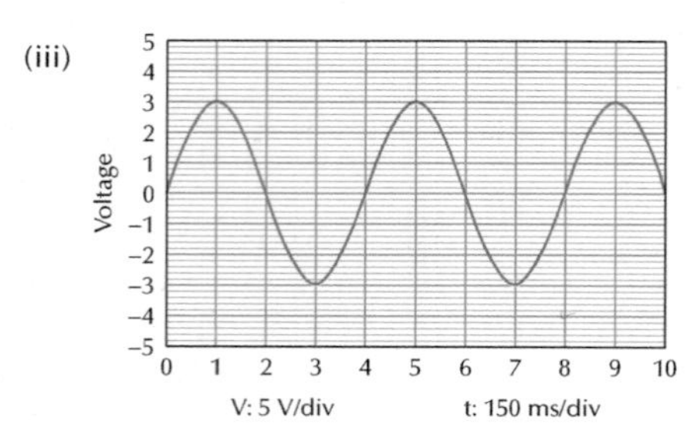

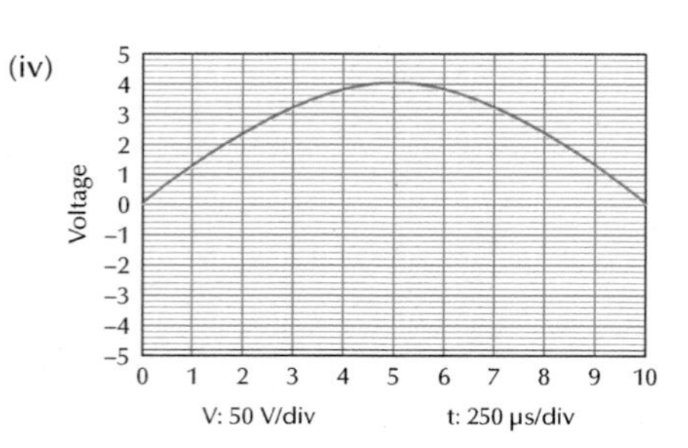

Fig 3.30

a) The peak voltage

b) The RMS voltage

c) The frequency

4 An engineer is investigating the electrical signal produced by a car alternator. He obtains the trace shown in Figure 3.31 on an oscilloscope where the time base setting is set to 5 ms/div and the voltage gain is set to 3 V/div.

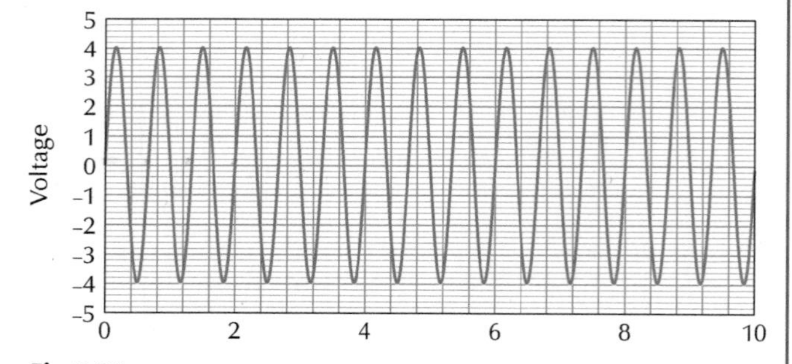

Fig 3.31

a) Determine the peak and RMS voltages of the signal.

b) Determine the frequency of the signal.

c) Suggest one change the engineer could make to allow for an easier calculation of the frequency of the signal.

5 If an oscilloscope were connected to the mains supply, an alternating voltage signal would be obtained. Assume the settings on the oscilloscope are 100 V/div and 5 ms/div.

a) Calculate the peak voltage of the UK mains supply.

b) State the frequency of the UK mains supply.

c) On graph paper, draw the trace that would be obtained for the UK mains supply with the above oscilloscope settings.

2 Current, potential difference, power and resistance

What you should know (National 5)

- Definition of potential difference as the voltage drop across a component
- Definition of current as the amount of charge flowing per unit time
- Properties of series circuits
- Properties of parallel circuits
- Resistance in series and parallel circuits

Learning intentions

- Be able to use relationships involving power, current, resistance and potential difference to analyse circuits, with calculations involving several steps
- Analysis of potential divider circuits

2.1 Circuits revision

Electrical circuits were introduced at National 5 level. They are reviewed here and are applied to more complex problems, leading to the Wheatstone bridge, a highly sensitive control circuit.

2.1.1 Series vs parallel

There are two basic set-ups for an electrical circuit: series, which has only one loop; and parallel, where there are multiple loops (two or more). These are illustrated in the circuit diagram in Figure 3.32.

The properties of the circuits are summarised in Table 3.1.

Series circuit	Parallel circuit
Current is the same at all points	Current splits into each branch
Voltage splits across each component	Voltage is the same across each branch

Table 3.1

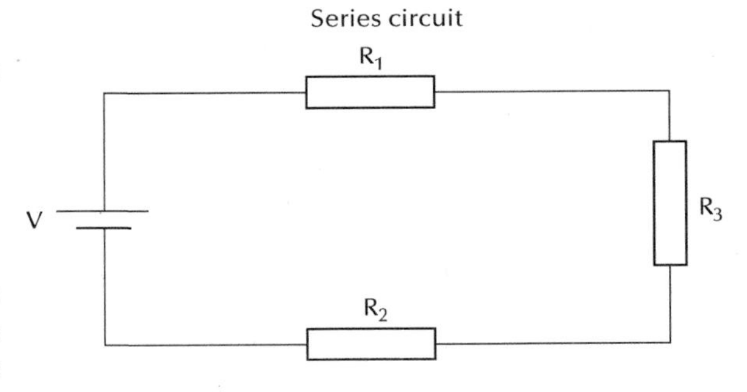

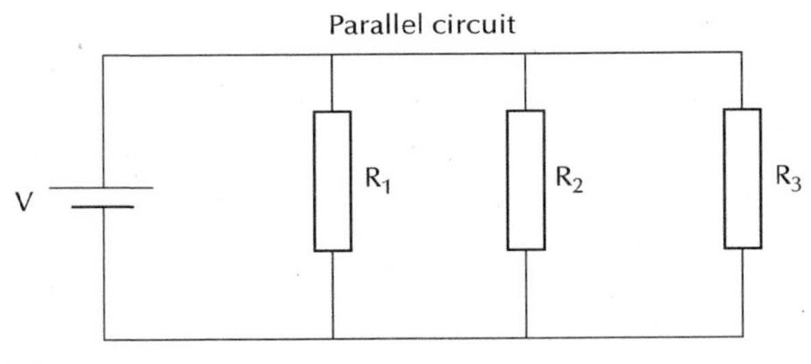

Fig 3.32

Series circuit

Consider the series circuit in Figure 3.32. The current has only one loop that it can flow around. This means that the current is the same at all points. The total potential difference across all resistors is equal to the source voltage, so:

$$I_T = I_1 = I_2 = I_3$$

However, the voltage across each component will be less than the source voltage. These individual voltages will add up to give the source voltage, so:

$$V_T = V_1 + V_2 + V_3$$

Parallel circuit

Consider the parallel circuit in Figure 3.32. The current can take different paths through each. The total current from the source will split across the different branches of the parallel circuit. This gives:

$$I_T = I_1 + I_2 + I_3$$

However, the voltage across each branch is the same as the potential difference across the power supply, which is the same between the top and bottom of each branch, so:

$$V_T = V_1 = V_2 = V_3$$

Exercise 2.1.1 Series vs parallel

1 For each of the circuits shown in Figure 3.33, calculate the values on the missing ammeters and voltmeters, assuming an ideal ammeter, i.e. no resistance.

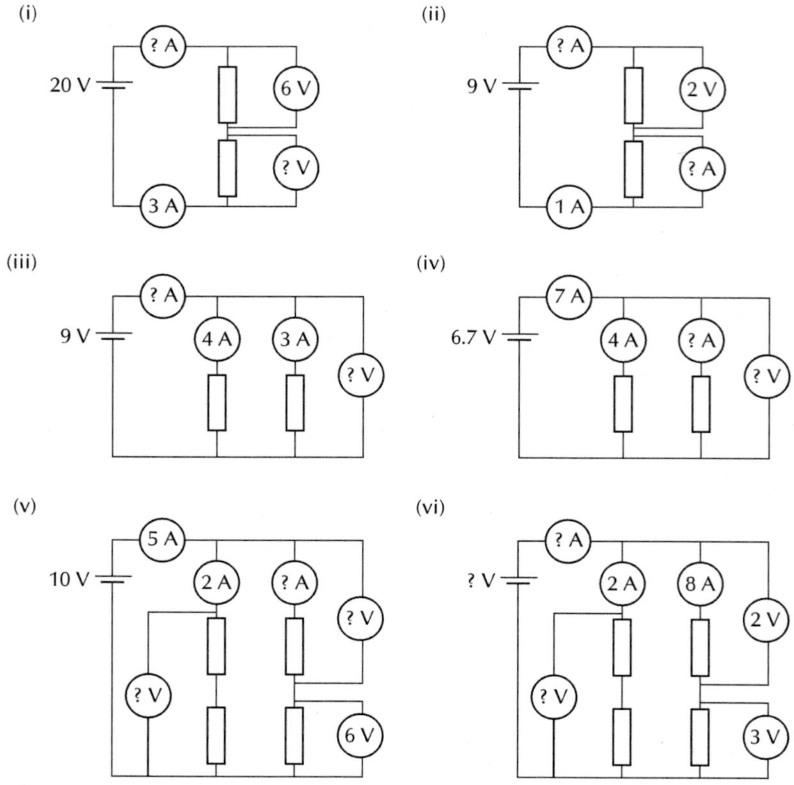

(i) 20 V, ? A, 6 V, ? V, 3 A

(ii) 9 V, ? A, 2 V, ? A, 1 A

(iii) 9 V, ? A, 4 A, 3 A, ? V

(iv) 6.7 V, 7 A, 4 A, ? A, ? V

(v) 10 V, 5 A, 2 A, ? A, ? V, ? V, 6 V

(vi) ? V, ? A, 2 A, 8 A, 2 V, ? V, 3 V

Fig 3.33

2.1.2 Revision: Ohm's law

Ohm's law was derived by experiment at National 5 level. It relates the current in a circuit to the voltage across a component. If a component is Ohmic (such as a resistor), it obeys Ohm's law and the voltage and current are directly proportional as shown in Figure 3.34.

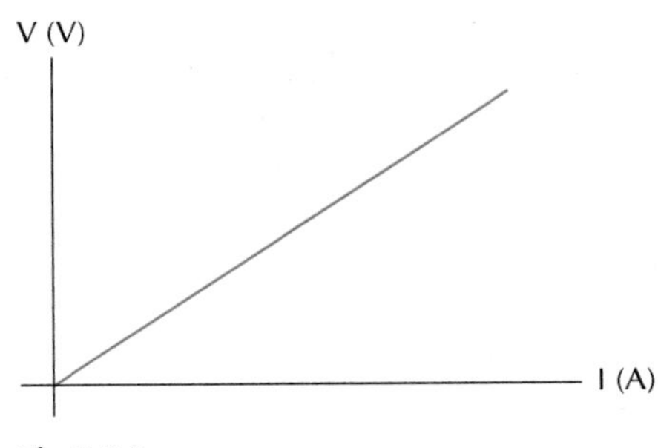

Fig 3.34

Mathematically, the proportionality can be written as:

$$V \propto I$$
$$V = kI$$

where k is a constant.

The constant here is the resistance of the component, which yields Ohm's law:

$$V = RI$$
$$V = IR$$

The gradient of the line in Figure 3.34 gives the resistance of the component.

> **Key point**
>
> Ohm's law links the current, voltage and resistance in a circuit:
>
> $$V = IR$$

2.1.3 Revision: total resistance

Circuits can be either series, parallel or a combination of both (known as a mixed circuit). However, any circuit can be broken down to just a power supply and a single series resistor. To do this, you need to be able to find the total resistance of a combination of resistors in series or parallel.

Series resistors

The relationship to find the total resistance for resistors in series can be derived by applying energy conservation. Consider the circuit shown in Figure 3.35.

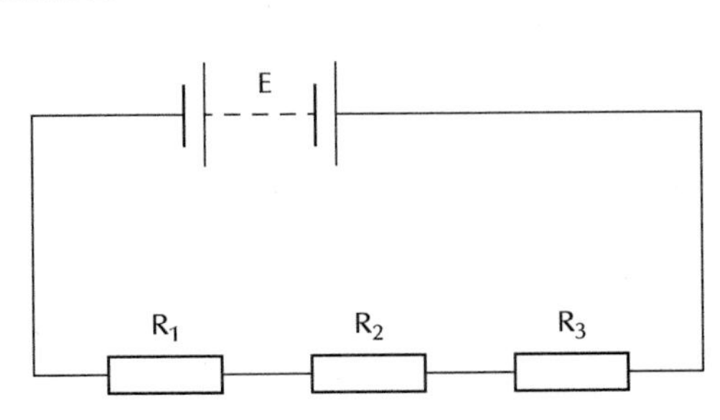

Fig 3.35

The energy supplied by the source (the EMF, electromotive force) is equal to the sum of the energy dissipated in the individual resistors in the circuit (remember that energy is linked to voltage):

$$E = V_1 + V_2 + V_3$$
$$E = IR_1 + IR_2 + IR_3$$

Modelling the three series resistors as one equivalent resistance, R_S, gives the circuit shown in Figure 3.36.

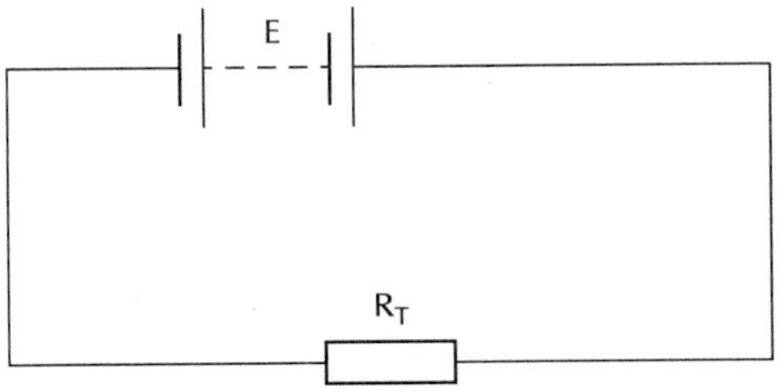

Fig 3.36

In this case:

$$E = IR_T$$

Setting the two equations equal to each other gives:

$$IR_T = IR_1 + IR_2 + IR_3$$

Cancelling out the current (because current is the same at all points in a series circuit, thus the same through each resistor):

$$R_T = R_1 + R_2 + R_3$$

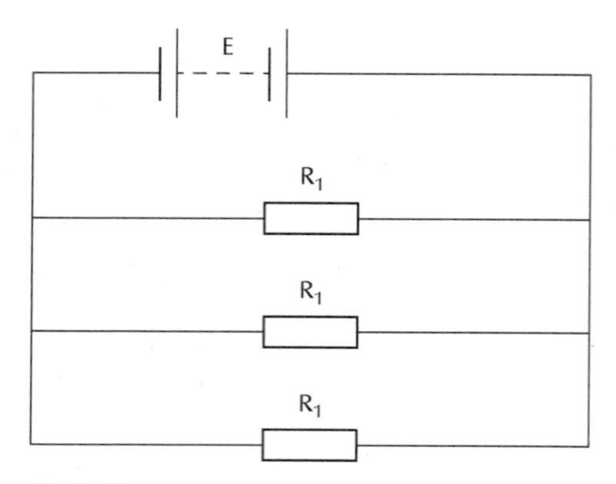

Fig 3.37

Parallel resistors

The principle of conservation of charge is used to derive the equation for total resistance of resistors connected in parallel. Conservation of charge states that the total charge leaving the battery in a given time must be equal to the sum of the charge through each of the resistors.

Consider the parallel circuit in Figure 3.37. For each resistor, the charge passing through it in a given time will be given by:

$$Q = It$$

The charge leaving the battery will be given by the same equation where I will be the total current. Conservation of charge gives:

$$Q_T = Q_1 + Q_2 + Q_3$$

Assuming a unit time, this can be written as:

$$I_T = I_1 + I_2 + I_3$$

Using Ohm's law, where E is the electromotive force (EMF) of the battery (same across each resistor), gives:

$$\frac{E}{R_S} = \frac{E}{R_1} + \frac{E}{R_2} + \frac{E}{R_3}$$

Cancelling out the E gives the equation for the total resistance in a parallel circuit:

$$\frac{1}{R_T} = \frac{1}{R_1} + \frac{1}{R_2} + \frac{1}{R_3}$$

Worked example

Calculate the total resistance of the combination of resistors shown in Figure 3.38 as well as the total current drawn from the source.

Here, there is a combination of series and parallel resistors. Start by finding the total resistance of the series combination of one leg:

$$R_T = R_1 + R_2$$
$$R_T = 10 + 20$$
$$R_T = 30\ \Omega$$

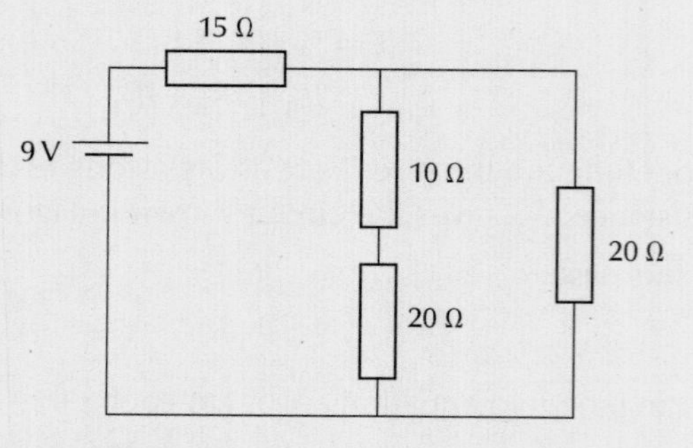

Fig 3.38

Now find the total resistance of the parallel combination,

$$\frac{1}{R_T} = \frac{1}{R_1} + \frac{1}{R_2}$$
$$\frac{1}{R_T} = \frac{1}{30} + \frac{1}{20}$$
$$\frac{1}{R_T} = 0.0833$$
$$R_T = \frac{1}{0.0833}$$
$$R_T = 12\ \Omega$$

This parallel combination is in series with the 15 Ω resistor, so the total resistance is given by:

$$R_T = R_1 + R_2$$
$$R_T = 12 + 15$$
$$R_T = 27\ \Omega$$

The total current from the source depends on the source voltage and the total resistance. Ohm's law can be used to solve for the total current:

$$V_T = IR_T$$

Substituting what you know and solving for the current gives:

$$I = \frac{V_T}{R_T}$$
$$I = \frac{9}{27}$$
$$I = 0.33A$$

Exercise 2.1.3 Total resistance and Ohm's law

1 Calculate the total resistance of the following combinations of resistors.

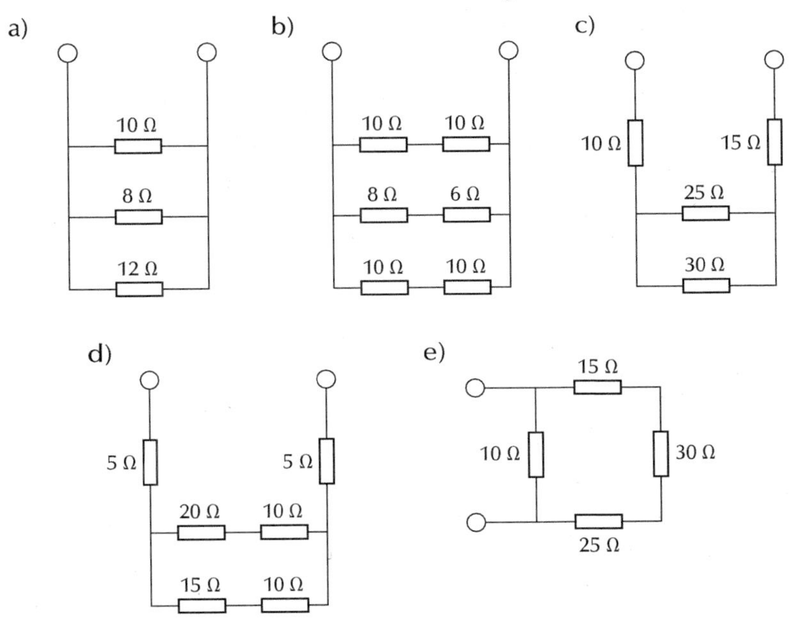

Fig 3.39

2 An engineer designs a circuit with three lamps, each with a resistance of 12 Ω. Each lamp has to receive a voltage of 20 V to operate correctly.

a) Design a circuit using a 20 V power supply that will allow the lamps to operate correctly.

b) Calculate the total resistance of the circuit.

c) Calculate the total current drawn from the power supply.

d) Calculate the current in each lamp.

(continued)

3 For each of the circuits shown in Figure 3.40, calculate the missing values on the ammeter or voltmeter.

a)

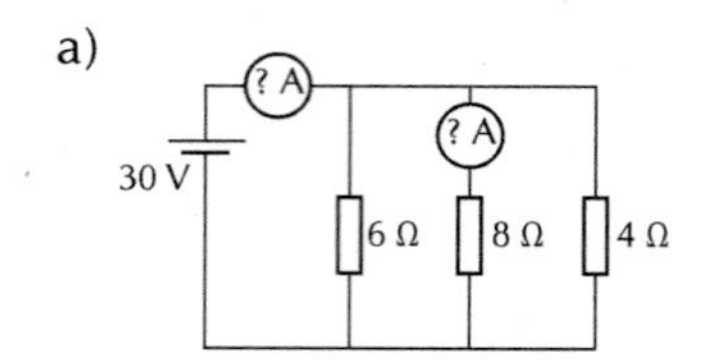

b)

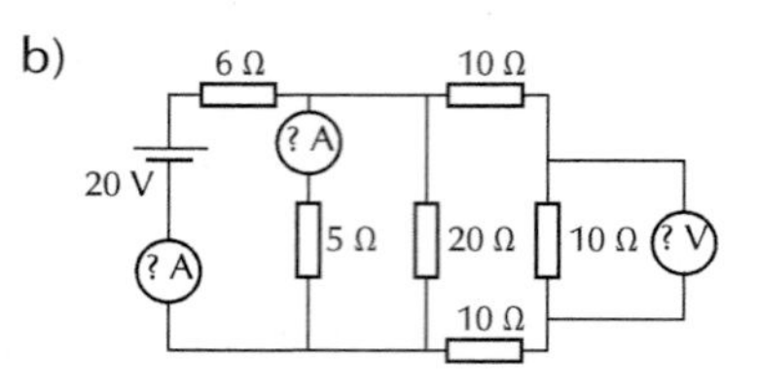

c)

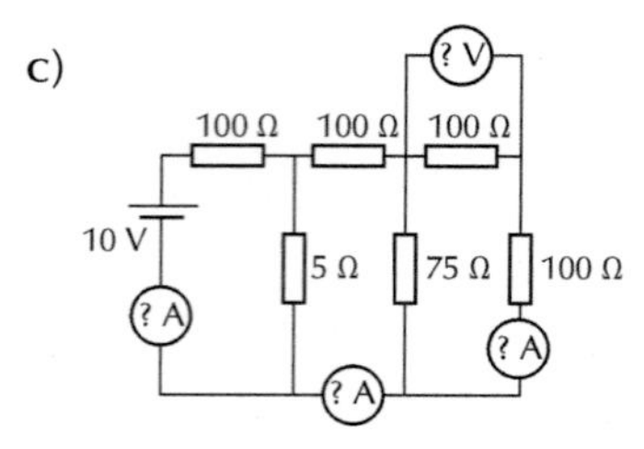

d)

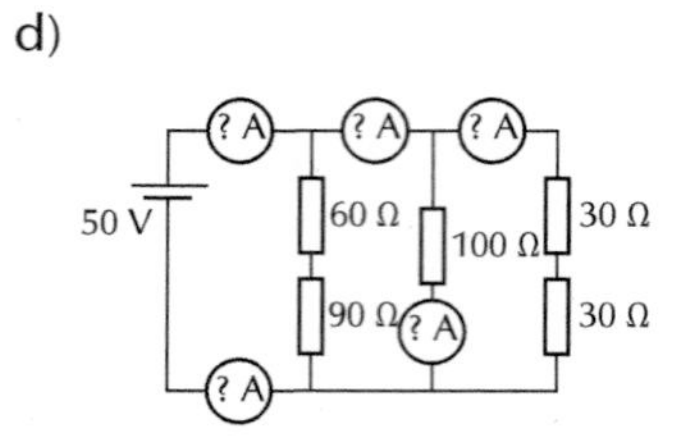

Fig 3.40

2.2 The potential divider

At National 5 level, a specific application of a series circuit called a potential divider was discussed in detail. The potential divider is revised here, and the extension to a new circuit called a Wheatstone bridge is considered.

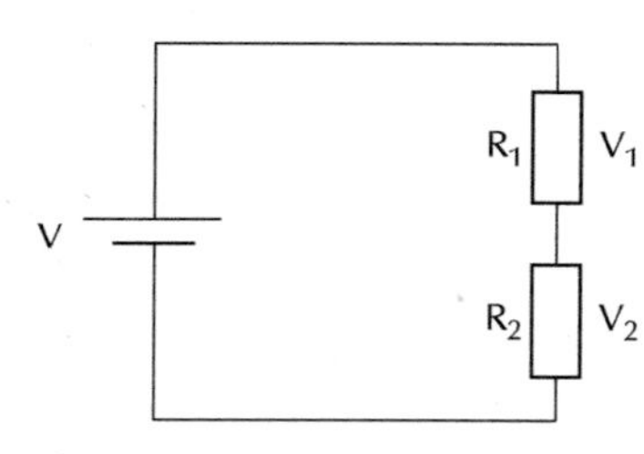

Fig 3.41

2.2.1 The potential divider

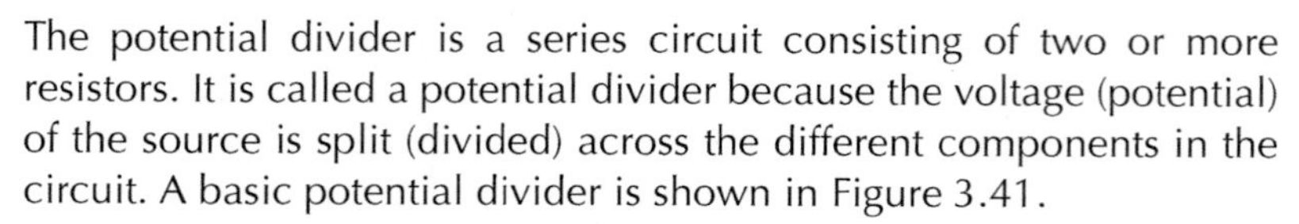

The potential divider is a series circuit consisting of two or more resistors. It is called a potential divider because the voltage (potential) of the source is split (divided) across the different components in the circuit. A basic potential divider is shown in Figure 3.41.

The rules for a series circuit, discussed above, apply to this circuit:

$$V_T = V_1 + V_2$$
$$R_T = R_1 + R_2$$

The current is the same at all points in a series circuit, so the current in each resistor must be equal:

$$I_1 = I_2$$

Applying Ohm's law gives the following:

$$\frac{V_1}{R_1} = \frac{V_2}{R_2}$$

This equation can be rearranged to give the familiar ratio equation for a potential divider (see National 5 book for detailed application):

$$\frac{V_1}{V_2} = \frac{R_1}{R_2}$$

In order to work out the voltage across a specific resistor when the supply voltage, V_S, is known, we can use the relationship,

$$V_2 = \left(\frac{R_2}{R_1 + R_2}\right) V_S$$

This equation was derived in the National 5 course text and detailed applications given there. This expression highlights that the greater the resistance of the resistor, the greater the share of the voltage it has.

2.2.2 The Wheatstone bridge

A Wheatstone bridge consists of two potential dividers connected in parallel as shown in Figure 3.42. Although not specifically part of the Higher course, the Wheatstone bridge is a key application of potential dividers.

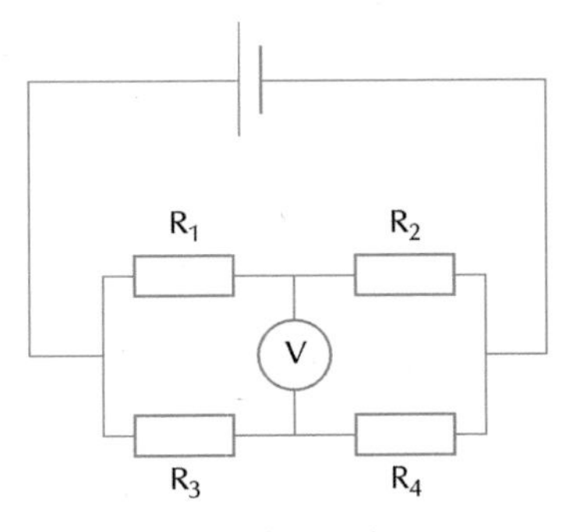

Fig 3.42

There are usually four components, which have associated resistances R_1, R_2, R_3, R_4. These can be resistors or components that change their resistance, such as a thermistor or light-dependent resistor (LDR).

As the two potential dividers are connected in parallel with each other, the voltage across each arm of the bridge is the same (and equal to the supply voltage). In other words, the voltage across R_1 and R_2 combined must equal the voltage across R_3 and R_4 combined:

$$V_S = V_1 + V_2$$
$$V_S = V_3 + V_4$$

The voltmeter in a bridge circuit measures the potential difference between the mid-point of both arms of the bridge. This will be the difference between the voltages across resistor R_2 and R_4.

$$V = V_4 - V_2$$

A Wheatstone bridge can either be balanced (where there is no potential difference between the mid-points) or unbalanced.

Balanced Wheatstone bridge

Consider the bridge circuit shown in Figure 3.43. The voltmeter forms the bridge between the two parallel potential dividers.

Consider each of the potential dividers in turn:

- Top: The voltage is split between resistors R_1 and R_2. The voltage at the mid-point, A, is equal to the voltage across resistor R_2 (V_2).
- Bottom: The voltage is split between resistors R_3 and R_4. The voltage at the mid-point, B, is equal to the voltage across resistor R_4 (V_4).

When the voltage at point A is equal to the voltage at point B, the potential difference across the bridge is equal to zero:

$$V = V_A - V_B$$
$$V = 0V$$

In this case the bridge is said to be balanced. In order for the bridge to be balanced, the top potential divider must split the voltage in the same ratio as the bottom divider. For this to be the case, the ratio of the resistors must be equal. This gives the condition for a balanced bridge:

$$\frac{R_1}{R_2} = \frac{R_3}{R_4}$$

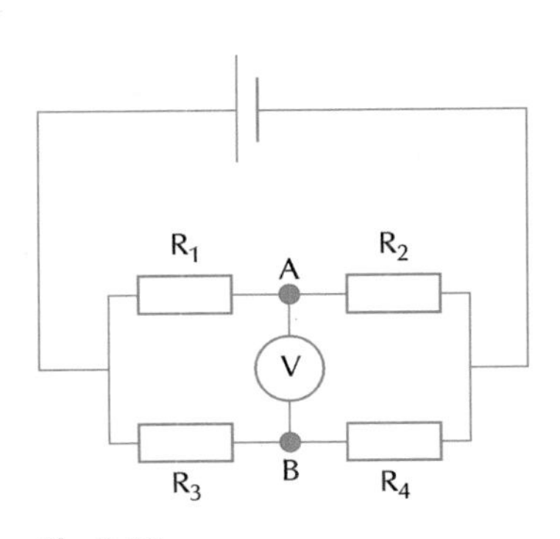

Fig 3.43

Unbalanced Wheatstone bridge

If the bridge is initially balanced, a small change to one of the resistors will cause the bridge to become unbalanced. This is represented in Figure 3.44 by a small change, ΔR:

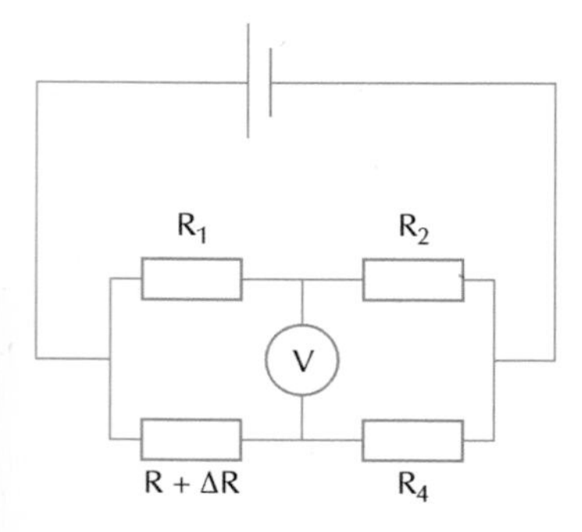

Fig 3.44

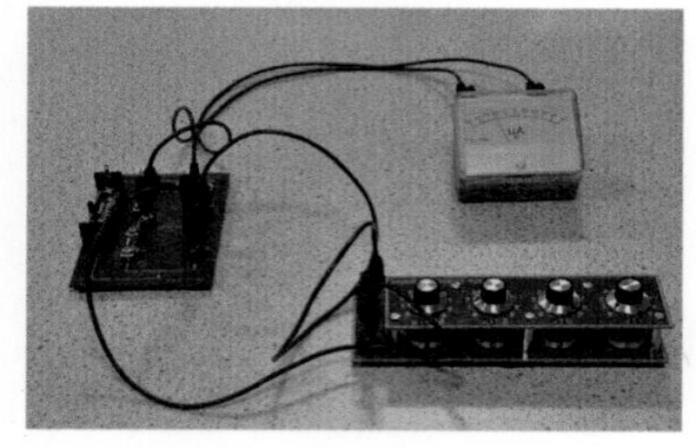

Fig 3.45

The reading on the voltmeter will change when the bridge is moved out of balance. The voltage reading on the voltmeter is directly proportional to the change in resistance:

$$V \propto \Delta R$$

Set up in this way, the Wheatstone bridge can be used to measure any physical change that causes a change in resistance. For example, using an LDR in the bridge circuit allows a measurement of changes in light level. These changes can be very small, as the bridge is a highly sensitive circuit. They can also be both positive and negative as Figures 3.45 and 3.46 show. The bridge allows a charge to flow either way depending on the potential difference across the bridge.

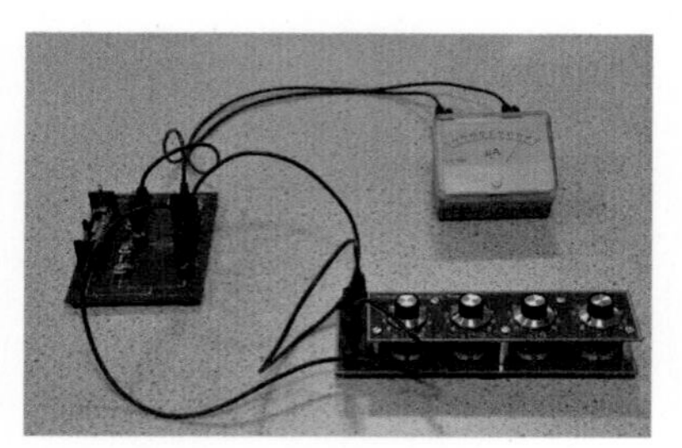

Fig 3.46

Experiment 2.2.2 Applying the Wheatstone bridge

This experiment examines applications of the Wheatstone bridge and how it can be used with different input components as a sensor.

Apparatus

- Wheatstone bridge circuit
- Power supply
- LDR
- Thermistor
- Microammeter
- Hot water
- Ice

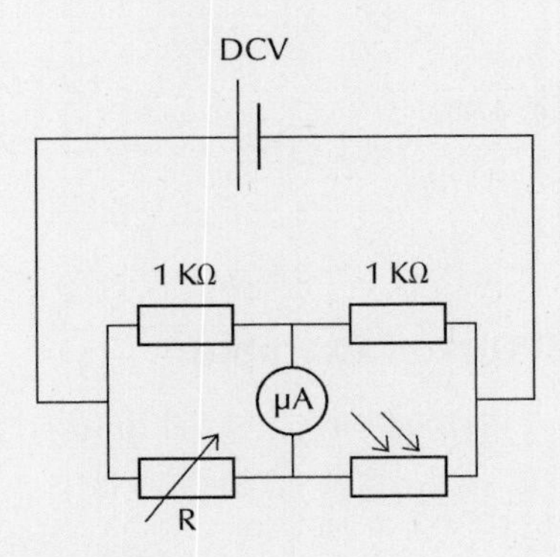

Fig 3.47

Instructions: light meter

1. Connect the circuit shown in Figure 3.47.
2. Balance the bridge with the LDR sitting on your bench by ensuring the current measured by the microammeter is 0.
3. Place the LDR in a bright place and measure the out-of-balance current.
4. Place the LDR in a dark place and measure the out-of-balance current.
5. Write a short report detailing what you have found out about the current when the LDR is placed in a light place, and when placed in a dark place. Comment on the direction of current.

Instructions: thermometer

1. Connect the circuit shown in the circuit diagram (Figure 3.48).
2. Place the thermistor into an ice/water mixture in a beaker (0°C).
3. Balance the Wheatstone bridge with the thermistor in a beaker of water and ice by adjusting the resistance of the variable resistor, R.
4. Place the thermistor in a beaker of boiling water (100°C) – measure the out-of-balance current obtained with the probe in boiling water.
5. Place the thermistor on your desk at room temperature and measure the out-of-balance current obtained.
6. Assuming that the out-of-balance current is directly proportional to the current obtained, use your results to calculate the room temperature.

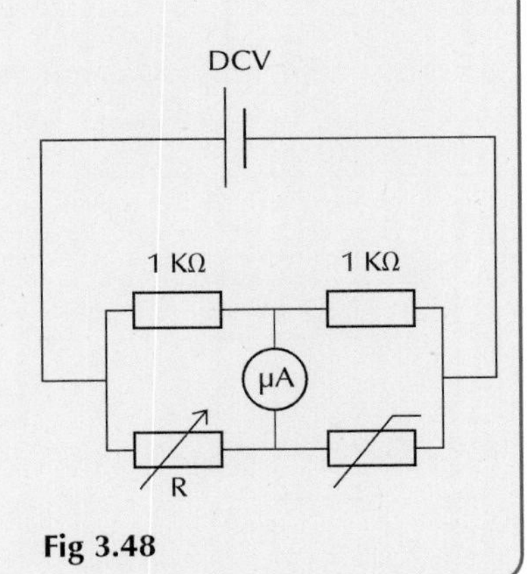

Fig 3.48

For the case of the unbalanced Wheatstone bridge, the voltage across the bridge can be found by calculating the voltage at points A and B in Figure 3.49, then finding the difference.

Both arms of the bridge are potential dividers so the potential divider equation can be applied:

$$V_1 = \left(\frac{R_1}{R_1 + R_2}\right)V_S$$

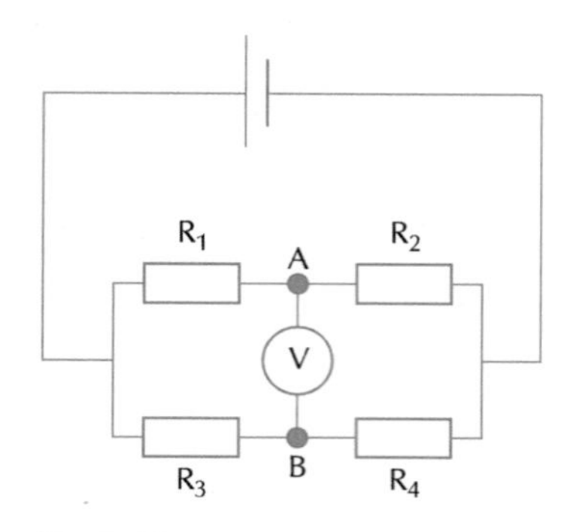

Fig 3.49

For the Wheatstone bridge shown in Figure 3.49:

- Point A: equal to the voltage left across R_2: $V_2 = \left(\frac{R_2}{R_1 + R_2}\right)V_S$
- Point B: equal to the voltage left across R_4: $V_4 = \left(\frac{R_4}{R_3 + R_4}\right)V_S$

The potential difference across the bridge is then given by:

$$V = V_4 - V_2$$

Worked example

Calculate the potential difference across the bridge in Figure 3.50.

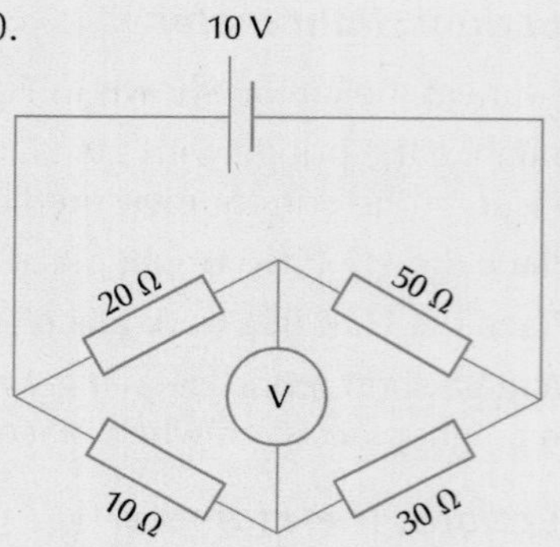

Fig 3.50

Start by finding the voltage at the top of the bridge:

$$V_1 = \left(\frac{R_2}{R_1 + R_2}\right)V_S$$

$$V_1 = \left(\frac{50}{20 + 50}\right) \times 10$$

$$V_1 = 7.14\ V$$

Now find the voltage at the bottom of the bridge:

$$V_1 = \left(\frac{R_4}{R_3 + R_4}\right)V_S$$

$$V_1 = \left(\frac{30}{10 + 30}\right) \times 10$$

$$V_1 = 7.50\ V$$

The difference between these two voltages gives the potential difference across the bridge:

$$V = 7.50 - 7.14$$

$$V = 0.36\ V$$

Exercise 2.2.2 The Wheatstone bridge

1 Calculate the missing voltage for each potential divider shown below.

a)

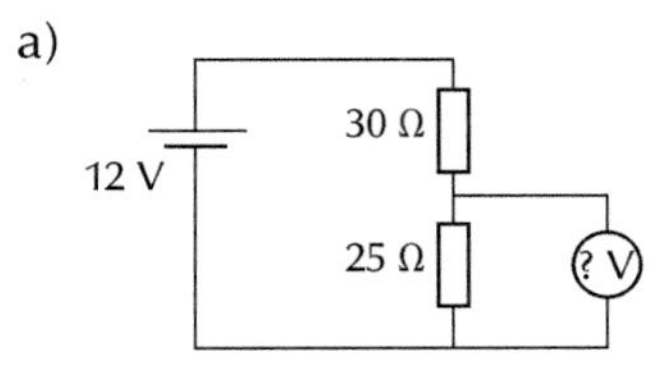

b)

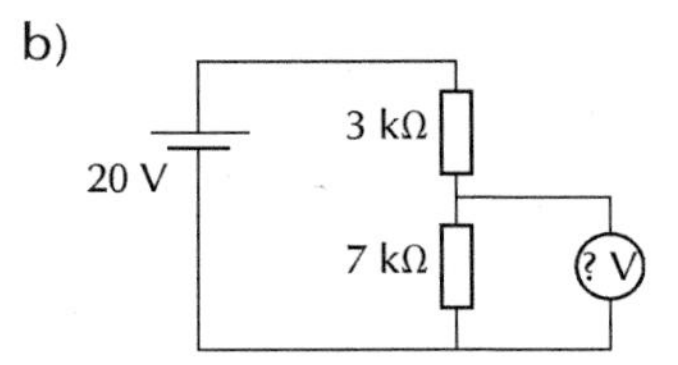

c)

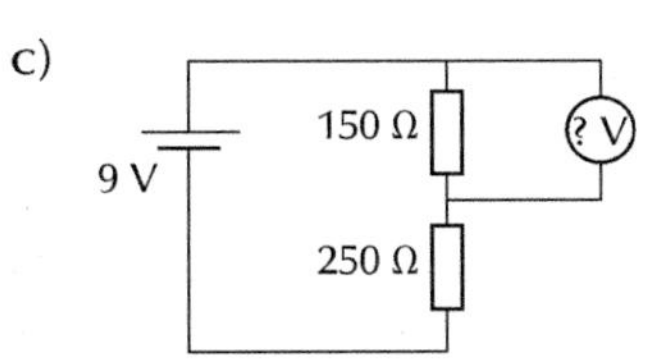

d)

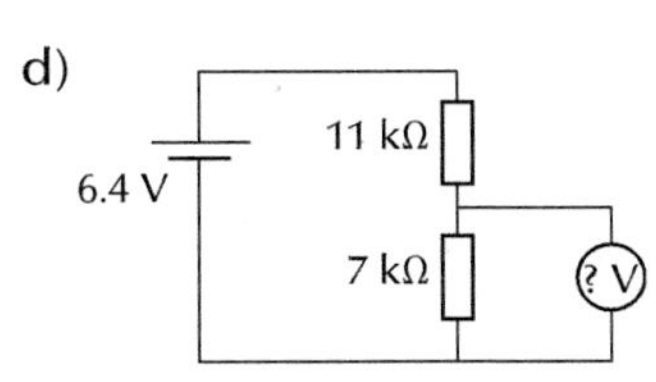

Fig 3.51

2 The following bridges are balanced. Find the value of the missing resistor in each case.

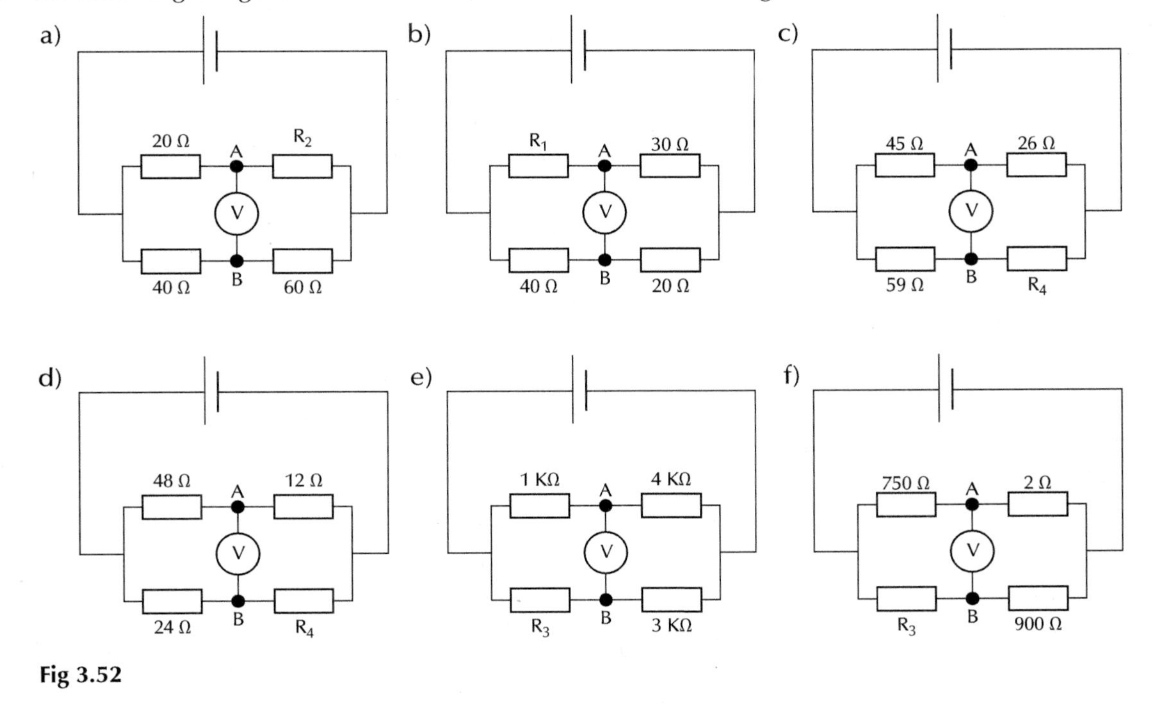

Fig 3.52

3 Calculate the potential difference across each of the bridges shown below.

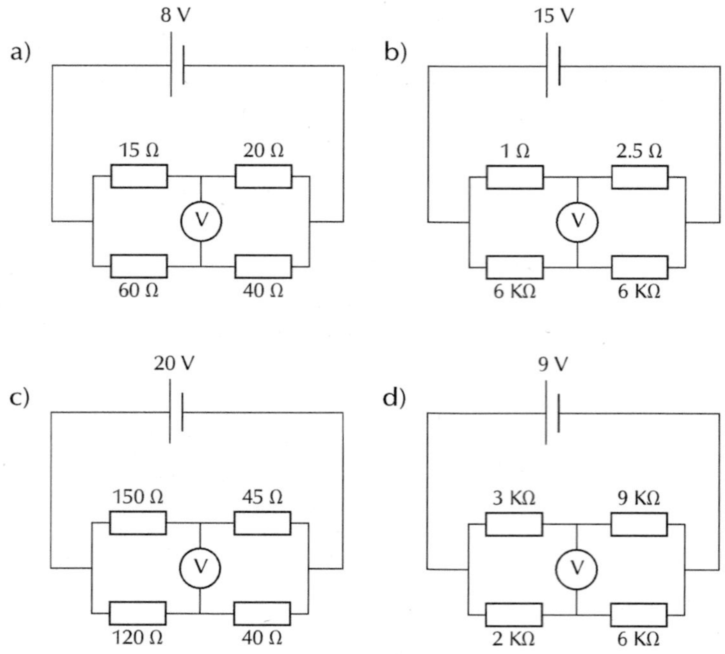

a) 8 V; 15 Ω, 20 Ω, 60 Ω, 40 Ω

b) 15 V; 1 Ω, 2.5 Ω, 6 KΩ, 6 KΩ

c) 20 V; 150 Ω, 45 Ω, 120 Ω, 40 Ω

d) 9 V; 3 KΩ, 9 KΩ, 2 KΩ, 6 KΩ

Fig 3.53

3 Electrical sources and internal resistance

You should already know (National 5)

- Electrical sources include batteries, the mains supply and power supplies
- Electrical sources supply electrical energy as either direct current or alternating current

Learning intentions

- Concept of electromotive force as the energy supplied to every coulomb of charge leaving the source
- Link between electromotive force and terminal potential difference as the 'actual' voltage the source supplies
- Comparing ideal supplies to real supplies, including the concept of lost volts, and the dependence of lost volts on the current drawn from the supply
- Determination of internal resistance of a source using graphical analysis

3.1 Sources of electricity

You have used electrical sources in your previous studies of physics. Typically, they take the form of batteries, or DC/AC power supplies in the lab. You have also used sources of electricity in your day-to-day life, for everything from lighting your home to cooking your food – the UK mains supply.

3.1.1 The battery

Batteries come in all shapes and sizes to fit different applications. If you compare the AA batteries you might find in a remote control to the Li-Ion battery for a laptop, for example, you will see a clear difference in their respective physical shapes (see Figures 3.54 and 3.55).

Fig 3.54

In addition to differences in shape and size, batteries also differ in terms of the voltage they deliver, and the maximum current they can deliver. Batteries must therefore be chosen to match their application. For example, a laptop battery can be rated at 19 V, but only needs to be able to supply a small current to run the various components of the computer. A car battery is rated at 12 V but needs to be able to supply a much larger current to turn over an engine during the starting

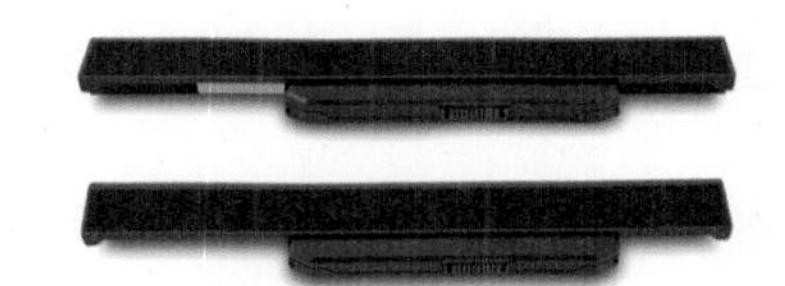
Fig 3.55

Key point

Batteries supply direct current (DC).

Fig 3.56

Key point

The UK mains supply is AC, 230 V RMS with a frequency of 50 Hz.

Fig 3.57

Fig 3.58

process. For this reason, the batteries will have a different chemical makeup that is suited to different uses.

Batteries do have several things in common, however. They all supply direct current (DC). Batteries produce electricity by means of a chemical reaction inside the cells that make them up. Different chemicals will give different voltages, maximum currents and charge and discharge times.

3.1.2 The wall socket

Mains electricity is in every home in the UK. We connect to it via sockets in the wall. Appliances can draw different currents from the mains supply depending on their resistance – this is linked to their power rating. The electricity from the mains is alternating current (AC). In the UK, the mains supply is 230 V RMS with a frequency of 50 Hz.

3.2 Internal resistance

If you read the case of a battery, it will tell you its voltage. For example, the battery in Figure 3.57 is a 9 V battery.

So far it has been assumed that this is the voltage that the battery will supply to a circuit. However, in practice this is not the case. The actual voltage delivered by the battery will be less because of energy losses within the battery, which are modelled as an internal resistance.

3.2.1 Ideal vs real

In the studies of electricity above, all electrical sources have been considered to be ideal. That means that the voltage stated by the source is the voltage that you have access to for the circuit. This voltage is assumed to remain constant, regardless of what is connected to the circuit. In practice, a real source will not supply the full voltage to the circuit. This is due to energy losses in the battery. The energy lost in the source will depend on what current the battery is supplying to the circuit.

You can see this effect in practice on a cold morning when starting a car. When you turn the key to start the engine (or push the starter button), the battery must supply a large current to turn the engine over. If the headlights are on at the same time, they will go dim when the starter motor is turning. This shows that the voltage from the battery has dropped when the starter motor is turning. The car battery is therefore far from an ideal source. The voltage from the battery changes depending on what the battery is powering at the time.

There are three important terms to become familiar with when discussing real and ideal sources. These are:

- Electromotive force (EMF, E): this is the amount of energy supplied to each coulomb of charge passing through the source. In the case of a battery, this energy is supplied by the chemical reactions taking place in the cell. It can only be measured across the terminals when the current drawn from the source is zero. In other words, it can only be measured when a very high resistance voltmeter is connected across the terminals of the source, with no additional components in the circuit (as shown in Figure 3.59).

- Terminal potential difference (TPD): this is the voltage available from the source when connected to an external load (for example, a lamp or a resistor). Like the EMF, it is measured across the terminals of the source. However, the TPD is the voltage measured when the source is connected to another component that draws current. The TPD will change depending on the current supplied by the battery.
- Lost volts: there is a difference between the EMF of a source and the TPD. The TPD is lower than the EMF because when the battery is connected to a source some energy is lost inside the battery. The difference between the EMF and the TPD is the lost volts. The greater the current drawn from the battery, the greater the lost volts.

$$E = TPD + LostVolts$$

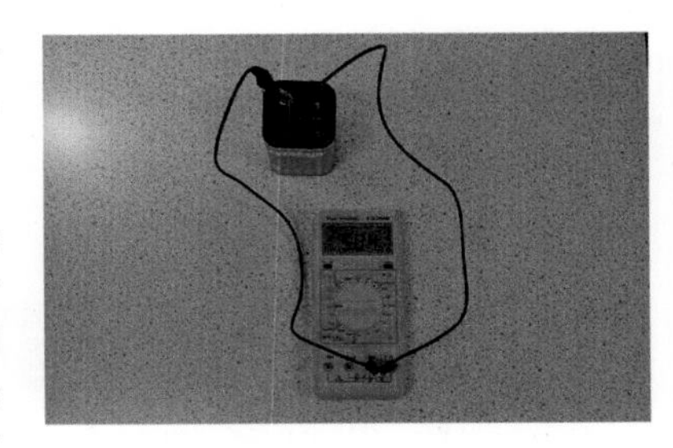

Fig 3.59

Key point

The terminal potential difference (available voltage) from a real source changes as the current drawn from the source changes. The difference between the TPD and the EMF is called the 'lost volts'.

3.2.2 Internal resistance

The concept of internal resistance is used to model the concept of energy being lost within a source of electricity (such as a battery). A typical source is therefore represented as the cell with a given EMF, E, and internal resistance, r, as shown in Figures 3.60 and 3.61.

The internal resistance is modelled to be in series with the cell. Like any other resistor in a circuit, there will be a voltage dropped across this resistor. This voltage is the lost volts – it is lost inside the cell so you do not have access to it. Changing the current drawing from the battery will change the voltage across the internal resistance according to Ohm's law:

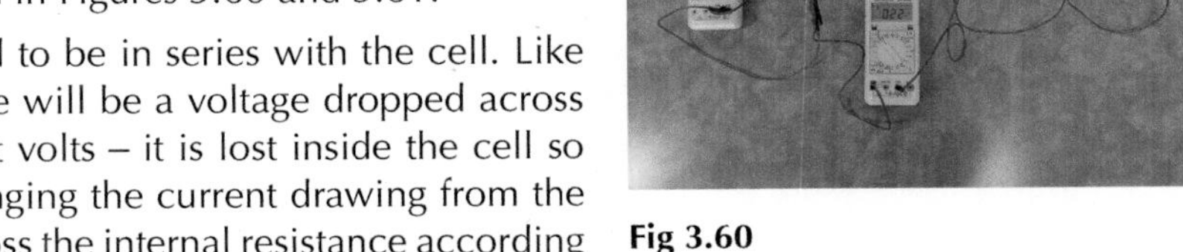

Fig 3.60

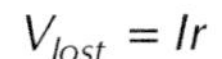

$$V_{lost} = Ir$$

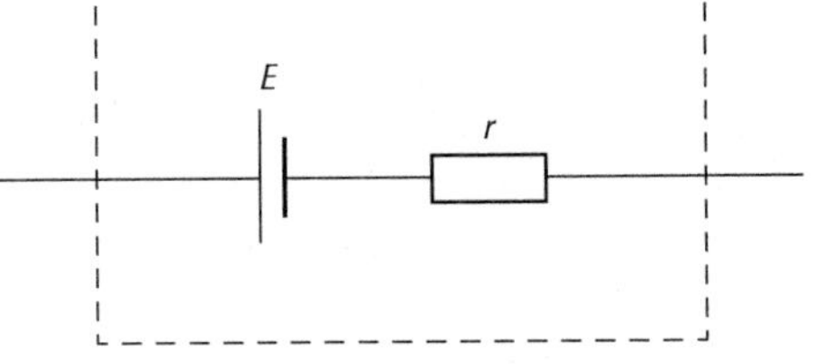

Fig 3.61

This highlights why the lost volts, and thus the TPD, changes with different currents drawn from the source. Consider the following situations:

- No current drawn: when no current is drawn from the source, the voltage across the internal resistance will be zero. This means that the lost volts will be zero and the TPD that is measured will be equal to the EMF:

$$EMF = TPD$$

- Small current drawn: if a large load resistor is connected to the source, a small current will be drawn (according to Ohm's law). This voltage will lead to a small voltage being dropped across the internal resistance. The lost volts will be small because,

$$LostVolts = Ir$$

- Large current drawn: if a small load resistance is connected to the power source, a large current will be drawn. A large voltage will be dropped across the internal resistance so lost volts will be large. In this case, the TPD will be low.

Experiment 3.2.2 Effects of resistance on TPD

This experiment investigates how the TPD from a source changes as the resistance connected to it is changed. This will result in a change in current drawn from the source.

Apparatus

- 6 V battery
- Three 6 V lamps
- DC ammeter
- DC voltmeter
- Connecting wires

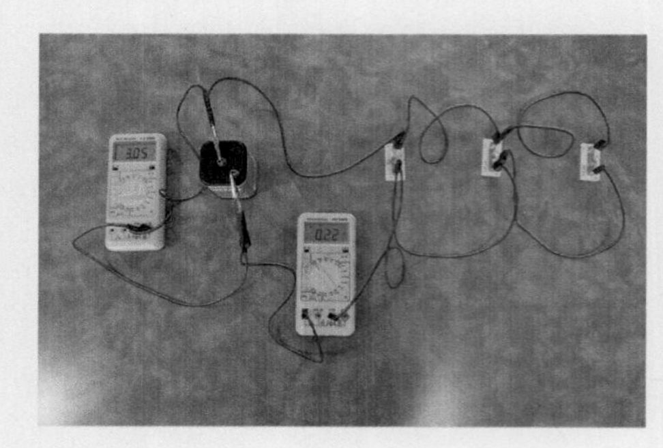

Fig 3.62

Instructions

1 Measure the EMF of the battery by connecting a voltmeter across the terminals of the battery with no components connected.

2 Build a series circuit (shown in Figures 3.63 and 3.64) containing a lamp and a battery. Connect a voltmeter across the terminals of the battery to measure the TPD and an ammeter in series to measure the current.

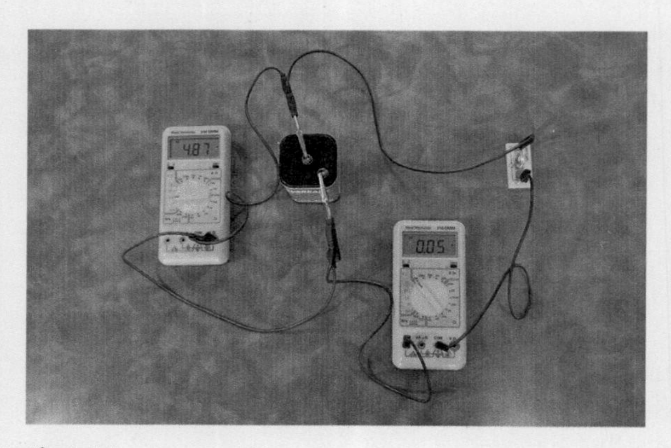

Fig 3.63

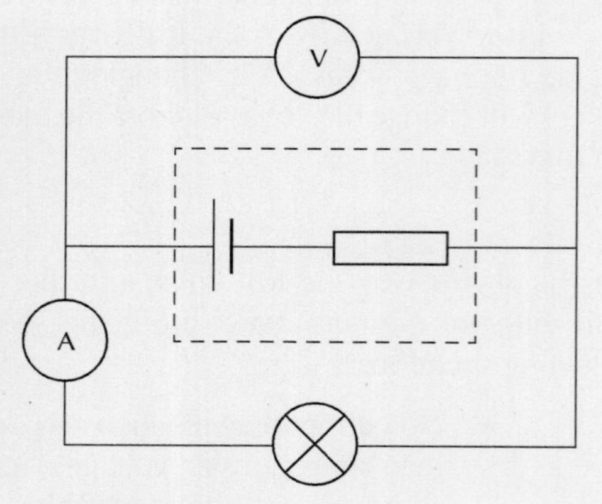

Fig 3.64

3 Repeat the above experiment adding a second lamp and then a third lamp as shown in the diagram below (Figure 3.66). Each time, measure the TPD using a voltmeter and the current drawn from the battery using the ammeter.

(continued)

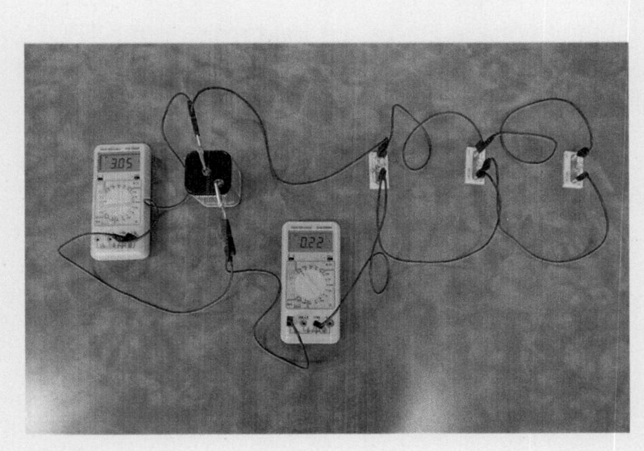
Fig 3.65

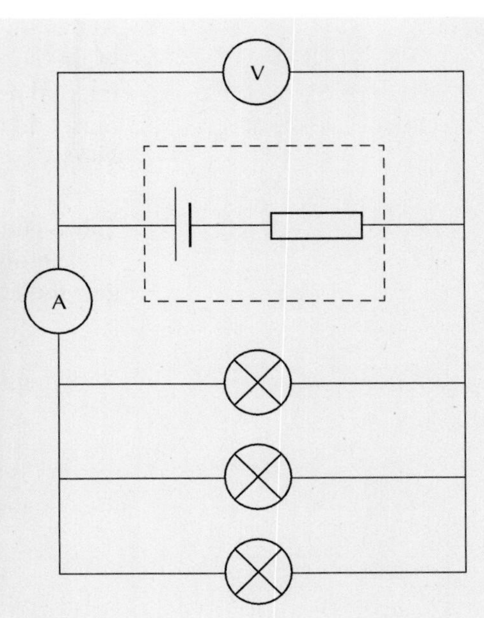

Fig 3.66

4 Record your results in a table similar to Table 3.2. Calculate the lost volts in each case using the equation,

$$LostVolts = E - TPD$$

Number of lamps lit	Current (A)	TPD (V)	Lost volts (V)
1			
2			
3			

Table 3.2

5 Calculate the internal resistance of the battery in each case. Use the current measured (the current through the internal resistance) and the lost volts (which is the voltage across the internal resistance) in the following equation (Ohm's law):

$$r = \frac{LostVolts}{I}$$

6 Write a short report detailing the results you have obtained. Include the following in your report:

- The effect of increasing the current drawn on the TPD.
- The effect of increasing the current drawn on the internal resistance.

The internal resistance is modelled as a series resistor. This means that problems involving internal resistance can be solved by modelling the circuits as series circuits. Key to a series circuit is that the voltage is split across the components in the circuit. Consider the circuit in Figure 3.67, showing a battery connected to a single load resistor with resistance R.

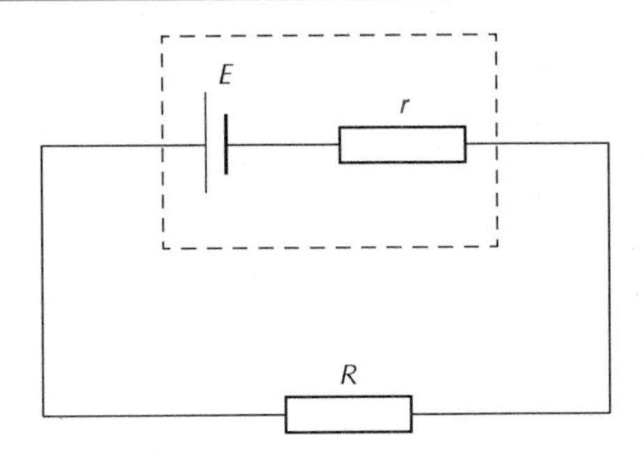

Fig 3.67

As this is a series circuit, the current is the same at all points. Knowledge of the current in the circuit is the key to solving problems involving internal resistance. The voltages across the components in the circuit must add up to give the total voltage:

$$V_S = V_1 + V_2$$

This means that the voltage across the load resistor (the TPD) and the voltage across the internal resistance (the lost volts) must add up to give the EMF:

$$EMF = TPD + LostVolts$$

Applying Ohm's law to each of the voltages gives:

$$E = IR + Ir$$

where I is the current in the circuit. The TPD is given the symbol V, so the equation becomes:

$$E = V + Ir$$

It is important to remember that this is simply a series circuit with two resistors.

Key point

The useful voltage from a real source is called the terminal potential difference (TPD), V. This changes depending on the current drawn from the source and obeys the following relationship:

$$E = V + Ir$$

Key point

When a voltmeter is connected across a source with no other components connected (no external load), the voltage measured by the voltmeter will be the EMF.

Key point

Energy losses within an electrical source are modelled as an internal resistance, which is a resistor in series with the voltage source as in Figure 3.68.

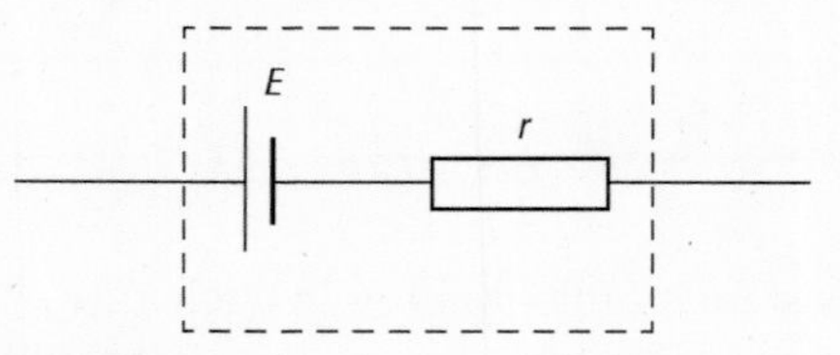

Fig 3.68

The cell has an EMF, E, which is the energy supplied to each coulomb of charge leaving the source. An EMF of 6 V means that 6 J of energy are supplied to every coulomb of charge. The internal resistance, r, causes a drop in voltage inside the source called lost volts.

Short circuit current

A battery can be short-circuited by connecting a wire between its positive and negative terminals. For an ideal source, this would lead to a near infinite current because the wire will have a very low resistance and the current is given by:

$$I = \frac{V}{R}$$

However, a real voltage source has an internal resistance. This will limit the amount of current the source can supply. The maximum current a source can supply is the short circuit current, which is given by:

$$I = \frac{E}{r}$$

where E is the EMF and r is the internal resistance. This equation is Ohm's law applied to a series circuit with a total resistance equal to the internal resistance. This circuit is shown in Figure 3.69.

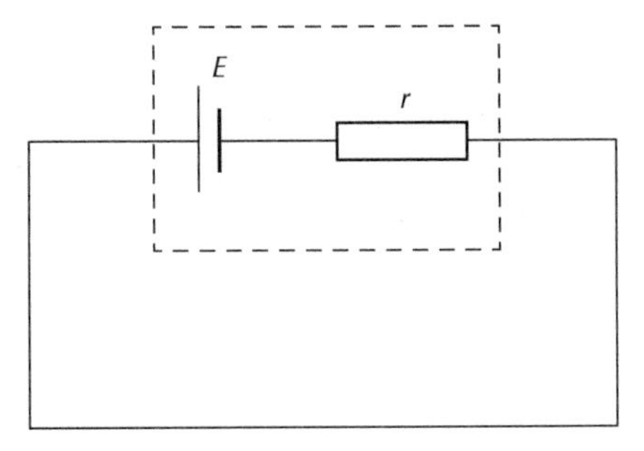

Fig 3.69

Worked example

A 9 V battery (battery with an EMF of 9 V) is connected to a fixed resistor of resistance 15 Ω as shown in Figure 3.70.

When connected to the resistor, the voltage measured by the voltmeter is 8.7 V.

a) Calculate the lost volts.

b) Calculate the current in the circuit.

c) Find the internal resistance of the battery.

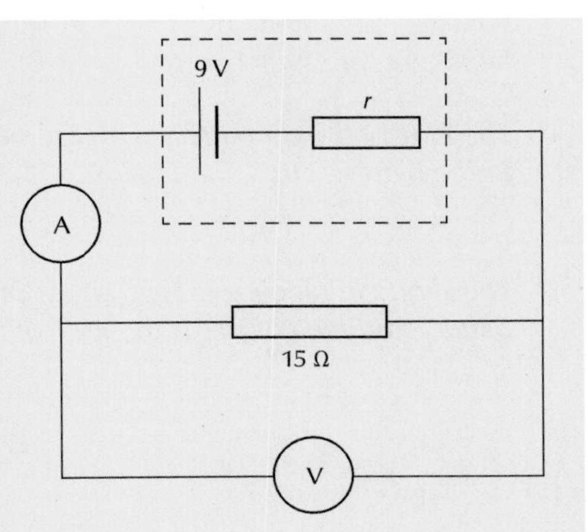

Fig 3.70

a) *The lost volts is equal to the difference between the EMF and the TPD:*

$$LostVolts = 9 - 8.7$$
$$LostVolts = 0.3\ V$$

b) *To find the current in the circuit, apply Ohm's law to the load resistor that has a known resistance:*

$$V = IR$$

The voltage across the resistor is the TPD, measured by the voltmeter in the circuit above. The resistor has a fixed resistance. Substituting these values and solving for current gives:

$$8.7 = 15 \times I$$
$$I = \frac{8.7}{15}$$
$$I = 0.58\ A$$

c) *The internal resistance can be found by applying Ohm's law again:*

$$V = Ir$$

The current in the internal resistance is the same as the current in the load resistor, calculated above. The voltage across the internal resistance is the lost volts. Substituting and solving for internal resistance gives:

$$0.3 = 0.58 \times r$$
$$r = \frac{0.3}{0.58}$$
$$r = 0.52\ \Omega$$

Worked example

A vehicle technician is investigating the internal resistance of a car battery. He connects a voltmeter across the terminals of the battery with no other components switched on and finds that the voltmeter reads 12.78 V.

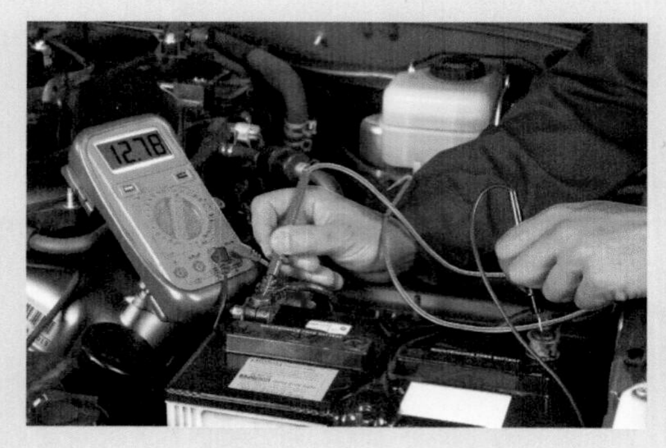

Fig 3.71

The car battery has an internal resistance of 0.7 Ω.

a) State the EMF of the car battery.

b) Calculate the short circuit current for the car battery.

c) The battery is connected to the headlights, which have a combined resistance of 22 Ω. Calculate the voltage measured by the mechanic when the headlights are switched on.

a) *The EMF of the car battery is measured by the voltmeter when no other components are connected to the battery. Here, the EMF is:*

$$E = 12.78\,V$$

b) *The short circuit current is given by Ohm's law, where the resistance is the internal resistance of the battery and the voltage is the EMF of the battery:*

$$I = \frac{E}{r}$$

$$I = \frac{12.78}{0.7}$$

$$I = 18.26\,\text{A}$$

c) *When the headlights are switched on, the total resistance in the circuit changes. As this is a series resistance, the total resistance can be found using:*

$$R_T = R_1 + R_2$$

Here, the new resistance is,

$$R = 22 + 0.7$$

$$R = 22.7\,\Omega$$

The current in the circuit can now be calculated using Ohm's law,

$$V = IR$$

$$12.78 = I \times 22.7$$

$$I = \frac{12.78}{22.7}$$

$$I = 0.56\,A$$

Ohm's law can be applied again to find the voltage measured by the mechanic, which is the voltage across the headlights (the TPD):

$$V = IR$$

$$V = 0.56 \times 22$$

$$V = 12.3\,V$$

Exercise 3.2.2 Internal resistance

1 State what is meant by electromotive force (EMF).

2 Electrical sources are not 100% efficient. How is the energy loss in an electrical source modelled? You may wish to draw a diagram to illustrate your answer.

3 A voltmeter is connected across a battery as shown in Figure 3.72.

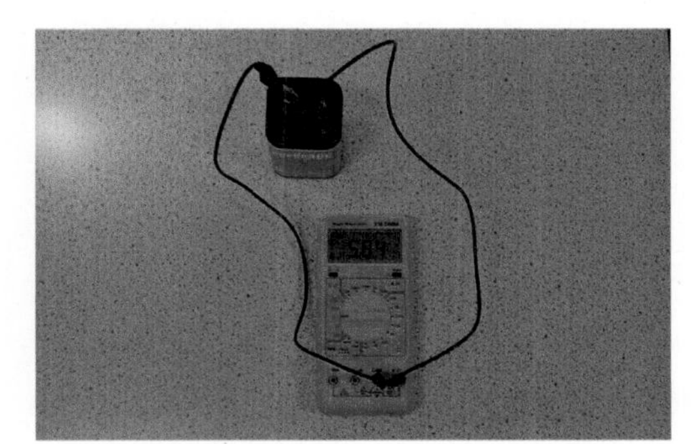

Fig 3.72

What is the voltmeter measuring in this example?

4 An AA battery has an EMF of 1.5 V and internal resistance of 1.5 Ω. It is connected to a torch lamp with a resistance of 9 Ω.

a) State what is meant by 'an EMF of 1.5 V'.

b) Calculate the current in the circuit when the battery is connected to the lamp.

c) Calculate the TPD when the battery is connected to the lamp.

d) Calculate the lost volts when the battery is connected to the lamp.

5 The internal resistance of a battery is being investigated. The battery is connected in series with a 15 Ω load resistor, switch and ammeter. A voltmeter is connected across the terminals of the battery as shown in the circuit diagram in Figure 3.73.

When the switch is open, the reading on the voltmeter is 5.9 V. When closed, the reading on the voltmeter drops to 4.6 V.

a) State the EMF of the battery.

b) Calculate the lost volts in the circuit when the resistor is connected to the battery.

c) Calculate the current when the switch is closed.

d) Calculate the internal resistance of the battery.

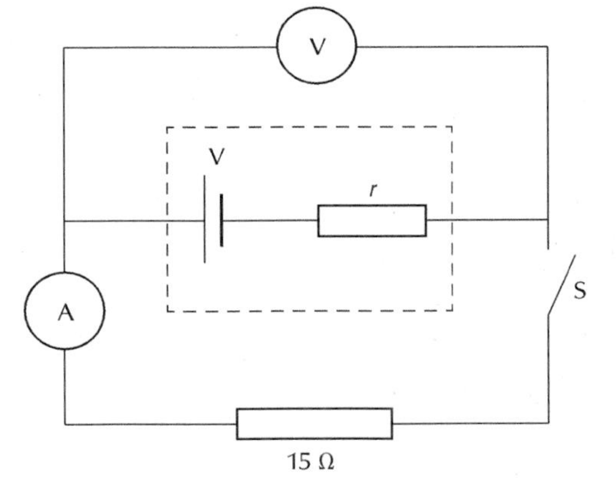

Fig 3.73

6 The terminals of a car battery are shorted and a current of 70 A is found. If the EMF of the car battery is 12.6 V, calculate the internal resistance of the battery.

7 A torch lamp is connected to a 9 V battery that has an internal resistance of 1.4 Ω. The torch lamp has a resistance of 8 Ω.

Fig 3.74

V

9 V

1.4 Ω

A

S

8 Ω

Fig 3.75

a) Calculate the current in the circuit when the switch is closed and the lamp lights.

b) Calculate the TPD when one lamp is connected to the battery, by closing switch S, as shown in the circuit diagram in Figure 3.75.

c) A second torch lamp is connected in parallel with the first as shown in Figure 3.76.

State and explain the effect on the current and TPD of connecting a second lamp in parallel to the battery.

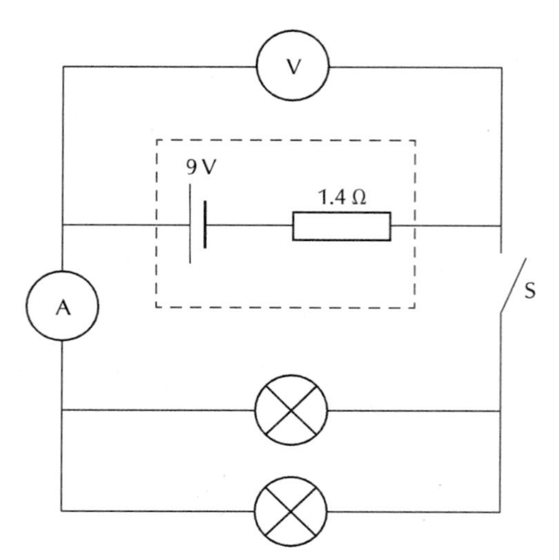

Fig 3.76

8 An experiment is being conducted to measure the internal resistance of a battery. Figure 3.77 shows the voltage and current measured from the battery when it is connected to a lamp, and when it is not connected.

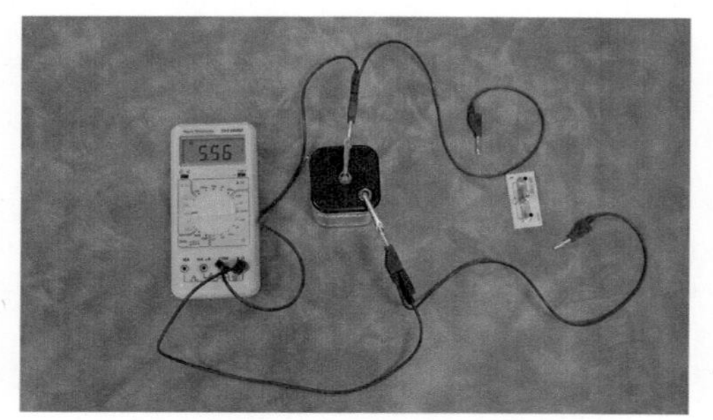

Voltmeter reading = 5.49 V;
Ammeter reading = 00.0 mA

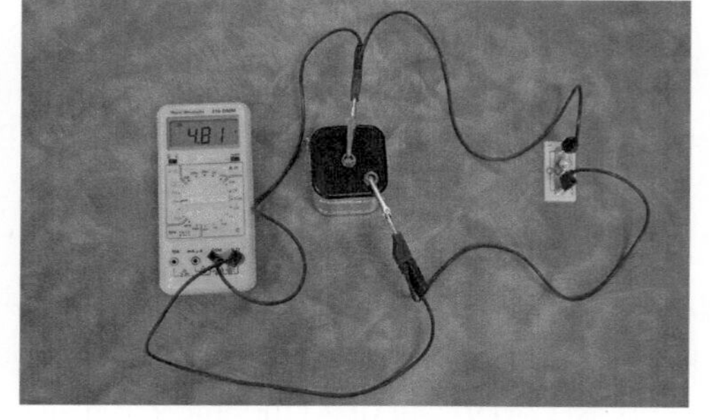

Voltmeter reading = 4.90 V;
Ammeter reading = 52.8 mA

Fig 3.77

Use this information to calculate the internal resistance of the battery.

(continued)

9 A remote control car uses four 1.5 V AA batteries connected in series to power its electric motor. Each of the batteries can be considered to be identical, with:

- EMF = 1.5 V
- Internal resistance = 0.2 Ω

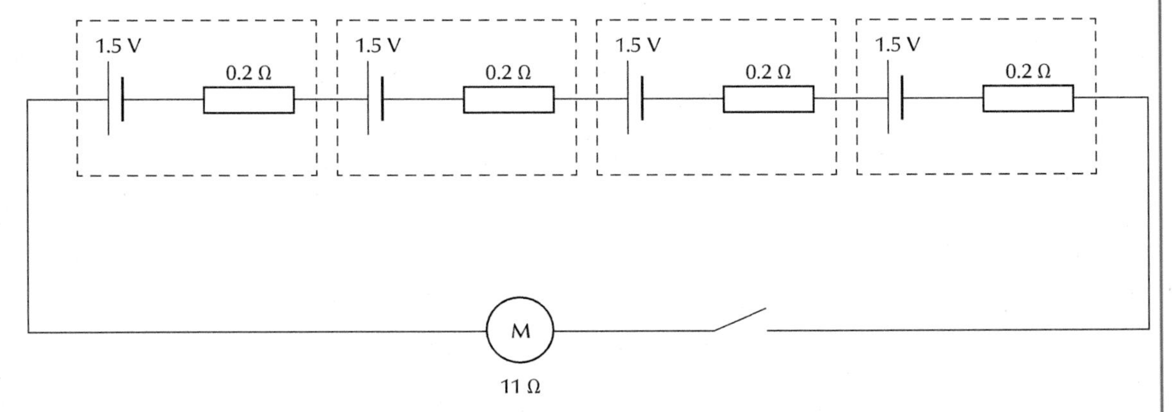

Fig 3.78

a) Calculate the total EMF of the four batteries connected in series.

b) Calculate the total resistance of the circuit when the motor is activated.

c) Find the current in the circuit when the motor is activated.

d) Find the lost volts when the motor is switched on.

3.2.3 Graphical determination of internal resistance

In an experiment, the TPD and the current in the circuit can be easily measured, using a set up shown in Figure 3.79.

If the resistance of the load resistor, R, is changed then this will lead to a change in current, lost volts and TPD. Plotting the observed TPD against current can be used to determine the EMF and internal resistance of the battery. Consider the equation for internal resistance:

$$E = V + Ir$$

This can be rearranged as follows:

$$V = E - Ir$$

$$V = -rI + E$$

Written in this form, it can be directly compared with the equation of a straight line:

$$y = mx + c$$

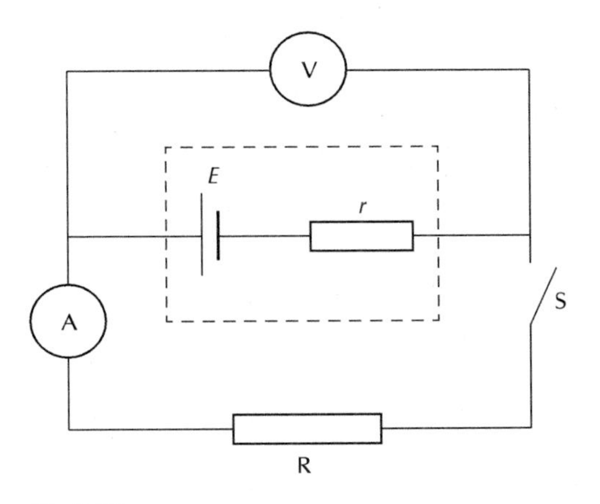

Fig 3.79

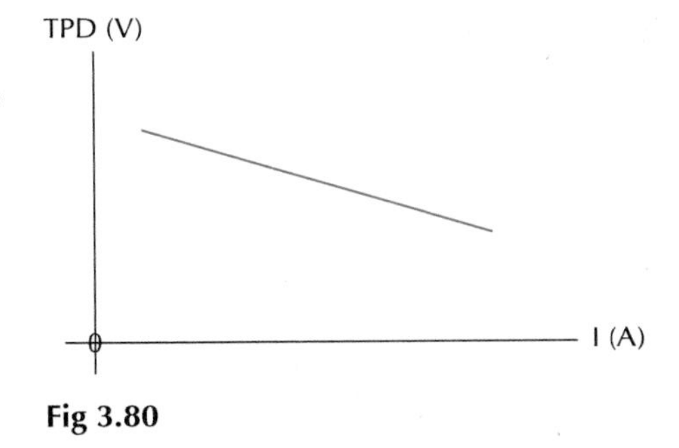

Fig 3.80

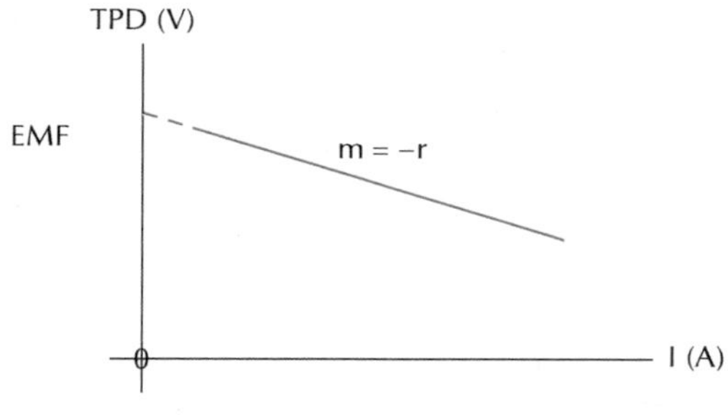

Fig 3.81

By comparing the two equations, you can see that the negative gradient of a graph of TPD vs current would give the internal resistance, and the y-intercept would give the EMF:

$$r = -m$$
$$E = c$$

The graph would look like Figure 3.80. Plotting this graph back to the y-axis will give the EMF, and finding the gradient will give the internal resistance (the negative signs cancel out), see Figure 3.81.

Experiment 3.2.3 Graphical determination of internal resistance

This experiment uses the measured TPD and current from a battery to find its EMF and internal resistance.

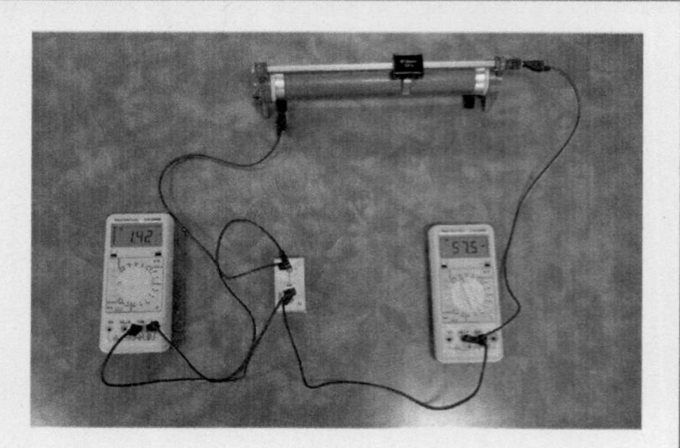
Fig 3.82

Apparatus

- Battery
- Ammeter
- Voltmeter
- Rheostat
- Connecting wires

Instructions

1 Set up the circuit as shown in the photograph and circuit diagram (Figures 3.82 and 3.83) – a series circuit containing a rheostat and battery with an ammeter to measure the current drawn and a voltmeter to measure the TPD.

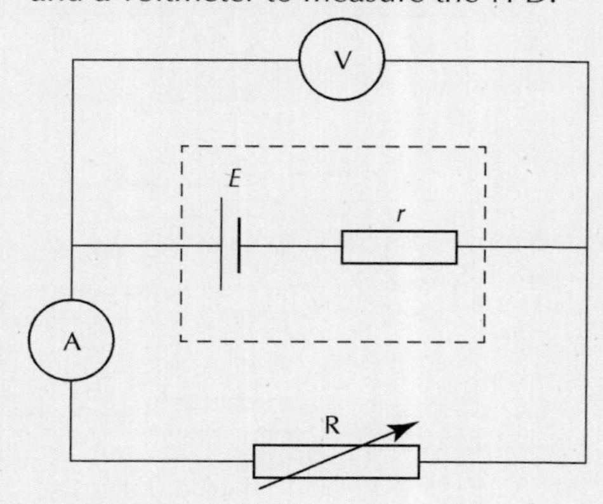

Fig 3.83

2 Construct a table similar to Table 3.3 to record your results:

TPD (V)	Current (A)

Table 3.3

(continued)

3 Adjust the rheostat until the desired current is drawn from the battery. Record the current drawn and the corresponding TPD measured by the voltmeter.
4 Repeat the above measurements of current drawn and corresponding TPD.
5 Plot a graph of the TPD vs current. Draw and fit a best fitting straight line to the graph.
6 Use the results above to determine the EMF and internal resistance of the battery.
7 Connect a voltmeter directly to the battery, with no other components to measure the EMF directly. How does this value compare with the results achieved by the above experiment? How does this value compare with the stated voltage of the battery?

Key point

A graph of TPD vs current drawn can be used to find the EMF and internal resistance of a battery as follows:

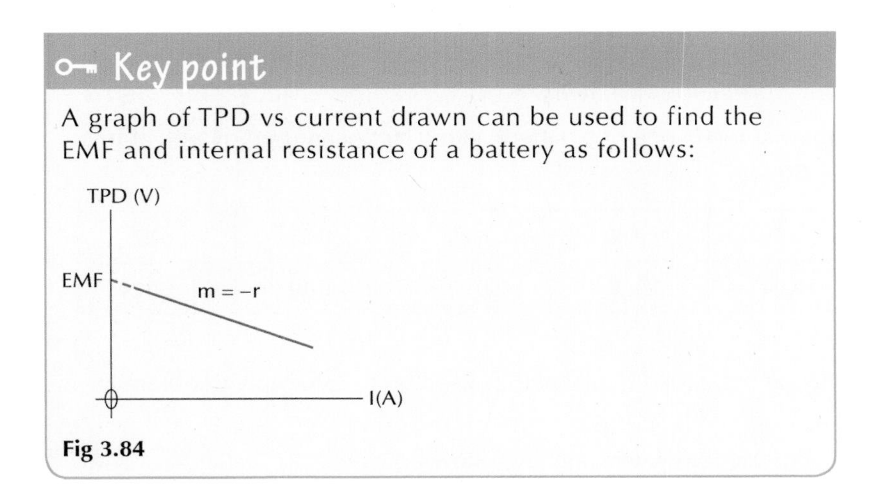

Fig 3.84

Exercise 3.2.3 Graphical determination of internal resistance

1 Students set up the experiment shown in the circuit diagram in Figure 3.85a to investigate the EMF and internal resistance of a battery.

One result taken from the experiment is shown in the photograph, Figure 3.85b.

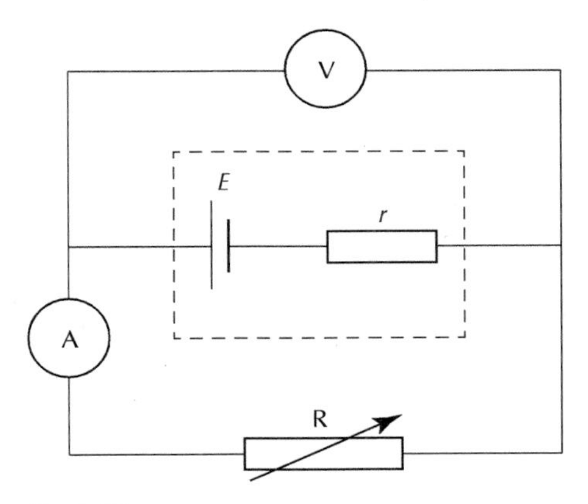

Fig 3.85a

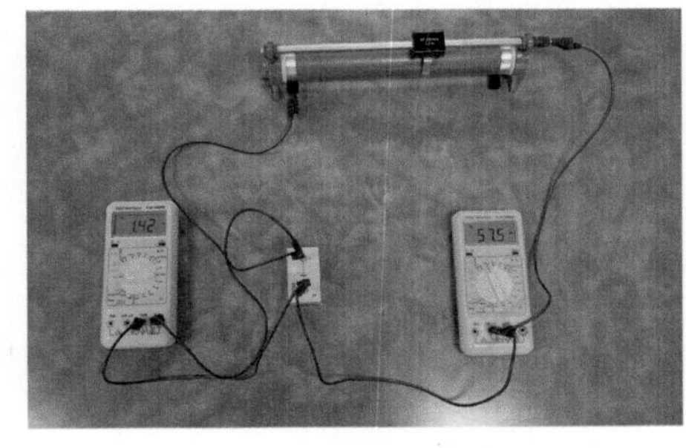

Voltmeter reading = 1.428 V;
Ammeter reading = 74.7 mA

Fig 3.85b

a) Calculate the resistance of the rheostat for the measurements made in Figure 3.85b.

b) The experiment was repeated for different load resistances. The measured values of the TPD and current drawn are shown in Table 3.4:

TPD (V)	1.346	1.383	1.398	1.411	1.444
Current (mA)	194.1	139.0	118.3	98.0	51.7

Table 3.4

(i) Plot a graph of TPD vs current for the above data. Remember to use the data from part a) as well!

(ii) Use your graph to determine the EMF and internal resistance of the battery.

(iii) Calculate the short circuit current for the battery.

2 Students are investigating the internal resistance of a battery after it has been used overnight in a torch. They obtain the following results.

Voltage (V)	1.09	0.93	0.80	0.69	0.57	0.47
Current (A)	0.02	0.04	0.06	0.08	0.10	0.12

Table 3.5

a) Draw a circuit diagram to show the experimental set-up used by the students to obtain the results in Table 3.5.

b) Plot a graph of these results and use it to determine the following:

(i) The EMF of the battery

(ii) The internal resistance of the cell

c) Calculate the short circuit current for this battery.

3 Students set up the experiment shown in Figure 3.86 to measure the internal resistance of a D-cell battery.

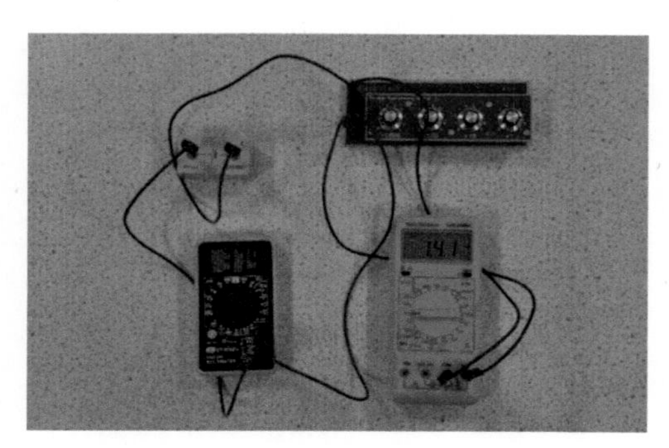

Fig 3.86

(continued)

The results from the experiment produce the following graph:

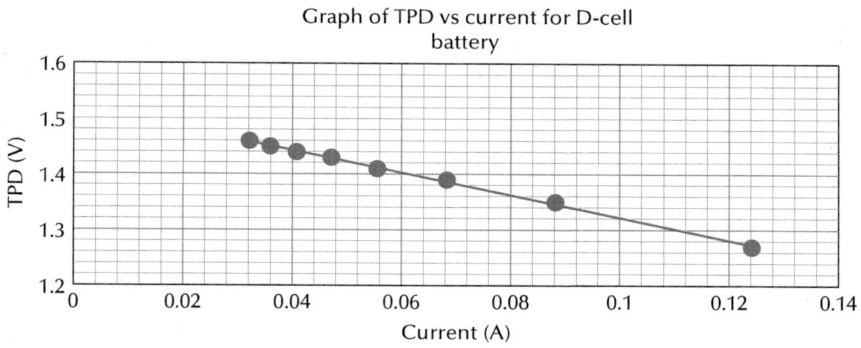

Graph of TPD vs current for D-cell battery

TPD (V)

Current (A)

Fig 3.87

The equation for the line was calculated using spreadsheet software on a computer to be

$$TPD = -2.02I + 1.52$$

a) Use the graph (and the equation given) to find the following:

(i) The EMF of the battery.

(ii) The internal resistance of the battery.

b) Calculate the short circuit current for the battery.

c) The battery is connected to a lamp that has a resistance of 6 Ω.

(i) Calculate the TPD across the lamp when connected to this battery.

(ii) Calculate the lost volts in the circuit.

(iii) Calculate the power dissipated across the lamp.

3.2.4 Load matching

The combination of internal resistance and load resistance forms a simple series circuit. This is shown in Figure 3.88.

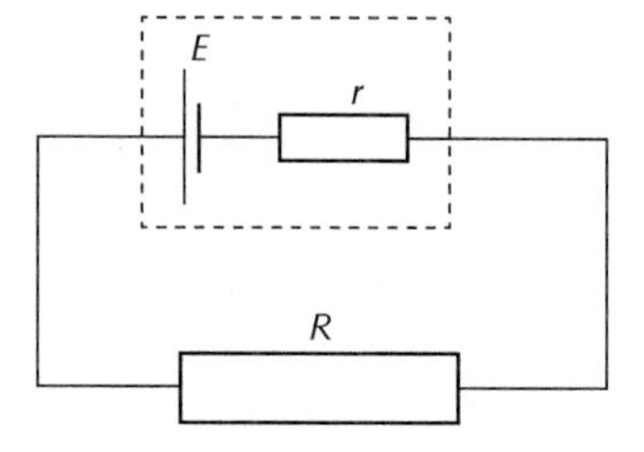

Fig 3.88

There is an optimum load resistance to get the maximum power transfer from the source. Consider having a source with an internal resistance of 20 Ω and EMF of 10 V, and then varying the load resistance. The power transferred to the load resistor will be given by the power equation:

$$P = I^2R$$

where the current is given by Ohm's law:

$$I = \frac{E}{R + r}$$

Changing the value of load resistance gives theoretical results shown in Table 3.6, calculated using the equations supplied.

Load resistance, R (Ω)	Current, I (A)	Power transferred, P (W)
0	0.5	0
5	0.4	0.8
10	0.33	1.11
15	0.29	1.22
20	0.25	1.25
25	0.22	1.23
30	0.2	1.2
35	0.18	1.16
40	0.17	1.11
45	0.15	1.07
50	0.14	1.02

Table 3.6

Plotting these results on a graph of load resistance vs power transferred gives the following:

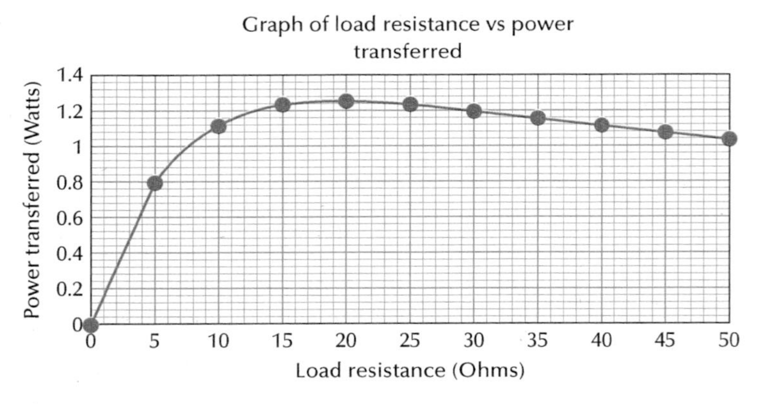

Fig 3.89

Key point

The maximum power is transferred to the load resistance when the load resistance is equal to the internal resistance of the source.

The graph shows that the maximum power transferred is when the load resistance is equal to the internal resistance of the source. This is known as load matching and is an important consideration when choosing electrical sources and designing circuits.

4 Capacitors

You should already know (National 5):

- A capacitor is a device that stores electric charge
- Capacitors can be used in time delay circuits

Learning intentions

- Definition of capacitance, linking the charge stores to the potential difference across a capacitor
- Total energy stored by a capacitor given by area under a potential vs charge graph
- Carry out calculations using relationships linking charge, capacitance, voltage and energy stored
- Analysis of charging and discharging graphs for both current and voltage, and how changing the resistance in the circuit affects this

4.1 Defining capacitance

Capacitors are components that are used in circuits to store electric charge and energy. At National 5 level, a capacitor was used in a circuit requiring a time delay. An example of this would be the courtesy lights in a car, where the lights stay on for a set time after the door has been closed.

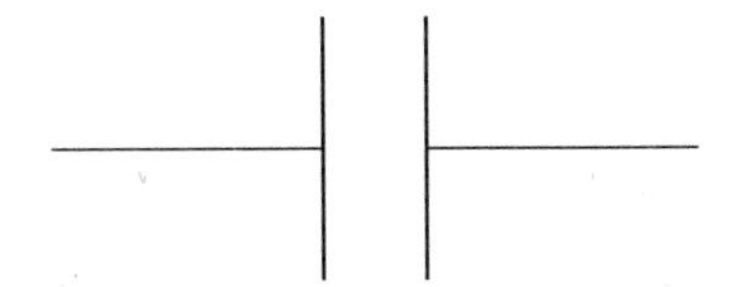

Fig 3.90

4.1.1 The capacitor

Capacitors store electric charge. The principle of the capacitor is based on there being two plates – one positively charged and the other negatively charged – which leads to energy being stored between the plates. They were initially designed as two parallel plates, which earned them the name 'parallel plate capacitors'. This gave rise to the circuit symbol used to represent a capacitor, which is shown in Figure 3.90.

One of the issues with a parallel plate capacitor is the amount of space required. This was overcome with an electrolytic design that looks like a Swiss-roll cut open. An electrolytic capacitor uses an electrolyte to replace one of its charged plates, and is able to occupy a much smaller volume for the same capacitance. A typical electrolytic capacitor design is shown in Figure 3.91.

Fig 3.91

4.1.2 Capacitance

The amount of charge stored by a capacitor depends on the voltage across the capacitor and also a property of the capacitor known as the capacitance, *C*. The greater the capacitance, the greater the charge stored for a given voltage.

Experiment 4.1.2 Determining capacitance

This experiment determines the capacitance of a capacitor and defines capacitance in terms of voltage and charge stored by a capacitor.

Apparatus

- Battery or DC power supply
- Capacitor
- Coulomb meter
- Voltmeter
- Resistor

Instructions

1. Build the circuit shown in the circuit diagram (Figure 3.92).
   The circuit consists of a DC power supply connected in series with the capacitor and a voltmeter in parallel to measure the voltage across the capacitor. A switch is used to switch between this and a second series circuit, which consists of a resistor and coulomb meter to measure the charge stored by the capacitor.
2. Close the switch to 'A' to fully charge the capacitor – the voltage across the capacitor when fully charged will be equal to the power supply voltage. Use the voltmeter to measure the voltage across the capacitor. Record the voltage used.

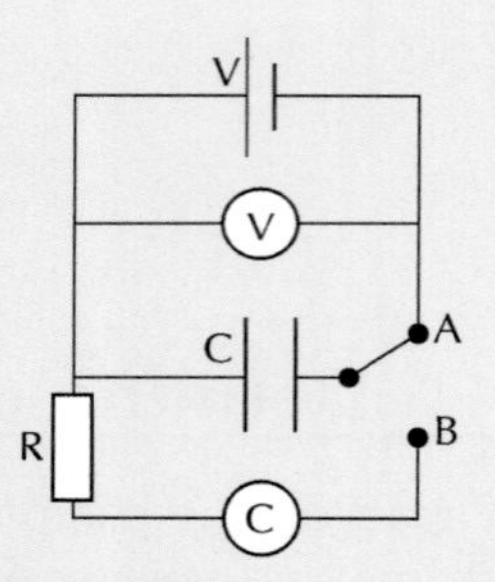

Fig 3.92

3. Once fully charged, move the switch to 'B' to discharge the capacitor through the resistor and coulomb meter. Use the coulomb meter to measure the amount of charge stored by the capacitor.
4. Use a table similar to Table 3.7 to record your results.

Potential difference (V)	Charge stored (C)	Charge ÷ voltage (CV^{-1})

Table 3.7

5. Repeat the experiment with different power supply voltages, recording your results.
6. Comment on the value of charge ÷ voltage for each of the voltages you have measured.
7. Plot a graph of voltage against charge stored.
8. Repeat the experiment with a capacitor with a different capacitance. Compare your results to those achieved for the first capacitor.

Defining capacitance

The capacitance of a capacitor, C, is defined as the charge stored per unit volt,

$$C = \frac{Q}{V}$$

Capacitance is measured in farads (F). The farad is a very big unit! Most capacitors have a capacitance measured in micro-farads (μF), or nano-farads (nF).

The greater the capacitance, the greater the charge stored per volt across the capacitor. If the voltage across a capacitor is plotted against the charge stored, a straight-line graph as shown in Figure 3.93 would be produced (as in the previous experiment).

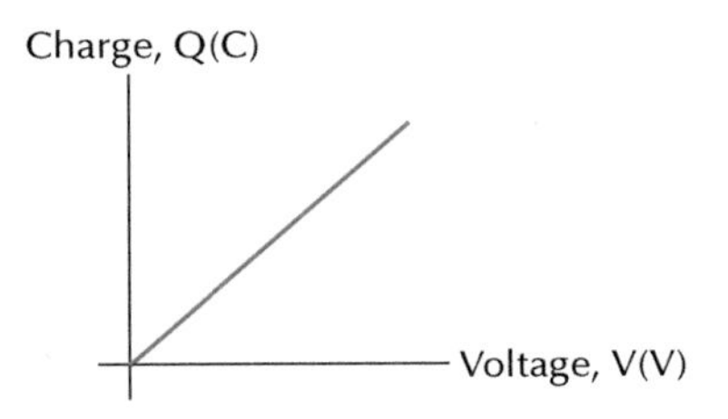

Fig 3.93

This shows that charge and voltage are directly proportional for a capacitor:

$$Q \propto V$$

This can be written as:

$$Q = kV$$

where k is a constant. Notice that this is equivalent to the equation for a straight line through the origin with Q on the y-axis and V on the x-axis. Here, the constant is the capacitance, thus:

$$Q = CV$$

which can be rearranged to give the definition for capacitance given above. The capacitance can therefore be obtained by finding the gradient of the line above.

Key point

The capacitance of a capacitor, C, is a measure of the charge stored, Q, per voltage, V, across the capacitor:

$$C = \frac{Q}{V}$$

Worked example

A capacitor can store 8 mC of charge when there is a voltage of 12 V across it. Calculate the capacitance of the capacitor.

The capacitance of the capacitor can be found by applying the definition of capacitance:

$$C = \frac{Q}{V}$$

Substituting what you know from the question and solving, you get:

$$C = \frac{8 \times 10^{-3}}{12}$$

$$C = 6.7 \times 10^{-4} F$$

$$C = 670\ \mu F$$

Fig 3.94

Exercise 4.1.2 Defining capacitance

1 A capacitor has a capacitance of 250 nF and is connected to a power supply with a voltage of 12 V.

a) What will the voltage across the capacitor be when it is fully charged?

b) Calculate the charge stored by the capacitor when it is fully charged.

c) The capacitor discharges in a time of 9s. Calculate the average discharge current.

2 A 400 μF capacitor stores 25 mC of charge. Calculate the voltage across the capacitor.

3 Students carry out an experiment to investigate the capacitance of a capacitor. They connect it to a DC power supply of 10 V and allow the capacitor to fully charge. The capacitor is then discharged through a coulomb meter, and is found to have stored 25 μC of charge.

a) How would the students be able to determine when the capacitor is fully charged?

b) Calculate the capacitance of the capacitor.

4 An engineer is designing a system that will supply power to a computer for a short time in the event of an electricity supply failure.

The power supply of the computer is rated at 20 V DC. The engineer calculates that at least 200 mC of charge needs to be stored for the computer.

a) Explain why a capacitor is an ideal component for use in this situation.

b) Calculate the minimum capacitance of capacitor required to meet the engineer's specification.

5 A capacitor can store 40 μC of charge with a voltage of 15 V across it.

a) Calculate the capacitance of the capacitor.

b) Calculate the charge stored by this capacitor if the voltage across it was 20 V.

c) What voltage would be required with this capacitor to store 190 μC of charge?

6 Students are investigating the properties of an electrolytic capacitor.

They connect it to a constant current source, which supplies a constant current of 50 mA to charge the capacitor. After a charging time of 30 s, the voltage across the capacitor is observed to be 15 V.

a) Calculate the charge stored by the capacitor after 30 s. (Hint – this is the charge supplied by the constant current of 50 mA.)

b) Now find the capacitance of the capacitor.

7 The capacitance of a capacitor is being determined by experiment. The capacitor is charged to a voltage of 12 ± 0.5 V by connecting it to a DC power supply. The capacitor is then discharged through a coulomb meter to measure the charge stored. The charge stored by this capacitor was measured five times for the same power supply voltage and the readings are shown below.

470 μC   485 μC   467 μC   471 μC   449 μC

a) Calculate the mean charge stored for the results above.

b) Calculate the approximate random uncertainty in charge stored.

c) Find the percentage uncertainty for both the charge stored and the voltage.

d) Use the results from this experiment to calculate the capacitance of the capacitor. Express your answer in the form *Capacitance ± Absolute Uncertainty*.

4.1.3 Charging and discharging a capacitor

When a capacitor is connected to a DC power supply, charge flows from the power supply to the plates of the capacitor.

Experiment 4.1.3a Charging a capacitor

This experiment investigates how the voltage across a capacitor changes as it is charged. The experiment also investigates how the charging current varies during the charging process.

Fig 3.95

Apparatus

- Capacitor
- DC power supply
- Wires
- Ammeter
- Voltmeter
- Fixed resistor (100 Ω)
- Switch
- Stopwatch

Instructions

1 Connect the circuit shown in the diagram and photograph in Figures 3.96 and 3.101.

Ensure the capacitor is fully discharged by checking that the voltage across it is zero volts. The capacitor can be discharged by connecting a wire to both terminals if required.

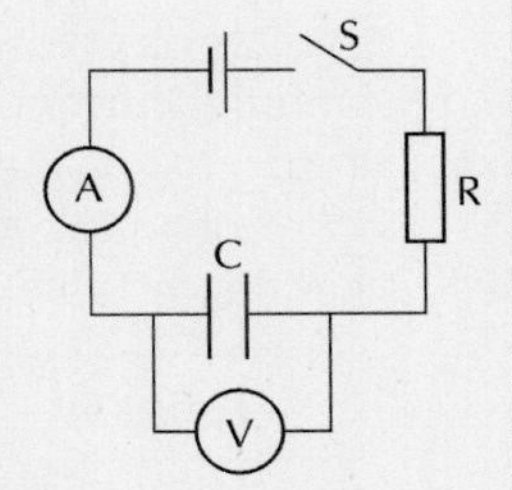

Fig 3.96

2 Close the switch and start the stopwatch. Record the current in the circuit and the voltage across the capacitor at either 10- or 30-s intervals depending on the rate at which the capacitor charges (this will depend on the capacitance of the capacitor – see later experiments). Use a table similar to Table 3.8 to record your results.

Time (s)	Voltage (V)	Charging current (A)
0		
10		
20		
30		
40		
50		
60		

Table 3.8

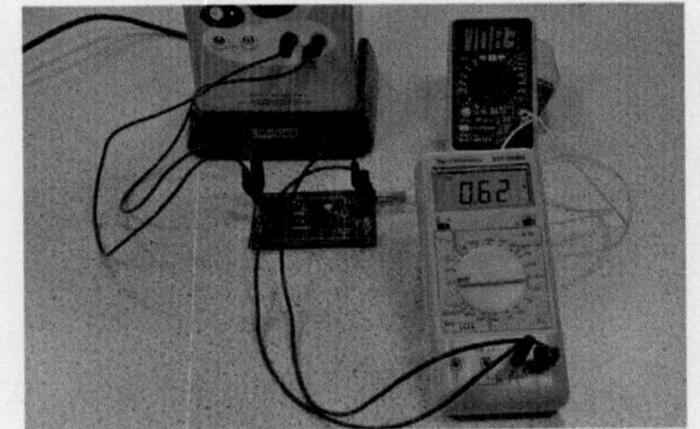

Fig 3.97

3 Plot a graph of the charging current vs time for the capacitor charging.

4 Plot a graph of the voltage across the capacitor versus time for the capacitor charging.

Experiment 4.1.3b Discharging a capacitor

This experiment investigates how the voltage across a capacitor changes as it discharges. The experiment also investigates how the discharging current varies during the discharging process.

Apparatus

- Capacitor
- DC power supply
- Wires
- Ammeter
- Voltmeter
- Fixed resistor (100 Ω)
- Switch
- Stopwatch

Instructions

1. Using the circuit shown in Experiment 4.1.3a, fully charge the capacitor.
2. Connect the fully charged capacitor into the circuit shown in Figure 3.98 (or simply flick the switch to discharge using the set-up in Figure 3.99).
3. Close the switch and start the stopwatch to record the time since the switch was closed. Record the values of the voltage across the capacitor and the current in the circuit using the voltmeter and ammeter, respectively. Depending on the capacitance of the capacitor, record the current and voltage at 10- or 30-s intervals. Record your results in a table similar to Table 3.9.

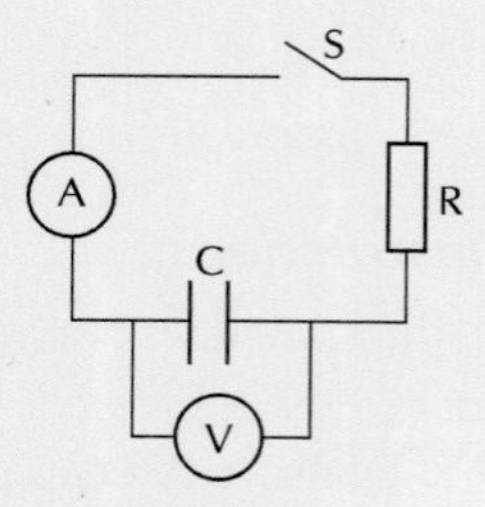

Fig 3.98

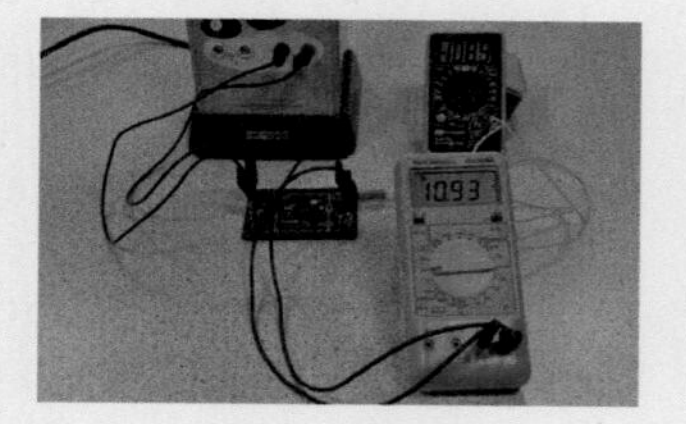

Fig 3.99

Time (s)	Voltage (V)	Discharging current (A)
0		
10		
20		
30		
40		
50		
60		

Table 3.9

4. Plot a graph of the discharging current versus time for the capacitor discharging.
5. Plot a graph of the voltage across the capacitor vs time for the capacitor discharging.

Charging a capacitor

When a capacitor is connected to a DC power supply, current will flow from the supply to the plates of the capacitor. Charge is stored by the capacitor as it builds up on the plates of the capacitor. Consider the circuit shown in Figure 3.100.

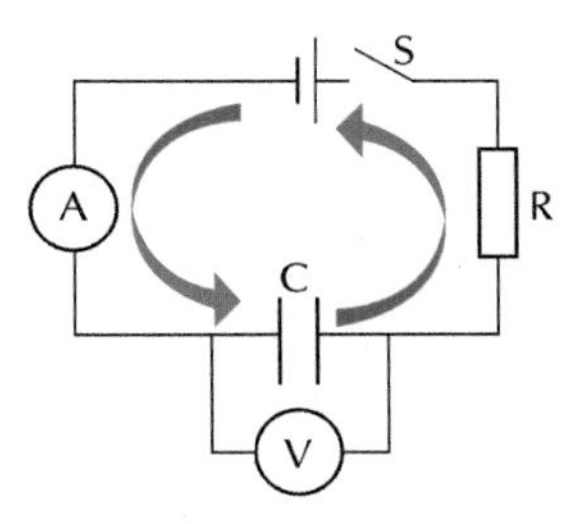

Fig 3.100

When the switch is closed, electrons will flow from the negative terminal of the power supply to the plate of the capacitor in the direction of the arrows shown. This will drive electrons off the other plate of the capacitor back to the positive terminal of the power supply, again in the direction of the arrow shown. This has the result of giving the plate of the capacitor an overall positive charge.

When negative charge is stored on the plate of the capacitor, it resists the flow of additional negative charge towards the plate. The more charge that is stored, the greater the resistance. This results in the current within the circuit reducing as the capacitor charges. Once there is sufficient charge on the plate, it will resist further charge flowing; at this stage, the capacitor is fully charged. This results in the current during the charging process varying as shown on the graph in Figure 3.101.

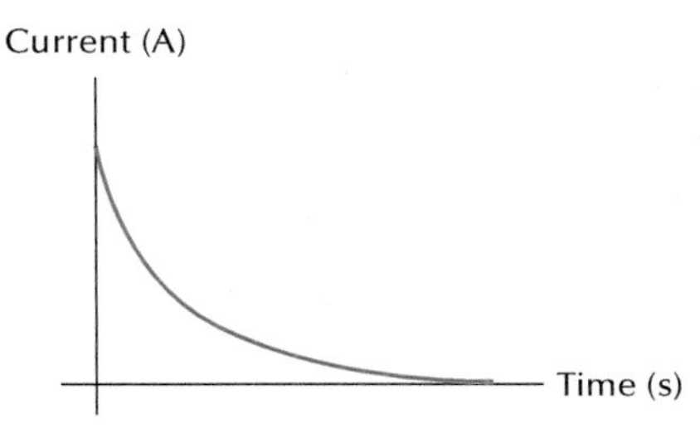

Fig 3.101

The initial charging current depends on the resistance of the resistor in the circuit, and is given by Ohm's law:

$$I = \frac{V}{R}$$

where V is the power supply voltage and R is the resistance of the resistor.

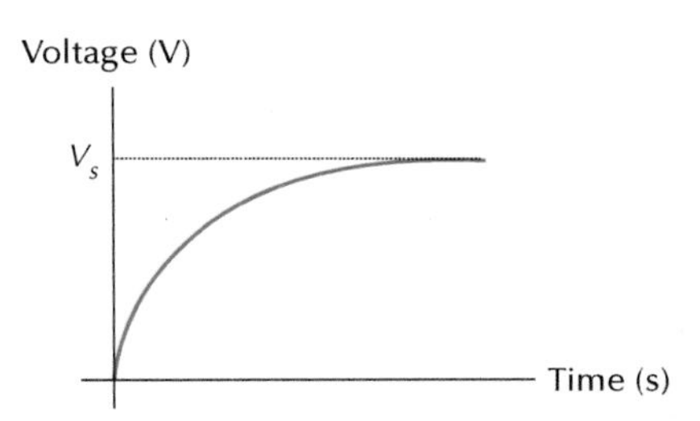

Fig 3.102

As the capacitor charges, the voltage across it increases. This reduces the potential difference between the supply and the capacitor, which reduces the current flowing. The amount of charge stored rises quickly during the initial stages of the charging process, when the current is large and it is easy to move charge onto the plates. As the capacitor becomes charged, the current drops, and the rate of increase of charge stored decreases. Eventually the capacitor is fully charged and no further transfer of charge takes place. This occurs when the voltage across the

Key point

When a capacitor charges, the current is high at first and decays to zero. The voltage across the capacitor starts at zero and increases quickly at first and then slowly until it reaches the power supply voltage. This results in the following current and voltage graphs for the charging of a capacitor:

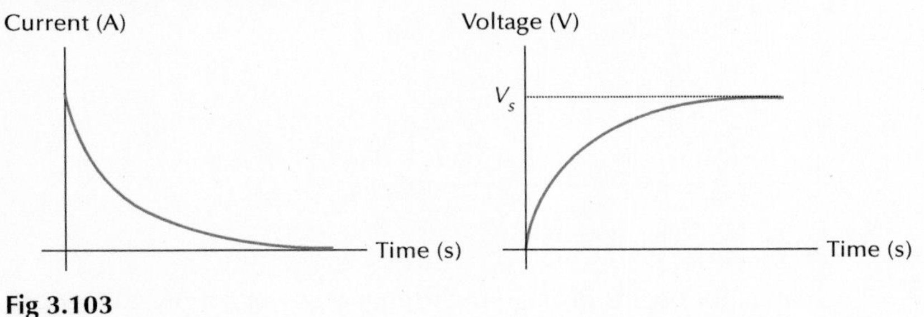

Fig 3.103

capacitor is equal to the supply voltage – the potential difference is zero, hence no current flows. This results in the graph shown in Figure 3.102 for the voltage against time for charging a capacitor.

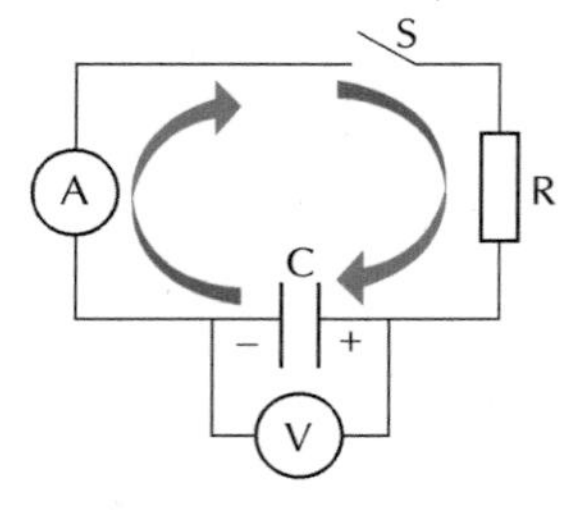

Fig 3.104

Discharging a capacitor

When a charged capacitor is connected to a circuit, it will discharge through the components in the circuit until there is no longer any charge stored. Consider a charged capacitor in the circuit shown in Figure 3.104.

When the switch is closed, electrons stored on the negative plate of the capacitor will flow round the circuit, creating a current. They will carry energy with them (see next section), which will dissipate in the resistor as heat. As the capacitor discharges, the current will decrease until the capacitor is fully discharged. The discharge current will be as follows:

> **Key point**
>
> The initial charging current when a capacitor is charging is given by Ohm's law:
>
> $$I = \frac{V_S}{R}$$
>
> Fig 3.105

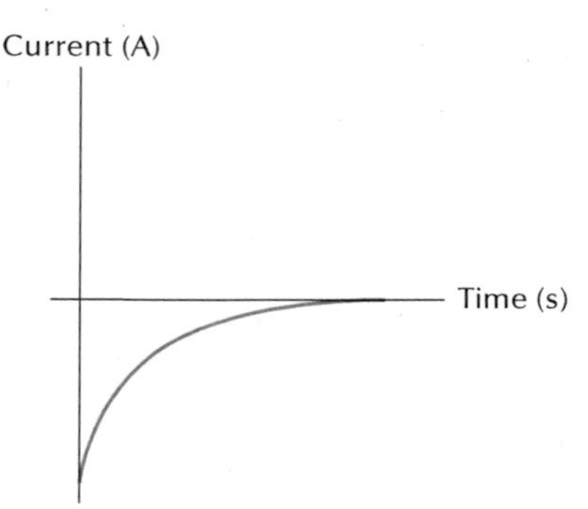

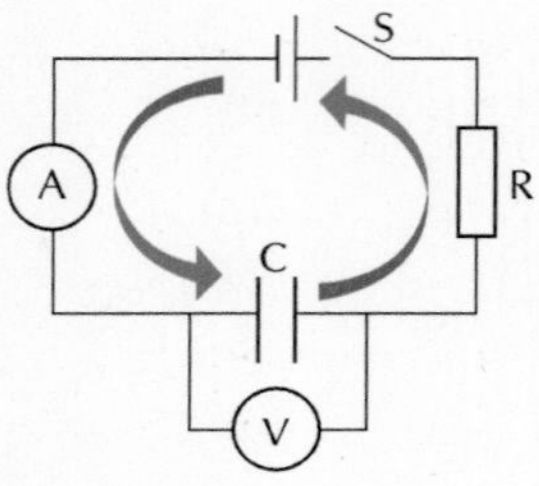

Fig 3.106

The initial discharge current will depend on the resistance of the resistor in the circuit and the initial voltage across the capacitor, which is the power supply voltage:

$$I = \frac{V_S}{R}$$

Notice that the discharging current here is negative. This represents the current in the opposite direction in the circuit, compared to when the capacitor was being charged.

The voltage across the capacitor will decrease as the capacitor discharges, falling quickly at first, and then more slowly as the current in the circuit decreases. This leads to the following graph for the voltage across a capacitor as it discharges:

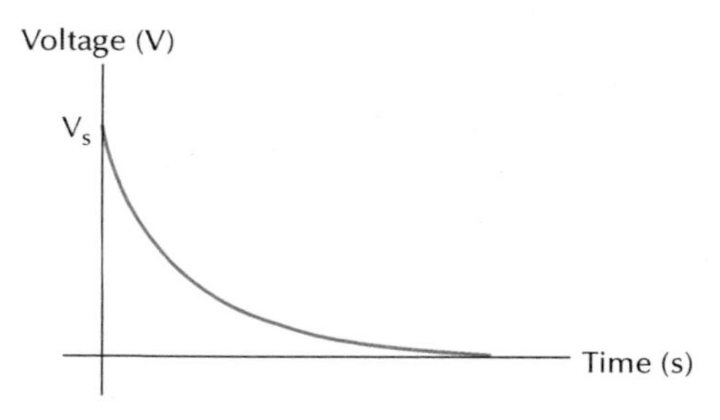

Fig 3.107

The capacitor begins discharging from the power supply voltage, V_S.

Key point

As a capacitor discharges, the current in the circuit starts high and decays to 0. The voltage across the capacitor starts high (at the power supply voltage) and decays to 0 when the capacitor is fully discharged. This leads to the following graphs for current and voltage when the capacitor is discharging:

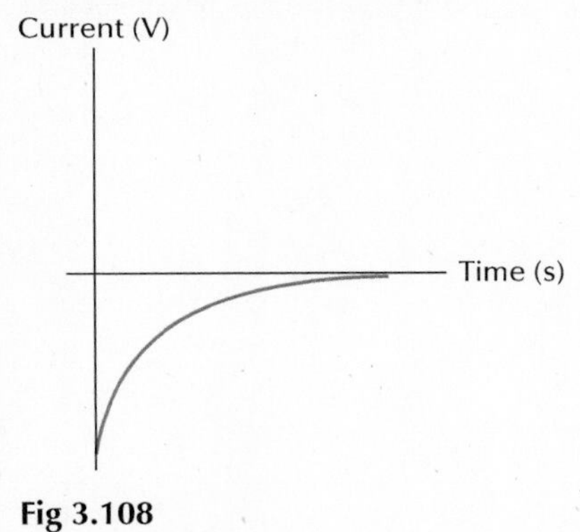

Fig 3.108

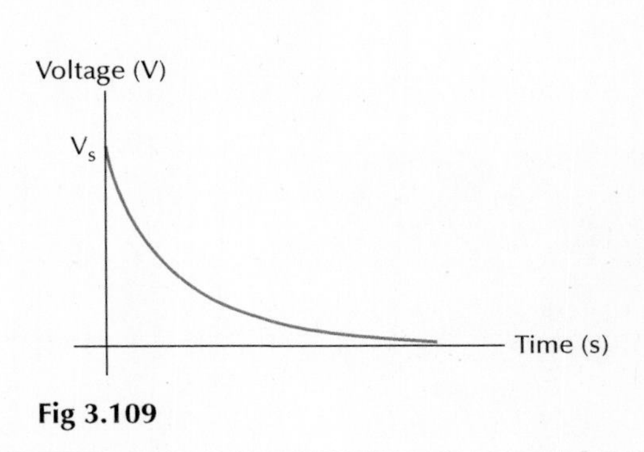

Fig 3.109

Effects of resistance and capacitance on charging and discharging

The above discussions show that the initial charging and discharging current depend on the resistance in the circuit according to Ohm's law:

$$I = \frac{V_S}{R}$$

and that the greater the capacitance of a capacitor, the greater the amount of charge stored for a given voltage. This comes from the definition of capacitance:

$$C = \frac{Q}{V}$$

which can be rearranged as follows to give the charge stored:

$$Q = VC$$

Changing the capacitance or resistance in the circuit will therefore have a significant effect on the charging and discharging times.

Key point

The initial discharge current depends on the power supply voltage used to charge the capacitor and the resistance through which the capacitor is dissipating its stored charge:

$$I = \frac{V_S}{R}$$

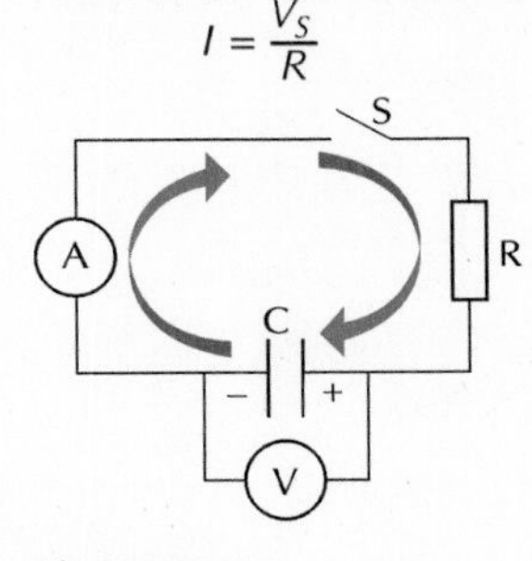

Fig 3.110

Experiment 4.1.3c Effects of resistance on the charging/discharging of a capacitor

This experiment investigates the effects of changing the resistance of the resistor in a charging circuit on the charging and discharging of a capacitor.

Apparatus

- Capacitor
- DC power supply
- Wires

- Ammeter
- Voltmeter
- Various resistors
- Switch
- Stopwatch

Instructions

1 Connect the circuit shown in Figures 3.111 and 3.112:

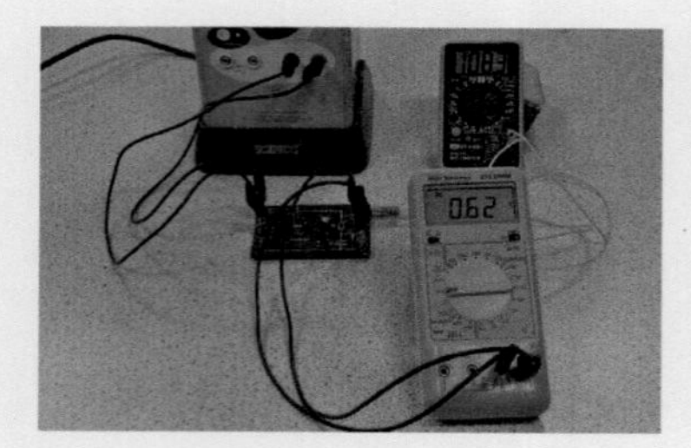

Fig 3.111

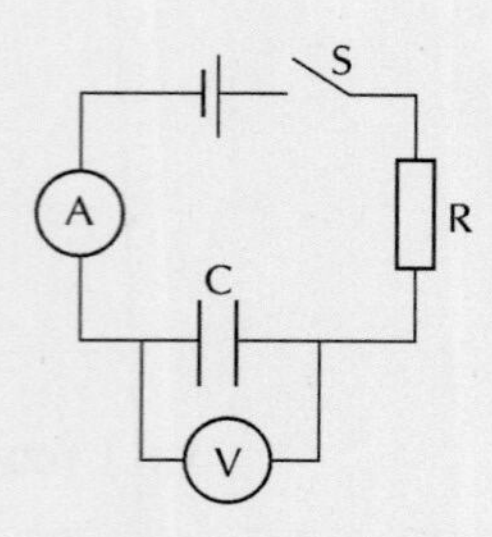

Fig 3.112

2 Ensure the capacitor is fully discharged by checking that the voltage across it is 0 volts. The capacitor can be discharged by connecting a wire to both terminals if required.

3 Close the switch and start the stopwatch. Record the current in the circuit and the voltage across the capacitor at 10-s intervals. Use a table similar to Table 3.10 to record your results.

Time (s)	Voltage (V)	Charging current (A)
0		
10		
20		
30		
40		
50		
60		

Table 3.10

4 Repeat steps 2 and 3 with the same capacitor (discharging it first), but with a different resistor in the circuit.

5 Plot a graph of the charging current vs time for the capacitor charging for the different resistors – you can use the same graph for all, carefully labelling which resistor is which.

6 Plot a graph of the voltage across the capacitor versus time for the capacitor charging. As above, the same graph can be used for all resistance values as long as the labelling is clear.

Experiment 4.1.3d Effects of capacitance on the charging/discharging of a capacitor

This experiment investigates the effects of changing the capacitance of the capacitor in a charging circuit on the charging and discharging of the capacitor.

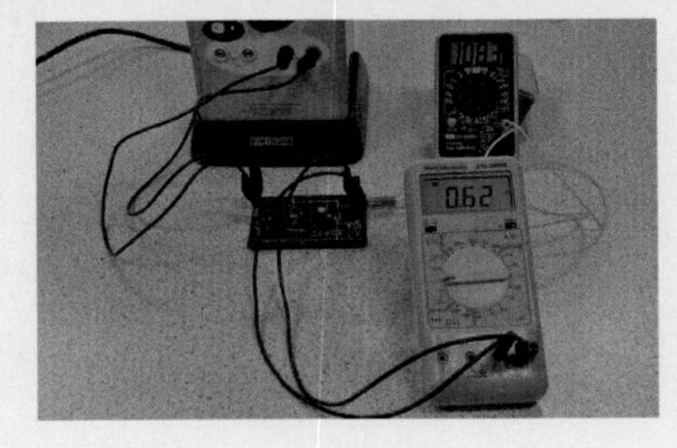

Fig 3.113

Apparatus

- Various capacitors
- DC power supply
- Wires
- Ammeter
- Voltmeter
- Resistor
- Switch
- Stopwatch

Instructions

1 Connect the circuit shown in Figures 3.114 and 3.115.

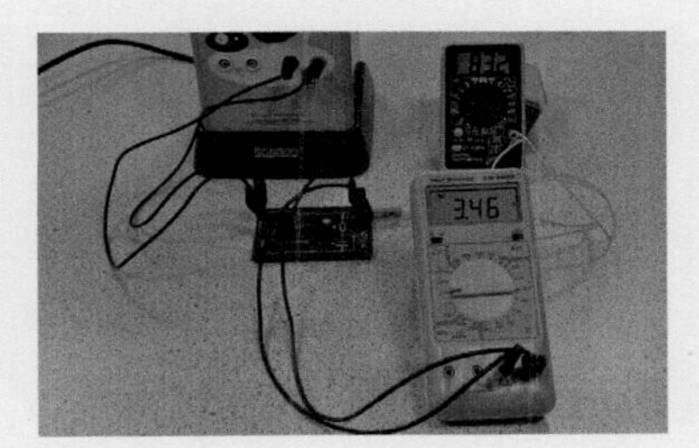

Fig 3.114

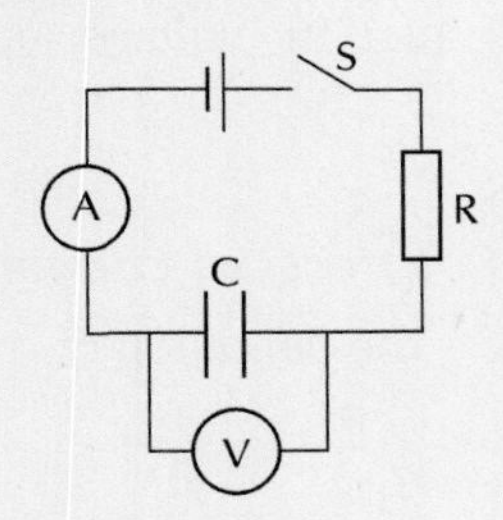

Fig 3.115

2 Ensure the capacitor is fully discharged by checking that the voltage across it is 0 volts. The capacitor can be discharged by connecting a wire to both terminals if required.

3 Close the switch and start the stopwatch. Record the current in the circuit and the voltage across the capacitor at 10-second intervals. Use a table similar to Table 3.11 to record your results.

4 Repeat steps 2 and 3 with a different capacitor (discharging it first), but with the same resistor in the circuit.

5 Plot a graph of the charging current vs time for the capacitor charging for the different capacitors – you can use the same graph for all, carefully labelling which resistor is which.

6 Plot a graph of the voltage across the capacitor vs time for the capacitor charging. As above, the same graph can be used for all resistance values as long as the labelling is clear.

Time (s)	Voltage (V)	Charging current (A)
0		
10		
20		
30		
40		
50		
60		

Table 3.11

Effects of resistance on capacitor charging

Changing the resistance of the resistor in a capacitor charging circuit will change the initial charging current and the length of time taken for the capacitor to fully charge. The greater the resistance, the smaller the initial charging current and the longer the time taken to charge. A capacitor will store a given amount of charge, but a smaller initial charging current will increase the time taken for its charge to flow to the capacitor. This is shown in the graphs in Figure 3.116.

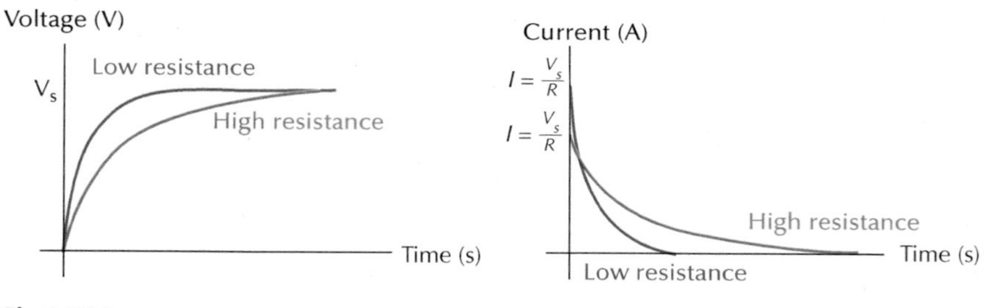

Fig 3.116

> **Key point**
>
> The resistance and capacitance of the circuit will affect the length of time taken for the capacitor to charge as follows:
>
> - Increase resistance → increase charging time (because initial charging current is lower).
> - Increase capacitance → increase charging time (because the capacitor stores more charge).

Remember that the initial charging current depends on the resistance and the voltage supply and is given by Ohm's law,

$$I = \frac{V_S}{R}$$

Effects of capacitance on capacitor charging

The definition of capacitance:

$$C = \frac{Q}{V}$$

shows that for a given supply voltage, the greater the capacitance, the greater the amount of charge the capacitor stores. If the resistance of the circuit is fixed, then the initial charging current will be the same. However, if a capacitor with a greater capacitance is used, it will take longer to charge because it stores more charge. This leads to the following current and voltage graphs for capacitors with a different capacitance:

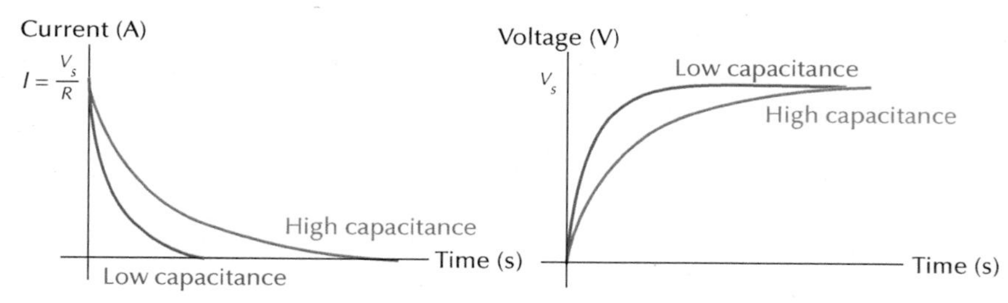

Fig 3.117

In summary, increasing either the resistance or the capacitance will result in an increase in the charging time.

Exercise 4.1.3 Charging and discharging a capacitor

1 Students wish to investigate the voltage and current across a capacitor as it charges. They have a capacitor, resistor, switch, ammeter, voltmeter and power supply. Draw a circuit diagram of the circuit they should set up in order to investigate the charging current and voltage.

2 Pupils have set up a circuit to measure the charging current and voltage across a capacitor as it charges, and have plotted the graphs shown in Figure 3.118.

The pupils have forgotten to include the axis labels on the graphs. Copy the graphs and include the axis labels (including units).

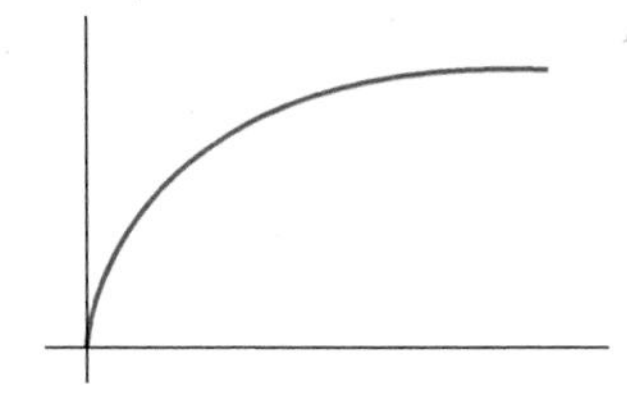

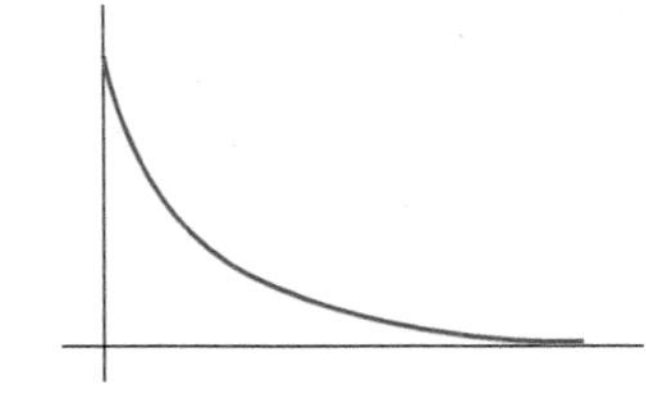

Fig 3.118

3 A 200 μF capacitor is connected in series with a 12 V power supply and a 5 kΩ resistor as shown in the circuit diagram in Figure 3.119.

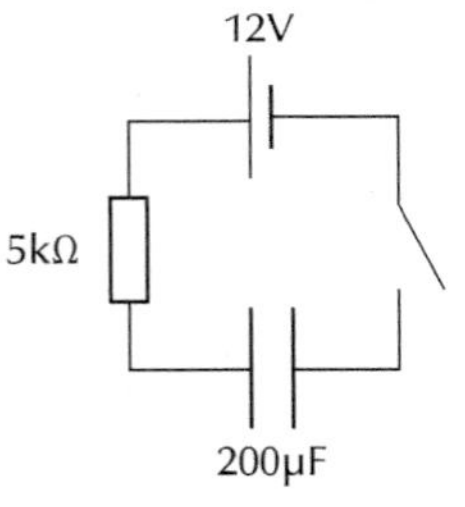

Fig 3.119

a) Calculate the initial charging current for the capacitor.

b) Calculate the charge stored in the capacitor when it is fully charged.

c) Plot a graph to show the current against time as the capacitor is charged. Include values on the current axis.

d) The capacitor is discharged. The resistor is replaced with a 10 kΩ resistor. How does the resistor change affect:

(i) The time taken to fully charge the capacitor?

(ii) The charge stored when the capacitor is fully charged?

e) Plot a second graph of charging current against time to show the effect of changing the resistance on the charging current.

4 A 1000 μF capacitor is connected to a 35 V DC power supply and allowed to fully charge. The capacitor is connected in series with the power supply and a 500 Ω resistor.

Fig 3.120

a) Calculate the initial charging current.

b) Plot a graph to show the charging current against time during the charging process.

c) Plot a graph to show the voltage across the capacitor against time during the charging process.

d) The capacitor is replaced with one of lower capacitance. How does this affect:

(i) The initial charging current?

(ii) The time taken to fully charge the capacitor?

4.2 Energy storage

Capacitors store electric charge. This means that capacitors store electrical energy. This energy can be released into a circuit and dissipated through components such as lamps and motors in the same way as energy from a battery.

4.2.1 Graphical definition of stored energy

If you charge a capacitor, it will store energy. A camera flash is a good illustration of this – energy is stored by a capacitor and when the switch to take the photograph is pressed, this energy is transformed rapidly into light energy by the lamp to give a bright flash.

Fig 2.121

The energy stored in the capacitor comes from the work done charging the capacitor. Consider the capacitor in the circuit shown in Figure 3.122.

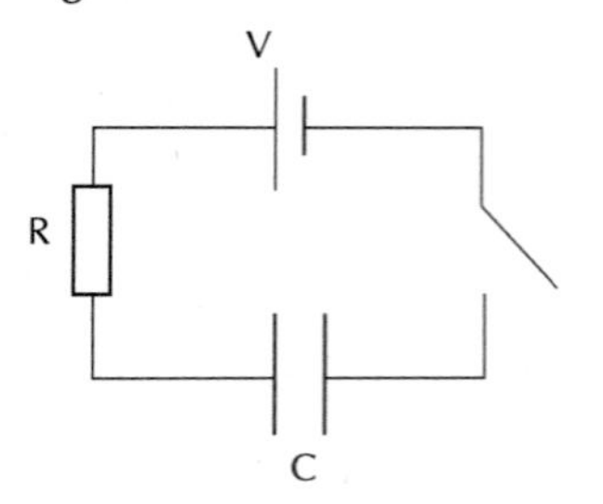

Fig 3.122

When the switch is closed, an electron will flow from the negative terminal of the power supply to the negative plate of the capacitor. This drives an electron off the positive plate to make it positively charged (see Figure 3.123).

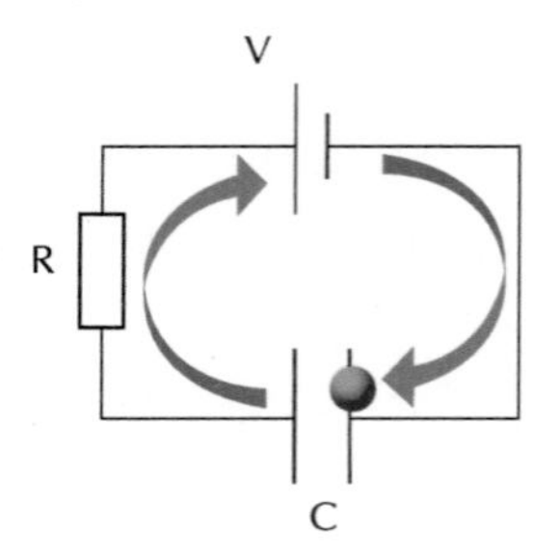

Fig 3.123

In order to move another electron to the negative plate of the capacitor, you have to do work against the charge already there because like charges repel. This work is done by the power supply:

$$W = QV$$

As the charge on the capacitor builds, the work required to move more charge to the capacitor increases. The voltage across the capacitor increases, as seen above. As the voltage is changing during the charging process, the above equation for work done will not work for the total energy stored by the capacitor. Instead, the energy stored will be given by the area under a charge vs voltage graph:

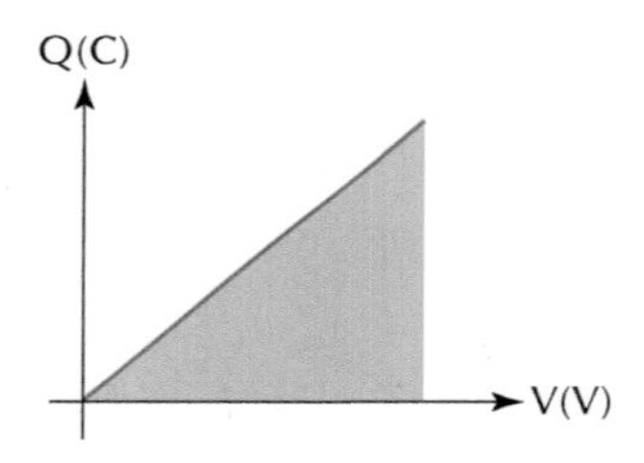

Fig 3.124

The area under this graph is given by:

$$Area = \frac{1}{2}QV$$

This leads to the energy stored by a capacitor being given by:

$$E = \frac{1}{2}QV$$

where Q is the charge stored and V is the voltage across the capacitor.

Key point

The energy stored by a capacitor that stores a charge, Q, when a potential difference, V, is across it is given by:

$$E = \frac{1}{2}QV$$

Experiment 4.2.1 Energy stored by a capacitor

This experiment investigates how the charge stored by a capacitor changes as the voltage across it increases. The relationship is used to find the energy stored by a capacitor.

Fig 3.125

Apparatus

- Constant current power supply
- Capacitor
- Stopwatch
- DC voltmeter

Instructions

1. Ensure the capacitor is fully discharged by checking the voltage across its terminals is zero. Note down the capacitance of the capacitor.
2. Connect the capacitor to a constant current supply and set the current to the desired charging current.
3. Connect a DC voltmeter across the capacitor to measure the voltage across it during the charging process. The set-up is shown in Figure 3.126.
4. Close the switch and allow the capacitor to charge. Use a stopwatch and the voltmeter to record the voltage across the capacitor at 10-s intervals during the charging process.
5. Record the voltage at 10-s intervals in a table similar to Table 3.12. The charge stored can be calculated using the equation:

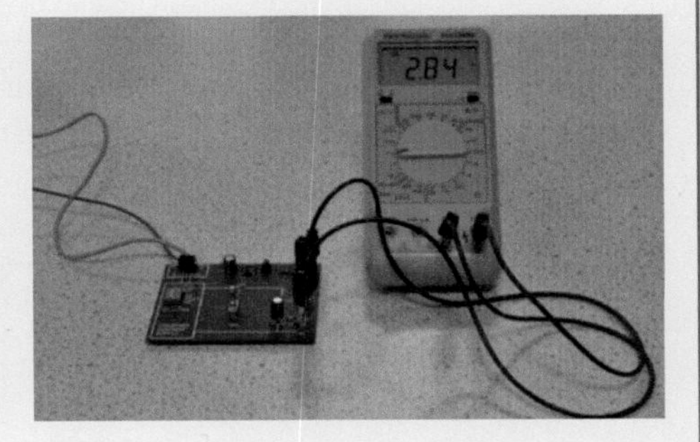

Fig 3.126

$$Q = It$$

where I is the current and t is the length of time.

Time (s)	Charge (Q = It) (C)	Voltage (V)
10		
20		
30		
40		
50		
60		

Table 3.12

The charging time and recording intervals can be adjusted to suit the particular charging current and capacitor chosen.

6. Plot a graph of charge stored against the voltage across the capacitor.
7. Use the graph above to calculate the energy stored by finding the area under a charge vs voltage graph.

4.2.2 Energy equations

The energy stored by a capacitor that stores a charge, Q, when a voltage, V, is across the plates is given by:

$$E = \frac{1}{2}QV$$

Using the definition for capacitance, two further equations for capacitance can be derived. Capacitance is defined as:

$$C = \frac{Q}{V}$$

Rearranging this for the charge stored gives:

$$Q = CV$$

This can be substituted into the equation for the energy stored by the capacitor to give:

$$E = \frac{1}{2}QV$$

$$E = \frac{1}{2}(CV)V$$

$$E = \frac{1}{2}CV^2$$

The definition of capacitance can also be rearranged for voltage, as follows:

$$V = \frac{Q}{C}$$

This can be substituted into the equation for the energy stored by the capacitor to give:

$$E = \frac{1}{2}QV$$

$$E = \frac{1}{2}Q\left(\frac{Q}{C}\right)$$

$$E = \frac{1}{2}\frac{Q^2}{C}$$

The choice of equation will depend on the variables that you have measured or have been given in the problem.

Key point

The energy stored by a capacitor is given by:

$E = \frac{1}{2}QV \quad E = \frac{1}{2}CV^2 \quad E = \frac{1}{2}Q^2$

Worked example

A capacitor is used to store energy in a desktop computer. In the event of power failure, the capacitor will discharge and keep the computer running for long enough to shut down properly. If the capacitor has a capacitance of 10 μF and charges to a voltage of 20 V, calculate the energy stored.

The voltage and capacitance are given in this problem, so the appropriate energy equation is:

$$E = \frac{1}{2}CV^2$$

Substitute the values given and solve for the energy. Take care to substitute the prefixes (e.g. μ) correctly:

$$E = \frac{1}{2}(10 \times 10^{-6})(20)^2$$

$$E = \frac{1}{2}(10 \times 10^{-6})(400)$$

$$E = 0.002\ J$$

Fig 3.127

Worked example

The lamp for a camera flash releases 4 J of energy in a time of 0.1 s. The energy is stored on a capacitor that is charged to a voltage of 12 V.

a) Calculate the power of the lamp.

b) Calculate the capacitance of the capacitor required to store energy for the lamp.

Fig 3.128

a) *The power of the lamp can be calculated using the definition of power:*

$$P = \frac{E}{t}$$

The energy is 4 J and it is transformed to light in a time of 0.1 s, so the power is:

$$P = \frac{4}{0.1}$$

$$P = 40\ W$$

b) *To find the capacitance of the capacitor, choose the appropriate energy equation for the information given. Voltage and energy are given, so choose:*

$$E = \frac{1}{2}CV^2$$

Substitute the known values and solve for the capacitance:

$$4 = \frac{1}{2}C(12)^2$$

$$4 = \frac{1}{2}(144)C$$

$$4 = 72C$$

$$C = \frac{4}{72}$$

$$C = 0.06\ F$$

Exercise 4.2.2 Energy storage by a capacitor

1 A 250 nF capacitor is connected to a 9 V DC power supply and allowed to fully charge.

a) Calculate the charge stored when the capacitor is fully charged.

b) Calculate the energy stored by the capacitor when fully charged.

2 A 70 μF capacitor stores 9 mC of charge.

a) Calculate the voltage across the capacitor when it is charged.

b) Calculate the energy stored by the capacitor.

3 An electrical engineer is choosing a capacitor for use in an energy storage circuit. She needs to store 15 mJ of energy when a voltage of 15 V is across the plates. Calculate the capacitance of the capacitor that the engineer should choose.

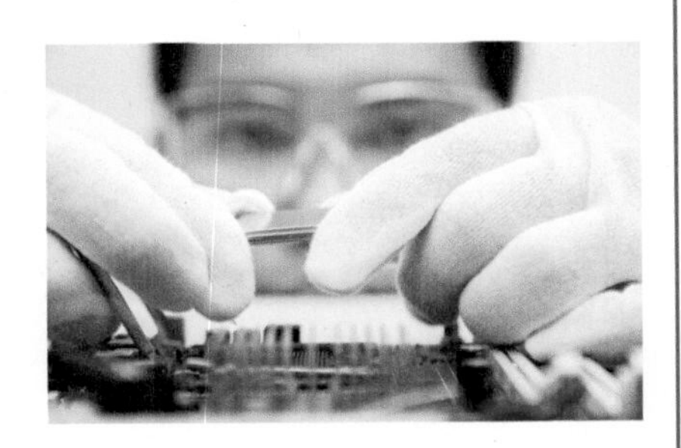

Fig 3.129

4 A solar buggy uses a solar cell to charge a capacitor. The capacitor stores electrical energy, which is then converted to kinetic energy to allow the buggy to move.

The capacitor fitted has a capacitance of 2.5 mF and charges until the voltage across it is 8 V.

a) Calculate the charge stored by the capacitor.

b) Calculate the energy stored by the capacitor.

c) The capacitor discharges through the motors in a time of 35 s. Find:

(i) The average discharge current.

(ii) The average power of the motor.

Fig 3.130

5 A computer's memory chips require power in the event of electricity supply failure, to ensure that they can be backed up before the computer switches off. The memory chips have to be kept supplied with electricity for 25 s to ensure they can fully save all data. They have a power consumption of 45 mW.

a) Calculate the electrical energy that must be stored in order to keep the memory chips supplied with electricity for 25 s. State one assumption you have made in this calculation.

b) A capacitor is used to store energy for the memory chips. The capacitor is charged to a voltage of 35 V.

(i) Calculate the capacitance of the capacitor required.

(ii) Calculate the charge stored by the capacitor.

(iii) Find the average discharge current through the memory chip.

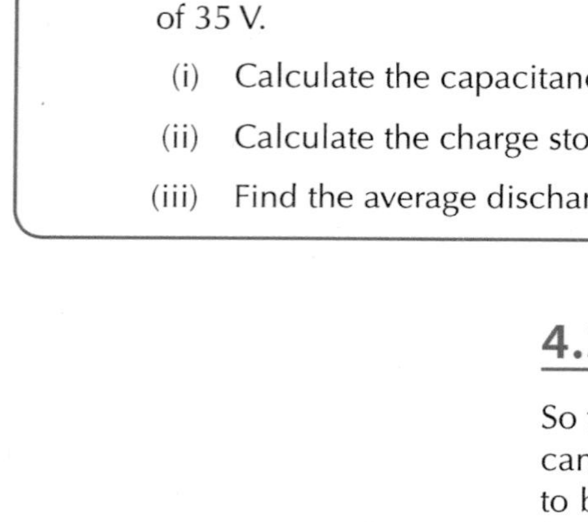

4.3 Capacitors in AC circuits

So far capacitors have been considered in DC circuits. However, they can also be applied to AC circuits, where their properties allow them to be used in a variety of different applications, including smoothing for DC power supplies, as well as part of filtering circuits.

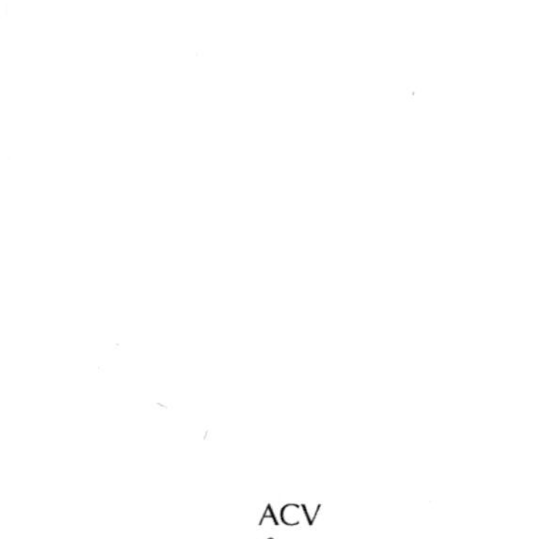

4.3.1 Frequency dependence

When connected to a DC power supply, a capacitor will fully charge up, and once fully charged the current in the circuit drops to zero. This is because it is no longer possible for additional charge to be pushed towards the plates of the capacitor; current stops.

Consider connecting the capacitor to an AC supply (Figure 3.131).

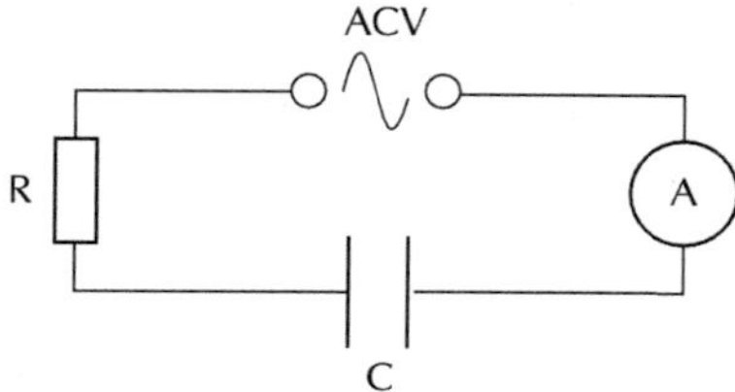

Fig 3.131

The current will begin flowing in one direction to charge the capacitor. When the current direction switches, it will discharge and then begin recharging the capacitor. If the frequency is high, the current never gets a chance to fall to a low value, so the average current will be large. For lower frequencies, the current can drop to lower values and thus the average current will be lower. This is shown on the graphs in Figures 3.132 and 3.133.

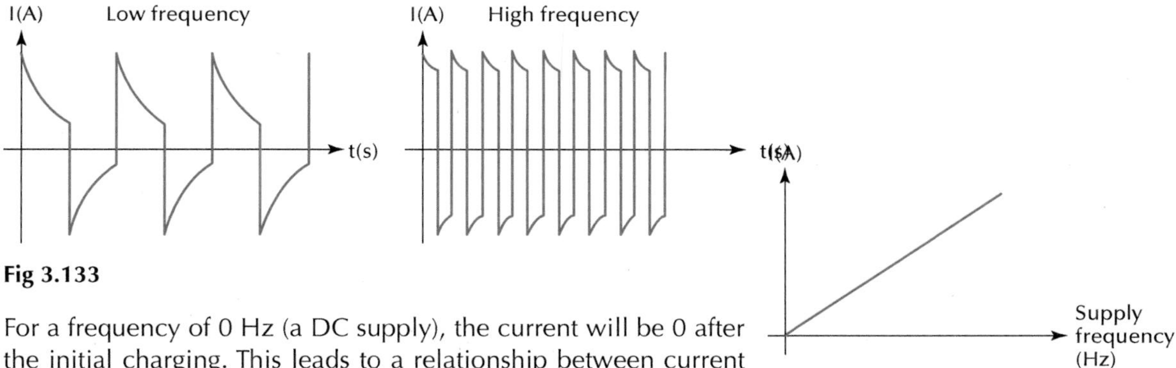

Fig 3.133

For a frequency of 0 Hz (a DC supply), the current will be 0 after the initial charging. This leads to a relationship between current and frequency shown in the graph in Figure 3.132.

Supply frequency (Hz)

Fig 3.132

Experiment 4.3.1 Determining the frequency response of a capacitor

This experiment investigates the frequency response of a capacitor by measuring the average current in a circuit as a function of the frequency of the supply voltage.

Apparatus

- AC voltage supply with variable frequency
- Capacitor
- Ammeter
- Resistor
- Oscilloscope

Instructions

1. Using the apparatus, set up the circuit shown in Figure 3.131, with the capacitor and resistor in series with the ammeter connected to a variable frequency AC supply.
2. Set the frequency of the supply to 10 Hz using an oscilloscope to check the frequency is accurate (do not rely on the dials of a signal generator). Measure the current in the circuit using an ammeter.
3. Change the frequency in the circuit and repeat the measurement above. Ensure the amplitude of the signal is kept constant by checking it using the oscilloscope.
4. Use a table similar to Table 3.13 to record your results.

Frequency (Hz)	Current (A)

Table 3.13

5. Plot a graph of the frequency of the supply against the current in the circuit.
6. Repeat the experiment using only a resistor in the circuit.
7. Write a short report that describes the experiment and the observed effect of supply frequency on the current in the circuit, when there is a capacitor and when there is only a resistor.

Key point

The supply frequency has no effect on the current in a circuit when there is only a resistor in series with the supply.

Key point

Capacitors act as a DC block. That is to say that they stop direct current from flowing. Once fully charged, the current in the circuit falls to zero and the capacitor has blocked the signal.

Key point

The greater the supply frequency, the greater the current in a circuit with a capacitor in series with the supply:

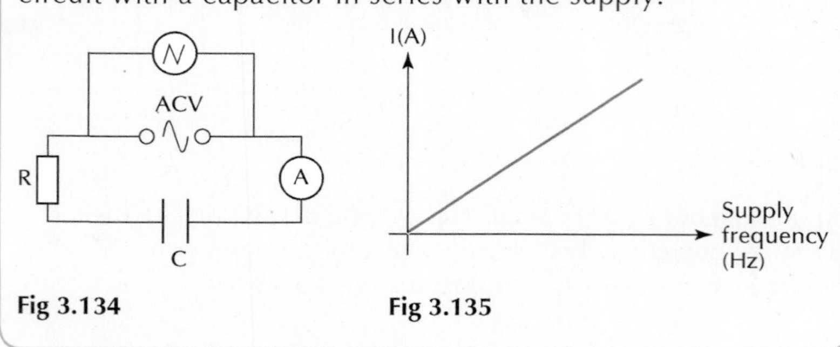

Fig 3.134

Fig 3.135

Fig 3.136

4.4 Practical applications

Capacitors have many applications in circuits. A computer system has many hundreds or thousands of capacitors in it, for example. They can also be used to supply a large amount of charge very quickly to a circuit; for example, when lighting a bright lamp for a short time (camera flash). They can also be used to smooth out noise to provide a 'clean' DC supply. Many of the bench power supplies in a lab use so-called smoothing capacitors for this purpose alongside a circuit known as a full wave rectifier. Two main applications of capacitors are discussed below.

4.4.1 Camera flash

If you have used an old film camera, you may have experienced waiting for the flash to become ready. During this time, there was often a high-pitched noise from the camera. The delay and noise were due to a capacitor charging and storing the large amount of energy needed to produce a very bright flash of light.

A circuit similar to Figure 3.137 is used for a camera flash.

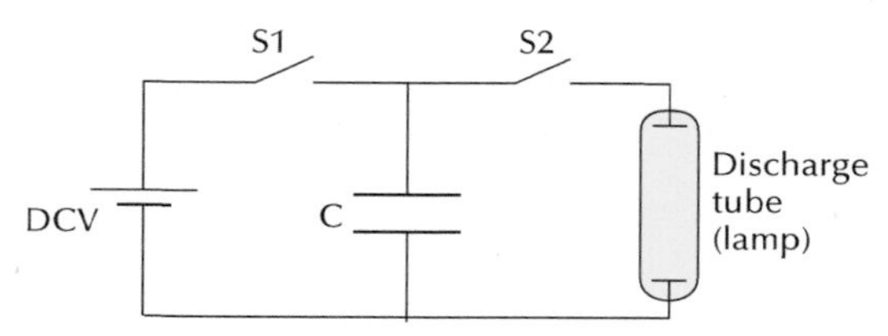

Fig 3.137

When switch S1 is closed, the capacitor charges up to a voltage equal to that of the power supply (DCV) voltage. The capacitor stores electric charge, so therefore stores electrical energy. When switch S2 is closed and S1 is opened, the capacitor discharges rapidly through the discharge tube. This means the energy stored is rapidly converted to light in the discharge tube, giving a bright (but short-lived) flash.

Exercise 4.4 Practical applications of capacitors

1 Students are investigating the effects of connecting a variable frequency power supply to a capacitor and a resistor. They set up the circuits shown in Figure 3.138.

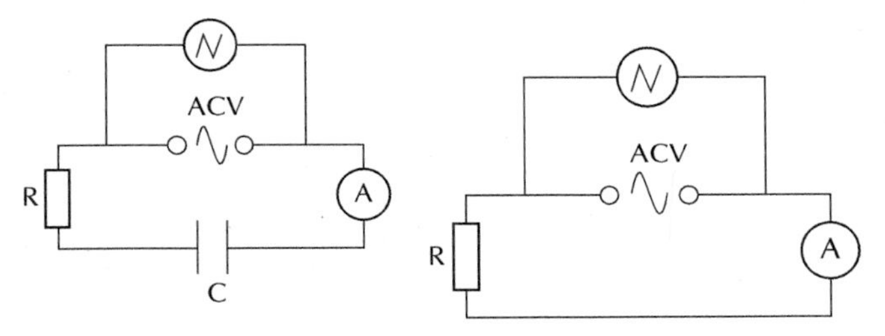

Fig 3.138

2 They vary the frequency and note the current for both circuits, and plot the results on a graph.

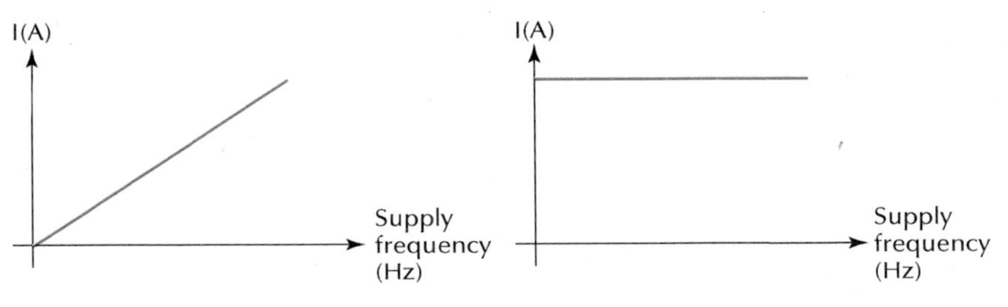

Fig 3.139

3 Copy both circuits and match them to their corresponding current vs frequency graphs.

4 A student writes the following statement about a capacitor: *'A capacitor can be used as a DC block.'*

Explain what is meant by this statement and describe how the capacitor can be used for this task.

5 A capacitor can be used in a camera to provide a large amount of energy to the camera flash in a short amount of time. The circuit shown in Figure 3.140 is used in a particular camera.

a) Describe how this circuit works to charge the capacitor and then discharge the capacitor through the lamp.

b) Calculate the maximum energy stored by the capacitor.

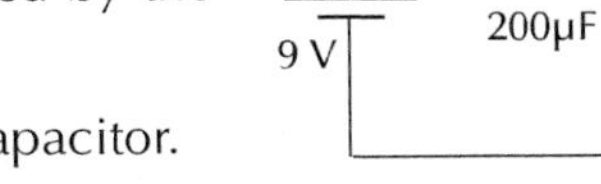

Fig 3.140

c) Calculate the charge stored by the capacitor.

d) The capacitor discharges through the lamp in a time of 3 ms. Calculate:

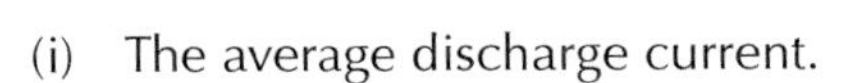

(i) The average discharge current.

(ii) The average power.

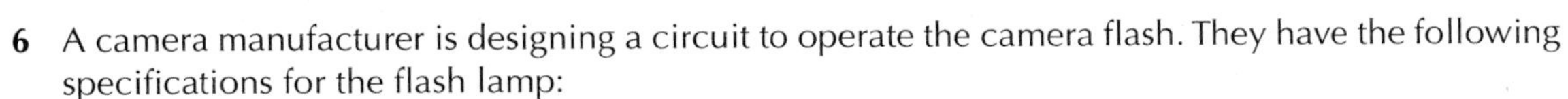

6 A camera manufacturer is designing a circuit to operate the camera flash. They have the following specifications for the flash lamp:

- Light energy in flash = 1.5 J
- Average power = 200 W
- Lamp efficiency = 65%

They design the circuit shown in Figure 3.141.

a) Calculate the minimum energy that must be stored by the capacitor.

b) Calculate the minimum capacitance for the capacitor in the circuit.

c) Calculate the discharge time for the capacitor.

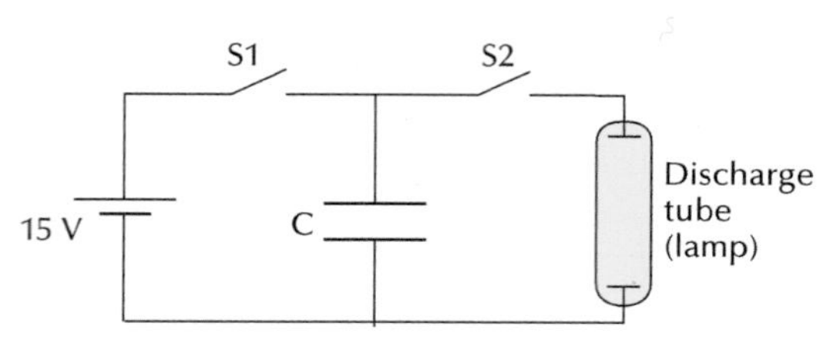

Fig 3.141

7 A lamp is connected in series with a capacitor to an AC power supply as shown in the circuit diagram in Figure 3.142.

a) There is an insulator between the plates of the capacitor. Explain why the lamp can still light up in the circuit above.

b) How does increasing the frequency of the supply affect the brightness of the lamp? Explain your answer.

c) The AC supply is replaced with a battery. Will the lamp still light? Explain your answer.

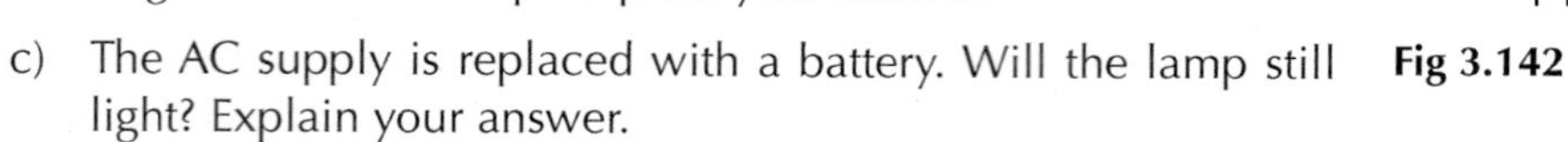

ACV

Fig 3.142

8 A courtesy light in a car is designed to illuminate the car interior when a door is opened or an interior switch is pressed. As part of a luxury feature, the courtesy light brightness slowly increases so as not to dazzle the driver or passengers at night.

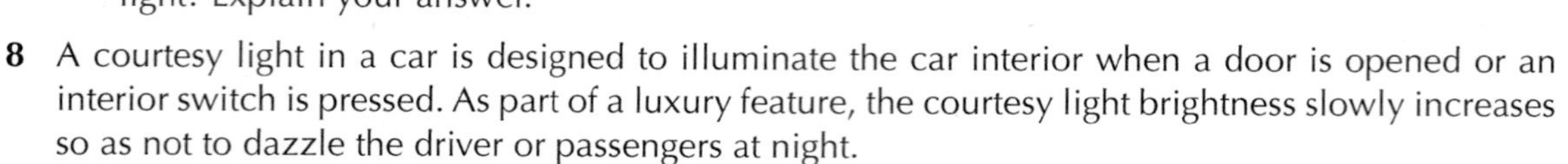

The manufacturer designs the circuit in Figure 3.143 to switch on the interior light lamps.

a) Use your knowledge of physics to explain why the courtesy lights do not illuminate at full brightness straight away.

b) The manufacturer finds that with the circuit shown in Figure 3.143, the lights come on too quickly. How could the length of time taken for the lights to illuminate at maximum brightness be changed? Give two solutions. (You may include additional components.)

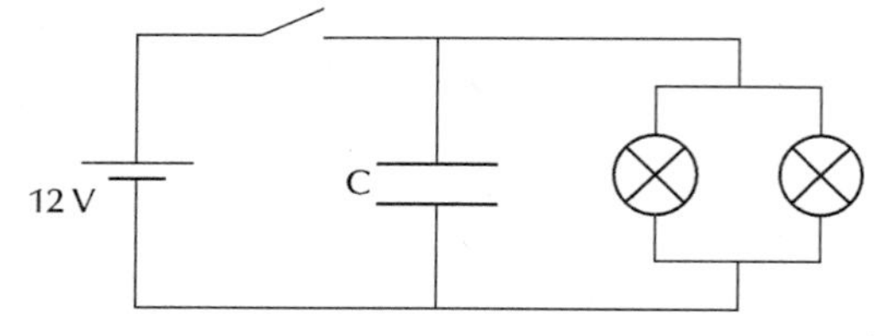

Fig 3.143

9 A sign on the outside of a restaurant uses a flashing neon lamp to show that it is open.

The neon sign will light when the voltage across the discharge tube has reached 150 V. It will remain lit so long as the voltage across the tube remains above 100 V. In order to make the neon light flash, an engineer designs the circuit shown in Figure 3.145:

Fig 3.144

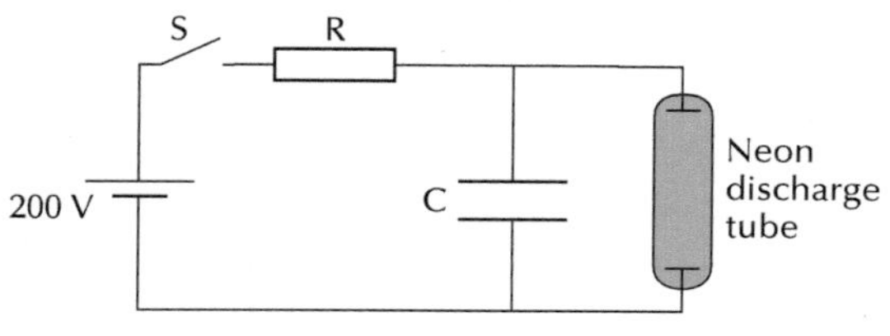

Fig 3.145

(continued)

a) Use your knowledge of physics to explain how the above circuit works to make the neon light flash.

b) The engineer initially chooses a resistor R with a resistance of 2 kΩ and a capacitor with capacitance 2 mF.

(i) Calculate the initial charging current for the capacitor.

(ii) Calculate the maximum energy stored by the capacitor when the neon light flashes.

c) State and explain two changes that could be made to the above circuit to reduce the flash rate of the neon light.

5 Semiconductors and p–n junctions

You should already know (National 5)

- Conductors allow electricity to flow; insulators prevent the flow of electricity
- A transistor consists of p–n junctions and can be used in switching circuits

Learning intentions

- Categorisation of solids into conductors, insulators and semiconductors
- Electrons occupy energy bands in a solid:
  - In a conductor, the highest energy band (conduction band) is not full so electrons can flow
  - In an insulator, the highest energy band (valence band) is full – gap to conduction band is large at room temperature, which prevents current from flowing
- In semiconductors, the gap between the valence and conduction bands is smaller at room temperature, allowing conduction in certain circumstances
- Two types of semiconductor – p-type and n-type
- P-type and n-type material being joined form a p–n junction – the electrical properties of this can be used in a variety of devices
- Solar cells are p–n junctions designed such that light on the p–n junction produces a potential difference across the junction: photovoltaic effect
- LEDs are forward-biased p–n junction diodes in which an energy gap between the conduction and valence band results in the emission of light

5.1 Conductors, insulators and semiconductors

Solids can be divided into three main categories according to their electrical properties. These are insulators that do not conduct electricity, conductors that do conduct electricity, and semiconductors that conduct electricity under certain circumstances.

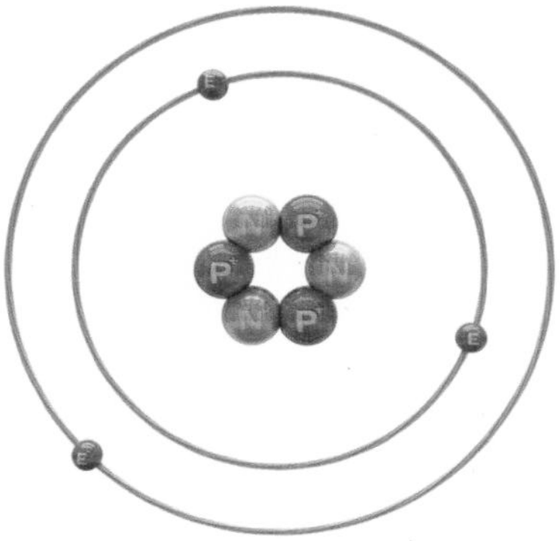

Fig 3.146

5.1.1 Band theory

In Area 2, the energy levels of individual atoms were discussed. In the Bohr model of the atom, electrons can orbit atoms only in very discrete orbits. This is shown in Figure 3.46 for lithium.

Each of the orbits represents an energy level that the electron can occupy. The electrons must occupy an energy level; they cannot exist between the bands. There is a maximum number of electrons that an energy level can have, which is governed by quantum mechanical processes and given by:

$$Max = 2n^2$$

where n is the number of the energy level. Working outwards from the centre of the atom, these maximum numbers are:

Energy level (n)	Maximum number of electrons
1	2
2	8
3	18
4	36
5	50

Table 3.14

Outer energy levels can contain a greater number of electrons. If these outer levels are not fully occupied, then the electrons are free to move.

Copper is a well-known electrical conductor. A typical copper atom is shown in Figure 3.147, which has 29 protons in the nucleus and 29 electrons orbiting.

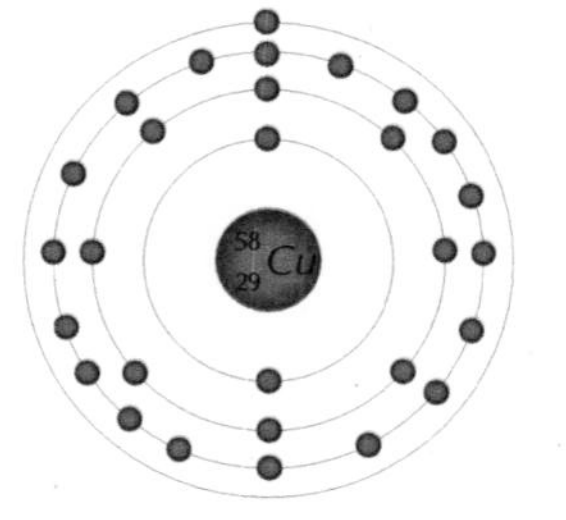

Fig 3.147

The inner three bands are fully occupied. The electrons are tightly bound and unlikely to move. However, there is a single electron in the fourth level. This electron, known as a valence electron, can be more easily removed from the atom.

Another known conductor of electricity is aluminium. Its atomic structure is shown in Figure 3.148.

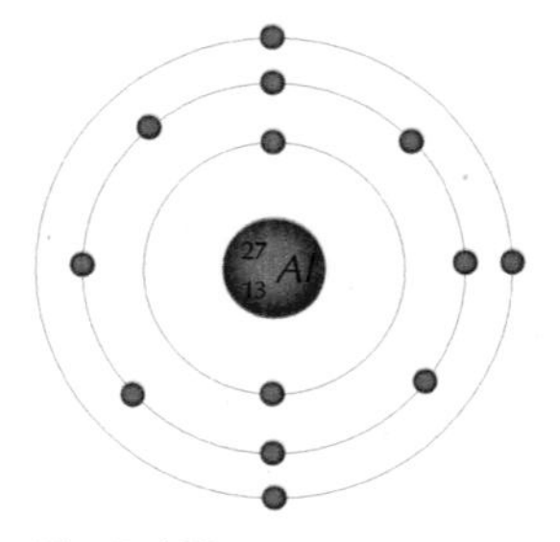

Fig 3.148

Here, the inner two energy levels are fully occupied. However, the third level, which has a capacity of 18 electrons, only contains 3 electrons. This means that these outer electrons are easy to move.

To view energy levels more easily, they are flattened out to give straight lines as shown in Figure 3.149 (see section on spectra: Area 2, Chapter 7).

E_4
E_3
E_2
E_1
E_0

Fig 3.149

In solids, there are many atoms interacting with each other, rather than atoms in isolation. When atoms interact, their outer energy levels will interact, and this leads to bands of available energies rather than discrete energy levels. This is demonstrated in the diagram illustrated in Figure 3.150.

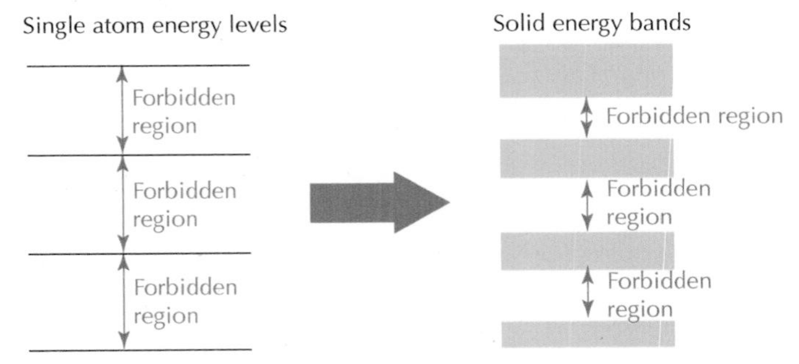

Fig 3.150

The valence band

The valence band is the outermost band that contains electrons. The valence band can be full, in which case electrons are not free to move,

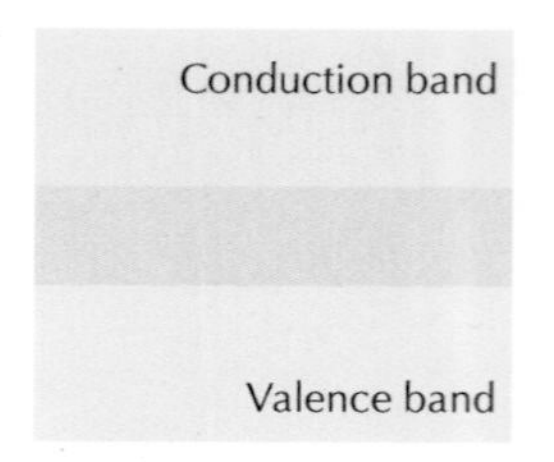

Fig 3.151

or it can be partially full. Electrons in the valence band are still bound to their atoms. The idea is the same as for single atoms, where the electrons in the outermost occupied energy level are called valence electrons.

The conduction band

Electrons with energies in the conduction band are free to move throughout the solid – their movement is the flow of electricity (an electric current).

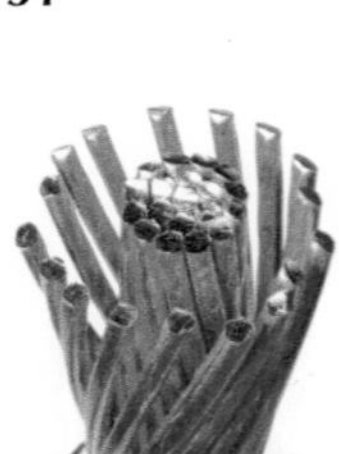

Fig 3.152

5.1.2 Conductors

A material that conducts electricity allows current to move easily through it. In other words, electrons are free to move within a conductor. For some conductors, the valence band is only part filled. This means that some electrons are free to move. In other conductors, the valence and conduction bands overlap, which also leads to there being free electrons to allow conduction. This is shown in Figure 3.151.

Conduction band

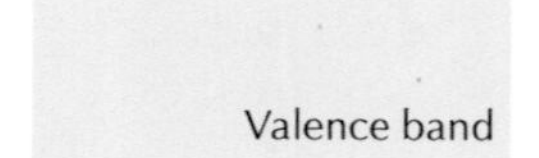

Valence band

Fig 3.153

5.1.3 Insulators

A material that is an insulator will not conduct electricity under normal circumstances. The valence band of an insulator is filled so there is no available space for an electron to move. The material is therefore not able to conduct electricity. The energy diagram in Figure 3.153 shows the energy bands for a typical insulator.

As the valence band is full, electrons must reach the conduction band in order to have space to move and conduct electricity. The large band gap between the valence and conduction bands prevents this from easily happening. Thermal energy at room temperature does not supply the electrons with enough energy to jump across the gap to the conduction band.

Insulators are used to prevent electricity from flowing. A typical use is the insulation around electrical cables, which is designed to prevent you from getting an electric shock from current-carrying wires. The wires are insulated with a material that does not conduct electricity so that you are shielded from potentially dangerous current being carried by the wire (see Figure 3.154).

Fig 3.154

5.1.4 Semiconductors

A semiconductor is a material that has an electrical conductivity between that of a conductor and an insulator. The gap between the conduction and the valence bands is much smaller, as shown in Figure 3.155.

Conduction band

Valence band

Fig 3.155

At absolute zero, there will be no electrons in the conduction band. As the temperature increases, electrons gain enough energy to transition to the conduction band. The material begins to conduct. The greater the temperature, the greater the number of electrons that make the jump, which lowers the resistance of the material by increasing its conductivity. Different materials will have a different band gap between the conduction and valence bands, which leads to different responses to changes in temperature.

Experiment 5.1.4 Semiconductors

This experiment investigates the effects of changing the temperature on the resistance of a thermistor (semiconductor) to demonstrate the effects of changing temperature on the electrons within the conduction and valence bands.

Apparatus

- DC power supply
- 1 kΩ fixed resistor
- Voltmeter
- Thermistor
- Connecting wires

Instructions

1. Build the potential divider circuit shown in Figure 3.156 consisting of a thermistor and a fixed resistor. The voltmeter is connected across the thermistor.

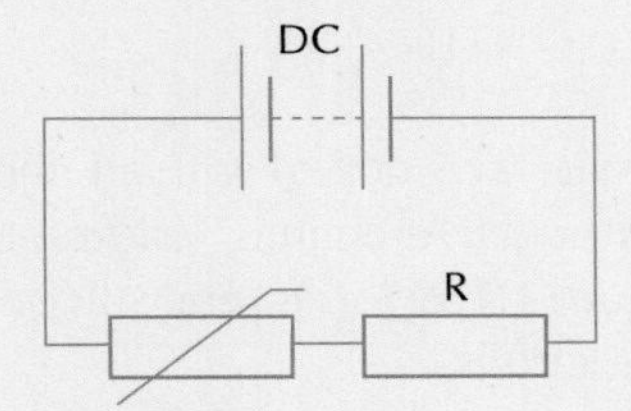

Fig 3.156

2. Immerse the thermistor in a bucket of ice (or cover the thermistor with an ice cube wrapped in a towel). Measure the voltage across the thermistor.
3. Increase the temperature of the thermistor and measure the voltage across it.
4. Write a short report about what you have found out. Remember to include the effect of changing temperature on the voltage across the thermistor (voltage is linked to the resistance of the thermistor). Do your results agree with the theory predicted above?

5.2 The p–n junction

Materials can be classified into three main categories according to their electrical conductivity, as discussed in the previous section. One of these classes is semiconductors. Semiconductors are insulators that have been doped with an impurity that allows them to conduct. The impurity is deliberately added, to either give an additional free electron or a missing electron which is considered as a positive 'hole'. This allows for more current to flow in the semiconductor, thus reducing its resistance. Doping must take place at the time the semiconductor crystal is being grown – you cannot simply add a doping material to a wafer of silicon, for example.

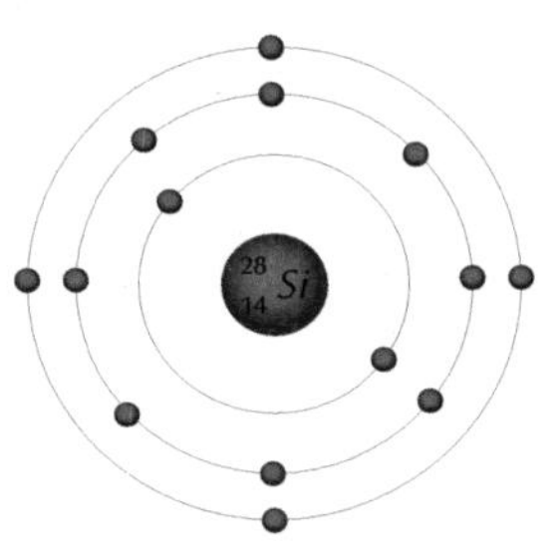

Fig 3.157a

Consider the silicon atom shown in Figure 3.157a. There are 14 electrons, which occupy the energy levels shown. The outer energy level (n = 3) has four valence electrons in it. When the atoms are brought together to form a solid, each silicon atom shares its four valence electrons with four other neighbouring silicon atoms. This is shown in Figure 3.157b, which shows a silicon lattice. Here, only the valence electrons are shown.

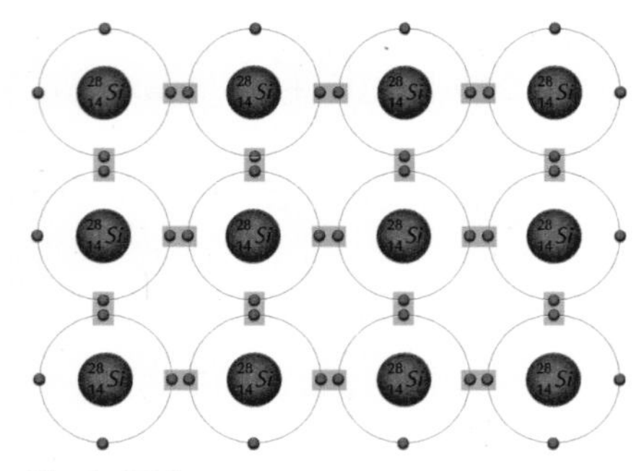

Fig 3.157b

Each of the four valence electrons form bonds (called covalent bonds) with neighbouring silicon atoms. This means that these electrons are not free to move easily through the solid. This leads to the familiar band structure for a semiconductor discussed above and shown here in Figure 3.158.

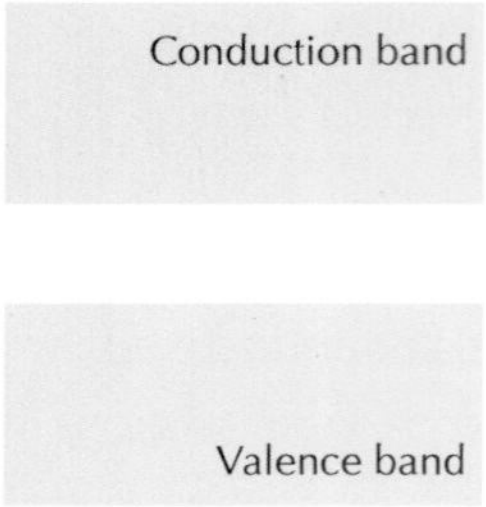

Conduction band

Valence band

Fig 3.158

The valence band is filled. The conduction band is closer to the valence band than with an insulator, and thermal excitation can excite electrons up to the conduction band to allow for conduction. However, the solid can also be doped with impurities to give extra electrons or missing electrons, which are much easier to move through the solid to give conduction. There are two possible doping options, which are discussed below.

5.2.1 n-type

In an n-type semiconductor, the material is doped with an impurity that gives an extra electron (extra negative charge, leading to the name *n-type*). This is shown in Figure 3.159 for doping silicon with arsenic, which has five valence electrons.

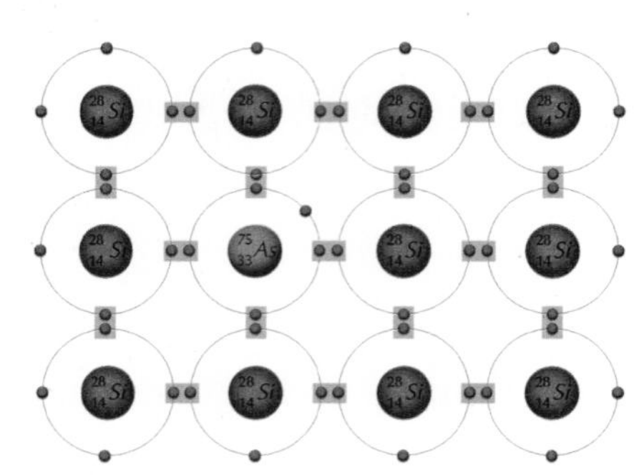

Fig 3.159

The extra electron is not bonded by a covalent bond and can therefore be easily moved through the solid. This extra free charge carrier increases the conductivity of the semiconductor.

The extra electrons from the arsenic atoms occupy levels close to the conduction band and are easily excited into the conduction band. The overall conductivity of the solid can be altered by changing the number of impurity atoms – the more impurity atoms there are, the greater the number of free electrons, which increases the conductivity.

5.2.2 p-type

In a p-type semiconductor, the material is doped with an impurity that has one less electron. This lack of a negative charge can be thought of as a positive charge, leading to the name 'p-type'. This is illustrated in Figure 3.160 with the doping of silicon with boron atoms, which have only three valence electrons.

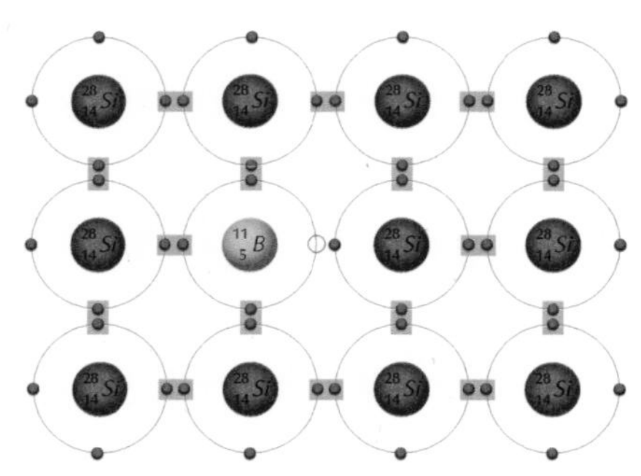

Fig 3.160

The missing electron is represented as a hole that electrons can move into. This allows the movement of charge throughout the semiconductor.

The impurity atoms give additional holes that are just above the valence band. Electrons can be easily promoted into these holes, leaving spaces in the valence band to allow conduction.

5.2.3 The p–n junction

One of the primary applications of p- and n-type semiconductors is the p–n junction. A p–n junction consists of a piece of p-type material placed in contact with a piece of n-type material as shown in Figure 3.161.

Fig 3.161

When these two types of material are brought close together, the resultant component will allow charge to flow in one direction but not the other. The resultant component is a diode, with the circuit symbol shown in Figure 3.162.

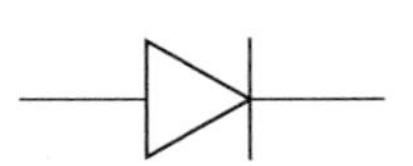

Fig 3.162

When p- and n-type material is brought together as shown, and no external potential difference is applied, some free-to-move electrons in the n-type material recombine with free holes in the p-type material. When this happens, the charge carriers near the boundary are no longer free. A region called the depletion layer is formed as shown in Figure 3.163.

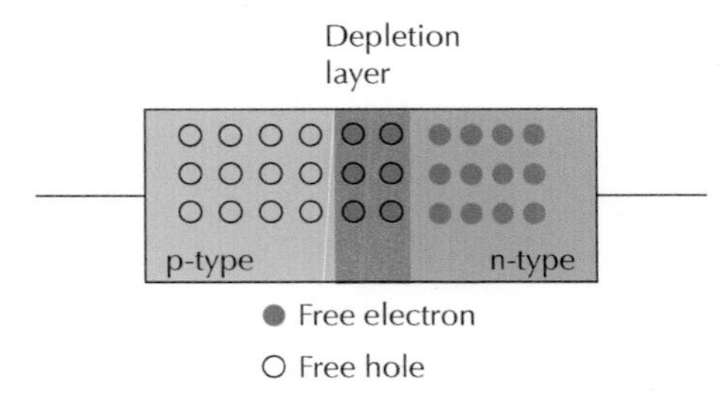

Fig 3.163

This induces a potential difference across the junction, which creates an electric field as discussed earlier in this area of study. In the n-type material, there is a shortage of electrons in the depletion layer compared to the rest of the material, giving an overall positive charge close to the boundary. Similarly, the shortage of holes in the p-type layer gives a negative charge (Figure 3.164).

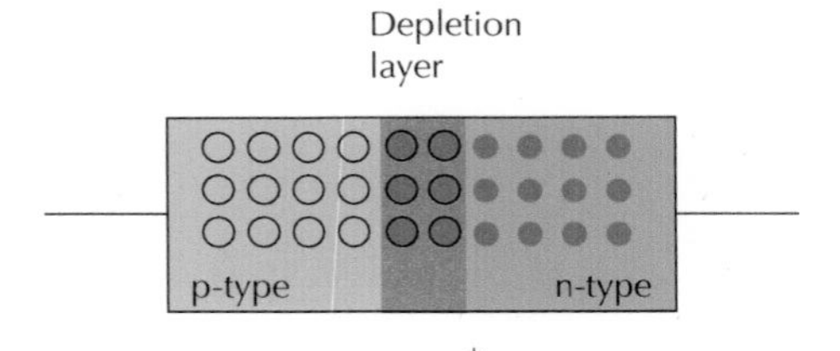

Fig 3.164

In the absence of an externally applied voltage, this electric field prevents any further flow of charges. On an energy level diagram, the p–n junction can be represented as shown in Figure 3.165.

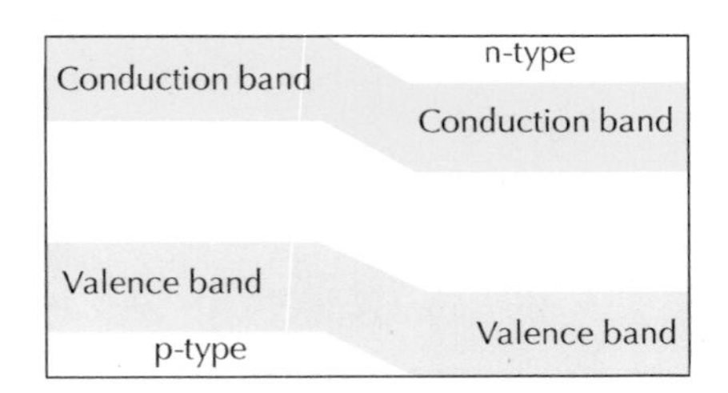

Fig 3.165

In the region of the depletion layer, the free electrons and holes have moved and recombined. There are no longer any free charge carriers in this region to move, so there is no current.

Forward bias

At National 5 level, the diode was introduced as a device that allowed current to move in only one direction. The diode would conduct current if it was forward biased. Here, the n-type material is connected to the negative terminal of a power supply and the p-type to the positive terminal, as shown in Figure 3.166.

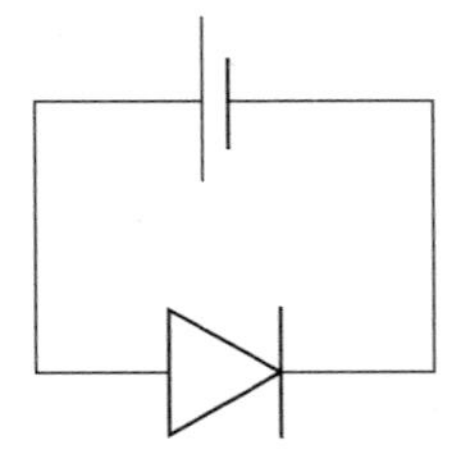

Fig 3.166

The potential difference of the power supply overcomes the electric field set up across the depletion region, allowing charge to flow. This can be visualised as the free electrons being attracted to the positive terminal, as shown in Figure 3.167. However, it is important to consider the effects in terms of energy levels. The applied potential difference reduces the electric field set up across the p–n junction. Electrons can now transition more easily into the conduction band of the p-type material, which allows for conduction across the diode.

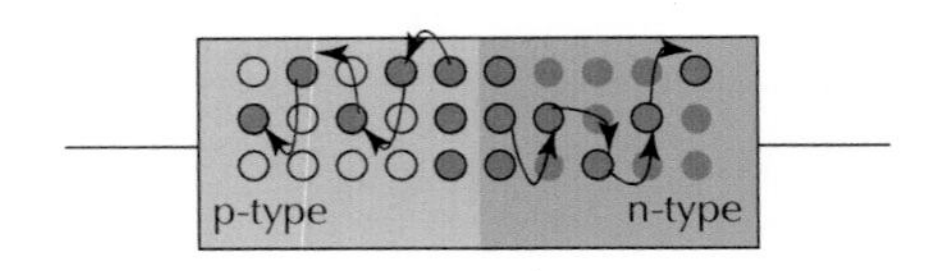

Fig 3.167

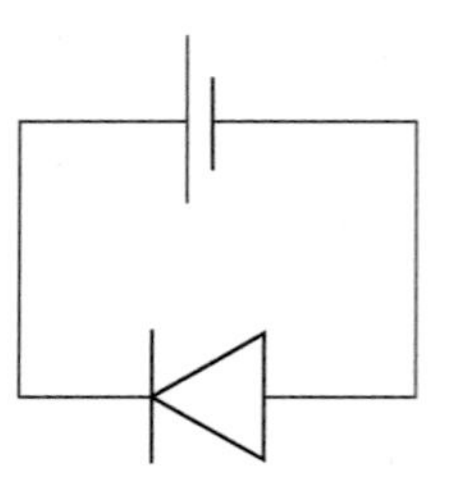

Fig 3.168

Reverse bias

When reverse biased, the diode will not conduct. The positive terminal of the battery is connected to the n-type material and the negative terminal to the p-type material as shown in Figure 3.168.

Here, the potential difference of the supply increases the electric field across the junction. This has the effect of preventing current movement by making it much more difficult for the electrons to reach the conduction band of the p-type material.

5.3 The LED

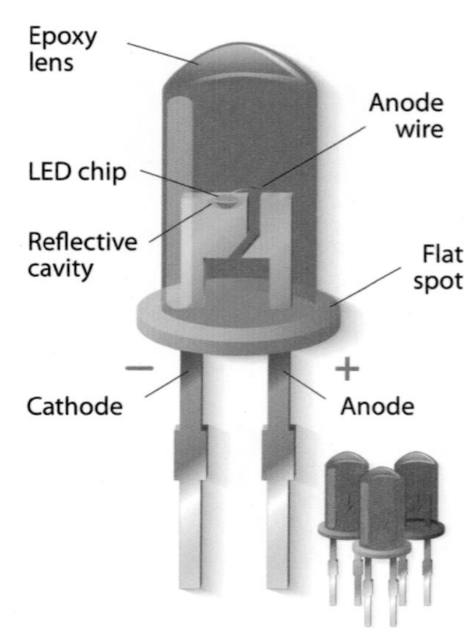

Fig 3.169

A light-emitting-diode (LED) is a device that converts electrical energy to light energy. They are characterised by low-temperature operation, and the ability to produce monochromatic light. A typical LED is shown in Figure 3.169.

The LED is a p–n junction diode that is forward biased. This means that electrons and holes are continuously recombining. With each recombination, a photon of energy is released.

Electrons gain energy from the applied voltage and move into the conduction band of the p-type material. An electron can then 'fall' into a hole in the valence band, losing energy, which results in the production of a photon of light. The colour of the photon will depend on the band gap between the conduction and valence band, which is different for different materials.

The voltage required to operate the LED will also depend on band gap – electrons require enough energy to reach the conduction band of the p-type. This will require a large enough voltage. As we have seen in the previous area of study the larger the band gap, the greater the energy of the photon and thus the greater the frequency of the photon. This means that LEDs that produce light at the blue/violet end of the spectrum typically require larger voltages to operate.

GO! Experiment 5.3 LED turn-on voltage

This experiment investigates the turn-on voltage of different LEDs and compares it to the colour of light emitted by the LED.

Apparatus

- Selection of LEDs
- Series resistor
- DC power supply
- Voltmeter

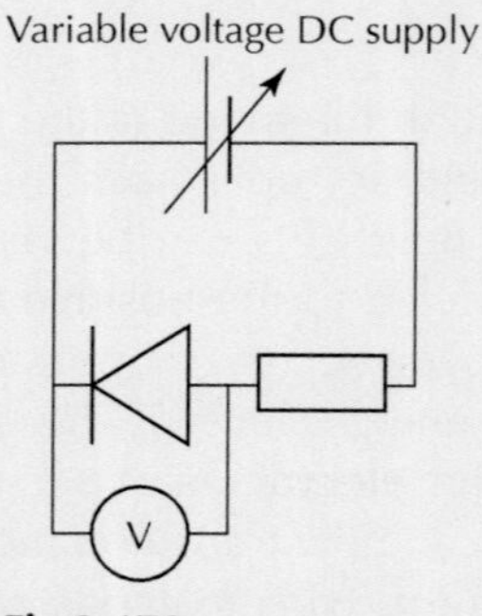

Fig 3.170

Instructions

1 Set up the series circuit shown in Figure 3.170 with the LED connected to a DC power supply (constant current) and series resistor.

(*continued*)

2 Gradually increase the supply voltage to the LED until you see it just emitting light and no more (a dark room will help). Note the turn-on voltage in a table similar to the one shown below.

LED colour	Turn-on voltage (V)

Table 3.15

3 Write a short report comparing the different colours to the turn-on voltage. Comment on the effect of increasing the photon energy (colours at the violet end of the spectrum) on the turn-on voltage.

5.3.1 The solar cell

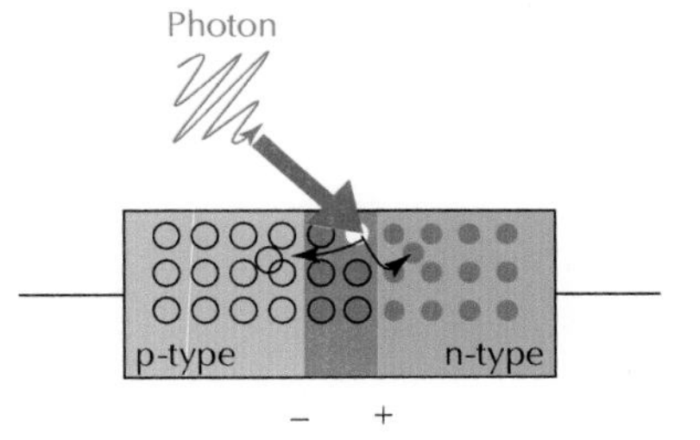

Fig 3.171

A p–n junction can also be used to produce electricity from light (convert light energy into electrical energy). This process can be seen to be very similar to the reverse process of the diode above. When photons with large enough energy are incident on the p–n junction, they split up electrons and holes to create free charge carriers. Photons with enough energy will promote electrons from the valence band into the conduction band. Electrons are then attracted to the n-type semiconductor, which sets up a potential difference across the cell.

This has the effect of increasing the potential difference across the diode. Light incident on the diode is producing a potential difference (voltage). Here, the diode is said to be operating in photovoltaic mode, and is often used in calculators as an additional power source to help sustain the life of the battery. It can be connected to a load as shown in Figure 3.172.

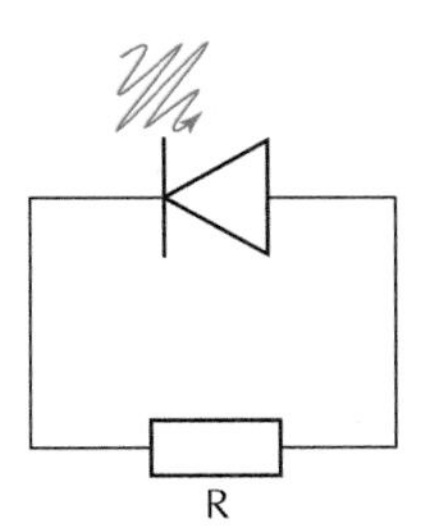

Fig 3.172

The current through the resistor is proportional to the number of photons incident on the junction. Hence, the current is proportional to the irradiance of light incident on the diode. This allows the diode to be used as a light sensor in addition to a renewable source of electricity.

5.3.2 The LDR

A p–n junction can be used in photovoltaic mode to produce electrical energy from light energy. The amount of energy produced is proportional to the light irradiance, allowing it to be used as a light sensor. A light dependent resistor (LDR) is another type of light sensor. The LDR was introduced at National 5 level as a device whose resistance changes according to light irradiance – the brighter the light, the lower the resistance:

Light Up Resistance Down (LURD)

An LDR is another example of a p–n junction. Contrary to the solar cell, where the p–n junction stands alone in the circuit, the LDR is a p–n junction connected to a power supply. The LDR is connected in reverse bias, which leads to a large depletion region as shown in the diagram below (and described above).

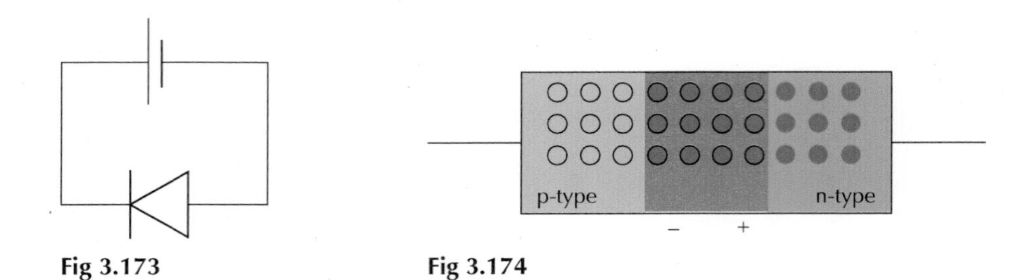

Fig 3.173

Fig 3.174

Normally in this situation, current does not flow in the circuit. However, when light is incident on the junction, electrons and holes are split apart (as with the solar cell). This leads to free charge carriers in the depletion region. The free charge carriers reduce overall resistance of the diode, allowing a current to flow. The conductivity of the diode is being changed, so this mode of operation is referred to as 'photoconductive mode'.

The greater the irradiance of the light, the greater the number of free charge carriers. This means that the resistance of the diode drops as the light irradiance increases. This effect was used in practice at National 5 level in circuits using LDRs in control circuits.

1. **Interpreting experimental data**
   - Variables
   - Graphical analysis
   - 'Unseen physics'
2. **Uncertainties**
   - Reading uncertainty
   - Random uncertainty
   - Percentage uncertainty

AREA 4
Experimental Physics

1 Interpreting experimental data

Experiments are carried out to investigate the relationship between variables. A simple example of this from National 5 and Higher level is the relationship between the unbalanced force and the mass:

$$F = ma$$

An experiment to investigate this is detailed in the National 5 course text where the unbalanced force is changed and the acceleration is measured. This experimental data would then be plotted on a graph to analyse it and determine a relationship between the force and mass.

1.1 Variables

In any experiment carried out, there are three main types of variable:

- Independent variable – this is the variable that you change as part of the experiment. In the example above, this would be the unbalanced force.
- Dependent variable – this is the variable that you measure as part of the experiment. It depends on the changes made to the independent variable. In the example above, this would be the acceleration.
- Constant variable – this is the variable that you ensure is kept constant as part of the experiment. In the example above, this would be the mass of the system.

1.2 Graphical analysis

We have already seen the use of graphs in science to represent experimental data. In physics, we typically plot a scatter graph and aim to find a line of best fit. This can be used to find the relationship between variables but also to determine a numerical value.

Consider the example above of investigating the effect of changing the unbalanced force on the acceleration of an object. Plotting the results on a graph would yield the result shown in Figure 4.1.

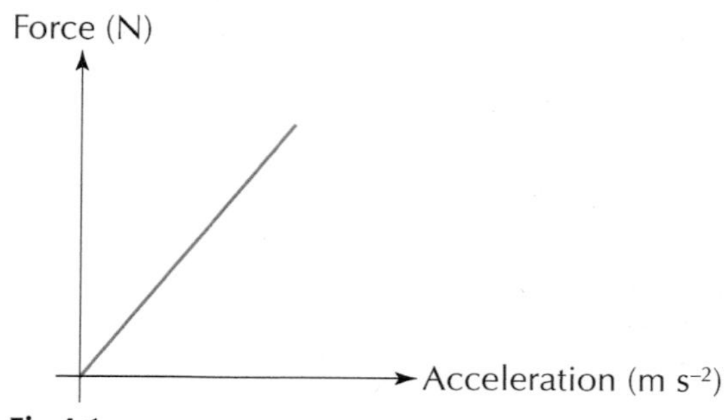

Fig 4.1

The straight line through the origin shows that unbalanced force is directly proportional to the acceleration:

$$F \propto a$$

However, we can also compare this with the equation of a straight line from mathematics:

$$y = mx$$

where m is the gradient of the line. Here, we have plotted force on the y-axis and acceleration on the x-axis so the equation of the straight line becomes

$$F = ma$$

Comparing with the equation for Newton's second law, we can see that the gradient of the line must equal the mass. The gradient of the line of best fit can be found by using the gradient formula

$$m = \frac{y_2 - y_1}{x_2 - x_1}$$

This can then be used to find the mass.

1.3 'Unseen physics'

We can also apply the graphical analysis technique to physics that we have not studied as part of the Higher course. The key idea is applying the skills of experimental data collection and graph plotting.

Consider the example of a simple pendulum. The period of the pendulum (length of time taken for one complete swing) is given by:

$$T = 2\pi\sqrt{\frac{l}{g}}$$

where l is the length of the pendulum and g is the acceleration due to gravity. Strictly, this is not part of the Higher course content but the skills of analysing a graph for this experiment are.

We can see from the relationship above that the period of the pendulum depends on its mass. Consider carrying out an experiment where the mass is changed (independent variable) and the period is measured (dependent variable).

Simply plotting period vs length would not give a straight line according to the equation above. However, if we square both sides of the above equation we yield

$$T^2 = \frac{4\pi^2}{g}l$$

This is the equation of a straight line where T^2 is plotted on the y-axis and l is plotted on the x-axis. The gradient of the resulting straight line would be

$$gradient = \frac{4\pi^2}{g}$$

This can then be used to find the acceleration due to gravity – a typical project experiment at Advanced Higher level.

2 Uncertainties

Whenever we make a measurement in an experiment, we are subject to a certain uncertainty in the measurement – we do not know the value we have measured exactly, to an infinite number of significant figures.

Consider using a ruler to measure the length of this textbook. A typical ruler is shown in Figure 4.2.

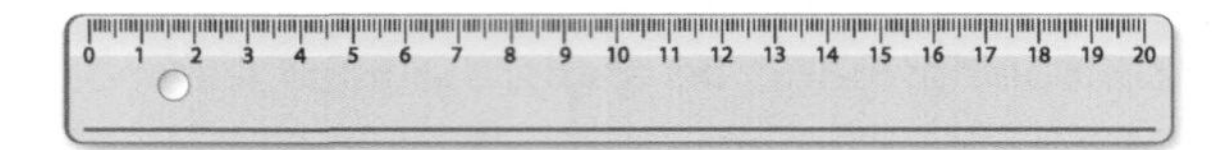

Fig 4.2

The smallest division on the ruler is millimetres. It is therefore not possible to measure the length of the textbook to an accuracy any greater than to the nearest half millimetre. There are also uncertainties in the measurement that you make. These uncertainties are discussed below.

2.1 Reading uncertainty

Whenever you make a measurement (take a reading), that measurement is subject to a reading uncertainty. The extent of this uncertainty depends on the instrument you have used to take the measurement.

If the instrument has an analogue scale (such as a ruler), then the reading uncertainty is plus or minus half of the smallest division. For example, say you measure the length of a pencil to be 75 mm using a ruler. The smallest division on the ruler is 1 mm, so the reading uncertainty is ± 0.5 mm. We would quote the measurement as 75 ± 0.5 mm.

If the instrument has a digital scale (such as a multimeter), then the reading uncertainty is plus or minus the smallest division. For example, say you measure the voltage across a resistor to be 2.54 V. This would have a reading uncertainty of ± 0.01 V and we would quote the measurement as 2.54 ± 0.01 V.

Clearly, the smaller the reading you make, the larger the impact of the reading uncertainty. The reading uncertainty can be considered as a percentage of the measured value – the greater the percentage, the greater the effect of the error. This is why you need to choose carefully the instrument you use to take the measurement with regard to the scale – you wouldn't chose a trundle wheel with centimetre divisions to measure the length of a sharpener that is only 2–3 cm long!

2.2 Random uncertainty

When carrying out an experiment, it is good practice to take multiple readings. For example, in the example of measuring acceleration for different applied forces, the acceleration would be measured three times if the results are concordant, and five times if not. The mean acceleration would then be calculated.

The spread of the results is an indication of the reliability of the experiment you are carrying out – a large spread in accelerations would point to issues with the experiment, while a small spread would give confidence in the measurements.

In mathematics, a measure of spread in data can be found by finding the standard deviation in the data. In physics, we use a simplified calculation to find the spread in data, which we call the random uncertainty. The random uncertainty in a set of data is given by:

$$\text{Random uncertainty} = \frac{\text{max. value} - \text{min. value}}{\text{number of results}}$$

2.3 Percentage uncertainty

The percentage uncertainty allows a comparison of the size of the error relative to the measurement being made. A percentage uncertainty is found by dividing the uncertainty by the measured result and then multiplying by 100 to convert to a percentage. This allows you to consider the relative size of the uncertainty compared to the measurement to assess whether it is significant or not.